Communications
in Computer and Information Science   2147

Rationale

The CCIS series is devoted to the publication of proceedings of computer science conferences. Its aim is to efficiently disseminate original research results in informatics in printed and electronic form. While the focus is on publication of peer-reviewed full papers presenting mature work, inclusion of reviewed short papers reporting on work in progress is welcome, too. Besides globally relevant meetings with internationally representative program committees guaranteeing a strict peer-reviewing and paper selection process, conferences run by societies or of high regional or national relevance are also considered for publication.

Topics

The topical scope of CCIS spans the entire spectrum of informatics ranging from foundational topics in the theory of computing to information and communications science and technology and a broad variety of interdisciplinary application fields.

Information for Volume Editors and Authors

Publication in CCIS is free of charge. No royalties are paid, however, we offer registered conference participants temporary free access to the online version of the conference proceedings on SpringerLink (http://link.springer.com) by means of an http referrer from the conference website and/or a number of complimentary printed copies, as specified in the official acceptance email of the event.

CCIS proceedings can be published in time for distribution at conferences or as post-proceedings, and delivered in the form of printed books and/or electronically as USBs and/or e-content licenses for accessing proceedings at SpringerLink. Furthermore, CCIS proceedings are included in the CCIS electronic book series hosted in the SpringerLink digital library at http://link.springer.com/bookseries/7899. Conferences publishing in CCIS are allowed to use Online Conference Service (OCS) for managing the whole proceedings lifecycle (from submission and reviewing to preparing for publication) free of charge.

Publication process

The language of publication is exclusively English. Authors publishing in CCIS have to sign the Springer CCIS copyright transfer form, however, they are free to use their material published in CCIS for substantially changed, more elaborate subsequent publications elsewhere. For the preparation of the camera-ready papers/files, authors have to strictly adhere to the Springer CCIS Authors' Instructions and are strongly encouraged to use the CCIS LaTeX style files or templates.

Abstracting/Indexing

CCIS is abstracted/indexed in DBLP, Google Scholar, EI-Compendex, Mathematical Reviews, SCImago, Scopus. CCIS volumes are also submitted for the inclusion in ISI Proceedings.

How to start

To start the evaluation of your proposal for inclusion in the CCIS series, please send an e-mail to ccis@springer.com.

Kangshun Li · Yong Liu

Editors

Intelligence Computation and Applications

14th International Symposium, ISICA 2023
Guangzhou, China, November 18–19, 2023
Revised Selected Papers, Part II

 Springer

Editors
Kangshun Li ⓘD
South China Agricultural University
Guangzhou, China

Yong Liu ⓘD
The University of Aizu
Fukushima, Japan

ISSN 1865-0929 ISSN 1865-0937 (electronic)
Communications in Computer and Information Science
ISBN 978-981-97-4395-7 ISBN 978-981-97-4396-4 (eBook)
https://doi.org/10.1007/978-981-97-4396-4

This Springer imprint is published by the registered company Springer Nature Singapore Pte Ltd.
The registered company address is: 152 Beach Road, #21-01/04 Gateway East, Singapore 189721, Singapore

If disposing of this product, please recycle the paper.

Preface

The 14th International Symposium on Intelligence Computation and Applications (ISICA 2023) was held on November 18–19, 2023 in Guangzhou, China, and served as a forum to present the current work of researchers and software developers from around the world as well as to highlight activities in the Intelligence Computation and Applications areas. It aimed to bring together research scientists, application pioneers, and software developers to discuss problems and solutions and to identify current and new issues in this area. ISICA 2023 received a total of 178 papers, all of which underwent peer review, and it ultimately accepted 82 papers.

These two-volumes features the most up-to-date research, organized in the following five parts. Section 1 explores the frontiers of evolutionary intelligent optimization algorithms. Section 2 focuses on the exploration of computer vision. Section 3 presents machine learning and its applications. Section 4 discusses big data analysis and information security. Section 5 covers some new Intelligent applications of computers. One of ISICA's missions is to explore how complex systems can inherit simple evolutionary mechanisms, and how simple models can produce complex morphologies.

On behalf of the Organizing Committee, we would like to warmly thank the sponsors, South China Agricultural University, Guangdong Key Laboratory of Big Data Analysis and Processing at Sun Yat-sen University, and Guangdong Polytechnic Normal University, which helped in one way or another to achieve our goals for the conference. We wish to express our appreciation to Springer for publishing the proceedings of ISICA 2023. We also wish to acknowledge the dedication and commitment of both the staff at Springer's Beijing office and the CCIS editorial staff. We would like to thank the authors for submitting their work, as well as the Program Committee members and reviewers for their enthusiasm, time, and expertise. The invaluable help of active members from the Organizing Committee, including Yan Chen, Lixia Zhang, Lei Yang, Wenxiang Wang, Shumin Xie, Jiaxin Xu, Tian Feng, Zifeng Jiang, Jiayu Zhang, Zhensheng Yang, Tao Lai, Ruolin Ruan, Shuizhen He, Junjie Wang, Mingchen Xie, Weicong Chen, Zhihao Zhou, Juhong Wu, Zhidong Zeng, Tianjin Zhu, Wensen Mo, Wenbin Xiang, Hassan Jalil, and Al-Daba Saqr in setting up and maintaining the online submission systems by Easy Chair, assigning the papers to the reviewers, and preparing the camera-ready version of the proceedings is highly appreciated. We would like to thank them personally for their help in making ISICA 2023 a success.

April 2024

Kangshun Li
Yong Liu

Organization

Honorary Chairs

Yuping Chen   Wuhan University, China
Yuanxiang Li   Wuhan University, China
Wensheng Zhang   Institute of Automation, Chinese Academy of
         Sciences, China

General Chairs

Kangshun Li   South China Agricultural University, China
Witold Pedrycz   University of Alberta, Canada
Jian Yin   Sun Yat-sen University, China
Yu Tang   Guangdong Polytechnic Normal University, China

Program Chairs

Yong Liu   University of Aizu, Japan
Kangshun Li   South China Agricultural University, China
Zhiping Tan   Guangdong Polytechnic Normal University, China
Yiu-ming Cheung   Hong Kong Baptist University, China
Jing Liu   Xidian University, China
Hailin Liu   Guangdong University of Technology, China
Hui Wang   Shenzhen Institute of Information Technology,
         China
Feng Wang   Wuhan University, China
Xuesong Yan   China University of Geosciences, Wuhan, China
Wenyin Gong   China University of Geosciences, Wuhan, China
Xuewen Xia   Minnan Normal University, China
Xing Xu   Minnan Normal University, China
Yinglong Zhang   Minnan Normal University, China

Local Arrangement Chairs

Zhijian Wu   Wuhan University, China
Yan Chen   South China Agricultural University, China

Publicity Chairs

Shunmin Xie   South China Agricultural University, China
Feng Wang   Wuhan University, China
Wei Li   Jiangxi University of Science and Technology,
                              China
Lixia Zhang   South China Agricultural University, China
Lei Yang   South China Agricultural University, China

Program Committee

Ehsan Aliabadian   University of Calgary, Canada
Rafael Almeida   University of Calgary, Canada
Ehsan Amirian   University of Calgary, Canada
Nik Bessis   University of Derby, UK
Yiqiao Cai   Huaqiao University, China
Zhangxing Chen   University of Calgary, Canada
Iyogun Christopher   University of Calgary, Canada
Guangming Dai   University of Calgary, Canada
Lixin Ding   Wuhan University, China
Ciprian Dobre   University Politehnica of Bucharest, Romania
Xin Du   Fujian Normal University, China
Christian Esposito   University of Salerno, Italy
Zhun Fan   Shantou University, China
Massimo Ficco   University of Campania Luigi Vanvitelli, Italy
Razvan Gheorghe   University Politehnica of Bucharest, Romania
Maoguo Gong   Xidian University, China
Zhaolu Guo   Jiangxi University of Science and Technology,
                              China
Tomasz Hachaj   Pedagogical University of Krakow, Poland
Guoliang He   Wuhan University, China
Jun He   Aberystwyth University, UK
Han Huang   South China University of Technology, China
Xiaomin Huang   Sun Yat-sen University, China
Ying Huang   Gannan Normal University, China

Contents – Part II

Machine Learning and Its Applications

Human Flow Prediction Model Based on Graph Convolutional Recurrent
Neural Network 3
   Hongwei Su and Maria Amelia E. Damian

Research on Text Classification Algorithm Based on Deep Learning 15
   Li Kangshun, Junjie Wang, and Wenbin Zhu

MobilenetV2-Based Network for Bamboo Classification
with Tri-Classification Dataset and Fog Removal Training 28
   Yan Chen, Dehao Shi, and Hongxing Peng

A Filter Similarity-Based Early Pruning Methods for Compressing CNNs 39
   Zifeng Jiang and Kangshun Li

Visualization of Convolutional Neural Networks Based on Gaussian Models ... 49
   Hui Wang and Tie Cai

A New Feature Selection Algorithm Based on Adversarial Learning
for Solving Classification Problems 56
   Xiao Jin, Bo Wei, Wentao Zha, and Jintao Lu

A Simulated Annealing BP Algorithm for Adaptive Temperature Setting 71
   Zi Teng, Zhixun Liang, Yuanxiang Li, and Yunfei Yi

Research on the Important Role of Computers in the Digital Transformation
of the Clothing Industry 95
   Ping Wang and Xuming Zhang

Multimedia Information Retrieval Method Based on Semantic Similarity 103
   Xuanyi Zong, Jingwen Zhao, Zhiqiang Chen, and Jinfeng He

Iterative Learning Control for Encoding-Decoding Method with Data
Dropout at Both Measurement and Actuator Sides 113
   Yongxian Chen and Yunshan Wei

A Domain Adaptive Segmentation Label Generation Algorithm
for Autonomous Driving Scenarios 127
   Kangshun Li and Tian Feng

Visualization Analysis of Convolutional Neural Network Processes 135
  Hui Wang, Tie Cai, Yong Wei, and Zeming Chen

Packet Performance Predictor Based on Graph Isomorphism Network
for Neural Architecture Search 142
  Yue Liu, Jiawang Li, Zitu Liu, and Wenjie Tian

Big Data Analysis and Information Security

Reversible Data Hiding Algorithm Based on Adaptive Predictor
and Non-uniform Payload Allocation 159
  Dan He

Research on Bayberry Traceability Platform Based on Blockchain 170
  Hongyu Xiao, Zihang Gao, Xiaojun Cui, and Nannan Zhao

Research on Satellite Navigation and Positioning Based on Laser Point
Cloud Data 179
  Yuming Sun and Hua Wang

Research and Application of System with Bayberry Blockchain Based
on Hyperledger Fabric 187
  Hongyu Xiao, Zihang Gao, Xiaojun Cui, and Nannan Zhao

A Multiparty Reversible Data Hiding Scheme in Encrypted Domain Based
on Hybrid Encryption 196
  Bing Chen, Lu Chai, Yong Wang, Jingkun Yu, and Wanhan Fang

Research on Smart Agriculture Big Data System Based on Spark
and Blockchain 208
  Yuming Sun and Hua Wang

Efficient Public Key Encryption Equality Test with Lightweight
Authorization on Outsourced Encrypted Datasets 214
  Chengyu Jiang, Sha Ma, and Hao Wang

Fake News Detection Model Incorporating News Text and User Propagation ... 225
  Shuxin Yang, Jiahao Li, and Weidong Huang

Data Analysis of University Educational Administration Information
Based on Prefixspan Algorithm 240
  Yiying Xu, Yi Liu, and Haili Yu

The Application and Exploration of Big Data in College Student
Information Management 253
    Jun Zhang and Yuanbing Wang

Research on Precision Marketing Strategy of Guangdong Characteristic
Products Enabled by Big Data in Rural 262
    Hua Wang and Yuming Sun

Research and Application of Offline Log Analysis Method for E-commerce
Based on HHS 270
    Haoliang Wang, Kun Hu, Lili Wang, and Jingtong Shang

Research on Communication Power of Cross-Cultural Short Video Based
on Qualitative Comparative Analysis 278
    Wen Meng and Chao Yu

Multi-recipient Public-Key Authenticated Encryption with Keyword
Search 287
    Kejin He, Sha Ma, and Hao Wang

Analysis of Information Security Processing Technology Based
on Computer Big Data 297
    Hua Wang and Fuyu Zhu

Intelligent Application of Computer

Research on the Quality Evaluation and Optimization of Ideological
and Political Education in Universities Driven by Artificial Intelligence 309
    Yuanbing Wang and Jun Zhang

Research on the Routing Protocol Algorithm Driven by the Dedicated
Frequency Points of the Internet of Things to Build a Network 319
    Lingwei Wang and Hua Wang

Artificial Intelligence in Intelligent Clothing: Design and Implementation 326
    Ping Wang and Xuming Zhang

Edible Oil Price Forecasting: A Novel Approach with Group Temporal
Convolutional Network and BetaAdaptiveAdam 337
    Lei Yang, Huade Li, Rui Xu, Zexin Xu, and Jiale Cao

Design and Application of a Teaching Evaluation Model Based
on the Theory of Multiple Intelligences 351
    Luyan Lai

Study on TNM Classification Diagnosis of Colorectal Cancer Based
on Improved Self-supervised Contrast Learning 360
  Tao Lai and Kangshun Li

Construction and Quality Evaluation of Learning Motivation Model
from the Perspective of Course Ideology and Politics 372
  Luyan Lai and Yongdie Che

The 3D Display System of Art Works Based on VR Technology 384
  Huyuan Lu, Qiner Xu, Zhiqiang Chen, and Beixin Zhong

Petrochemical Commodity Price Prediction Model Based on Wavelet
Decomposition and Bayesian Optimization 394
  Lei Yang, Rui Xu, Huade Li, and Zexin Xu

Mutate Suspicious Statements to Locate Faults 407
  Guangsheng Zhan, Shi Cheng, and Jinbao Zhang

Iterative Learning Control with Variable Trajectory Length in the Presence
of Noise 419
  Yuangao Yan, Xixian Tan, and Yunshan Wei

Quality Control Model of Value Extraction of Residual Silk Reuse Based
on Improved Genetic Algorithm 430
  Qi Ji, Mingxing Li, and Chao Shen

Key Technology and Application Research Based on Computer Internet
of Things 442
  Lingwei Wang and Hua Wang

Research on the Innovative Application of Computer Aided Design
in Environmental Design 450
  Lei Wang

Research on Intelligent Clustering Scoring of English Text Based
on XGBOOST Algorithm 459
  Zhaolian Zeng, Wanyi Yao, Jia Zeng, Jiawei Lei, Feiyun Chen,
  and Peihua Wen

Machine Learning-Assisted Optimization of Direction-Finding Antenna
Arrays 476
  Qing Zhang, Miao Gong, Gouqiong Li, Xinyu Ma, Yiheng Chen,
  Fei Zhao, and Sanyou Zeng

Author Index 487

Contents – Part I

Frontiers of Evolutionary Intelligent Optimization Algorithms

An Improved NSGA-II Algorithm with Markov Networks 3
  Yuyan Kong, Jintao Yao, Juan Wang, Peiquan Huang, and Zhenzhen Qiu

An Improved Particle Swarm Optimization Algorithm Combined with Bat
Algorithm 18
  Hongyu Xiao, Nannan Zhao, Zihang Gao, and Xiaojun Cui

An Improved Whale Optimization Algorithm Combined with Bat
Algorithm and Its Applications 26
  Xiaofeng Wang, Jian'ou Wang, and Chanjuan Lin

Fusion of Nonlinear Inertia Weight and Probability Mutation for Binary
Particle Swarm Optimization Algorithm 39
  Jiayu Zhang and Kangshun Li

An Evolutionary Algorithm Based on Replication Analysis for Bi-objective
Feature Selection 49
  Li Kangshun and Hassan Jalil

Improved Particle Swarm Algorithm Using Multiple Strategies 62
  Yunfei Yi, Zhiyong Wang, and Yunying Shi

A Reference Vector Guided Evolutionary Algorithm with Diversity
and Convergence Enhancement Strategies for Many-Objective
Optimization 73
  Lei Yang, Yuanye Zhang, and Jiale Cao

Research on Mine Emergency Evacuation Scheme Based on Dynamic
Multi-objective Evolutionary Algorithm 88
  Furong Jing, Hui Liu, and Yanhui Zang

An Adaptive Dynamic Parameter Multi-objective Optimization Algorithm 101
  Yu Lai and Lanlan Kang

Adaptive Elimination Particle Swarm Optimization Algorithm
for Logistics Scheduling 113
  Kexin Lin, Wei Li, and Yuqi Ou

A Modified Two_Arch2 Based on Reference Points for Many-Objective
Optimization 125
    Shuai Wang, Dong Xiao, Futao Liao, Shaowei Zhang, Hui Wang,
    Wenjun Wang, and Min Hu

Floorplanning of VLSI by Mixed-Variable Optimization 137
    Jian Sun, Huabin Cheng, Jian Wu, Zhanyang Zhu, and Yu Chen

A Multi-population Hierarchical Differential Evolution for Feature
Selection 152
    Jian Guan, Fei Yu, and Zhenya Diao

Research on State-Owned Assets Portfolio Investment Strategy Based
on Improved Differential Evolution 165
    Dong Ji and Dandan Cui

A Particle Swarm Optimization Algorithm with Dynamic Population
Synergy 178
    Qianqian Dong, Wei Li, and Fufa He

Preference-Based Multi-objective Optimization Algorithms Under
the Union Mechanisms 192
    Yi Zhong and Lanlan Kang

Auto-Enhanced Population Diversity with Two Options 207
    Yangcong Ou, Ming Yang, and Jing Guan

Exploration of Computer Vision

Robot Global Relocation Algorithm Based on Deep Neural Network
and 3D Point Cloud 223
    Yan Chen, Zhengying Li, and Wenbin Qiu

Safety Zone and Its Utilization in Collision Avoidance Control of Industrial
Robot 231
    Yongcai Zhang, Yih Bing Chu, and Tian Jiang

Entropy of Interval Type-2 Fuzzy Sets and Its Application in Image
Segmentation 247
    Jianqiao Shen, Haijun Qian, and Huabei Nie

An Improved Algorithm for Facial Image Restoration Based on GAN 254
    Jibo Zhang, Jia Yuan, Dongbo Zhang, and Lu Xiang

A Study of PyTorch-Based Algorithms for Handwritten Digit Recognition 266
    Kangshun Li, Mingchen Xie, and Xuhang Chen

A High-Quality Video Reconstruction Optimization System Based
on Compressed Sensing 277
    *Yanjun Zhang, Yongqiang He, Jingbo Zhang, Zhihua Cui,
    and Xingjuan Cai*

Conv and Efficient Multi-Scale Attention Module for YOLOv5 292
    Xuan Guo and Weidong Huang

BRA-YOLO: Object Detection Algorithm with Bi-Level Routing
Attention for YOLOv5 302
    Xing Huang and Weidong Huang

A Pest Detection Algorithm Based on Improved YOLO 312
    Kangshun Li, Shuizhen He, and Jiancong Wang

Improving Interactive Differential Evolution for Cartoon Face Image
Combination 326
    Bo Tang, Fei Yu, Qingrong Ou, Bang Liang, and Jian Guan

Mask Reconstruction Augmentation and Attention Aggregation for Stereo
Matching 340
    Zhaokui Li, Zhongxin Yang, Jinen Zhang, and Jinrong He

Machine Learning and Its Applications

Construction of an Intelligent Salary Prediction Model and Analysis of BP
Neural Network Applications 357
    Xuming Zhang, Ling Peng, and Ping Wang

Human Action Recognition Classification Based on 3D CNN Deep
Learning 369
    Li Kangshun, Tianjin Zhu, and Hangchi Cheng

A KNN Algorithm Based on Mixed Normalization Factors 388
    Hui Wang, Tie Cai, Yong Wei, and Jiahui Cai

Modified Carnivorous Plant Algorithm Based on Lévy Flight
for Optimizing the BP Model 395
    Chen Ye, Peng Shao, and Shaoping Zhang

CR-IFSSL: Imbalanced Federated Semi-Supervised Learning with Class
Rebalancing 409
   Yutong Xie, Haiyan Liang, Xianmin Wang, Jing Li, Ziyu Cheng,
   Siming Huang, Feng Liu, and Li Guo

Kernel Fence GAN: Unsupervised Anomaly Detection Model Based
on Kernel Function 420
   Lu Niu and Shaobo Li

A News Recommendation Approach Based on the Fusion of Attention
Mechanism and User's Long and Short Term Preferences 429
   Yi Xiong

Research of the Three-Dimensional Spatial Orientation for Non-visible
Area Based on RSSI 443
   Huabei Nie, Jianqiao Shen, Haihua Zhu, Ani Dong, Yongcai Zhang,
   and Yi Niu

Collaborative Filtering Recommendation Algorithm Based on Improved
KMEANS 451
   Xuesong Zhou, Changrui Li, and Jia Shi

Emotion Analysis of Weibo Based on Long Short-Term Memory Neural
Network 463
   Li Kangshun, Weicong Chen, and Yishu Lei

Author Index 471

Machine Learning and Its Applications

Human Flow Prediction Model Based on Graph Convolutional Recurrent Neural Network

Hongwei Su[1,3] and Maria Amelia E. Damian[2(✉)]

[1] Graduate School, University of the East, Manila, Philippines
[2] Faculty College of Engineering, University of the East, Manila, Philippines
mariaamelia.damian@ue.edu.ph
[3] China Urban Construction Design and Research Institute CO., LTD., Guangzhou, China

Abstract. At present, many cities have launched real-time "congestion degree", "traffic index" and other travel reference indicators, according to which urban residents can reasonably choose to travel. However, these real-time monitoring travel reference indicators do not predict the future, but only respond to the past situation. Therefore, it is necessary to study the prediction of urban crowd flow, and thus become the focus of research.

The existing research focuses on dividing cities into grids and predicting the flow of people in each grid area in the next period. However, considering the rapid change of urban crowd flow, measures such as crowd density control require a long response time and the specific location of peak traffic, this paper uses the structure of the graph to divide regions, and innovatively constructs the regional geographic adjacency diagram and regional flow diagram, and uses the graph convolutional cyclic neural network to model the spatial and temporal relationship of regional human flow. In this paper, the public data set of a city taxi is selected for experimental verification and analysis. The experimental results show that the proposed method improves the prediction accuracy of regional passenger flow and reduces the root-mean-square error by 0.04, which is better than other existing schemes.

Keywords: Graph Convolutional Recurrent Neural Network · Traffic Prediction · Long Short-Term Memory Network

1 Introduction

The prediction of urban crowd flow plays an important role in the realization of fine urban management, and it is very necessary to conduct in-depth research on it. However, the accurate prediction of urban crowd flow has always been a very challenging problem. In particular, it is still in its infancy to build a prediction model of future continuous multi-time human flow in multiple areas and locations of cities. This is mainly due to the following difficulties in the prediction of urban crowd flow:

Accurate prediction models need to model the spatiotemporal correlation of the human flow in all regions or sites at various periods, but the high mobility of urban

K. Li and Y. Liu (Eds.): ISICA 2023, CCIS 2147, pp. 3–14, 2024.
https://doi.org/10.1007/978-981-97-4396-4_1

population leads to the complex spatiotemporal correlation of the human flow in all regions or sites.

Urban crowd flow changes rapidly, and scheduling strategies such as crowd density control and resource allocation require a long response time and the exact location of peak traffic. Methods that target specific urban locations, such as traffic stations, and predict future continuous multi-time traffic flow need to be further studied.

The flow of people in various areas and stations of the city changes regularly, and there are periodicity in working days and weekends, but most of the time is at a low level, and there are not many areas and moments of traffic peak. When most of the samples are at the low value level, the model tends to output low predicted values to satisfy the low total error achieved by most of the samples. This led to an "underestimation" of forecasts at the peak.

In recent years, artificial intelligence technology represented by deep learning has made breakthrough progress in image processing, natural language processing, graph data mining and other fields [1]. Compared with traditional machine learning, the method based on deep learning has two outstanding advantages: First, it has outstanding feature learning ability and avoids cumbersome feature engineering, such as shallow automatic learning of detailed features such as edges and textures, and deep learning of abstract features related to categories [2]. Second, the scale and performance of the problems it handles are qualitatively improved compared with the past, such as high-precision pixel-level image semantic segmentation, which can simultaneously output the category judgment results of each pixel of the image [3]. Therefore, how to apply deep learning to a series of temporal and spatial series prediction such as urban crowd flow prediction [4] has attracted the attention of researchers. However, due to the particularity of the spatiotemporal series data of urban crowd flow, how to build a prediction model based on deep learning that can simultaneously model both spatiotemporal and spatial information, how to design a network structure that considers relevant factors of cities and can integrate geospatial information, and how to improve the prediction performance by using the unique periodicity of spatiotemporal series have become the focus of research.

2 Research Status

The traffic prediction problem can be divided into two categories, the first type takes the road section as the research object, and the second type takes the region as the research object. At present, the problem of regional traffic prediction is becoming more and more important in smart cities. Different from the road section flow problem, the regional flow problem is more concerned with the flow situation in the city, in terms of areas. Literature [5] models the distribution and flow of people in a region. Literature [6] models the migration of people between cities. In the prediction of regional human flow, the main problem needs to be solved is how to model the relationship between regional flow and time change, as well as the logic in space. The former is mainly reflected in the smoothness and periodicity of time. Some researchers use nonlinear models such as neural networks for time series prediction [7], and some use linear models such as Kalman filter for time series prediction [8]. Autoregressive Moving Average (ARMA) plays a fundamental role in time series estimation models. Autoregressive Integrated Moving Average model (ARIMA) based on ARMA is widely used in time series prediction.

In addition to the time rule, the spatial rule of regional human flow is equally important. Some professionals obtain the distribution of functional areas of cities by clustering according to the connections between regional human flow, and then estimate the human flow by means of kernel density estimation [9]. Recently, with the rise of deep neural networks, the prediction methods based on deep neural networks have been more and more applied to the prediction of regional pedestrian flow. Literature [10] uses Convolution Neural networks (CNN) to speculate on the flow of people in a region, and establishes a corresponding model for the actual changes of the flow of people through the convolutional method in mathematics. On this basis, the flow of people in a region in the following period of time is predicted. In addition, some experts use Neural networks to model from the perspective of space, and then use Recurrent Neural Network (RNN) to model the time rule to predict the flow of people [11]. Some researchers designed the Deep Spatio-Temporal Residual Networks (ST-ResNet) [12, 13], which divided the urban area according to the grid, represented the human flow as tensor data, and used convolutional neural networks to explore the spatial rules. Then some features are selected manually and put into the channel information center to study the time rule

For the problem of regional pedestrian flow prediction, the previous method has certain limitations:

(1) The traditional regional pedestrian flow forecasting method has disadvantages in the exploration of spatial laws;
(2) Traditional machine learning methods require artificial design attributes and cannot fully describe space-time laws;
(3) Convolutional neural networks can only be used to cut regions into mesh-like inputs. In other words, convolutional operation can only be used for standardized input, without considering the characteristics of real geography. Real road network information rarely presents regular shape, and it is more accurate to describe it with the structure of the graph.

Therefore, in order to predict the regional human flow more accurately, the regional geographic adjacency diagram is constructed according to the adjacency relationship between regions, and the corresponding diagram is constructed according to the actual situation of human flow in different regions. Then, the spatial and temporal attributes of regional human flow were modeled through the heterogeneous graph to explore the mutual influence of regional human flow.

3   Human Flow Prediction Model Based on Graph Convolutional Recurrent Neural Network

Firstly, some problems are formally defined.

Regional human flow is a quantity that describes the flow relationship between regions. In order to better describe the regional human flow, we define the region and the regional human flow as follows:

Definition 1: Urban area. The geographical space of the city is divided into multiple disjoint areas, and the set of divided areas is denoted R, $R = \{r_1, r_2, r_3 \ldots r_n\}$.

Definition 2: Regional traffic. During the period t, the traffic entering a city area r is denoted by $f_{r,t}^i$, that is, the number of people entering area r during the period t. During the period t, the traffic flowing out of a city area r is denoted as f_t^o, that is, the number of people flowing out of area r during the period t. It is synthesized into a binary group ($f_{r,t}^i$, f_t^o), denoted as f_t^r, that is, the human flow of area r in the time period t.

Definition 3: Prediction of regional traffic. For the specified area r and the given time period t, the regional pedestrian flow prediction of the predicted area r during the time period t is f_t^r.

Due to the large number of people and strong mobility, it is difficult to obtain accurate information about the flow of people. Therefore, taxi data should be selected as the data for research. Taxi is one of the important ways for people to travel. The data of passengers getting on and off the taxi are introduced, which covers a large number of population flows in the city and is a kind of sparse sampling of the flow of people in urban areas. Therefore, the study of regional taxi passenger traffic can also reflect the actual change of regional passenger traffic. Compared with other information, it can show the spatiotemporal dependence of human flow to a greater extent.

Based on the adjacency relationship between regions, the map of regional geographic adjacency relationship is constructed, and the corresponding chart is constructed by using the flow of people between regions. Then through the heterogeneous map to model, in order to reflect the relationship between the flow of people in relevant areas. First, define the regional geographic adjacency diagram and regional flow diagram as follows:

Definition 4: Regional geographic adjacency diagram. The undirected regional geographic adjacency graph Gn = (vn, En, wn) is constructed with elements in urban area R as vertices, where vn is the vertices in urban area R divided according to road network, En is the edge set, and its construction mode is as follows: If r_i and r_j in region R are adjacent, then vertex Ri and vertex Rj in region geographic adjacency graph G are connected to form an undirected edge. wn is an adjacency matrix, if there is an edge between two nodes, the corresponding position value is 1, otherwise it is 0.

Definition 5: Regional flow diagram. Taking elements in urban region R as vertices, the flow graph $G_f^t = (Vf, E_f^t, cW_f^t)$ of the directed graph is constructed, where v_f is the vertices in urban region R divided by road network. E_f^t is the set of edges in the period t, constructed as follows: in the period t, if the number of flows from vertex Ri to vertex Rj is greater than the threshold δ, then a directed edge is connected from vertex Ri to vertex Rj. w_f^i is a weighted adjacency matrix whose weight is the sum of the number of people flowing from vertex Ri to vertex Rj in the time period t.

In order to deal with the spatial dependencies of different cases, the weighted sum of the k power of the Laplace transform is calculated as a spectral transform. This is based on Laplacian's KTH power exactly supported by the k-hop neighbor, representing the diffusion of traffic at different scales. Computing the KTH Laplacian matrix is computationally expensive, so the Chebyshev polynomial expansion is applied for an efficient approximation, as shown in formula 1:

$$y^t = g_w(L)x^t = u \sum_{k=0}^{k-1} w_k \wedge k\varphi^k u^t x^t \approx \sum_{k=0}^{k-1} w_k T^k \wedge x^t \qquad (1)$$

where gw (L) is a learning filter based on the Laplace matrix. The parameter is the vector of the Chebyshev coefficient, and $T^k \wedge x^t$ is the value of the Chebyshev polynomial of order k in $\wedge$xt. This approximation reduces the computational cost of filtering from $O(|n|^2)$ to $O(K|\varepsilon|)$.

Based on the graph convolutional neural network (GCN), the matrix in the graph convolutional neural network is introduced Multiplication is replaced by the graph convolution operation $*$ G in formula 1, incorporating spatial dependencies into the gated cycle unit. The graph convolution operation is applied to both input and hidden states to obtain the graph convolution gated cyclic unit. In order to calculate the gradient, the graph convolution gated cyclic unit is superimposed, and the recursion of the fixed number of steps T is expanded, and the backpropagation in time is used. The relevant formula is as follows:

$$r_t = \sigma\left(w_f * GX_t + U_r * GH_{t-1} + B_r\right) \qquad (2)$$

$$U_t = \delta(W_u * GX_t + U_r * GH_{t-1} + B_u) \qquad (3)$$

$$C_t = tanh(w_c * GX_t + U_c * G(r_t \cdot H_{t-1} + B_c)) \qquad (4)$$

$$h_t = U_t \cdot H_{t-1} + (1 - U_t) \cdot C_t \qquad (5)$$

Graph convolutional recurrent neural networks can be used to model the spatial and temporal dependence of regional human flow. We found that cities are divided into functional areas according to their functional characteristics, and people move between them. On the map of geographical adjacency, we stack the traffic in k time periods in the latest period according to relevant requirements. In detail, given the regional geographic proximity graph Gn and the historical regional pedestrian flow data, the most likely regional pedestrian flow value in the future period is inferred. We record the number of people at the vertex r_i of the region at time t as $f^{r_i}{}_t$, and concatenate the number of people at each vertex in the region to obtain the input matrix $X^R{}_t$ of the regional geographic adjacencies graph Gn at time t, and take the number of people at time t−k, t−k + 1,..., t−2, t−1 as historical data to deduce the prediction result $H^n{}_t$ of the regional adjacencies graph at time t.

Similarly, regional flow diagram G_f is used to model the spatial and temporal dependence of regional human flow. In the regional flow graph, the crowd flow from vertex R_i to vertex R_j is described. According to the regional flow relationship, the future regional flow rate can also be predicted. With the help of the graph convolutional recurrent neural network, the corresponding model is built according to the actual situation of local flow. As with the geographical adjacency graph Gn, the prediction result m of the regional flow graph at time t is deduced H_t^f. Finally, it will be combined H_n^t with H_t^f the fully connected network to obtain the predicted result of the final regional human flow $H_t = \delta(Wd[H_t^n, H_t^f])$.

4 Experimental Verification

In order to estimate the regional human flow, this chapter is based on recurrent neural network. In order to verify the experimental effect, we collected the taxi information set released by NYC-Taxi, which mainly included the movement tracks of all vehicles in a city, as well as the number of people involved at that time, cost and other information. In this information set, a city is divided into 265 zones by different sections, and each zone is numbered as the area number for boarding and unloading passengers. We used the same method when dividing the flow of people in the region, and collected the relevant information for the whole year of 2018 to serve as the information set for this training and prediction process. In accordance with the method described in the previous section, the data is divided every two hours, and the number of guests in each period of each area is counted as the inbound and outbound traffic of the area.

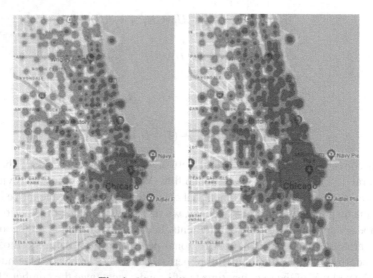

Fig. 1. Map of all taxi areas in a city

Figure 1 shows the map of all taxi areas in a certain city, that is, the division of areas in the experiment. Each area is numbered in the table, and the corresponding name of the queried area is attached to the query table.

Tables 1 and 2 give a brief overview of the data set and list the key information of one of the records, which we will then preprocess and format into the dimensions needed to train the graph convolutional recurrent neural network.

Table 1. Data set situation table

List	Content
Dataset	Xcitytaxi
Data type	Taxi passenger data
Time span	Full Year 2020
Total data volume	400 million

Table 2. Sample description table

List	Sample
Carrying time	2022-09-11 12:01
Alighting time	2022-09-11 12:30
Passenger area number	38
Drop off area number	71
Number of passenger	2
Travel mileage	35.3
Aggregate amount	50

Figure 2 is the connection diagram of the geographic adjacency relationship diagram for one of the large regions. The adjacent regions are connected to form the topology structure on the right, and then the corresponding adjacency matrix is installed according to the way introduced before.

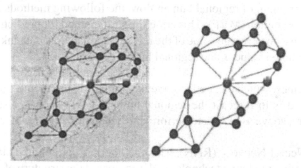

Fig. 2. Schematic diagram of geographical adjacency

For the data set of regional passenger flow mentioned above, the records of passengers boarding the train within the same two hours are extracted, and then these data items are superimposed on the number of passengers according to the area number r_i of passengers boarding the train, and the outgoing flow f_t^o of the area r_i is obtained.

Then, the passenger alighting record in this time slice is extracted, and the number of passengers is superimposed according to the area number r_i of the passenger alighting, as the inbound flow f_t^r of the area r_i. Thus, the binary group (f_t^i, f_t^o) is obtained, and the number of people in the recorded area $\int ti$ during the period t is f_t^r, that is $f_t^r = (f_t^i, f_t^o)$.

For the data required by the regional flow chart mentioned above, it is constructed in the following way: Select the boarding time, construct the regional flow chart every two hours, and screen out the required data items according to the time period t. Next, for the filtered data, record its boarding area number as R_i and getting off area number as R_j. Connect a directed edge from node R_i to node R_j. If this edge does not exist previously, the new weight is the number of passengers in this record; if this edge already exists, the new number of passengers is superimposed to the weight. Because of the construction in this way, some edges may have low weights, which means that the frequency of some activities is low, which will make the edges we build very dense, so we set the threshold δ, after each time slice construction, according to the threshold δ re-screening, if the weight of the edge is greater than δ, the edge is retained, otherwise the edge is removed. Finally, 60% of them are taken as training set to learn the model, 10% as verification set to adjust the hyperparameter, and 30% as test set to verify the model ability.

Evaluation index: The prediction problem of regional pedestrian flow is essentially a regression problem, so we use the average.

Root error (RMSE) to act as a performance measure. The calculation is as follows:

$$RMSE = \sqrt{\frac{1}{N} \sum_{i=1}^{N} (X_i - \overline{X}_i)^2} \tag{6}$$

where N is the sample size, X_i and $\overline{X}_i$ are the true value and the predicted value respectively.

The taxi data set used in this paper is unstructured data divided according to the road network. In order to simultaneously model the spatiotemporal properties, we choose the graph convolutional recurrent neural network. In order to verify the superiority of this structure in the problem of regional human flow, the following methods are compared:

Mean-Value-Function (MVF): This is the simplest method for predicting the regional human flow. We use the average value of the regional human flow data ink periods before time t as the predicted value of the regional human flow at time t.

Exponential Moving-Average (EMA): This is a commonly used method in timing forecasting schemes, which can better solve the prediction of smooth timing problems. Our ultimate goal is to forecast the regional human flow, which is also a time series forecast in nature, so we use EMA to perform the benchmark method of regional human flow forecast.

Recurrent Neural Network (RNN): This method has certain advantages at present and has better timing modeling technology. We compared the structure of many recurrent networks and finally used 2-layer Long short-term memory network (LSTM) to predict the regional traffic flow. And as one of our benchmark solutions. Differential regression moving Model (ARIMA): This method can predict unstable sequences. We also chose ARIMA as our benchmark solution.

Since the partitioning method is inconsistent with the partitioning method of convolutional neural network, the method related to convolutional network is not compared here. In addition to the comparison with the previous benchmark method, it is also compared with some of its own networks that remove some of the design, the scheme is as follows:

Regional adjacent-Graph convolutional neural network (Negihbour-GCN, NGCN): Only the regional geographic adjacency diagram is used, and the first k time slices at time t are spliced into input data, and the graph convolutional neural network is used for prediction to obtain the predicted value of regional passenger flow.

Regional adjacent-flow graph-Graph convolutional neural network (Negihbour-FlowGCN, NFGCN): It integrates regional geographic adjacency diagram and regional flow diagram, hoping to learn cross-regional transfer relationship from the flow diagram and predict the regional human flow.

Negihbour-GCRNN (NGCRNN): Only the regional geographic adjacency graph is used to model and predict the regional human flow.

Regional adjacent-flow Graph-Graph convolutional recurrent neural networks (NegihbourFlowGCRNN, NFGCRNN): This is the method used in this paper, based on a predictive model of graph convolutional recurrent neural networks. By integrating the regional adjacency graph and flow graph, the local similarity and frequent flow pattern are used to predict the regional human flow from time and space.

5 Experimental Results and Analysis

In terms of taxi data, we used the above method to conduct experiments. Table 3 and Fig. 3 show the prediction effect of regional inbound and outbound traffic. The method we use is far better than other schemes. By integrating the two types of connected graphs and modeling the spatial and temporal dependence relationship at the same time, we improve the prediction accuracy and error energy of regional human flow Down to about 0.03.

Table 3. Prediction effect of regional flow standard

Means	Regional discharge RMSE	Zone inlet flow RMSE
MVF	0.1676	0.1684
EMA	0.1194	0.1206
RNN	0.0811	0.0806
ARIMA	0.1068	0.1102
NGCN	0.0734	0.0761
NFGCN	0.0521	0.0439
NGCRNN	0.0487	0.0476
NFGCRNN	0.0319	0.0327

By comparing the final model proposed in this chapter with the benchmark method mentioned above and the RMS error chart, we can find that the simplest MVF algorithm has a high error, which may be due to the fact that the mean value cannot model the regional human flow well. EMA and ARIMA algorithms have been greatly improved compared with MVF algorithms. After we introduced the long short-term memory network, the modeling ability was greatly improved, and the time rule of data could be extracted by LSTM, and the RMSE was reduced to about 0.08.

Next, we introduce two kinds of heterogeneous graphs respectively. First, we fuse the regional geographic adjacency graph and the graph convolutional network, and use the first k time slices at time t for prediction. In the same way, we fuse the regional flow graph. The results show that the RMSE of NGCN method is similar. Finally, we integrate two types of graphs in the graph convolutional neural network at the same time, that is, the NFGCN algorithm, and use the time window k to predict the regional human flow at time t. Compared with a single graph, the RMSE is smaller because the regional flow graph is integrated and can extract the cross-regional movement rule.

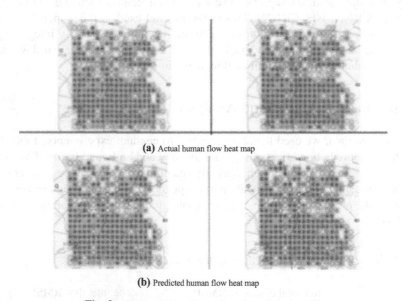

(a) Actual human flow heat map

(b) Predicted human flow heat map

Fig. 3. Comparison of actual and predicted traffic

When we experiment the convolutional network of graphs, we find that when the size of time window k is different, its prediction effect is different. Therefore, we find that it is not enough to model the timing rule only with the time window, so we design the graph convolutional recurrent neural network. First of all, try to use only the geographical adjacency diagram of the region, which is the NGCRNN algorithm. Experiments show that the error is lower than that of graph convolutional network. Finally, the method NFGCRNN in this paper integrates regional geographic adjacency diagram and regional flow diagram to model the spatial and time-dependent relationship, which improves the effect of regional human flow, and the RMSE error can be reduced to about 0.03.

6 Conclusion

In this paper, the proposed algorithm is verified and compared by experiments, and the experimental results fully demonstrate the superiority of the model, and its RMSE can be reduced to about 0.03 in the end, which is significantly improved compared with other models.

This chapter mainly conducts the prediction of regional human flow. Through analysis, it is found that the existing prediction of regional human flow is based on the network format, and the data form is regular data of matrix, so as to explore the spatio-temporal dependence. But in real life, there are many scenarios with non-European space data, and we need to model both time and space laws on these data. Therefore, in order to solve this problem, this paper uses the data of taxi loading and unloading, integrates the regional geographic adjacency diagram and regional flow diagram, and uses the graph convolutional recurrent neural network to predict the regional passenger flow. Finally, we collected data from the publicly available dataset NYC-Taxi, on which the predicted RMSE of the model proposed in this chapter is only 0.03, which is better than all other benchmark schemes.

References

1. Song, C., Lin, Y., Guo, S., Wan, H.: Spatial-temporal synchronous graph convolutional networks: a new framework for spatial-temporal network data forecasting. Proceedings of the AAAI Conference on Artificial Intelligence **34**(1), 914–921 (2020)
2. Fang, X., Huang, J., Wang, F., et al.: Constgat: contextual spatial-temporal graph attention network for travel time estimation at baidu maps. In: Proceedings of the 26th ACM SIGKDD International Conference on Knowledge Discovery & Data Mining, pp. 2697–2705 (2020)
3. Zhang, J., Zheng, Y., Qi, D., et al.: Predicting citywide crowd flows using deep spatiotemporal residual networks. Artif. Intell. **259**, 147–166 (2018)
4. Yao, H., Tang, X., Wei, H., et al.: Modeling Spatial-Temporal Dynamics for Traffic Prediction (2018)
5. Shi, C., JinBao, Z., Zhan, G., et al.: Circuit implementation of respiratory information extracted from electrocardiograms. J. Datab. Manage. **33**(2) (2022)
6. Liu, P., Zhang, Y., Kong, D., et al.: Improved spatio-temporal residual networks for bus traffic flow prediction. Appl. Sci. **9**(4), 615 (2019)
7. Wang, H., Su, H.: Star: a concise deep learning framework for citywide human mobility prediction. In: Proceedings of 2019 20th IEEE International Conference on Mobile Data Management (MDM), pp. 304–309. IEEE (2019)
8. Lin, Z., Feng, J., Lu, Z., et al.: DeepSTN+: context-aware spatial-temporal neural network for crowd flow prediction in metropolis. Proceedings of the AAAI Conference on Artificial Intelligence **33**(1), 1020–1027 (2019)
9. Pan, Z., et al.: Urban traffic prediction from spatio-temporal data using deep meta learning. In: Proceedings of the 25th ACM SIGKDD International Conference on Knowledge Discovery & Data Mining, pp. 1720–1730 (2019)
10. Zheng, C., Fan, X., Wang, C., Qi, J.: Gman: a graph multi-attention network for traffic prediction. Proceedings of the AAAI Conference on Artificial Intelligence **34**(01), 1234–1241 (2020)

11. Yao, H., Tang, X., Wei, H., Zheng, G., Li, Z.: Revisiting spatial-temporal similarity: a deep learning framework for traffic prediction. In: Proceedings of the AAAI conference on artificial intelligence **33**(01), 5668–5675 (2019)
12. Boucher, M.: Transportation electrification and managing traffic congestion: the role of intelligent transportation systems. IEEE Electrification Magazine **7**(3), 16–22 (2019)
13. Pavlyuk, D.: Feature selection and extraction in spatiotemporal traffic forecasting: a systematic literature review. Eur. Transp. Res. Rev. **11**(1), 1–19 (2019)

Research on Text Classification Algorithm Based on Deep Learning

Li Kangshun[1,2(✉)], Junjie Wang[1], and Wenbin Zhu[1]

[1] College of Computer Science, Guangdong University of Science and Technology,
Dongguan 523808, China
likangshun@gdust.edu.cn

[2] College of Artificial Intelligence, Dongguan City University, Dongguan 523808, China

Abstract. Short texts are characterized by their brevity, lack of standardization, and strong contextual dependency. Therefore, improving the accuracy of short text classification is a significant research topic. The use of deep learning models to effectively mine the semantic information of short texts for better text classification results warrants further exploration and research. In this paper, we propose a TextRCNN-ECA model based on the ECA attention mechanism to address the issue of weak feature extraction ability in the BiLSTM model. Firstly, our model utilizes BiLSTM to capture contextual feature information in the short text, which is then concatenated with the original word vectors to retain and control the transmission of contextual information effectively. Secondly, we employ max pooling to obtain the vector feature with the highest relevance to the text. Additionally, we introduce the ECA attention mechanism to assign weights to the text feature vectors, allowing for the identification of more critical features within the text. Experimental results demonstrate that TextRCNN-ECA exhibits a superior ability to capture key information in text.

Keywords: Text classification · Convolutional neural network · Recurrent neural network · attention mechanism

1 Introduction

1.1 Research Background and Significance

Short text classification plays an important role in the efficient utilization of short text data. The study of text classification began in the 1960s. With the development of machine learning, researchers have found some efficient techniques and applied them to text classification tasks. In short text classification task, deep learning technology can extract deeper text features, so as to better help people extract valuable information from short text data. In view of the strong feature expression ability of deep learning, based on the advanced deep learning theory and on the basis of training word vector representation with nested word embedding model, how to use deep learning model to better mine the semantic information of short text, so as to achieve better text classification effect is still a topic worthy of further exploration and research.

K. Li and Y. Liu (Eds.): ISICA 2023, CCIS 2147, pp. 15–27, 2024.
https://doi.org/10.1007/978-981-97-4396-4_2

1.2 Research Status at Home and Abroad

With the continuous progress of artificial intelligence technology, text classification algorithms based on machine learning methods have been proposed successively [7, 9, 10]. However, in the pre-processing stage, machine learning algorithms mainly rely on manual methods to obtain sample features, and the classification results are limited by feature extraction to some extent. On the other hand, when faced with complex scenarios such as massive data and multiple semantics, the time complexity of machine learning algorithms is relatively high, which is slightly lacking in practical applications.

Recurrent Neural Networks (RNN) with short-term memory ability can effectively represent contextual semantic information when processing text data with sequence characteristics [2]. Rao [3] used word vector and LSTM (Long Short Term Memory [4] to capture contextual dependencies in text, improving the accuracy of text classification. Attention mechanism is one of the common improvement methods in the innovation of deep learning algorithms, which can enhance the ability of models to extract key features [5]. Zhou used LSTM network and attention mechanism to solve cross-language emotion classification task [6].

1.3 The Main Research Content of This Paper

To address the issue of poor feature extraction in the BiLSTM model, we propose the TextRCNN-ECA model, which incorporates the Efficient Channel Attention (ECA) mechanism. Firstly, the model utilizes BiLSTM to capture long-distance feature information within the text. This information is then concatenated with the original word vectors, enabling effective retention and control of the transmission of contextual features. Subsequently, the maximum pooling operation is employed to obtain vector features that exhibit the highest correlation with the text.

2 Introduction to Relevant Theories and Techniques

2.1 Text Classification Task Overview Text Classification Task Overview

Text classification is a basic and typical task in natural language processing. Its purpose is to automatically classify text resources according to pre-set rules and knowledge laws, and it is also an effective method to deal with text information overload. Common applications include news classification, public opinion analysis, mail filtering, and so on (Fig. 1).

2.2 Text Preprocessing

Text preprocessing is particularly important in the whole process of text classification, and it is a very key part. The processing result has a great influence on the final text classification effect. The specific operations of text preprocessing are generally related to specific tasks, among which the commonly used operations include: corpus cleaning, Chinese word segmentation and word stop and so on.

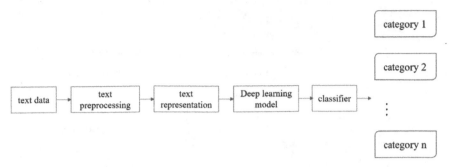

Fig. 1. Text classification process diagram based on deep learning.

2.3 Text Representation

For a computer to understand text data, the text data must be transformed into a mathematical vector pattern. Word embedding is a technique for embedding text data into a vector space in the form of a mathematical vector. Each word vector in the vector space represents the semantic information of the corresponding word.

3 TextRCNN-ECA Model Based on BiLSTM

In order to improve the effect of BiLSTM model on text classification task, this chapter designs TextrCNN-ECA model based on TextRCNN and channel attention mechanism ECA. Secondly, it introduces the attention mechanism of ECA, its motivation and its reformed structure. The data set used in this paper is then introduced, analyzed, and processed. Then, in the experimental section, we discuss the positive effects of attention mechanism and maximum pooling operation on the text classification effect of BiLSTM model, and the optimal parameter combination of TextRCNN-ECA model. Finally, by comparing different text classification models, it is proved that TextRCNN-ECA model has better text classification effect.

3.1 TextRCNN-ECA Model Text Classification

The structure of the TextRCNN-ECA model is shown in Fig. 2.

3.2 Data Processing and Word Embedding Layers

Word embedding is a technique for representing words as low-dimensional dense vector forms on a continuous space. Assuming that the sentence length after length normalization is n, the word embedding model finally outputs a word vector matrix $W \in Rn \times e$, where the dimension of a word corresponding to a word vector is e.

The word embedding model in this paper adopts the pre-trained Word2Vec model, which is the word embedding model of "Sogou News pre-trained word + word 300d". In the process of TextRCNN-ECA model training, the word embedding layer will use

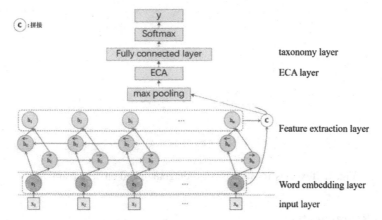

Fig. 2. TextRCNN-ECA model structure diagram.

the training set in this paper to fine-tune the Word2Vec model, so as to generate the word vector matrix that is more suitable for the data set in this paper.

Before the text data is entered into the word embedding layer, it needs to be standardized for text data, as shown in Table 1.

Table 1. Text data standardization process.

Text data standardization process
(1) Segmentation of text and labels according to the format of the data set
(2) Divide the text data by characters and input it into the list
(3) According to the length of the list, determine whether the text data reaches the length n set by the model
(4) If not, use the < pad > character to expand the length of the list to n; If the length of the list is greater than n, the first n bits of the list are taken as input
(5) Convert the characters in the list to the corresponding numeric index according to the word list
(6) Repeat the above steps to finally convert all the text data into a list form

The word embedding layer finally converts the One-hot encoding in the list into Word2Vec word vector matrix and outputs it by looking up the table.

3.3 Feature Extraction Layer

The feature extraction layer of TexTRCNN-ECA model adopts the structure of TextRCNN model. RNN can capture context information better than CNN, which is conducive to capturing long-distance text dependencies. TextRCNN uses RNN to replace the convolutional layer in TextCNN. The model takes the original word vector and context as its text feature vector, and uses the middle layer of bidirectional RNN to obtain the context information on both sides. Then the context information is combined with the original word vector to form a new text feature vector, and then input into the pooling layer.

In TextRCNN's original paper [11, 12], the RNN model used is different from the traditional RNN model, and its RNN model formulas are as follows: Eq. (1) and Eq. (2).

$$c_l(w_i) = f(W^{(l)}c_l(w_{i-1}) + W^{(sl)}e(w_{i-1})) \tag{1}$$

$$c_r(w_i) = f(W^{(r)}c_r(w_{i+1}) + W^{(sr)}e(w_{i+1})) \tag{2}$$

The traditional RNN model formula is such as Eq. (3) and Eq. (4).

$$c_l(w_i) = f(W^{(l)}c_l(w_{i-1}) + W^{(sl)}e(w)) \tag{3}$$

$$c_r(w_i) = f(W^{(r)}c_r(w_{i+1}) + W^{(sr)}e(w)) \tag{4}$$

In this article, TextRCNN adopts the traditional RNN model structure, namely formula (3) and formula (4).

The RNN model portion of the TextRCNN model can use LSTM or other RNN variants. In this chapter, BiLSTM model is selected as the RNN model in TextRCNN model.

3.4 ECA Layer

Wang constructed a lightweight channel attention module ECA in 2019 [1], which effectively reduces model complexity while maintaining model performance through appropriate cross-channel interaction. The ECA attention mechanism module structure is shown in Fig. 3.

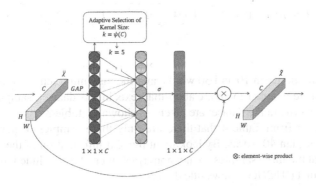

Fig. 3. ECA attention mechanism module structure diagram.

Wang argued that in the case of fixed grouping, the channel dimension is proportional to the size of the convolution kernel k [1]. Considering that the dimension of the channel is generally an integer power of 2, the calculation formula between the size of the convolution kernel k and the dimension of the channel C is shown in Eq. (5).

$$k = \varphi(C) = \left| \frac{\log_2(C)}{\gamma} + \frac{b}{\gamma} \right|_{odd} \tag{5}$$

3.5 Classification Layer

The classification layer consists of a fully connected layer and a classifier. The fully connected layer expands the text feature vector obtained from the previous layer after ECA attention mechanism weighting into a one-dimensional vector, which provides input to the classifier, and the number of neurons is determined by the number of target categories. The classifier selects the Softmax activation function to calculate the probability distribution of the text in each category and get the final text classification result.

3.6 Experimental Design and Result Analysis

3.6.1 Experimental Data Set Introduction and Analysis

The data set used in this chapter is the THUCNews data set generated after screening and filtering the historical data of Sina news RSS subscription channels from 2005 to 2011.For the collated THUCNews news title data set, the statistics of its news title length are plotted, as shown in Table 2.

Table 2. THUCNews Statistics on the length of news headlines.

Header text length statistics	count	mean	std.	min.	25%	50%	75%	max.
amount	836070	19.473	4.094	2	17	20	23	48

It can be seen from Table 3 that the title text of the sample in THUCNews news title data set is mostly between 10 and 30 words, and the maximum length of the title is 48. In order to analyze whether there are abnormal texts in this dataset, samples with titles longer than 40 words in the dataset are taken, as shown in Table 3.

It can be seen from Table 4 that there are only three samples of news texts with headlines longer than 40 words. By looking at the text information of the samples, you can confirm that the three samples are not abnormal. It can be concluded that there is no abnormal text in THUCNews news title data set.

In general, in the problem of text multi-classification, the proportion of each sample category has a great influence on the classification effect. Therefore, the distribution of sample categories should be considered before the task of text multi-classification.

For the THUCNews news title dataset, the quantity and proportion of each news category are shown in Table 4.

Table 3. Title text that is longer than 40 words.

caption text	category	length for heading
Hilton MaldivesResort&Spa Rangali	household	48
Final Q2 personal consumption expenditures quarterly price rate 1.4% Previous: 1.3% estimated: 1.3%	stock	42
Us October core Producer Price Index m/m –0.6% Previous: –0.1% estimated: 0.1%	stock	42

Table 4. The number and proportion of each news category in the THUCNews headline data.

serial number	category	quantity	proportion
0	finance and economics	37098	0.0444
1	lottery	7588	0.0091
2	hous	20050	0.024
3	stock	154398	0.1847
4	househole	32586	0.039
5	education	41936	0.0502
6	technology	162929	0.1949
7	society	50849	0.0608
8	fashion	13368	0.016
9	current politics	63086	0.0755
10	sports	131600	0.1574
11	constellation	3578	0.0043
12	game	24373	0.0292
13	recreation	92631	0.1108

It can be seen from the statistical results in Table 5 that there are a total of 14 news categories in THUCNews headline data set, among which science and technology news has the largest number, with a total of 162,929 data samples, accounting for nearly 20%. It was followed by stock news and sports news. The number of lottery news and horoscope news is relatively small, both of which are less than 10,000. From the distribution diagram of news categories, it can be seen more directly that there is an obvious sample imbalance in THUCNews news title data set. Therefore, in order to keep the data distribution of training set, verification set and test set as uniform as possible, the data set is first randomly scrambled when the training set, verification set and test set are divided, and then the data set is divided by hierarchical sampling. In this paper, the partition ratio of training set, verification set and test set is 7:1.5:1.5. After partitioning, the training set contains 585,249 samples, while the verification set and test set contain 125,410 and 125,411 samples respectively.

3.6.2 Contrast Model

(1) TextRNN model

First, input the text X = [x1, x2,..., xn] into the word embedding layer to calculate the word vector E = [e1, e2,..., en]. Then put the word input vector E to BiLSTM network, take the forward and reverse LSTM hide status to nod on the final time step (" h "_" n ")→ and ←(" h "_" n "), then joining together get the "h" _ "n" = "[" (" h" _ "n") → "; ("h" _"n") ←"]", and then through the full connection layer and Softmax activation function to get the final text classification result (Fig. 4).

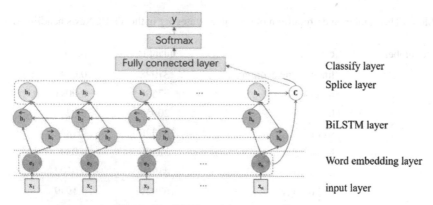

Fig. 4. TextRNN model structure diagram.

(2) TextRNN-Attention model

The introduction of attention mechanism into the BiLSTM model can effectively improve the accuracy of model feature extraction [6].

Output H = [h1, h2,... on BiLSTM hidden layer using attention mechanism, hn], assuming q is a query vector, the correlation between q and hi is calculated by scoring function, and then Softmax function is used for normalization processing. Finally, the attention distribution of q on each hi a = [a1, a2,..., an] is obtained. Taking a$_i$ as an example, the relevant calculation formula is shown in Eq. (6).

$$a_i = \text{Softmax}(s(h_i, q)) = \frac{\exp(s(h_i, q))}{\sum_{i=1}^{n} \exp(s(h_i, q))} \tag{6}$$

Finally, the output H of the BiLSTM hidden layer is weighted and summed according to the attention distribution, as shown in Eq. (7).

$$context = \sum_{i=1}^{n} a_i \cdot h_i \tag{7}$$

The calculation method of the scoring function in this chapter is shown in Eq. (8).

$$s(h, q) = h^{\text{T}} \cdot q \tag{8}$$

(3) TextCNN model

    TextCNN uses one-dimensional convolution to obtain n-gram representation of text features. TextCNN can flexibly process various sequence information of text by adjusting the size of convolution kernel, thus improving the text classification ability of the model.

(4) DPCNN model

The DPCNN model is a deep convolutional model, as shown in Fig. 5.

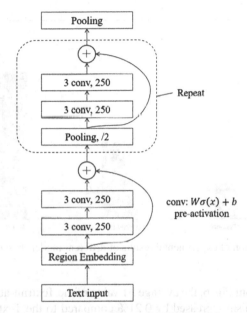

Fig. 5. DPCNN model structure diagram DPCNN model structure diagram.

The macro average F1 value is defined as Formula (9).

$$Macrof - \frac{1}{n} \sum_{i=1}^{n} \frac{2 \times TP_i}{2 \times TP_i + FP_i + FN_i} \tag{9}$$

The loss function is defined in the form shown in Eq. (10).

$$L(y_i, f(x_i; \theta)) = - \sum_j w_j y_{ij} \log f(x_i; \theta)_j, \; w_j = \frac{\frac{1}{n_j}}{\sum_j \frac{1}{n_j}} \tag{10}$$

3.6.3 Experimental Design and Result Analysis

The experiments in this section obtained the optimal evaluation indexes of each model on the test set through multiple rounds of training and model tuning.

(1) In order to illustrate the influence of Attention mechanism and maximum pooling operation on the BiLSTM text classification results on the dataset in this paper, the classification results of Textrnn-attention model and TextRCNN model are compared with the classification results of TextRNN model.

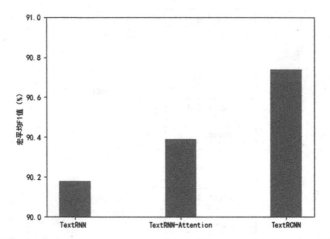

Fig. 6. Comparison of experimental results of different models based on BiLSTM.

As can be seen from Fig. 6, the average F1 value of the Textrnn-attention model with the Attention mechanism increased by 0.21% compared to the TextRNN model without the attention mechanism. The TextRCNN model with the addition of the maximum pooling layer improved the macro average F1 value by 0.56% compared to the TextRNN model. This shows that both the attention mechanism and the maximum pooling operation can make the model capture better text feature vectors, and thus make the model's text classification better. It can also be observed from Fig. 6 that in the THUCNews news title dataset in this paper, the maximum pooling operation improves the BiLSTM model better than the attention mechanism.

(2) As can be seen from Table 5, when the number of LSTM layers is 1, the macro average F1 value of the TextRCNN-ECA model is significantly higher than that of the LSTM layers is 2, indicating that for the TextRCNN-ECA model, the classification effect of the model gradually deteriorates with the increase of the number of LSTM layers. In addition, because the training speed of LSTM is relatively slow, and the measurement of time factor, the number of LSTM layers is 1. At this time, the classification effect of the model is better and the efficiency is higher.

Table 5. Table of experimental results of different parameter combinations of TextRCNN-ECA.

Parameter combination (batch size, number of LSTM layers, number of hidden LSTM layers)	Macro average F1 value (%)
(128, 1, 128)	90.28
(128, 1, 256)	90.61
(128, 1, 512)	90.45
(256, 1, 128)	90.00
(256, 1, 256)	**91.46**
(256, 1, 512)	90.95
(512, 1, 128)	90.58
(512, 1, 256)	90.80
(512, 1, 512)	91.06
(128, 1, 128)	82.22
(128, 2, 256)	82.12
(128, 2, 512)	82.37
(256, 2, 128)	83.96
(256, 2, 256)	83.37
(256, 2, 512)	83.35
(512, 2, 128)	84.88
(512, 2, 256)	83.32
(512, 2, 512)	84.54

(3) In order to illustrate the advantages of TextrCNN-ECA model in text classification, TextRNN, Textrnn-attention, TextRCNN, TextCNN, DPCNN and other common text classification models were compared with the classification results of TextrCNN-ECA model.

The parameter Settings of TextRCNN-ECA model in this experiment are shown in Table 6.

The experimental results are shown in Fig. 7.

Table 6. TextRCNN-ECA model parameters.

Name of parameter	Parameter values
text length	40
Word vector dimension	300
Pre-trained word vectors	sgns.sogou.char
LSTM Number of layers	1
LSTM Number of hidden layers	256
activation function	ReLU
optimizer	Adam
learning rate	1e-3
Epoch number	20
Batch size	256
Number of batches processed in the early stop method	1500

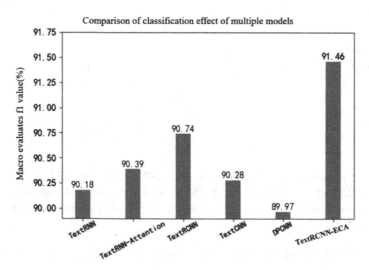

Fig. 7. Comparison of classification effect of multiple models.

4 Summary and Prospect

To solve the problem of poor feature extraction in BiLSTM model, TextRCNN-ECA model is proposed in Sect. 3.1. The model uses BiLSTM to obtain the context information in text and spliced it with the original word vector, so as to effectively retain and control the transmission of the context feature information. Then the vector features with the highest correlation with the text are obtained through the maximum pooling operation. Then ECA attention mechanism is introduced to weight the text feature vectors filtered by the maximum pooling operation, so as to obtain the more critical features in the text.

In the experiment, the effects of attention mechanism and maximum pooling operation on the results of BiLSTM text classification on the THUCNews news title dataset were discussed, and the effects of three parameters, namely, training batch size, LSTM layer number and LSTM hidden layer number, on the TextRCNN-ECA model were discussed. The optimal parameter combination of TextRCNN-ECA model is found. Finally, the comparison experiment proves that TextRCNN-ECA has better performance of text classification.

References

1. Wang, Q., Wu, B., Zhu, P.F., et al.: ECA-net: efficient channel attention for deep convolutional neural networks. IEEE/CVF Conf. Comput. Vis. Pattern Recogn. **2019**, 11531–11539 (2020)
2. Elman, J.L.: Finding structure in time. Cogn. Sci. **14**, 179–211 (1990)
3. Rao, A., Spasojevic, N.: Actionable and political text classification using word embeddings and lstm. ArXiv (2016). abs/1607.02501
4. Hochreiter, S., Schmidhuber, J.: Long short term memory. Neural Comput. **9**(8), 1735–1780 (1997)
5. Bahdanau, D., Cho, K., Bengio, Y.: Neural machine translation by jointly learning to align and translate. CoRR (2014). abs/1409.0473
6. Zhou, P., Shi, W., Tian, J., et al.: Attention-based bidirectional long short-term memory networks for relation classification. In: Proceedings of the 54th Annual Meeting of the Association for Computational Linguistics (vol. 2: Short Papers) (2016)
7. Colas, F., Brazdil, P.: Comparison of SVM and some older classification algorithms in text classification tasks. In: IFIP AI (2006)
8. Johnson, R., Zhang, T.: Deep pyramid convolutional neural networks for text categorization. In: Annual Meeting of the Association for Computational Linguistics (2017)
9. Trstenjak, B., Mikac, S., Donko, D.: KNN with TF-IDF based framework for text categorization. Proc. Eng. **69**, 1356–1364 (2014)
10. McCallum, A., Nigam, K.: A comparison of event models for naive Bayes text classification. In: AAAI Conference on Artificial Intelligence (1998)
11. Lai, S., Xu, L., Liu, K., et al.: Recurrent convolutional neural networks for text classification. In: AAAI Conference on Artificial Intelligence (2015)
12. Li, C., Zhou, A., Yao, A.: Omni-dimensional dynamic convolution. ArXiv (2022). abs/2209.07947

MobilenetV2-Based Network for Bamboo Classification with Tri-Classification Dataset and Fog Removal Training

Yan Chen, Dehao Shi[✉], and Hongxing Peng

School of Mathematics and Informatics, South of China Agricultural University,
Guangzhou 510642, China
shi382760724@stu.scau.edu.cn

Abstract. The problem of low recognition accuracy of the front and back sides of bamboo slices has been a challenge in traditional bamboo processing, which has hindered the large-scale promotion of fully automated production equipment. In this paper, a classification recognition network based on MobilenetV2 was proposed to tackle this problem. The materials for the front and back sides of processed bamboo were collected on-site in the factory. The bamboo slices were initially recorded using a camera, and then the images were captured by capturing frames from the recorded videos. The classification accuracy reached 98%-99% after training on three different datasets in a three-classification experiment. Nevertheless, the recognition accuracy may decrease in actual production and processing environments, due to the presence of dust. To overcome this issue, two types of datasets were selected for screening: clear and unclear. The recognition accuracy of MobileNetV2 training on the unclear dataset showed a noticeable decline. However, the overall recognition accuracy can reach 96%-97% after training the dehazed MobileNetV2 network.

Keywords: Classification · CNN · MobelNetV2 · Deep Learning · Industrial Applications of Bamboo

1 Introduction

The production of bamboo filaments is a critical process for manufacturing bamboo products. Bamboo filaments are made by processing bamboo slices, which are in turn made from bamboo. The process starts by cutting the bamboo into tubes, approximately two meters long. The tubes are then split into bamboo slices by machines. Finally, the bamboo slices are fed into machines to be processed into bamboo filaments. In traditional bamboo filaments processing, each step is carried out by manual operation of the machine to process the bamboo continuously, including the step of manually placing the front side of the bamboo slice upward and feeding it into the processing machine.

This production mode requires a large amount of labor for repetitive tasks. In order to improve efficiency and reduce costs, machines that automatically feed bamboo slices have been developed to replace manual labor. The automated machine is able to pick up bamboo slices from the material hopper one by one and place them with the front side facing up into the processing machine for final processing into bamboo filaments. The primary task of the automated phase of this process is to recognize the front and back sides of the bamboo slices to determine whether they should be flipped. Currently, the recognition of the front and back sides of bamboo slices is mostly determined by setting color thresholds based on the RGB three-parameter recognition method. However, this method is affected by environmental factors, resulting in low accuracy rates, which has prevented it from being widely adopted. With the development of computer vision and artificial intelligence algorithms, new directions and ideas have emerged, providing new opportunities for the classification recognition of bamboo strip front and back sides.

Artificially-designed neural networks such as MobileNet [1, 3], ThunderNet [4], ShuffleNet [2, 5], and SqueezeNet [6] have achieved remarkable results. The MobilenetV2 model used in this paper has been successfully applied in many applications. Mohamed Almghraby et al. [7] have applied MobilenetV2 in real-time detection of whether a person is wearing a mask, achieving a verification accuracy of up to 98%. Qian Xiang et al. [8] have designed and trained the MobilenetV2 network model for a fruit-picking robot, accurately classifying various fruits.

This paper specifically addresses the issue of low recognition accuracy for bamboo slices. By collecting and compiling various types of training and testing sets for bamboo slices, MobilenetV2-based neural network models were adjusted and optimized in conjunction with dehazing networks. The datasets were extensively cross-examined, and a model was ultimately obtained that could effectively recognize the front and back sides of bamboo slices in complex environments. This model is capable of accurately identifying bamboo strips that have lost their original color due to contamination by dust and mud in a fast and efficient manner, even in dusty environments.

2 Method

The overall architecture of the model is shown in the figure below. First, the image is processed by the dehazing network, in this case the GCAnet. The network uses multiple convolution and deconvolution modules and applies residual structures to dehaze the image. After dehazing, the image undergoes a one-dimensional convolution to upsample the image. Then, the image enters the Bottleneck module for convolutional operation. In the process of stepwise downward convolution, residual structures are used to add feature maps of the same size. After 17 layers of Bottleneck, there is another round of convolution and pooling operations. Finally, the feature map is flattened into a one-dimensional vector with N categories, which is output with the Softmax function (Fig. 1).

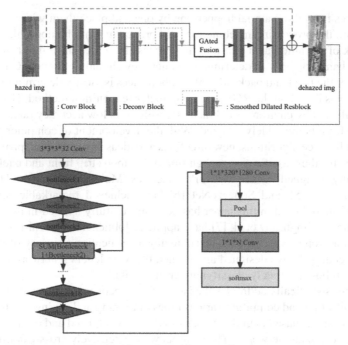

Fig. 1. Network architecture diagram.

2.1 Dehaze Network

GCAnet is a new end-to-end gated context aggregation network for image dehazing, which is trained by directly learning the residual between the original and the hazy image. The network uses smooth extended convolutions to avoid grid-like artifacts and uses a gating subnetwork to integrate features from different levels. First, the network encodes the image features and then incorporates context information to fuse features of different levels. Next, the decoded feature map is obtained by solving the residual map and added to the hazy image to achieve dehazing.

The network model consists of three convolutional blocks as an encoder, a deconvolutional block and two convolutional blocks as a decoder, and a smooth dilated residual block in the middle to learn contextual information. Meanwhile, a gated fusion network is used to fuse features at different levels. Finally, the network outputs the decoded feature map, obtains the residual, and adds it to the hazy image to achieve the dehazing effect.

2.2 MobilenetV2

After the dehazing network, the image is input into MobilenetV2 for model training. MobilenetV2 is a lightweight network proposed in 2019 that adopts depthwise separable convolution used in MobilenetV1 and introduces inverted residual to improve network performance. The specific network model of MobilenetV2 is shown in Fig. 2. The network mainly consists of a 3x3 convolutional network, 17 Bottleneck linear bottleneck layers, and a 7x7 pooling layer, followed by Softmax for classification. In Bottleneck, when s = 1, i.e. stride = 1 and the output size is the same as the input size, inverted residual structures are applied. When s = 2, inverted residual structures are not used.

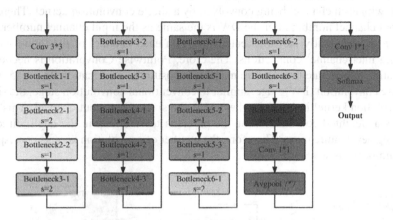

Fig. 2. Structure of MobilenetV2 model.

Bottleneck.
The structure of Bottleneck is shown in Fig. 3. This structure is divided into two types. The first type is an inverted residual structure with a stride of 1, which is different from the traditional residual structure that first reduces dimensionality and then increases dimensionality. This structure first uses pointwise convolutions for dimensionality expansion, applies depthwise convolutions on the expanded feature maps, and finally uses pointwise convolutions for dimensionality reduction. The second type of structure has a stride of 2 and does not use inverted residual structures. These two types of Bottlenecks are alternately used in the 17 bottleneck stacking process to form the main part of the MobilenetV2 network.

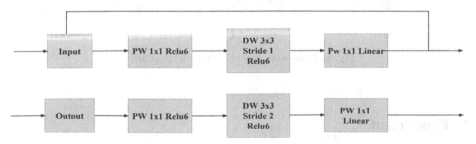

Fig. 3. Bottleneck **network structure.**

Depth-Wise Convolution and Point-Wise Convolution.

Depthwise Convolution and Pointwise Convolution both belong to depthwise separable convolution. The principle diagram of them is shown in Figs. 4 and 5. In Fig. 4, in the depthwise convolution (DW) stage, each convolution kernel only works on a single channel, with each channel being convolved by a single convolution kernel. Therefore, the output channel number after the DW is the same as the input channel number. This operation independently performs calculations on each channel of the input, without extracting inter-channel connections. Therefore, Pointwise Convolution is needed for combination to generate new feature maps with positional information. In Fig. 5, in the pointwise convolution (PW) stage, similar to conventional convolution stage, the size of the convolution kernel is 1x1xM, where M is the size of the input channel. This layer performs a weighted combination on the previous feature map in the depth direction, generating new feature maps. Therefore, the number of convolution kernels corresponds to the number of new feature maps.

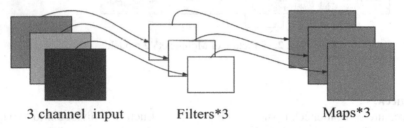

3 channel input   Filters*3   Maps*3

Fig. 4. Depth-wise convolution model diagram.

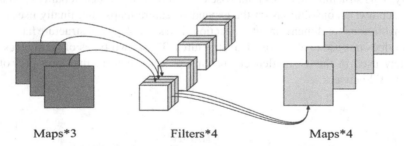

Maps*3   Filters*4   Maps*4

Fig. 5. Point-wise convolution model diagram.

3   Experiment

3.1   Evaluation Indicators

In this experiment, Top-1, Top-2, Recall, and Precision were used as the main evaluation indicators.

Top-1 Indicator.
The Top-1 and Top-2 indicators are the probability that the value of the correct category in the final network output vector is in the first or top 2 rankings.

Recall and Precision Indicators.
First, we introduce the concept of TP TN FP FN

TP (True Positives) means that the sample is divided into positive samples, and is divided correctly.

TN (True Negatives) means that the sample is divided into negative samples and is correctly divided.

FP (False Positives) is the fact that the sample is negative, but is considered positive. FN (False Negatives) means that the fact is a positive sample, but is considered a negative sample.

Then Recall is the probability that the sample considered by the model to be correct and indeed true is the probability of all correct samples. The formula is Recall = TP/(TP + FN). And Precision is the probability that the samples considered by the model to be correct and indeed correct are the probability of all samples considered by the model to be correct. The formula is P recision = TP/(TP + FP).

3.2 Construction of the Dataset

The bamboo data in this study were collected by shooting from an overhead view and recording videos. The frames of the video were processed manually to obtain positive and negative sides of bamboo and the background images, which mainly include three categories: bamboo front, bamboo back, and background. Two datasets were created based on the presence of bamboo nodes and the type of background images, with clear and unclear versions of each dataset. The video frames in the datasets were manually processed to distinguish between the positive and negative sides of bamboo and the background. This was done to verify the neural network's ability to recognize bamboo in complex scenes where image quality may vary.

Deep learning relies heavily on large datasets. To avoid the phenomenon of model overfitting during training, the dataset created in this study used methods of random rotation, cropping, and flipping for data augmentation (Table 1, Figs. 6 and 7).

Table 1. Amount of images in each dataset.

Categories	Amount
Front	540
Front side with bamboo node	400
Back	460
Back side with bamboo node	400
Background	890

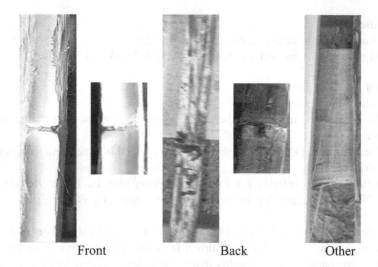

Front   Back   Other

Fig. 6. Clear image dataset.

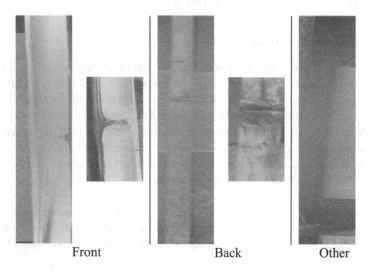

Front   Back   Other

Fig. 7. Unclear image dataset.

3.3 Result and Discussion

The experiment utilized 224*224 pixel-sized images to train and test a MobilenetV2 network with a dehazing module. A dynamic learning rate was employed, with the training process gradually decreasing the learning rate from $1 \times e^{-3}$ to $0.01 \times e^{-3}$ (i.e., 0.01 times the maximum learning rate). With a BatchSize of 64, the model was effectively trained to improve training speed and convergence. Additionally, the Adam optimizer was used to further accelerate the training process.

To avoid the network from classifying images that are not bamboo into either the positive or negative class, a third "background" class was added in the three-class network. The three classes are "positive," "negative," and "background." This helps ensure that non-bamboo images are not misclassified by the network during training.

Experiment in Ideal Environment.
Table 2 displays two types of datasets, wherein the images are all clear, representing training and testing in ideal conditions. The Top-1 accuracy in both datasets is remarkably high, reaching 98% ~ 99%, with recall rates at 99.6% or higher. The second dataset, included to address issues where networks tend to learn primarily based on color features, incorporates bamboo nodes to enable networks to learn the positive and negative shapes of bamboo nodes. This intends to reduce instances where factory-produced bamboo undergoes discoloration or other changes in color, leading to incorrect identification.

Table 2. Experiment in ideal environment.

Category	Top-1	Top-2	Recall	P
raw dataset	98.92%	100%	99.84%	99.6%
adding bamboo joint images	99.06%	100%	99.6%	99.61%

Training with an Unclear Data Set.
In Table 3, all indicators of the first type of training dataset showed favorable performance in the test set with non-clear images, but the accuracy was not high in the test set with clear images. As for the second type of training dataset, since the training set was clear, it only works well in the test set with clear images but not in non-clear environments. Therefore, it can be concluded that neither of these two training datasets can achieve good performance simultaneously on both non-clear and clear test sets, which means a single network cannot correctly classify both clear and non-clear images. Hence, a dehazing algorithm is introduced to pre-process the dataset, resulting in uniformly dehazed images for training experiments.

Experimental Results of Dehazing Algorithm.
In Table 4, the training dataset undergoes haze removal through a GCAnet algorithm, followed by training using MobilenetV2. Top-1 is an important indicator for determining the model's performance, and combined with Table 3, the accuracy of the models trained with dehazed images shows a slight improvement compared to models trained without dehazing, using the same dataset. However, only when performing dehazing operations on the test set, could the model achieve good performance, indicating that the dehazing algorithm played a certain role. Nonetheless, the results also show that using dehazing algorithms on clear image datasets still resulted in poor performance, indicating that a single model cannot be used universally for detection.

Table 3. Training results for unclear and clear datasets.

Training set	Test set	Category	Top-1	Top-2	Recall	P
Unclear training set	unclear test set	raw dataset	98.52%	100%	99.84%	99.6%
		adding bamboo joint images	96.77%	100%	99.45%	99.28%
	clear test set	raw dataset	90.19%	99.8%	98.31%	98.02%
		adding bamboo joint images	89.22%	99.53%	95.66%	95.75%
Clear training set	unclear test set	raw dataset	85.85%	99.46%	93.44%	96.20%
		adding bamboo joint images	80.12	99.26	88.73	92.53
	clear test set	raw dataset	98.92%	100%	99.84%	99.6%
		adding bamboo joint images	99.06%	100%	99.6%	99.61%

Table 4. The experimental results on the dehazed training set.

Training set	Test set	Category	Top-1	Top-2	Recall	P
Dehazed unclear training set	unclear test set	raw dataset	98.52%	100%	99.84%	99.6%
		adding bamboo joint images	96.77%	100%	99.45%	99.28%
	dehazed unclear test set	raw dataset	90.19%	99.8%	98.31%	98.02%
		adding bamboo joint images	89.22%	99.53%	95.66%	95.75%
	dehazed clear test set	raw dataset	85.85%	99.46%	93.44%	96.20%
		adding bamboo joint images	80.12	99.26	88.73	92.53

The Training with Both Dehazed and Clear Datasets Combined.
Finally, a new dataset that combines both clear and non-clear images after dehazing was introduced for training and a new model was proposed. Table 5 shows the experimental results, indicating a significant improvement in accuracy for the dehazed clear test set and dehazed non-clear test set. The Top-1 accuracy is generally above 95%, and both precision and recall are above 98%. Ultimately, the proposed model demonstrates good performance on both clear and hazy test sets after dehazing, providing a stable detection network suitable for complex environments.

Table 5. Final experiment results.

Training set	Test set	Category	Top-1	Top-2	Recall	P
Combined dehazed and clear training set	dehazed unclear test set	raw dataset	94.74%	99.93%	98.18%	98.45%
		adding bamboo joint images	95.49%	99.80%	98.06%	98.21%
	dehazed clear test set	raw dataset	97.84%	100%	99.64%	99.46%
		adding bamboo joint images	96.83%	100%	99.50%	99.30%

4 Conclusion

To better adapt to the high-dust environment in factories, various datasets were used for training in this study. By testing various models trained with different datasets, the best performing model for this environment was finally identified. The training scheme for this model includes dataset creation and dehazing algorithm application. Regarding dataset creation, this study prepared multiple types of datasets for experimentation. During training, it was found that the network relied too heavily on color features to identify the front and back sides of bamboo due to excessive color feature learning. To address this issue, an equal number of images with bamboo nodes were inserted into the training set to encourage the network to learn more about bamboo's shape characteristics rather than solely relying on color features. Regarding dehazing algorithms, GCAnet was selected as the primary dehazing algorithm. The resulting dataset comprised dehazed images combined with clear images and was twice the size of the original dataset. A larger dataset allows for better convergence of the network and mitigates overfitting. The model trained using this dataset achieved high recognition accuracy on foggy images after dehazing and delivered good results on clear images after dehazing. It thus demonstrated a good recognition effect in complex environments.

References

1. Howard, A.G., et al.: Mobilenets: efficient convolutional neural networks for mobile vision applications. In: Proc. of the IEEE Conf. on Computer Vision and Pattern Recognition, pp. 432–445 (2017)
2. Zhang, X., Zhou, X., Lin, M., Sun, J.: Shufflenet: an extremely efficient convolutional neural network for mobile devices. In: Proc. of the IEEE Conf. on Computer Vision and Pattern Recognition, pp. 6848–6856 (2018)
3. Sandler, M., Howard, A., Zhu, M., Zhmoginov, A., Chen, L.: MobileNetV2: inverted residuals and linear bottlenecks. In: Proc. of the IEEE Conf. on Computer Vision and Pattern Recognition, pp. 4510–4520 (2018)
4. Qin, Z., et al.: ThunderNet: towards real-time generic object detection on mobile devices. In: Proc. of the IEEE Int'l Conf. on Computer Vision, pp. 6718–6727 (2019)
5. Ma, N., Zhang, X., Zheng, H., Sun, J.: ShuffleNet V2: practical guidelines for efficient cnn architecture design. In: Proc. of the European Conf. on Computer Vision, pp. 116–131 (2018)

6. Landola, F.N., et al.: SqueezeNet: AlexNet-level accuracy with 50x fewer parameters and < 0.5 MB model size. In: Proc. of the 5th Int'l Conf. on Learning Representations (2017)
7. Almghraby, M., Elnady, A.O.: Face mask detection in real-time using mobilenetv2. Int. J. Eng. Adva. Technol. **10**(6), 104–108 (2021)
8. Xiang, Q., Wang, X., Li, R., et al.: Fruit image classification based on mobilenetv2 with transfer learning technique. In: Proceedings of the 3rd international conference on computer science and application engineering, pp. 1–7 (2019)

A Filter Similarity-Based Early Pruning Methods for Compressing CNNs

Zifeng Jiang and Kangshun Li[✉]

College of Mathematics and Information, South China Agricultural University,
Guangzhou 510642, Guangdong, China
likangshun@sina.com

Abstract. There are a large number of redundant features in deep neural networks, the pruning method based on redundant filters can identify redundant features and prune them. However, these methods require a long-term pre-training, which greatly increases the computational cost. Inspired by the early-bird tickets, this paper proposes an early pruning algorithm based on filter similarity. First, the C-SGD optimizer is used in training to move the filter towards the cluster center. At the same time, we propose an Intra-Cluster Similarity Metric to measure the degree of redundancy in a cluster, when the similarity reaches the specified threshold, it means that the filters in the cluster are redundant and the network is stable, and then the redundant filters can be safely removed. Our method can greatly re-duce the training time and compress the network, and the pruned model can achieve the accuracy of the original model after retraining. Our algorithm pruned ResNet-56 and was able to reduce 60.8%FLOPs on CIFAR-10 with 0.16% accuracy degradation. For ResNet-50, it can reduce FLOPs by 55.7%, TOP1 accuracy is only reduced by 1.37%, and the time required for training can be greatly reduced.

Keywords: Network Pruning · Filter pruning · Cosine similarity · Early pruning · Feature correlation · Deep learning

1 Introduction

In recent years, convolutional neural Network (CNN) has made great development and has become the mainstream method in the field of computer vision. It has achieved good results in image classification [1], object detection [2], instance segmentation [3] and other visual fields. However, convolutional neural networks can achieve such superior performance at the cost of a huge number of model parameters and high training cost, which brings great pressure to the storage and calculation of the model. For example, the classical image classification model ResNet-50 model [4] has about 25.6 M parameters and requires more than 100 MB of space to store. When processing an image with a resolution of 224×224, it requires 3.8G FLOPs (floating point operations), which greatly limits the deployment of convolutional neural networks in the real world, especially in embedded devices.

In order to solve the deployment difficulties of convolutional neural networks, the academic community has proposed a variety of model compression methods, such as

K. Li and Y. Liu (Eds.): ISICA 2023, CCIS 2147, pp. 39–48, 2024.
https://doi.org/10.1007/978-981-97-4396-4_4

low-rank decomposition [7, 10], parameter quantization [8, 14], network pruning [8, 9, 12, 13], etc. Network pruning can be divided into two categories: unstructured pruning and structural pruning. Among them, structured pruning has been widely used because it can compress the model and accelerate the inference speed at the same time.

Among them, the similarity-based pruning method as a newly proposed method is receiving more and more attention [15]. However, as Fig. 1, the existing pruning methods based on similarity require a long pre-training period to distinguish redundant features, which greatly limits the application of this method. Recently, You H et al. [13] proposed an early-bird lottery algorithm that can greatly reduce the time required for the pre-training phase. Therefore, we propose to combine the early bird lottery algorithm with the C-SGD optimizer and propose an early pruning algorithm based on filter similarity. The main contributions of this paper are as follows:

1. We introduce the idea of early bird algorithm into similarity-based pruning methods and propose an improved filter pruning algorithm, which can perform pruning at the early stage of network training.
2. Proposed an Intra-Cluster Similarity Metric (**ICSM**) to calculate the similarity of filters in a cluster.
3. Experiments show that our method can greatly reduce the training time of pruning algorithm and improve the application efficiency of pruning algorithm.

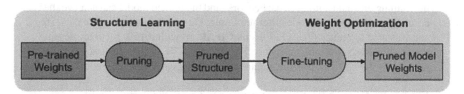

Fig. 1. Traditional network pruning flow.

2   Related Work

2.1   Similarity Based Pruning Methods

It is a well-known fact that deep neural networks are often over parameterized and there are a lot of redundant features in the network. Based on this fact, Ayinde B O et al. [16] proposed a pruning method based on redundant features of neural networks. The redundant features are clustered by clustering algorithm, and the redundant features in each cluster are simply removed to prune the network, which can accelerate the inference speed of the network. RoyChowdhury et al. [17] proposed that cosine similarity can be used to measure the similarity of two filters in the network. Ding X et al. [18] went further and proposed an improved Centripetal SGD optimizer, which can make the similarity filter move closer to its cluster center in the training process, so that the filter has a higher degree of redundancy. However, these methods require a long time of pre-training to make the filter redundant, which greatly limits the practical application of pruning algorithms.

2.2 Other Pruning Methods

In addition to similarity based pruning methods, there are regularization based pruning methods [20, 21] and magnitude based pruning methods [19, 22]. Regularization based pruning methods add a regularization term to the objective function to make a part of weights in the network tend to 0 after a certain training, and finally prune the network by removing the weights close to 0. Liu Z et.al [20] found the extensive use of BN layer in CNN and proposed to add a regularization term for BN layer parameters to the objective function. By regularizing the scaling factors in the BN layer, a part of the scaling factors can be made to go to 0, which can be easily pruned. The magnitude based pruning method pruning according to the magnitude of the weights in the network, such as APoZ [19], which measures the redundancy according to the number of zero values in the output feature map, and pruning is carried out by iteratively pruning the channels with the most zero values.

2.3 Early Bird Ticket

Haoran You et al. [13] suggested that it is possible to find the winning Lottery Ticket [23, 24] early in network training and based on this, they proposed the concept of Early Bird Ticket. Consider a randomly initialized dense network $f(x;\theta)$, f is able to achieve accuracy f_{acc} and minimum loss f_{loss} by SGD optimizer in i-th iterations. Based on this, consider a subnetwork of $f' = f(x;m \odot \theta)$, m is a mask for the network f. The early-bird lottery hypothesis states that there is a mask m such that the subnetwork $f(x;m \odot \theta)$ after t-th iteration training its accuracy f'_{acc} can reach the accuracy of original network f_{acc}, where $t << i$. The specific method is to perform a pre-pruning at the end of each round of training, and calculate the similarity between the subnetworks obtained in each round. When the similarity of the subnetworks becomes stable, it is considered that $f(x;m \odot \theta)$, that is, the winning ticket has been found. Because the early-bird lottery algorithm [13] can find the winning ticket in the early stage of training, which can greatly save training time, and this can alleviate the problem of long pre-training of the pruning algorithm, we consider introducing the idea of early-bird lottery into the similarity-based pruning algorithm.

3 Methods

3.1 Formulation

As a classical network, the main structure of convolutional neural network is composed of convolutional layer and fully connected layer. The pruning algorithm is generally aimed at the filter in the convolutional layer. Let i be the index of each layer, each filter F_j^i in layer i-th can be represented as a $u^i \times v^i$ convolution kernel, $F_j^i \in R^{u^i \times v^i}$, then the parameters of all filters in the i-th layer can be represented as $K^i \in R^{u^i \times v^i \times c^i \times c^{i-1}}$, c^i is the number of filters in the i-th layer. At the same time, the batch normalization layer of the i-th layer can be represented by $\mu^i, \sigma^i, \gamma^i, \beta^i$. Then all parameters of the i-th layer can be represented as $P^i = (K^i, \mu^i, \sigma^i, \gamma^i, \beta^i)$. The similarity-based pruning method calculates the similarity between filters sim (F_1^i, F_2^i) by some method, and when sim $(F_1^i, F_2^i) > \tau$, τ is the similarity threshold, then the filters F_1^i or F_2^i are considered redundant and only one of them is kept. Let the retained filters of i-th layer be R^i then the i-th layer obtained by pruning is $P_{R^i}^i = (K_{R^i}^i, \mu_{R^i}^i, \sigma_{R^i}^i, \gamma_{R^i}^i, \beta_{R^i}^i)$. In the similarity-based pruning method, to calculate the similarity between filters, we simply expand the convolution kernel of $u^i \times v^i$ into a $(u^i*v^i) \times 1$ feature vector V. We can calculate the similarity between two filters using the cosine similarity:

$$sim(F_1^i, F_2^i) = \frac{V_1 \cdot V_2}{|V_1|*|V_2|} \qquad (1)$$

We use mask distance [13] to measure the similarity between two subnetworks, consider A complete network A, the subnetwork obtained by pruning is A^*, set A binary mask m, the corresponding mask of the pruned channels in the subnetwork A^* is set to 0, and the unpruned channels in the subnetwork A^* are set to 1, so that we can represent the structure of the subnetwork A^* according to the mask m. By calculating the Hamming distance of the mask m (i.e., the mask distance), we can count the number of channels through which the two subnetworks differ to get their similarity. So the similarity between subnetworks can be expressed as

$$sim(A_1^*, A_2^*) = Hamming\ Distance(m_1, m_2) \qquad (2)$$

3.2 Filter Similarity Visualizetion

The C-SGD algorithm uses the C-SGD optimizer to make the filters in the same cluster closer to each other, and pruning is carried out after a certain amount of iterative training to make the filters close enough. However, this method lacks the means to judge the clustering degree of the filter, which leads to a long pruning process. We believe that a long pruning training is not necessary because the redundancy of the network is already balanced early in the training. Therefore, we first use a visual method to observe the changes of the intra-cluster similarity during the clustering process.

For further verification, we also use cosine similarity to measure the similarity between filters. Cosine similarity is a measure of the degree of similarity between two vectors, which determines the similarity of two vectors by calculating the cosine value of the Angle between them. The cosine similarity can be calculated by formulation (1), and we can compute the cosine similarity between the two filters in layer i and the average similarity of the filters in cluster C:

$$SIM\left(F_j^{(i)}, F_k^{(i)}\right) = \frac{\left(F_j^{(i)} \cdot F_k^{(i)}\right)}{\left(\|F_j^{(i)}\|*\|F_k^{(i)}\|\right)} \qquad (3)$$

$$Average\ Cosine\ SIM\ (C) = \frac{2}{|C|(|C|-1)} \sum_{j \in C} \sum_{k \in C} SIM\ (K_{:,:,:,j}^{(i)}, K_{:,:,:,k}^{(i)}) \qquad (4)$$

According to Eq. 3.2.4, we visualize the average cosine similarity of each network layer of the ResNet-56 network, and the results are shown in Fig. 2.

It is obvious that the average cosine similarity of the filter clusters increases rapidly at the beginning of the network training, which means that the filters become similar to

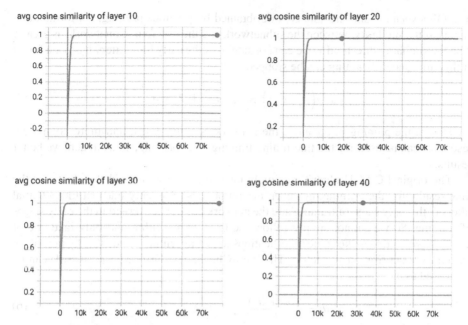

Fig. 2. Curve of average cosine similarity of ResNet56 convolutional layers 10, 20, 30, and 40. The X-axis represents the number of iterations and the Y-axis represents the average cosine similarity

each other and the redundancy between the filters increases. After about 5000 iterations, the rise in cosine similarity quickly approaches 1 and becomes stable, indicating that the filters within the same cluster have become very similar. This further validates our hypothesis that the network quickly becomes redundant and reaches equilibrium when optimized with the C-SGD optimizer, so we can prune the network at the early phase of network training.

3.3 Filter Similarity Based Early Pruning Methods

C-SGD is an improved SGD optimizer, which can make the filter move closer to its cluster center during the training process. C-SGD first simply flattens each filter into a feature vector, and then uses the k-means clustering algorithm [25] to cluster N^i feature vectors into $N^i*(1-p)$ clusters according to the pruning ratio p. After a certain number of iterative training, the feature vectors in a cluster collapse to a point, and the other feature vectors in the cluster are redundant. Pruning is done by keeping only one of the filters. However, the problem of this method is that it requires a long time of iterative training, and the time efficiency is not as good as other pruning methods. Therefore, we consider introducing the idea of early-bird lotteries, which can quickly identify the winning lotteries based on the similarity between subnetworks at the early stage of C-SGD training, and quickly perform pruning instead of having to carry out a long iterative training.

For the complete network A, after each epoch of training, we perform pre-pruning to obtain the subnetwork $\mathbf{A}^*$, and use the mask distance to calculate the similarity sim(

A*, A)between A* and the subnetwork obtained by previous pruning. As formulation eq. (5), if the similarity between the subnetwork and the first four subnetworks is greater than the similarity threshold τ, we can consider that the early bird lottery has been found at this time and the training can be stopped.

$$sim\left(A_i^*, A_{i-j}^*\right) > \tau, \quad j = 1, 2, 3, 4 \tag{5}$$

Due to the pruning in advance, the network is compressed, and more computing resources can be used for the fine-tuning training after pruning, so as to achieve better results.

The original C-SGD algorithm simply takes one of the filters in the cluster as the final result after the training, however, our purpose is to quickly find out the potential filters in the early stage of training and the network has not converged in the early stages of training. So we cannot simply follow the C-SGD algorithm and only take one of them, so we need to make some modifications to the algorithm. We do this by proposing an Intra-Cluster Similarity Metric. The Intra-Cluster Similarity Metric (ICSM) can be expressed as θ:

$$\theta = \frac{\Sigma sim(F_c, F_j)}{n} \quad j = 1, 2, \ldots, n \tag{6}$$

where n is the number of filters within a cluster and F_c is the cluster center of that cluster.

In the process of training, the ICSM in each cluster is monitored, and the redundancy of the filters in the cluster is also taken into account. Only when both the intra-cluster similarity metric and the subnetwork similarity satisfy the threshold, the winning ticket is considered to have been found and the true pruning is performed. The specific process of our proposed algorithm is as follows:

Algorithm 1: Filter Similarity–Based Early Pruning Methods

Input: full network **A**, pruning ratio **p**, max train epoch **T**, subnetwork similarity threshold τ_s and ICSM threshold τ_θ

Output: pruned subnetwork **A***

Calculate the number of filter cluster n according to **p**

Clustered filters of each layer using the k-means clustering algorithm

for epoch t = 1, ⋯, **T do**

Train the network **A** use C-SGD optimizer, make the filter move closer to the cluster center.

Calculate the Redundancy of network L_θ by formulation (6)

Do pre-pruning, get subnetwork **A***

Calculate the similarity between subnetworks L_N by formulation (5)

if $L_N > \tau_s$ and $L_\theta > \tau_\theta$ **then**

Stop training, save the subnetwork **A***

end for

---

4 Experiment

4.1 Setting

We next conduct experiments using networks with different structures and pruning methods to show the effectiveness of the proposed algorithm on the classification task. All experiments are performed on CIFAR10 and CIFAR100 datasets. The CIFAR-10 dataset contains 50 000 training images and 10 000 test images divided into 10 classes. We conduct experiments using two networks with different architectures, Resnet and VGGnet, to test the ability of the proposed algorithm on different network architectures.

In the network training phase, we use C-SGD optimizer to train the network, with batch size 64 and hyperparameter ε is $3 * 10^{-3}$. At the same time, in order to speed up the filter to the cluster center, we set the initial learning rate of the optimizer to 0.1, and the learning rate decay by 0.1 on the 80th and 120th epochs. We run five tests for each set of experiments to reduce chance. Pruning is performed when the subnetwork structure similarity L is larger than the threshold τ and the network filter redundancy is smaller than the threshold τ. In the pruning phase we simply keep only the first filter in each cluster.

After the pruning is completed, we fine-tune the subnetwork for 160epochs using the SGD optimizer with an initial learning rate of 0.1, and the learning rate decays by 0.1 at the 80th and 120th epochs, respectively, and the rest of the training parameters are the same as in the training phase.

4.2 Result and Analysis

As shown in Table 1, our proposed algorithm can work normally on a variety of networks, and is close to or even better than the existing filter similarity based pruning algorithms in terms of pruning rate and accuracy. At the same time, because we apply the early bird lottery to the algorithm, our algorithm requires far less training time than other algorithms, and is superior to other algorithms in time efficiency (Table 2).

On ResNet-56 and VGG16, both C-SGD and our method achieve high FLOPs reduction while maintaining high accuracy. This indicates that these two methods strike a good balance between pruning efficiency and model performance. It achieves good results on CIFAR-10 dataset and ImageNet dataset. Our algorithm pruned ResNet-56 and was able to reduce 60.8%FLOPs on CIFAR-10 with 0.16% accuracy degradation. For ResNet-50, it can reduce FLOPs by 55.7%, TOP1 accuracy is only reduced by 1.37%, and the time required for training can be greatly reduced.

Table 1. Experimental results under CIFAR10 dataset.

Network	Method	Base Top1	Pruned Top1	FLOPs %
ResNet-56	Redundant Filter-based Pruning [16]	92.07%	91.57%	51.6%
	C-SGD [18]	93.56%	93.60%	60.85%
	HRank [5]	93.52%	93.17%	50.0%
	Network Slimming [20]	93.34%	93.26%	44.9%
	Channel pruning [6]	92.8%	91.8%	50.0%
	Ours	93.56%	93.40%	60.85%
Vgg16	Redundant Filter-based Pruning	93.42%	92.94%	48.5%
	C-SGD	93.48%	93.46%	60.85%
	HRank	93.42%	93.30%	50.0%
	Network Slimming	93.48%	93.15%	44.9%
	Channel pruning	92.80%	92.43%	50.0%
	Ours	93.48%	93.35%	59.1%

Table 2. Experimental on ResNet-50 under ImageNet dataset.

Network	Method	Base Top1	Pruned Top1	FLOPs %
ResNet-50	C-SGD	75.33%	74.54%	55.7%
	HRank	76.15%	74.98%	43.7%
	DCP	76.01%	74.95%	55.7%
	Channel pruning	76.15%	72.30%	33.2%
	Ours	75.33%	73.96%	55.7%

5 Conclusion

We introduce the idea of early bird lottery and propose an early pruning algorithm based on filter similarity. At the same time, we also propose an Intra-Cluster Similarity Metric to measure the redundancy of filters in a cluster. When both the intra-cluster filter metric and the similarity between subnetworks meet the threshold is, we consider that the subnetwork at this time is the winning lottery ticket and can be pruned safely. Experiments on the cifar10 and cifar100 dataset show that our algorithm can prune convolutional networks well, and the final test accuracy can reach the level of a variety of similarity-based pruning methods. At the same time, since our algorithm prune at the early stage of training, it can greatly reduce the training time and improve the efficiency of the pruning algorithm in practical use.

References

1. Krizhevsky, A., Sutskever, I., Hinton, G.E.: Imagenet classification with deep convolutional neural networks. Commun. ACM **60**(6), 84–90 (2017)
2. Girshick, R., Donahue, J., Darrell, T., et al.: Rich feature hierarchies for accurate object detection and semantic segmentation In: Proceedings of the IEEE Conference on Computer Vision and Pattern Recognition, pp. 580–587 (2014)
3. Long J, Shelhamer E, Darrell T. Fully convolutional networks for semantic segmentation In: Proceedings of the IEEE Conference on Computer Vision and Pattern Recognition, pp. 3431–3440 (2015)
4. He, K., Zhang, X., Ren, S., et al.: Deep residual learning for image recognition In: Proceedings of the IEEE Conference on Computer Vision and Pattern Recognition, pp. 770–778 (2016)
5. Lin, M., et al.: Hrank: filter pruning using high-rank feature map. In: 2020 IEEE/CVF Conference on Computer Vision and Pattern Recognition (CVPR), pp. 1526–1535 (2020)
6. He, Y., Zhang, X., Sun, J.: Channel pruning for accelerating very deep neural networks. In: International Conference on Computer Vision (ICCV), vol. 2, p. 6 (2017)
7. Denton, E.L., Zaremba, W., Bruna, J., LeCun, Y., Fergus, R.: Exploiting linear structure within convolutional networks for efficient evaluation. In: NIPS (2014)
8. Han, S., Pool, J., Tran, J., Dally, W.: Learning both weights and connections for efficient neural network. In: NIPS, pp. 1135–1143 (2015)
9. Liu, L., Huang, Q., Lin, S., et al.: Exploring inter-channel correlation for diversity-preserved knowledge distillation In: Proceedings of the IEEE/CVF International Conference on Computer Vision, pp. 8271–8280 (2021)
10. Sainath, T.N., Kingsbury, B., Sindhwani, V., Arisoy, E., Ramabhadran, B.: Low-rank matrix factorization for deep neural network training with high-dimensional output targets. In: (ICASSP), pp. 6655–6659. IEEE (2013)
11. Jaderberg, M., Vedaldi, A., Zisserman, A.: Speeding up convolutional neural networks with low rank expansions In: Proceedings of the British Machine Vision Conference. BMVA Press (2014)
12. Luo, J.-H., Wu, J., Lin, W.: ThiNet: a filter level pruning method for deep neural network compression. In: 2017 IEEE International Conference on Computer Vision (ICCV), pp. 5068–5076 (2017)
13. You, H., Li, C., Xu, P., et al.: Drawing early-bird tickets: toward more efficient training of deep networks In: International Conference on Learning Representations (2019.2020)
14. Rastegari, M., Ordonez, V., Redmon, J., et al.: Xnor-net: Imagenet classification using binary convolutional neural networks. In: European conference on computer vision. Springer, Cham, pp 525–542 (2016)
15. Vadera, S., Ameen, S.: Methods for pruning deep neural networks. IEEE Access **10**, 63280–63300 (2022)
16. Ayinde, B.O., Inanc, T., Zurada, J.M.: Redundant feature pruning for accelerated inference in deep neural networks. Neural Netw. **118**, 148–158 (2019)
17. RoyChowdhury, A., Sharma, P., Learned-Miller, E., Roy, A.: Reducing duplicate filters in deep neural networks. In: NIPS Workshop on Deep Learning: Bridging Theory and Practice, pp 1–7 (2017)
18. Ding, X., Ding, G., Guo, Y., et al.: Centripetal sgd for pruning very deep convolutional networks with complicated structure. In: Proceedings of the IEEE/CVF Conference on Computer Vision and Pattern Recognition, pp. 4943–4953 (2019)
19. Hu, H., Peng, R., Tai, Y.-W., Tang, C.-K.: Network trimming: A data-driven neuron pruning approach towards efficient deep architectures. arXiv preprint arXiv:1607.03250 (2016)

20. Liu, Z., Li, J., Shen, Z., et al.: Learning efficient convolutional networks through network slimming In: Proceedings of the IEEE International Conference on Computer Vision, pp. 2736–2744 (2017)
21. Zhuang, T., Zhang, Z., Huang, Y., et al.: Neuron-level structured pruning using polarization regularizer. Adv. Neural. Inf. Process. Syst. **33**, 9865–9877 (2020)
22. Guo, Y., Yao, A., Chen, Y.: Dynamic network surgery for efficient DNNs. In: Proceedings of the 30th Internatonal Conference on Neural Information Processing Systems, pp. 1387–1395 (2016)
23. Frankle, J., Carbin, M.: The Lottery Ticket Hypothesis: Finding Sparse, Trainable Neural Networks In: International Conference on Learning Representations (2018)
24. Frankle, J., Dziugaite, G.K., Roy, D., et al.: Linear mode connectivity and the lottery ticket hypothesis. In: International Conference on Machine Learning, pp. 3259–3269. PMLR (2020)
25. Likas, A., Vlassis, N., Verbeek, J.J.: The global k-means clustering algorithm. Pattern Recogn. **36**(2), 451–461 (2003)

Visualization of Convolutional Neural Networks Based on Gaussian Models

Hui Wang and Tie Cai[✉]

Shenzhen Institute of Information and Technology, Shenzhen 518172, China
cait@sziit.edu.cn

Abstract. Convolutional neural networks (CNN) have made breakthrough progress in tasks such as image classification, target detection and scene recognition. The trained model has excellent automatic feature extraction and prediction performance, and can provide users with end-to-end "input-output" form Solution. However, due to distributed feature encoding and increasingly complex model structures, people have never been able to accurately understand the internal knowledge representation of the convolutional neural network model and the underlying reasons that prompted it to make specific decisions. We can find that the output and input of CNN with an appropriate prior over the weights and biases is a Gaussian process (GP). For the CNN, it is a signal Gaussian model. So, we can use single Gaussian model to explain the computational process of CNN. We show the GP distribution and computational process over the weights and biases. The experiment results show that this analysis method can give a good visualization effective.

Keywords: Convolutional visualization · Gaussian process · weights and biases

1 Introduction

Convolutional visualization refers to understanding the working principle and learned features of a convolutional neural network by visualizing the filters and feature maps in the network. The commonly used convolutional visualization methods include: visualization of feature maps, visualization of convolutional kernels, visualization of heatmaps, and visualization of gradients. Convolutional visualization can help deep learning researchers better understand the concepts and principles of convolution, thereby better designing and optimizing convolutional neural networks. Through visualization, researchers can have a clearer view of each step in convolution operations, including input, convolution kernel, convolution operation, and output, thereby better understanding the essence and function of convolution. In addition, convolutional visualization can help researchers better understand advanced concepts in convolutional neural networks, such as pooling and batch normalization, in order to better design and optimize deep learning models.

However, there are certain limitations behind this advantage of CNN. On the one hand, people still cannot better understand the internal knowledge representation of CNN

K. Li and Y. Liu (Eds.): ISICA 2023, CCIS 2147, pp. 49–55, 2024.
https://doi.org/10.1007/978-981-97-4396-4_5

and its accurate semantic meaning. Even model designers have difficulty answering what features CNN has learned., the specific organizational form of features and the importance measurement of different features, etc., causing the diagnosis and optimization of CNN models to become empirical and even blind repeated trials, which not only affects the model performance, but may also leave potential loopholes; another On the other hand, real-life applications based on CNN models have been deployed in large numbers in daily life, such as face recognition, pedestrian detection and scene segmentation. However, for some special industries with low risk tolerance, such as medical, finance, transportation, military and other fields. The issues of interpretability and transparency have become major obstacles to its expansion and deepening. These fields have strong practical needs for deep learning models such as CNN, but due to model security and interpretability issues, they are still unable to be used on a large scale. The model may make some common sense errors in practice, and cannot provide the reasons for the errors, making it difficult for people to trust its decisions.

Based on the interpretability theory principles, existing neural architecture search(such as CNN) algorithms can be broadly classified into three different categories: Reinforcement learning interpretability model based on tree architecture, visualization [1, 2], gradient-based neural architecture search algorithms [3–5]. Specifically, the proposed algorithms often consume heavy computational resources due largely to the inexact reward resulted from reinforcement learning techniques [6]. The algorithms based on gradient are better than the algorithms based on reinforcement learning. Unfortunately, the gradient-based algorithms require constructing a supernet in advance, which also demands specialized expertise. The CNN algorithms solve the CNN problems by exploiting Evolutionary Computation (EC) techniques. Specifically, EC is a class of computational paradigms such as Genetic Algorithms (GA) [7], Genetic Programming (GP) [8], and Particle Swarm Optimization (PSO) [9] solving challenging optimization problems by simulating the evolution of biology or swarming social behavior [10–14]. An initial population of candidate solutions is generated and iteratively updated by the tree structure for generations. Each image sample in each generation needs to be evaluated to obtain the process image feature value. As a result, the population will gradually increase in fitness. Compared to the CNN algorithms based on reinforcement learning. Compared with the gradient-based CNN algorithms, CNN algorithms are often fully automatic, and they can produce an appropriate CNN without any human intervention [15, 16].

However, the exited analysis methods have some shortness, such as the inaccurate analysis results and inaccurate recognition effective. So, we use single Gaussian model to explain the computational process of CNN. We show the GP distribution and computational process over the weights and biases. The experiment results show that this analysis method can give a good visualization effective.

2   Review of CNN and Single GP

We will define the relationship between CNN and single GP by Central Limit Theorem. We review the single-hidden CNN before moving to the multi-layer CNN.

2.1 CNN

With the introduction of VGGNet [8] and GoogleNet [9], researchers have found that by deepening the number and width of convolutional neural networks, the network can extract more deep information, which is helpful for tasks such as image detection and classification. A large number of cascaded CNN and fused CNN methods have appeared in this field of expression recognition. In 2015, Yu et al. [10] designed a cascade CNN structure, combined with an optimization algorithm that automatically assigns cascade CNN weights, and won the second place in the EmotiW2015 competition for expression recognition based on static images with a recognition accuracy of 61.29% on the SFEW2.0 data set, it has been proved through experiments that the CNN architecture has excellent performance in the field of expression recognition. The article proposes two loss functions of the cascade network, namely the maximum likelihood loss function eq. (1) and the folding loss function eq. (2):

$$\min_{\omega} - \sum_{i=1}^{N} \log \sum_{k=1}^{K} P_k(y_i|X_i)\omega_k + \lambda \sum_{k=1}^{K} \omega_k^2$$

$$s.t. \sum_{k=1}^{K} \omega_k = 1, \omega_k \geq 0, \forall k \tag{1}$$

$$\min_{\omega} \sum_{i=1}^{N} \sum_{y \neq y_i} \left[1 - \frac{\sum_{k=1}^{K} (P_k^{i,y_i} - P_k^{i,y})\omega_k}{\gamma} \right] + \lambda \sum_{k=1}^{K} \omega_k^2$$

$$s.t. \sum_{k=1}^{K} \omega_k = 1, \omega_k \geq 0, \forall k \tag{2}$$

where N is the number of samples in the training set, $P_k(y_i|X_i)$ represents the probability of correct prediction of the i-th sample, K is the number of trained networks, ω_k is the weight of the k-th network, $P_k^{i,y_i} \triangleq P_k(y|X_i)$ represents the i-th The probability that the sample is predicted to be the y-th label, λ, γ are hyperparameters.

2.2 Single GP

When the sample data Y is univariate, the Gaussian distribution follows the probability density function below:

$$P(x|\theta) = \frac{1}{\sqrt{2\pi\sigma^2}} \exp(-\frac{(x-\mu)^2}{2\sigma^2}) \tag{3}$$

where, μ is the data mean), σ is the data standard deviation.

3 The Visualization of CNN Based on Single GP

We will simulate the change process of parameters and weight as a number of GP distribution. Then, we will compute the probability distribute by GP distribution. We have provided an alternative derivation in terms of Bayesian marginalization over pooling layer.

Suppose that z_j^{l-1} is a GP, identical and independent for every j, After $l-1$ steps, the CNN network computes

$$z_i^l(x) = b_i^l + \sum_{j=1}^{N_l} W_{ij}^l x_j^l(x), \, x_j^l(x) = \phi(z_j^{l-1}(x)) \qquad (4)$$

In fact, these recurrence relations have appeared in other contexts. They are exactly the relations derived in the mean field theory of signal propagation in fully-connected random neural networks. There are also many researchers trying to design convolutional neural network architecture and loss specifically for expression recognition. Function, a sparse batch normalization CNN model referred to as SBN-CNN consist of sparse batch normalization convolution neural network. The SBN-CNN model is improved based on VGGNet. In the shallow convolution layer Use large convolution kernels, combined with batch normalization technology and dropout technology to avoid possible problems during model training.

4   Experiment Results Analysis

We will verify the effective of the proposed analysis method. We simulated the process of visualization analysis in convolutional. Based on this, we used image data set to test the proposed. Will show the visualization presentation. The multi-layer CNN is a fully-connected network with identical width at each layer. To verify the effective of GP predictions, training is on the mean squared error loss. We formulate classification as regression, which have given a most results. In the study, single GPs were chosen to be rectified unlinear units and hyperbolic tangent. Class labels were encoded as a one-hot, zero-mean, regression target, such as (i.e., entries of -0.1 for the incorrect class and 0.9 for the correct class).

Explainable AI (XAI) research is committed to explaining artificial intelligence models in a way that humans can understand, so that humans can understand the internal operating logic and decision-making results of the model, and provide convenience for troubleshooting and widespread use of the model. Visualization and Research work on interpreting neural network models has attracted increasing attention. The European Parliament introduced provisions on automated decision-making in the General Data Protection Regulation (GDPR), stipulating that data subjects have access to automated decision-making the right to interpret information involved. In addition, in 2019, the High-level Expert Group on Artificial Intelligence proposed ethical guidelines for trustworthy artificial intelligence. Although there are different legal opinions on these provisions, there is general agreement on the implementation of such the necessity and urgency of principles. The National Institute of Standards and Technology (NIST) released 4 principles on XAI in August 2020: Provability (explanation results can be proven by evidence), Usability (explanation results can be understood by users of the model and meaningful to users), accuracy (explanation results must accurately reflect the model's operating mechanism), and limitation (explanation results can identify situations that are not suitable for its own operation).

Fig. 1. The visualization analysis result of proposed method.

The experiments results are shown as Fig. 1. Explainable AI research is committed to explaining artificial intelligence models in a way that humans can understand, so that humans can understand the internal operating logic and decision-making results of the model, and provide convenience for troubleshooting and widespread use of the model. Visualization and Research work on interpreting neural network models has attracted increasing attention. The European Parliament introduced provisions on automated decision-making in the General Data Protection Regulation, stipulating that data subjects have access to automated decision-making the right to interpret information involved. In addition, the High-level Expert Group on Artificial Intelligence proposed ethical guidelines for trustworthy artificial intelligence. Although there are different legal opinions on these provisions, there is general agreement on the implementation of such the necessity and urgency of principles. The National Institute of Standards and Technology released 4 principles on explainable AI: Provability (explanation results can be proven by evidence), Usability (explanation results can be understood by users of the model and meaningful to users), accuracy (explanation results must accurately reflect the model's operating mechanism), and limitation (explanation results can identify situations that are not suitable for its own operation). Models trained on historical data sets may lead to the introduction of many biases that are not easily detectable. Biased rules may be deeply hidden in the trained model, and these biased rules may be regarded as general rules. The opacity of the model Sex hides these potential problems, so it becomes very difficult to judge whether the model results are fair when it comes to certain issues (such as gender, race etc.). In addition to social and moral issues, the opacity of the model Sexuality also affects issues such as accountability, product safety and industrial segregation of responsibilities.

5 Conclusions

To overcome the problem that convolutional neural network has made breakthrough progress in tasks such as image classification, target detection and scene recognition. We proposed that the output and input of CNN with an appropriate prior over the weights and biases is a Gaussian process (GP). For the CNN, it is a signal Gaussian model. So, we can use single Gaussian model to explain the computational process of CNN. We show the GP distribution and computational process over the weights and biases. The experiment results show that this analysis method can give a good visualization effective. Future work may involve evaluating the NNGP on a cross entropy loss using the approach. Training used the Adam optimizer with learning rate and initial weight/bias variances optimized over validation error using the Google Vizier hyperparameter tuner. Dropout was not used. In future work, it would be interesting to incorporate dropout into the NNGP covariance matrix using an analysis approach.

Acknowledgement. This work was supported by the Guangdong Basic and Applied Basic Research Foundation under Grant No. 2022A1515011447, the Shenzhen Fun damental Research fund under No. Grant 0220820010535001, the National Natural Science Foundation Youth Fund Project of China under Grant No. 62203310, Shenzhen Institute of Information Technology Key Laboratory Project under Grant No. SZIIT2023KJ005.

References

1. Kaelbling, L.P., Littman, M.L., Moore, A.W.: Reinforcement learning: a survey. J. Artif. Intell. Res. **4**, 237–285 (1996)
2. Zoph, B., Le, Q.V.: Neural architecture search with reinforcement learning. arXiv preprint arXiv:1611.01578 (2016)
3. Liu, H.,. Simonyan, K., Yang, Y.: Darts: Differentiable architecture search. arXiv preprint arXiv:1806.09055 (2018)
4. Baeck, T., Fogel, D.B., Michalewicz, Z. (eds.): Handbook of Evolutionary Computation. CRC Press (1997). https://doi.org/10.1201/9780367802486
5. Real, E., Moore, S., Selle, A., Saxena, S., Suematsu, Y. L., Tan, J., Le, Q.V., Kurakin, A.: Large-scale evolution of image classifiers. In: International Conference on Machine Learning, pp. 2902–2911. PMLR (2017)
6. Sun, Y., Xue, B., Zhang, M., Yen, G.G.: Evolving deep convolutional neural networks for image classification. IEEE Trans. Evol. Comput. **24**(2), 394–407 (2019)

7. Mitchell, M.: An Introduction to Genetic Algorithms. The MIT Press (1998). https://doi.org/10.7551/mitpress/3927.001.0001
8. Banzhaf, W., Nordin, P., Keller, R. E., Francone, F. D.: Genetic Programming: an Introduction: On the Automatic Evolution of Computer Programs and Its Applications. Morgan Kaufmann Publishers Inc. (1998)
9. Kennedy, J., Eberhart, R.: Particle swarm optimization. In: Proceedings of ICNN'95-International Conference on Neural Networks, vol. 4, pp. 1942–1948. IEEE (1995)
10. Sun, Y., Yen, G.G., Yi, Z.: Igd indicator-based evolutionary algorithm for many-objective optimization problems. IEEE Trans. Evol. Comput. **23**(2), 173–187 (2018)
11. Deb, K., Pratap, A., Agarwal, S., Meyarivan, T.: A fast and elitist multiobjective genetic algorithm: Nsga-ii. IEEE Trans. Evol. Comput. **6**(2), 182–197 (2002)
12. Sun, Y., Yen, G.G., Yi, Z.: Improved regularity model-based eda for many-objective optimization. IEEE Trans. Evol. Comput. **22**(5), 662–678 (2018)
13. Jiang, M., Huang, Z., Qiu, L., Huang, W., Yen, G.G.: Transfer learning-based dynamic multiobjective optimization algorithms. IEEE Trans. Evol. Comput. **22**(4), 501–514 (2017)
14. Sun, Y., Yen, G.G., Yi, Z.: Reference line-based estimation of distribution algorithm for many-objective optimization. Knowl. Based Syst. **132**, 129–143 (2017)
15. Sun, Y., Xue, B., Zhang, M., Yen, G.G.: Completely automated cnn architecture design based on blocks. IEEE Trans. Neural Netw. Learn. Syst. **31**(4), 1242–1254 (2019)
16. Elsken, T., Metzen, J.H., Hutter, F.: Neural architecture search: a survey. The J. Mach. Learn. Res. **20**(1), 1997–2017 (2019)
17. Lee J, Bahri Y, Novak R, et al. Deep neural networks as Gaussian processes. arXiv preprint arXiv:1711.00165 (2017)

A New Feature Selection Algorithm Based on Adversarial Learning for Solving Classification Problems

Xiao Jin[1], Bo Wei[1,2(✉)], Wentao Zha[1], and Jintao Lu[1]

[1] School of Computer Science and Technology (School of Artificial Intelligence),
Zhejiang Sci-Tech University, Hangzhou 310018, China
{202130504101,2021329621010,2021329621140}@mails.zstu.edu.cn,
weibo@zstu.edu.cn
[2] Longgang Research Institute, Zhejiang Sci-Tech University, Longgang 325000, China

Abstract. As a data preprocessing technique, the objective of feature selection (FS) is to eliminate irrelevant and redundant features. In recent years, evolutionary computation (EC) has shown great potential in solving FS problems. However, most of the existing EC-based FS approaches still face some shortcomings. To address these issues, a novel FS algorithm based on adversarial learning is proposed in this paper. In this work, an inter-population adversarial learning strategy is proposed. It enables two subswarms to compete and learn from each other in the evolutionary process. In addition, an interval-based flipping strategy is proposed. In this strategy, some features of a certain interval are flipped. A novel surrogate model is employed to pre-evaluate the candidate solutions, aiming to reduce the computational cost. Experiments on 10 UCI datasets indicate that the proposed SSAPSO is able to select smaller subsets of features in most cases with high classification accuracy.

Keywords: Particle swarm optimization · Salp swarm algorithm · Feature selection · Swarm intelligence

1 Introduction

With the rapid advancement of big data, high-dimensional data has become increasingly prevalent in various fields. The high-dimensional data often contains great value, but also contains many redundant or even irrelevant features. These features not only greatly increase the runtime of machine learning algorithms, but can also degrade their performance. FS technology plays a crucial role in removing redundant or irrelevant features and selecting an optimal subset of features. It has been extensively applied in dimensionality reduction [1], fault detection [2], cancer classification [3], sentiment analysis [4], and other domains.

K. Li and Y. Liu (Eds.): ISICA 2023, CCIS 2147, pp. 56–70, 2024.
https://doi.org/10.1007/978-981-97-4396-4_6

FS approaches mainly include the following: filter approaches, wrapper approaches, and embedding approaches. The filter approach evaluates candidate feature subsets based on metrics such as correlation between features and labels or features [5]. The wrapper approach can effectively evaluate the feature subset by introducing a learning algorithm [6, 7]. The embedded approach uses FS as part of model training and can be automated during training [8]. Among these three approaches, the wrapper and embedded approaches are more likely to achieve higher performance because they utilize learning algorithms.

The huge search space of high-dimensional dataset makes traditional algorithms and exhaustive search methods no longer applicable. However, evolutionary computing (EC) has shown significant advantages in solving FS problems because of its excellent global search ability. In numerous EC technologies, salp swarm algorithm (SSA) [9] and particle swarm optimization (PSO) [10] are utilized to solve FS problems due to their simplicity and efficiency. However, the current FS approach still suffers from some shortcomings. For instance, the topology of the algorithm results in the population always overutilizing the same neighborhood and ignoring more potential regions. Moreover, the valuable information carried by the data is not used effectively.

In this work, a new algorithm (SSAPSO) based on adversarial learning is used to solve the FS problem. In SSAPSO, the salp swarm and the particle swarm will compete against each other and learn from each other. The introduction of an interval-based flipping strategy can also enhance the explore ability of the algorithm. Furthermore, the main contributions are presented below.

a) A novel inter-population adversarial learning strategy is proposed. In this strategy, the salp swarm and particle swarm perform adversarial competition and learning. The structure of multiple learning exemplars enables the above two sub-swarms to exchange information. Such a design can effectively prevent the population from falling into local optimum.

b) An interval-based flipping strategy based on the correlation between features and labels is proposed. This strategy partitions the features into different intervals based on relevant information, and then the roulette wheel strategy is used to select an interval. Some features of the selected interval will be flipped. The elite solutions will be employed to generate candidate solutions by executing the strategy.

c) A new surrogate model is proposed to pre-evaluate candidate solutions. The evaluation results of the previous iteration are employed to construct the surrogate model, which reduces the computational cost.

The rest of the sections are arranged as follows: Sect. 2 briefly reviews related work on PSO and SSA. The proposed SSAPSO will be elaborated in Sect. 3. In Sect. 4, we compare SSAPSO with representative algorithms on the UCI datasets. Section 5 is the conclusions and future directions.

2 Background and Related Work

2.1 Feature Selection

Given a D dimensional dataset with N instances, the task of the FS approach is to find d features from the dataset and achieve the minimum classification error rate. It means that the FS approach needs to remove redundant and irrelevant features. For example, in a general classification problem, the objective function $F(\cdot)$ is concerned with the classification error rate and the size of the feature subset, which is calculated by Eq. (1).

$$minF(X) \quad s.t. \ X = \{x_1, x_2, \ldots, x_j, \ldots, x_D\} \tag{1}$$

where $j \in \{1, 2, \ldots, D\}$ and $x_j \in \{0, 1\}$; $x_j = 1$ means that the jth dimension feature is selected; Otherwise, the feature is not added to the final feature subset.

2.2 Canonical PSO

The PSO is a well-known metaheuristic algorithm for simulating bird swarm behavior [10]. In PSO algorithm, suppose a population is searching in a D dimensional decision space, the position vector of the ith particle can be expressed as $X_i = (x_{i,1}, x_{i,2}, \ldots, x_{i,D})$, and its velocity vector is denoted by $V_i^t = (v_{i,1}, v_{i,2}, \ldots, v_{i,D})$. The personal historical best position (pbests) and the global best position (gbest) are employed to update the positions of the particles, and the update formula is shown in Eqs. (2) and (3).

$$v_{i,j}^{t+1} = \omega v_{i,j}^t + c_1 r_1 \left(pbest_{i,j}^t - x_{i,j}^t \right) + c_2 r_2 \left(gbest_j^t - x_{i,j}^t \right) \tag{2}$$

$$x_{i,j}^{t+1} = x_{i,j}^t + v_{i,j}^{t+1} \tag{3}$$

where c_1 and c_2 are acceleration coefficients of learning. r_1 and r_2 are two random numbers from uniform distribution. ω is the inertia weight varying with the number of iterations t.

2.3 Canonical SSA

The SSA is an emerging algorithm recently proposed by Mirjalili et al. [9]. SSA was inspired by simulating the group behavior of salps as they move and forage in the ocean. The role of salp in the SSA is divided into leader and follower. The leaders can search near the food source, while the followers gradually move closer to the leaders. The Eq. (4) is employed to update the leader.

$$x_j^{t+1} = \begin{cases} F_j^t + c_1 \left((ub_j - lb_j)c_2 + lb_j \right) & if \ c_3 \geq 0.5 \\ F_j^t - c_1 \left((ub_j - lb_j)c_2 + lb_j \right) & else \end{cases} \tag{4}$$

where F is the food source. ub and lb control the upper and lower boundaries of the decision space, respectively. t is the number of evolutions of the current population. c_2 and c_3 are two random numbers from uniform distribution. The expression is as follows:

$$c_1 = 2e^{-\left(\frac{4l}{L}\right)^2} \tag{5}$$

where l represents the evolution times of the population; L is the upper limit of evolution times.

The follower updates its position as presented by Eq. (6).

$$x_{i,j}^{t+1} = \frac{1}{2}\left(x_{i,j}^t + x_{i-1,j}^t\right) \tag{6}$$

where $i \geq 2$ and x_j^i represents the position of the ith follower salp in the jth dimension.

2.4 FS Based on PSO and SSA in Classification Problems

PSO-based approaches have been extensively employed for solving classification problems in recent years. A novel feature selection approach is proposed in [11]. In the local search strategy of this approach, two operators search in the neighborhood, which improves the performance of the algorithm. Ke Chen et al. proposed a new FS approach (CUS-SPSO). In CUS-SPSO, the correlation-guided update strategy can make full use of correlation information, which guides particles to more potential regions [12]. In [13], an explicit representation of particle is proposed. At the same time, a new update model and an adaptive strategy to change the particle size are proposed.

In addition, several variants of SSA have been successively employed to solve the FS problem. Balakrishnan et al. introduced an novel SSA that aims to enhance the ability of the salp to explore different areas by updating its position with Levy flight [14]. Aljarah proposed a binary SSA [15] with asynchronous update rules and a new leadership structure. The algorithm constructs multiple sub-chains, different sub-chains use different update strategies. A salp swarm algorithm based on quadratic interpolation and local escape operator (LEO) is proposed [16]. The algorithm focuses on the quadratic interpolation of the optimal search agent to improve the solution performance of the algorithm.

A new hybrid metaheuristic algorithm SSA-FGWO is proposed in [17]. The basic idea of SSA-FGWO is to use grey wolf optimization (GWO) to improve SSA. Firstly, First, the strategy in SSA is employed to update the positions of the leaders. Secondly, the strong exploration strategy of GWO was used to update the positions of the followers. This hybrid design can improve the classification performance of SSA-FGWO. A novel gravitational search algorithm for Gaussian PSO (GPSOGSA) [18] is proposed. It contains an absolute Gaussian parameter that helps to improve the local search ability for identifying local attractors. The local attractor strives to reach the global optimal position with a good convergence rate. In [19], The spiral-shaped mechanism is introduced into the PSO algorithm (HPSO-SSM). The spiral mechanism is used as the exploitation operator to effectively promote the classification accuracy of the HPSO-SSM.

3 Proposed Approach

3.1 Overview of the SSAPSO

An overview of the proposed SSAPSO algorithm is given in Algorithm 1 for details. In SSAPSO, the representation and range of solutions are consistent with the canonical PSO algorithm. In addition, a threshold of 0.5 is employed to decide whether to select a feature.

3.2 Inter-population Adversarial Learning Strategy (IALS)

Both PSO and SSA have certain limitations when used to solve FS problems. In this paper, we propose an inter-population adversarial learning strategy (IALS). In this strategy, the population is divided into salp swarm and particle swarm. The two swarms are updated by different update methods. Specifically, after the followers in the salp swarm compete with the particles, the losers will learn the information of the winners. The two subswarms not only compete but also learn from each other, which fully promotes communication and improves the search ability of the population.

Algorithm 1: Overview of SSAPSO

Input: Dataset, Maximum number of iterations T, Population size N, Boundary of particle positions *boundary*, Elite archives E, Set $t = 1$
Output: The optimal feature subset

1 Initialize the population ;
2 Evaluate the fitness values of particles in the initial population ;
3 Update the *pbests* and the *gbest* of the initial population ;
4 Calculate the feature weights according to ReliefF ;
5 Normalize feature weights ;
6 Features are divided into three feature intervals: *Low*, *Mid*, and *High* ;
7 **while** $t \leq T$ **do**
8 | Update the positions of particles according to Algorithm 2 ;
9 | Evaluate the fitness values of particles in the current population ;
10 | Update the *pbests* and the *gbest* ;
11 | Update Elite archive E ;
12 | Elite solutions are flipped several times to generate candidate solutions ;
13 | A surrogate model is used to evaluate candidate solutions ;
14 | The N candidate solutions with better fitness values are screened, and KNN real evaluation is performed on them ;
15 | The gbest is updated if the fitness value of the candidate solution is better than the gbest ;
16 | $t = t + 1$;
17 **end**
18 **return** The optimal feature subset ;

The leaders of the salp swarm search near the food source, which enables SSA to have strong exploitation ability. In the IALS, the update method of leaders does not change, while the followers need to conduct adversarial learning with the particle swarm in addition to the original update method. The positions of the followers are updated in the way shown in Eq. (7) and (8).

$$x_{1,j}^{t+1} = \begin{cases} \left(x_{1,j}^{t} + bestleader_{j}^{t+1} + winp_{j}^{t}\right)/3 & \text{if } x_{1}^{t} \text{ is the loser} \\ \left(x_{1,j}^{t} + bestleader_{j}^{t+1}\right)/2 & \text{else} \end{cases} \tag{7}$$

$$x_{i,j}^{t+1} = \begin{cases} \left(x_{i,j}^{t} + x_{i-1,j}^{t} + winp_{j}^{t}\right)/3 & \text{if } x_{i}^{t} \text{ is the loser} \\ \left(x_{i,j}^{t} + x_{i-1,j}^{t}\right)/2 & \text{else} \end{cases} \tag{8}$$

where the *bestleader* is the leader with the best fitness value after updating; x_1 is the first follower who moves closer to the leader; *winp* is the corresponding winning particle of salp in the competition.

On the other hand, when the canonical PSO selects the optimal feature subset in the high-dimensional feature space, the particles will overutilize the same neighborhood and fall into the local optimum. Therefore, in IALS, each particle may have a different learning exemplar, so that the population can maintain better diversity in the evolutionary process. The positions of losers and winners in the particle swarm are updated in ways shown in Eqs. (9) and (10), respectively.

$$\begin{cases} v_{i,j}^{t+1} = \omega v_{i,j}^{t} + C_1 R_1\left(wins_{j}^{t} - x_{i,j}^{t}\right) + C_2 R_2\left(gbest_{j}^{t} - x_{i,j}^{t}\right) \\ x_{i,j}^{t+1} = x_{i,j}^{t} + v_{i,j}^{t+1} \end{cases} \tag{9}$$

$$\begin{cases} v_{i,j}^{t+1} = \omega v_{i,j}^{t} + C_1 R_1\left(pbest_{j}^{t} - x_{i,j}^{t}\right) + C_2 R_2\left(bestleader_{j}^{t+1} - x_{i,j}^{t}\right) \\ x_{i,j}^{t+1} = x_{i,j}^{t} + v_{i,j}^{t+1} \end{cases} \tag{10}$$

where the *bestleader* is consistent with the above. *wins* is the corresponding winning salp of particle in the competition. The losers in the competition learn information about the *gbest* and *wins* to explore more promising regions. R_1 and R_2 are random numbers from 0 to 1.

In this way, two subswarms can learn valuable information from each other in a competition. It means that IALS can effectively improve the search ability of the population and aims to search for more promising solutions. IALS is shown in Algorithm 2.

Algorithm 2: IALS

Input: Population size N, Boundary of particle positions *boundary*, Current population *pop*, The *gbest*, *pbests*, Fitness values of individuals in the population *Fit*

Output: Updated population *popnext*

1 **for** $i = 1$ *to* $2 * N/5$ **do**
2 | Update the positions of leaders in the salp swarm according to Eq. (4) ;
3 **end**
4 Evaluate the fitness values of the leaders and find the best leader *bestleader* ;
5 The salp swarm and the particle swarm perform a random pairwise adversarial competition ;
6 **for** $i = 2 * N/5 + 1$ *to* $7 * N/10$ **do**
7 | **if** *i=2*N/5+1* **then**
8 | | Update *i*th follower according to Eq. (7)
9 | **else**
10 | | Update *i*th follower according to Eq. (8)
11 | **end**
12 **end**
13 **for** $i = 7 * N/10 + 1$ *to* N **do**
14 | **if** *ith particle is the loser in the competition* **then**
15 | | Update *i*th particle according to Eq. (9)
16 | **else**
17 | | Update *i*th particle according to Eq. (10)
18 | **end**
19 **end**
20 **return** *popnext* ;

3.3 An Interval-Based Flipping Strategy (IFS)

For a dataset, it usually contains valuable information. Therefore, filter approach can use measures such as correlation to select feature subsets. Moreover, the features with very high or very low correlation values are easy to determine whether they should be selected, while features with moderate correlation are more difficult to decide [12]. In this paper, an interval-based flipping strategy (IFS) is proposed. Based on the correlation information, the original features are divided into three intervals: *Low*, *Mid*, and *High*. In this strategy, one of the intervals is selected as the flipping interval by the roulette wheel method, and then part of the features in this interval are flipped. The detailed description of the proposed interval-based flipping strategy is as follows.

ReliefF has been widely used as an algorithm to describe the correlation between features and labels [20, 21]. ReliefF algorithm is used to calculate the weights of features, and the weight value of a feature depends on its ability to distinguish neighboring instances. If the value between the two closest instances in different classes is very different, the weight value of the corresponding feature will be increased. Instead, the weight of the feature will be decreased. Features with feature weights above the threshold t_1 and below the threshold t_2 are classified as *High* and *Low*, respectively, and the remaining features are classified as *Mid*. According to Eq. (11), the roulette wheel is used to select the flipping interval.

$$Interval_F = \begin{cases} Low & if\ rand\ \leq Num(Low)/Num(All) \\ High & if\ rand\ \geq (Num(Low) + Num(Mid))/Num(All) \\ Mid & else \end{cases} \quad (11)$$

where $Interval_F$ represents the flipping interval; $Num(Low)$ and $Num(Mid)$ represent the number of features in *Low* and *Mid*, respectively; $Num(All)$ represents the number of features in the original feature set. As a result, features with moderate correlations are more likely to be flipped.

In the proposed IFS, the individuals with top $N/5$ fitness values after each real evaluation are included in the elite archives. IFS is then used several times to flip elite individuals, which results in a large number of promising particles. Figure 1 shows an example of elite individuals generating new positions through IFS. Given a dataset with 10 dimensions of features. The original position vector of the elite individual is $X = \{0.35, 0.46, 0.12, 0.81, 0.76, 0.68, 0.57, 0.43, 0.75, 0.62\}$, and the features of this dataset are divided into *Low*, *Mid*, and *High* intervals based on the feature weights. Assuming that the *Mid* interval is selected by the roulette wheel method, some features in the *Mid* interval will be flipped. The position vector of the candidate solution can be obtained by IFS, that is, $X^* = \{0.35, 0.46, 0.88, 0.81, 0.76, 0.32, 0.43, 0.75, 0.62\}$. In this way, the local search ability of the population can be enhanced by searching the neighborhood candidates of the elite solutions.

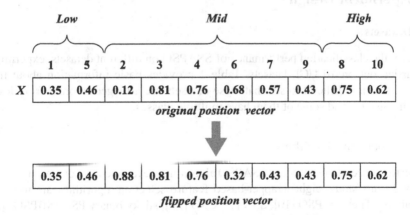

Fig. 1. Illustration of the proposed IFS

3.4 Pre-evaluation with Surrogate Model

The fitness evaluation of a large number of candidate particles will greatly increase the running time of the SSAPSO, so a new surrogate model is used to pre-evaluate the fitness value of candidate particles. Inspired by [12, 22], the k-Nearest Neighbor (KNN) algorithm was selected to construct the surrogate model. The huge training set is often the main reason for the long training time of the model. To avoid expensive computation, we take the positions of the population in the previous iteration and their corresponding real fitness values as the surrogate training set.

Specifically, the steps to construct the surrogate model to estimate the fitness value of a candidate solution are as follows. The positions of the population in the previous iteration are used as the surrogate training set. The model then computes the Euclidean distance between the candidate solution and each instance in the surrogate training set. Finally, the fitness values of the three closest instances are selected to estimate the fitness values of the candidate solutions. Then the pre-evaluated fitness value of the candidate solution is displayed in Eq. (12).

$$F_{pre} = \frac{F_1 * d_3 + F_2 * d_2 + F_3 * d_1}{d_1 + d_2 + d_3} \tag{12}$$

where F_1, F_2, and F_3 are the true fitness values of the three nearest neighbors to the evaluated candidate solution, respectively. d_1, d_2, and d_3 are the distances of the three nearest neighbors to the evaluated candidate solution, respectively. F_{pre} represents the fitness value predicted by the surrogate model. In this way, the nearest neighbor to the candidate solution will receive a higher weight. The predicted fitness values of these candidate solutions are ranked, and then the more optimal N candidate solutions are authentically evaluated.

4 Experiment Design

4.1 Datasets

To verify the classification performance of SSAPSO on different datasets, experiments are carried out on 10 UCI datasets. Table 1 provides basic information about these datasets. It is not difficult to see that these datasets are arranged in order from low to high dimensions, and some of them have multiple classes.

4.2 Comparison Algorithms

The performance of the proposed SSAPSO algorithm is evaluated by comparative experiments. In this study, eight wrapper-based feature selection algorithms are used as a comparison. They are PSO [10], SSA [9], CSO [23], sticky binary PSO (SBPSO) [24], PSOEMT [25], and MIRFDE [26]. It is worth noting that the classification method using all features (Full) is also included in the comparison experiment.

4.3 Experiment Settings

In all experiments, each algorithm was run 30 times independently and the average score was taken. For each experiment, 70% of the instances in the dataset are randomly selected as the training set and the remaining instances are used as the test set. Furthermore, considering that some datasets have fewer instances, ten-fold cross-validation is used

Table 1. Datasets

Datasets	Features	Instances	Classes
WallRobot	24	5456	4
German	24	1000	2
Ionosphere	34	351	2
Chess	36	3196	2
Movementlibras (Mov)	90	360	15
Musk1	166	476	2
Madelon	500	2600	2
Isolet	617	1560	26
MultipleFeatures (MFS)	649	2000	10
COIL20	1024	1440	20

to evaluate the classification error rate on the training set. KNN is used to evaluate the feature subset and k is set to 3. Finally, the feature subset selected by algorithm is evaluated on the test set.

FS should not only control the classification error rate but also limit the size of the feature subset. Therefore, Eq. (13) is employed to evaluate the selected feature subset.

$$minFit(x) = \alpha \times error + (1 - \alpha) \times \frac{Num(selected)}{Num(all)} \tag{13}$$

where α is the weight that determines the importance of the classification error rate (*error*) in the evaluation. *Num(selected)* refers to the size of the obtained feature subset and *Num(all)* refers to the size of the original feature set. Therefore, the original bi-objective problem becomes a minimization problem. It is worth noting that the α is set to 0.9 in this experiment.

4.4 Parameter Settings

The parameters in the SSAPSO algorithm are set as follows: The number of individuals N in the population is set to 50. The upper limit of the number of iterations T is set to 100. As shown in Table 2, the parameter settings of all the compared algorithms are consistent with those in the literature.

4.5 Experimental Results and Analysis

Table 3 shows the average classification accuracy (%) of SSAPSO and 6 comparison algorithms running 30 times on the test set. The results shown in bold in the table indicate that the corresponding algorithm achieved the best results. However, W/T/L means the comparison results between SSAPSO and other algorithms. In general, it can be seen from Table 3 that SSAPSO achieves the highest average classification accuracy

Table 2. Parameter settings of the 6 compared algorithms

Algorithms	Parameter settings
PSO	$c_1 = c_2 = 1.49445, w = 0.9 - 0.4 * (t/T)$
SSA	$c_1 = 2e^{-(4l/L)^2}$
CSO	$\phi = 0.1, \lambda = 0.5$
SBPSO	$i_s = 4/N, \alpha = 2, ustkS = 8 \times T/100$
PSOEMT	$\delta = 0.7, \rho = 0.05, rmp = 0.6, m = 10$
MIRFDE	$F = 0.5, CR = 0.2, \beta = 0.3$

on 7 datasets out of the 10 datasets. On the Ionosphere, SSAPSO is 1.33% lower than PSOEMT. On Mov, MIRFDE is nearly 3% ahead of SSAPSO; SSAPSO is 2.14% lower than SBPSO on Musk1. In addition, it is easy to see that SSAPSO achieves better results on high-dimensional datasets. This means that IALS and IFS of SSAPSO can effectively strengthen the search ability of the algorithm.

Table 4 illustrates the number of features selected by SSAPSO compared with the other six algorithms on all 10 datasets. As can be seen in Table 4, the proposed SSAPSO method selects the fewest features on the Madelon and COIL20 datasets. In general, SSAPSO has a similar number of selected features as PSO, SSA, CSO, and SBPSO on most datasets, and the dimensionality reduction ability of SSAPSO is better than MIRFDE but worse than PSOEMT. SSAPSO selects about 40% of the features in the original feature set on Musk1, Mov, and Isolet datasets. However, on the datasets Wall-Robot, Madelon, and COLI20, less than 20% of the features in the original feature set are selected. This shows that the proposed SSAPSO algorithm can reduce the number of selected features and improve the classification accuracy.

Table 3. Comparison of SSAPSO and 6 representative algorithms in terms of average classification accuracy (%)

Datasets	Full	PSO	SSA	CSO	SBPSO	PSOEMT	MIRFDE	SSAPSO
WallRobot	84.89	94.44	91.51	94.35	93.91	89.60	84.55	**94.76**
German	62.94	69.19	69.08	69.08	63.11	60.79	62.84	**69.44**
Ionosphere	82.74	88.16	87.68	88.44	87.99	**90.63**	82.82	89.30
Chess	93.86	96.60	94.40	96.71	94.84	89.41	94.78	**97.65**
Mov	75.48	75.77	74.75	75.86	75.80	59.10	**79.12**	76.14
Musk1	85.23	84.31	84.10	84.01	**88.90**	78.19	83.93	86.76
Madelon	56.79	58.85	55.93	59.18	61.26	77.41	54.16	**85.79**
Isolet	79.49	80.17	77.60	79.87	80.49	77.98	80.42	**80.92**
MFS	97.00	97.61	97.56	97.68	97.20	96.95	97.00	**97.76**
COIL20	97.89	98.23	97.83	98.22	98.00	93.25	97.69	**98.50**
W/T/L	10/0/0	10/0/0	10/0/0	10/0/0	9/0/1	9/0/1	9/0/1	–

Table 4. Comparison of SSAPSO and 6 representative algorithms in terms of numbers of selected features

Datasets	Full	PSO	SSA	CSO	SBPSO	PSOEMT	MIRFDE	SSAPSO
WallRobot	24.00	4.20	6.47	4.17	4.00	**1.00**	21.97	4.40
German	24.00	8.70	9.73	9.73	11.14	**7.00**	22.00	11.77
Ionosphere	34.00	8.63	11.87	8.90	9.37	**6.00**	32.00	8.93
Chess	36.00	13.67	17.47	14.63	24.17	**6.00**	33.89	23.40
Mov	90.00	31.70	34.57	33.03	53.49	**29.00**	76.91	34.53
Musk1	166.00	71.90	71.30	72.13	68.31	**29.00**	127.31	72.80
Madelon	500.00	240.00	220.27	238.53	227.97	20.00	325.23	**12.80**
Isolet	617.00	292.90	258.60	291.87	371.63	**118.00**	389.69	267.03
MFS	649.00	268.23	247.67	273.60	268.71	**161.00**	404.60	336.47
COIL20	1024.00	417.57	394.87	432.30	408.86	144.00	618.11	**130.63**

Due to space limitations, Table 5 shows the comparison of the classification accuracy of SSAPSO and its variants on the four datasets. SSAPSO1 is the updated model consistent with the canonical SSA with the canonical PSO but with IFS and surrogate model. Furthermore, SSAPSO2 uses IALS to update the positions of the population but does not adopt IFS and surrogate model. It can be seen that on the low-dimensional datasets German and Ionosphere, SSAPSO achieves higher classification accuracy than SSAPSO1 and SSAPSO2. This means that IALS can effectively enhance the search ability of the algorithm and avoid fast convergence of the algorithm. In addition, Fig. 2 shows the classification accuracy variation trend of SSAPSO and its variants on the four datasets. It is easy to see that the search ability of SSAPSO1 and SSAPSO on the training set is relatively close, which means that the strong local search ability of IFS can find more promising solutions in the neighborhood. On the datasets Ionosphere and Isolet, SSAPSO is slightly inferior to SSAPSO1 in the early iteration stage. However, IALS can maintain population diversity, which leads SSAPSO to achieve higher classification accuracy in the end.

Table 5. Comparison of SSAPSO performance with SSAPSO variants

Datasets	SSAPSO1	SSAPSO2	SSAPSO
German	69.03	69.32	**69.44**
Ionosphere	88.33	89.21	**89.30**
Musk1	85.97	83.64	**86.76**
Isolet	80.43	80.17	**80.92**

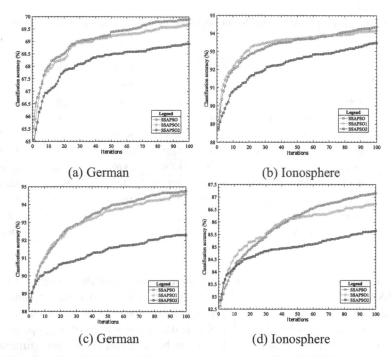

(a) German   (b) Ionosphere

(c) German   (d) Ionosphere

Fig. 2. The changing trend of classification accuracy achieved by SSAPSO and its variants on the four datasets

5 Conclusion

In this paper, a new FS approach based on SSA and PSO (SSAPSO) is used to solve the classification problem. To improve the search ability of SSAPSO, a novel interpopulation adversarial learning strategy is proposed. An interval-based flipping strategy based on the correlation between features and labels is employed to search in the neighborhood of elite solutions. A new surrogate model is proposed to predict the fitness values of candidate solutions, aiming to reduce the computational cost.

The performance of the SSAPSO is evaluated by comparing it with 6 representative algorithms. Experimental results show that, in most cases, SSAPSO can achieve higher classification accuracy based on reducing a certain number of features. This is because the proposed inter-population adversarial learning strategy and interval-based flipping strategy can significantly promote the search ability of the SSAPSO, and even in the face of high-dimensional datasets, SSAPSO can also achieve good results.

In future work, we will focus on finding a new topology to promote the classification accuracy of the proposed algorithm on high-dimensional datasets.

Acknowledgment. This work was supported by the Basic Public Welfare Research Project of Zhejiang Province (LGF22F020020), the National Natural Science Foundation of China (61806204), Scientific Research Starting Foundation of Zhejiang Sci-Tech University (20032309Y), and the Research Fund Project of Zhejiang Sci-Tech University Longgang Research Institute (LGYJY2023003).

References

1. Wang, Z., Liang, S., Xu, L., Song, W., Wang, D., Huang, D.: Dimensionality reduction method for hyperspectral image analysis based on rough set theory. Eur. J. Remote Sens. **53**(1), 192–200 (2020)
2. Lee, C.Y., Wen, M.S.: Establish induction motor fault diagnosis system based on feature selection approaches with MRA. Processes **8**(9), 1055 (2020)
3. Sowan, B., Eshtay, M., Dahal, K., et al.: Hybrid PSO feature selection-based association classification approach for breast cancer detection. Neural Comput. Applic. **35**(7), 5291–5317 (2023)
4. Chen, J., Chen, Y., He, Y., et al.: A classified feature representation three-way decision model for sentiment analysis. Appl. Intell. **52**, 7995–8007 (2022)
5. Solorio-Fernandez, S., Martínez-Trinidad, J.F., Carrasco-Ochoa, J.A.: A supervised filter featureselectionmethodformixeddatabasedonspectralfeatureselectionandinformation-theory redundancy analysis. Pattern Recogn. Lett. **138**, 321–328 (2020)
6. Espinosa, R., Jiménez, F., Palma, J.: Multi-surrogate assisted multi-objective evolutionary algorithms for feature selection in regression and classification problems with time series data. Inf. Sci. **622**, 1064–1091 (2023)
7. Jain, R., Xu, W.: Artificial Intelligence based wrapper for high dimensional feature selection. BMC Bioinformatics **24**(1), 392 (2023)
8. Khaire, U.M., Dhanalakshmi, R.: Stability of feature selection algorithm: a review. J. King Saud Univ.-Comput. Inform. Sci, **34**(4), 1060–1073 (2022)
9. Mirjalili, S., Gandomi, A.H., Mirjalili, S.Z., et al.: Salp Swarm Algorithm: a bio-inspired optimizer for engineering design problems. Adv. Eng. Softw. **114**, 163–191 (2017)
10. Kennedy, J., Eberhart, R.: Particle swarm optimization. In: ICNN'95-International Conference on Neural networks, vol. 4, pp. 1942–1948. IEEE (1995)
11. Moradi, P., Gholampour, M.: A hybrid particle swarm optimization for feature subset selection by integrating a novel local search strategy. Appl. Soft Comput. **43**, 117–130 (2016)
12. Chen, K., Xue, B., Zhang, M., Zhou, F.: Correlation-guided updating strategy for feature selection in classification with surrogate-assisted particle swarm optimization. IEEE Trans. Evol. Comput. **26**(5), 1015–1029 (2021)
13. Qu, L., He, W., Li, J., Zhang, H., et al.: Explicit and size-adaptive PSO-based feature selection for classification. Swarm Evol. Comput. **77**, 101249 (2023)
14. Balakrishnan, K., Dhanalakshmi, R., Khaire, U.M.: Improved salp swarm algorithm based on the levy flight for feature selection. J. Supercomput. **77**(1), 1239912419 (2021)
15. Aljarah, I., Mafarja, M., Heidari, A.A., Faris, H., Zhang, Y., Mirjalili, S.: Asynchronous accelerating multi-leader salp chains for feature selection. Appl. Soft Comput. **71**, 964979 (2018)
16. Qaraad, M., Amjad, S., Hussein, N.K., Elhosseini, M.A.: An innovative quadratic interpolation salp swarm-based local escape operator for large-scale global optimization problems and feature selection. Neural Comput. Appl. **34**(20), 17663–17721 (2022)
17. Qaraad, M., Amjad, S., Hussein, N.K., Elhosseini, M.A.: Large scale salp-based grey wolf optimization for feature selection and global optimization. Neural Comput. Appl. **34**(11), 8989–9014 (2022)
18. Kumar, S., John, B.: Anovel gaussian based particle swarm optimization gravitational search algorithm for feature selection and classification. Neural Comput. Appl. **33**(19), 12301–12315 (2021)
19. Chen, K., Zhou, F.Y., Yuan, X.F.: Hybrid particle swarm optimization with spiral-shaped mechanism for feature selection. Expert Syst. Appl. **128**, 140–156 (2019)

20. Robnik-Šikonja, M., Kononenko, I.: Theoretical and empirical analysis of ReliefF and RReliefF. Mach. Learn. **53**, 23–69 (2003)
21. Eiras-Franco, C., Guijarro-Berdiñas, B., Alonso-Betanzos, A., Bahamonde, A.: Scalable feature selection using ReliefF aided by locality-sensitive hashing. Int. J. Intell. Syst. **36**(11), 6161–6179 (2021)
22. Xue, Y., Tang, T., Pang, W., Liu, A.X.: Self-adaptive parameter and strategy based particle swarm optimization for large-scale feature selection problems with multiple classifiers. Appl. Soft Comput. **88**, 106031 (2020)
23. Gu, S., Cheng, R., Jin, Y.: Feature selection for high-dimensional classification using a competitive swarm optimizer. Soft. Comput. **22**, 811–822 (2018)
24. Nguyen, B.H., Xue, B., Andreae, P., Zhang, M.: A new binary particle swarm optimization approach: momentum and dynamic balance between exploration and exploitation. IEEE Trans. Cybern. **51**(2), 589–603 (2019)
25. Chen, K., Xue, B., Zhang, M., Zhou, F.: An evolutionary multitasking-based feature selection method for high-dimensional classification. IEEE Trans. Cybern. **52**(7), 71727186 (2020)
26. Hancer, E., Xue, B., Zhang, M.: Differential evolution for filter feature selection based on information theory and feature ranking. Knowl.-Based Syst. **140**, 103–119 (2018)

A Simulated Annealing BP Algorithm for Adaptive Temperature Setting

Zi Teng[1], Zhixun Liang[2], Yuanxiang Li[3], and Yunfei Yi[2(✉)]

[1] School of Mathematics and Statistics, Central South University, Changsha 410083, Hunan, China
[2] School of Big Data and Computer, Hechi University, Yizhou 546300, Guangxi, China
gxyiyf@163.com
[3] School of Computer Science, Wuhan University, Wuhan 430072, Hubei, China

Abstract. The backpropagation (BP) algorithm is so far, the most effective algorithm for neural network learning and training. This algorithm is based on the fast gradient descent method in optimization theory, which is essentially a deterministic local optimization algorithm. In this paper, for improving its local optimization, enhancing its global optimization potential ability is discussed by combining with the simulated annealing mechanism. The BP algorithm is utilized as the core component under the framework of simulated annealing algorithm, and the Metropolis criterion is used as the heuristic strategy. Integrating advantages of the BP algorithm and the simulated annealing algorithm, An adaptive simulated annealing based BP algorithm is designed. In this algorithm there two adaptive strategies, one is the setting of annealing temperature, and the other is the random perturbation of connection weight and bias value. The adaptive temperature setting can simultaneously reflect changes of the loss function value, and the adaptive perturbation is related to the network complexity and loss function value. By using the small batch gradient descent method used in machine learning commonly, the update of the annealing temperature is synchronized with the input sample batch, which follows up the search state of the algorithm running. The perturbed renewal of those parameters of the neural network is based on the current network configuration to avoid blind random perturbation. The algorithm is tested with the Fashion-MNIST and CIFAR-10 datasets, which are widely used in the current deep learning research field. The classification and recognition accuracies of the algorithm exceed 99% both. These results are better than the best results reported currently for the two datasets. In addition, the adaptive BP algorithm is repeated 30 times with these two datasets respectively. The statistical analysis results show the robustness and stability of the algorithm.

Keywords: BP algorithm · simulated annealing algorithm · adaptive temperature setting · pattern recognition

1 Introduction

Deep learning methods based on deep neural networks are an important direction in the new generation of artificial intelligence research. Notably, applications of these methods in pattern recognition have led to remarkable progress. As a result, theoretical methods and applied researches on new generation artificial intelligence are experiencing a period of rapid growth.

However, researches on methods of sample-based supervised learning have primarily focused on the back propagation of error (BP) algorithm to training neural networks, from early shallow to contemporary deep networks, and the role of this algorithm remains unchallenged. For example, the posterior learning method, which is based on the Bayesian full probability formula, and the stochastic gradient method, which delays gradient vanishing, and so on, they just adjust the learning rate and change the direction of gradient descent based on the BP algorithm. In recent years, swarm intelligence algorithms, such as nature-inspired and evolutionary algorithms, have been thoroughly studied and applied. Thus, these algorithms are now being combined with BP algorithms and introduced into deep neural networks and deep learning, to allow adjustments of network structure and learning parameters in the BP algorithm [1–3].

The simulated annealing algorithm(SAA) is a classical nature-inspired computing algorithm by imitating the physical annealing process, but it is not a swarm search algorithm. The simplicity and speed of this algorithm are attributed to its single-point search approach. The core idea of the SAA is to simulate the process of a metal being cooled slowly (annealing) to its lowest energy state after being heated, with a clear physical background. It could overcome the shortcomings of the local search method with a random jump-trap strategy, expecting to search for the global optimal solution of an optimization problem [1]. Currently, the simulated annealing strategy has been integrated into the design and application of swarm intelligence algorithms [4–11]. The BP algorithm is derived from the gradient descent method in optimization theory and is essentially a deterministic local optimization algorithm. In this paper, the simulated annealing strategy is used to improve the local search characteristics of the BP algorithm for searching for the global minimum or better local minimum of the loss function.

A key parameter in the SAA is the temperature. The temperature decreased slowly, directly affects the performance of solving. In fact, the SAA eventually falls into a local search when the temperature is very low. In classical setting methods and common applications, the temperature decreases monotonically with the annealing process, so at some moments the temperature is forced increasing, that is referred to as tempering. However, the change of temperature in tempering does not relevant to the optimization objective and cannot reflect the search behaviour of the algorithm in real time. Considering the physical background of the SAA, we then modelled and analysed the convergence and time complexity based on the theory of gas kinematics and statistical physics. Finally, we derived a formula for estimating the temperature with the objective function value. Based on this formula, a temperature adaptive setting method was proposed, which is valuable of practical generalizing and applications. The temperature set by this method is closely related to the objective function and changes synchronously with the function value. The experimental tests and analyses showed that the temperature has a decreasing trend in the whole, and follows up the change of the objective function value synchronously, which just reflects the search behaviour of the SAA in real time.

Therefore, we will generalize the theoretical results above to combine with the BP algorithm in this paper. To improve the optimization performance of the BP algorithm, the adaptive temperature setting method is naturally integrated into the BP algorithm so that enabling the potential of global optimization. Now deep learning methods are not in the manner of inputting samples one by one because of a large number of samples, but in batches for saving computing time. Embedding the simulated annealing mechanism into the BP algorithm, the annealing process will be carried out along with sample batches. That is, the temperature varies after current batch of samples are dealt with by the neural network. Then a simulated annealing BP algorithm with adaptive temperature setting is designed.

In Sect. 2, the basics of the BP algorithm and SAA are introduced, especially the adaptive annealing temperature setting method. In Sect. 3, the adaptive simulated annealing based BP algorithm will be designed in detail, and the adaptive update strategy for the values of connection weights and bias of the neural network is also described. In Sect. 4, test experiments and analysis are illustrated. The Fashion-MNIST and CIFAR-10 datasets, which widely applied in current deep learning researches, are explained. Comparative analysis and statistical analysis are both made, and results show the excellent performance of the adaptive simulated annealing BP algorithm. The fifth part gives conclusion and discussion, summarizing the work of this paper and presenting some directions for further researches.

2 BP and Simulated Annealing Algorithms

2.1 BP Algorithm

In the 1970s and 1980s, training algorithms in multilayer neural networks were supported by improving computing power of computers. As a result, neural networks that had not been used in decades once again captured the attention of artificial intelligence researchers, particularly in the realm of efficient learning and training algorithms. Following the BP algorithm coming into being, neural networks have emerged as one of the main fields of artificial intelligence [12]. At present, computing and storage capabilities of computer systems have been greatly improved. Neural networks have been developed into deep network architectures and entered the era of deep learning. The development of neural network researches directly promotes the research and development of a new generation of artificial intelligence.

The BP algorithm uses the method of reverse conduction error to guide a neural network in performing sample learning and training, to solve complex classification and pattern recognition problems. The learning and training processes iteratively adjust the network parameters, such as the connection weight and bias values, by using the gradient descent method of optimization theory, so that optimizing the loss function to the minimum and obtaining the best parameter configuration of the network.

The left side of Fig. 1 is a schematic diagram of a feedforward neural network with multiple hidden layers (for the sake of simplicity, one hidden layer is used to represent multiple hidden layers), and the neurons between the adjacent layers are fully connected and assigned connection weight values. There are no connections among neurons in the same layer or between neurons across layers.

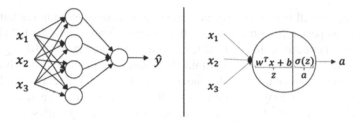

Fig. 1. Schematic diagram of the structure of a fully connected feedforward neural network.

Let the network have q hidden layers, the input layer have d neurons, the output layer have l neurons, the connection weight of the ith neuron in the input layer and the hth neuron in the adjacent hidden layer be w_{ih}, and the connection weight of the hth neuron in the hidden layer and the jth neuron in the output layer be w_{hj}. Samples in the training set are input into the input layer. In the hidden and output layers, the input of a neuron is given by the inner product of the vector of the neighbouring neuron values in the preceding layer and the vector of their corresponding connection weights. The neuron values are determined by a transfer function, which is the sum of the activation function acting on the inner product and b, a bias value. Let the training set be $D = \{(x_1, y_1), (x_2, y_2), \ldots, (x_m, y_m)\}$, of which $x_i \epsilon \mathbb{R}^d$, $y_i \in \mathbb{R}^l$. For sample $(x_k, y_k)(k = 1 \ldots m)$, assuming that the output at the output layer is $\hat{y}_k = (\hat{y}_1^{\,k}, \hat{y}_2^{\,k}, \ldots, \hat{y}_l^{\,k})$, $\hat{y}_j^{\,k} = a_j = \sigma(z_j)$, $z_j = \beta_j + \theta_j$, $\beta_j = w^T x$ (shown on the right side of Fig. 1), where σ is the activation function, β_j is the input value of the jth neuron in the output layer, and is the bias value of the th neuron in the output layer. The bias represents the intercept of the transfer function, which simulates the threshold effect of the neuron. The activation function σ is generally the *sigmoid* function, which performs a nonlinear transformation on the input. In addition, in a hidden layer h, the bias of each neuron is denoted as γ_h.

Let the loss function of the neural network for sample (x_k, y_k) be $E_k = f(y_j^{\,k}, \hat{y}_j^{\,k})$. The loss function can be the mean square error, the cross entropy function, which is used in this paper, and so on. During the learning and training processes of the network, all parameters (the connection weight w_{ij} and bias values θ_j), are simply recorded as a vector W, then, the update formula is expressed as:

$$W \leftarrow W + \Delta W \tag{1}$$

Given a learning rate η, the gradient descent method is used to adjust the parameters. Firstly, the update increments shown as formula (2), of the connection weight are calculated between the output layer and its adjacent hidden layer, because values of the loss function on the output layer are known.

$$\Delta W_{ij} = -\eta \frac{\partial E_k}{\partial W_{hj}} \tag{2}$$

Then, the process of gradient descent goes on from the output layer to the input layer. This process advances layer by layer in reverse, updating all parameters by gradient descent optimization until convergence. That is, the core idea of the BP algorithm [13].

When inputting the first sample, computing inner product, transformation, backpropagation of error, gradient descent optimization and parameter update are performed, and so on, this iteration continues until all samples are input and processed. This is referred to as completing a cycle or one epoch of learning and training. Then, the next epoch of learning and training is conducted again. Initially, as a small threshold, an allowable error of loss function between two epochs is set as the stopping condition, and the learning and training processes are carried out repeatedly until the stopping condition is met. Then, the BP algorithm terminates.

In the era of big data, sample sets have become large-scale datasets, and the sample number is continuously increasing. Thus, it is impossible to input samples and update parameters one by one as an epoch. It is now common to divide a sample set into a training set and a test set, and then the training set are divided into a number of batches. In an iterating step, samples in a batch are input, and the sum of the output errors is calculated as the loss function of those samples in the batch. Therefore, the learning and training processes are performed in the batch manner [13, 14]. To prevent overfitting, which occurs when the neural network learns too much from the training set and results in poor generalizability to non-training set, the loss function should be regularized. It is common and practice to add the squared sum of parameters, referred to as the regularization term, into the objective function. The resulting objective function becomes the weighted sum of two parts. Assuming that each batch contains p samples, and $num = 1, \ldots, batch$, there is:

$$loss_{num} = \lambda \frac{1}{p} \sum_{k=1}^{p} E_k + (1 - \lambda) \sum_i w_i^2 \tag{3}$$

where $0 < \lambda < 1$.

Minimizing the objective function (3), the BP algorithm is briefly described as follows.

Algorithm 1 BP (backpropagation of errors)
algorithm
1. INITIALIZE
2. BEGIN
3. Input sample set $D = \{(x_k, y_k)\}_{k=1}^m$, the learning rate η;
4. Divide the training set from D into $bath$ groups (batches), and the number of samples in each group is roughly equal;
5. $epoch := 0, E\_0 := 0, Stop := 0$, and set the stopping criterion tolerance ε;
6. Randomly initialize the neural network connection weights W_{ij} and bias values θ_j in the range (0, 1);
//For simplicity, these values are recorded as W
7. END
8. NET-LEARNING&TRAINING
9. BEGIN
10. WHILE !Stop DO
11.   BEGIN
12.   FOR $num := 1\ to\ batch$ DO
13.   Enter the num batch of sample groups, compute the network output and loss function $loss_{num}$;
14.   Error backpropagation and gradient descent optimization to update parameters W;
15.   ENDFOR ($batch$)
16. $epoch := epoch + 1$;
17.   Calculate the total error of the training set after parameter update E_{min};
18.   IF $|E_{min} - E_0| < \varepsilon$ THEN $Stop := 1$;
ELSE $E_0 := E_{min}$;
19.   END (WHILE)
20.   Save connection weights and bias values W, output the number of learning training epochs, the minimum value $loss_{min}$ of the loss function;
21.   END

The other part of the sample set is the test set, which is used to test the learning and training effect of the BP algorithm. For brevity, the test part is not listed in Algorithm 1.

In 1985, David Rumelhart et al. further developed the theory and application of the BP algorithm [14]. In recent years, it has been applied widely to learn and train with large-scale datasets to different neural networks, especially the deep neural networks. In applications, it is a usual expectation to enhance the performance and robustness of the BP algorithm. However, there are two basic problems considering its optimization mechanism. The first problem is that the gradient descent is often slowly called the gradient vanishing for complex and deep neural networks in the middle and later periods of the BP algorithm executing. The second is that, once the neural network parameters are initialized the BP algorithm can only converge to a local minimum of the loss function, as a result, the parameter setting of the network represents a local optimal solution. And yet the BP algorithm itself lacks a mechanism to escape from the local optimal solution. The first problem is one of the hotspots of current deep learning methods, such as attention mechanisms and stochastic gradient descent [15, 16]. In this paper, the second problem is to address. By integrating the simulated annealing mechanism into the BP algorithm, a combined algorithm would be of the capacity of escaping local optimal solutions and the potential of converging to the global optimal solution.

2.2 Simulated Annealing Algorithm

The SAA is a classical heuristic algorithm that imitates the physical annealing process. The idea originates from the analogy between optimization problem solving and the solid annealing process for improving local search methods with a stochastic perturbation mechanism. Now, simulated annealing strategies have been integrated into the designations and applications of swarm intelligence algorithms. The SAA compares the objective function value of the optimization problem to the energy of the physical system and simulates the process in which the physical system tends to the minimum energy state. The random jump-trap mechanism is called the Metropolis criterion whose theoretical basis is the Boltzmann distribution in equilibrium statistical physics. Giving an objective function f, i represents the current solution of the problem and j, the new solutions generated from i, which are analogous to the states of the physical system and the transitions between states. The SAA states as follows.

Algorithm 2 SAA(Simulated Annealing Algorithm)
1. INITIALIZE
2. BEGIN
3. $k := 0; i := l_0; L_k := L_0;$
4. REPEAT
5. FOR $n := 1$ to L_k DO
6. BEGIN
7. GENERATE (j from i);
8. IF $f(j) \leq f(i)$, THEN $i := j$;
9. ELSE IF
10. $exp\{[f(i) - f(j)]/T_k\} > random[0,1)$
    THEN $i := j$;
11. ENDFOR
12. $k := k + 1$;
13. UPDATE (T_k);
14. CALCULATE (L_k);
15. Until the stop criterion
16. END

Lines 9–10 expresses the Metropolis criterion, where random[0,1] represents a random number in the interval[0,1]. The exponential function is the ratio of two Boltzmann distributions describing two equilibrium states of the physical system. T_0 is the initial temperature set in initializing the algorithm, T_k is the temperature as a function of the annealing step k, and L_k is the number of inner loop at step$k$, which was referred to as the Markov chain length in early theoretical analysis. During the inner loop at temperatureT_k, the SAA simulates the transition process among states of the system going on times. That is, at a loop step a new solution is generated, if it is worse than the current solution, the Metropolis criterion determines whether the solution is accepted as the current solution. When the new solution is accepted, it is referred to as the jump-trap step.

T_k and L_k are two important parameters in the SAA, which directly affect the search performance of the algorithm [17, 18]. It has been found that is of an upper bound [19–21], through numerous experimental and applied studies, as well as our earlier time-complexity analysis, so a suitable integer can be chosen because an excessively large L_k leads to invalid searches with too long inner loop. In the middle and later stages of the annealing process, the integer can be smaller. In line 13 of SAA, cooling occurs slowly according to the temperature updating formula, that is, the annealing. The classical temperature update formula is set by a simple decreasing serial, $T_{k+1} = \beta T_k$, where β is referred to as the decreasing factor, which is a value in the interval (0.8,1) to ensure slow cooling. While annealing slowly to a very low temperature, the Metropolis criterion can no longer jump trap, poor solutions are hardly accepted and SAA falls into a local search. Therefore, the tempering strategy was proposed, and the SAA is improved to an annealing and tempering algorithm. To do so, a mechanism of monitoring and observing the annealing process must be set during the middle and late search stages of the algorithm. The mechanism helps to determine whether the algorithm is trapped in a local search. If it is, then the temperature would be raise at certain moments of the annealing, called the tempering, to enable the process escaping the local search. This improvement is an important advancement in the SAA researches [20, 21].

2.3 Adaptive Temperature Setting

The temperature setting method mentioned above is referred to as a fixed setting method. Once the temperature T_0, the factor β and the termination temperature T_{min} are determined. The annealing temperature changes in a slowing process of step-by-step, which is not related to behaviours and running states of the algorithm, and nor to changes of the objective function value, but controls the search process of the algorithm. So a tempering strategy was presented, it tracks and inspects search states of the algorithm. Several tempering moments are embedded into the search process artificially, ensuring that the algorithm performs several turns tempering without changing its established process. But the tempering value was set according to a rule of thumb and is not related to changes of function values. That is, changes of temperature cannot reflect operating states of the algorithm, nor the historical information of the algorithm searching process.

We have made a time complexity analysis on the relaxation model established for the SAA in the early stage, a theoretical estimation formula for the annealing temperature was derived [22].

$$T_k = \frac{1}{3}(f_{k-1} - f_k) \tag{4}$$

This formula directly represents the temperature with the objective function value obtained by the algorithm and shows that the magnitude of the temperature value is equivalent to the magnitude of the difference of function values between two adjacent annealing processes. But the estimation has only theoretical significance, when solving practical problems, $f_{k-1} - f_k$ will fluctuate randomly. Thus, this formula cannot be used directly as a temperature setting.

However, the direction and method of temperature setting can be found from the Formula (4). For the two typical types of numerical and non-numerical optimization problems, then, we derived the corresponding adaptive temperature setting methods [22]. In this paper, it is to discuss the combination of SAA with BP algorithm that is regarded as a numerical algorithm. The setting formula for numerical optimization problems is given below.

$$T_{k+1} = \alpha |f_{k-1} - f_k| + \sum_{i=2}^{k-1} \beta_i (\alpha |f_{i-1} - f_i|) \tag{5}$$

The cumulative sum in the formula is referred to as the regularized compensation term of the linear superposition, which represents the influence of the previous temperature on the current temperature, β_i is referred to as the inertia factor, and it can be gotten that $\beta_i = \beta^{k-i+1}$. The larger the β value is, the greater the influence of inertia is. In the algorithm running, f_k can be replaced by f_{kmin}, which is the currently optimal value of the function after performing k times annealing searches. α is the weight factor, and it indicates the influence degree of the function value on the temperature. Experimental analysis of previous researches showed that $\alpha = 10$ is a moderate value.

Setting Formula (5) not only utilizes the current results but also utilizes the historical information of the annealing process, that is, the historical information is used to estimate and correct the annealing temperature. When the difference of objective function values between adjacent annealing searches is small, the inertia term plays a major role in the temperature setting, which is equivalent to classical conventional cooling, and the algorithm performs a local search near the current solution. While the difference changes in a large extent, the former term becomes the main term, and the inertia term plays an auxiliary role. Therefore, the setting strategy maintains high annealing temperatures in the early stages of algorithm running, expanding the search range, and makes small adjustments by the inertia term in the later stages focusing on local optimization. The first half controls the detection behaviors of the algorithm, and the second half controls its mining or optimizing behaviors. Furthermore, from middle stages to later, searching states fluctuate, and annealing searches exhibit adaptive tempering behaviors so that having an opportunity to escape from a local optimum trap.

3 Simulated Annealing Based BP Algorithm

As mentioned above, the combination of nature-inspired computing and heuristic algorithms with the BP algorithm is one of the important directions of current artificial intelligence research. However, those researches have focused mostly on optimizing structures of deep neural networks, adjusting parameters of algorithms such as the learning rate in the BP algorithm, improving the gradient descent direction to delay gradient disappearance [23], and so on. In this section, combination of Algorithm 1 and Algorithm 2 is discussed for improving the optimization approach of the BP algorithm and finding more promising initial solutions. Under the framework of SAA, the BP algorithm is integrated into, in which the BP algorithm acts as the core algorithm and the Metropolis criterion as the heuristic strategy. And then, the simulated annealing based BP algorithm is designed.

3.1 Adaptive Temperature Setting

The primary task of the SAA is to provide a method for generating a new solution from the current solution according to the characteristics of the problem. When the structure of the neural network is determined, the target of the BP algorithm is to minimize the loss function. This is achieved by optimizing the connection weight and bias (for simplicity, these values are collectively referred to as parameters and expressed as W). The objective function is shown in Formula (3). Here and thereafter, the cross-entropy loss function is used, for it is widely used in classification and deep learning research and gaining effectiveness.

$$loss = \frac{1}{p} \sum_{k=1}^{p} [-y_k \cdot \ln \hat{y}_k - (1 - y_k) \cdot \ln(1 - \hat{y})_k] \tag{6}$$

where represents the number of samples. For sample $x, y \in \mathbb{R}^l$, is the corresponding true label value vector, $\hat{y} \in \mathbb{R}^l$, which is the predicted label value vector outputting by the neural network. The two vectors are normalized, and the dot product in the sum represents the vector inner product.

According to the batch processing method in Algorithm 1, if the parameters of the current neural network W are determined, the network output and function *loss* can be calculated. Considering the complexity of the problem and the network, each component of the current solution is randomly perturbed, but related to the network and the function *loss*, simply expressed as formulas (7),

$$W \leftarrow W + \Delta W \tag{7}$$

And eq. (8),

$$\Delta W = 10^{-2(co+q)} \times loss \times random(-1, 1) \tag{8}$$

where co represents the number of convolutional layers in the deep neural network, and q is the number of hidden layers in the network.

Values ΔW are related to the network because a slight perturbation on the complex network configuration may cause the value of function *loss* to fluctuate extensively. To ensure the stability of the solving process, values change inversely with the network complexity. Those values are also correlated to the value of function *loss*. This idea is inspired from the temperature setting method discussed in Sect. 2.3. That is, ΔW changes synchronously with function *loss*, because the ideal minimum value of the *loss* is 0. As the *loss* increases, the perturbation of parameters can also be larger, maintaining the global optimization potential, when $loss \rightarrow 0$, then $\Delta W \rightarrow 0$, the algorithm keeps local searches, and converges quickly.

3.2 Algorithm Description

In contrast to pure optimization problems, training neural networks poses additional challenges due to the complexity of the network structure, especially in the case of modern deep neural networks, which can be increasingly complex. Up to now, theoretical analysis and applications have indicated that the BP algorithm is one of effective learning algorithms by training neural networks. Therefore, the BP algorithm and the SAA can be combined to exploit their strengths and complement their advantages for bring out the best in each other.

In the new algorithm, utilizing simulated annealing process as the framework, the mini-batch gradient descent method will be embedded into the framework. It is followed closely by the input batch of samples to update the annealing temperature. For input samples of a batch, the gradient descent performs firstly, and the corresponding loss function value is calculated, to obtain the current solution (connection weight and bias values), which is the basis for parameter perturbation. At temperature T_k with a batch of samples, the BP algorithm runs L_k times, and also, those parameters are adaptively perturbed L_k times. This combination not only allows the BP algorithm to potentially achieve global optimization but also makes the random perturbation being reliable, avoiding blind searches. Therefore, the parameter perturbation method given by formulas (7) and (8) are referred to as adaptive perturbation strategies.

Combining these two algorithms, the new algorithm will exhibit detection behavior in the early period, and mining and optimization behavior in the later period. These behaviors are similar to the "two E" (exploration and exploitation) mechanism in the evolutionary algorithm. Having explained, a formal description of the adaptive simulated annealing based BP algorithm can be obtained, as given in Algorithm 3.

Algorithm 3 ASA-BP (adaptive simulated annealing based BP algorithm)

1. INITIALIZE
2. BEGIN
3. Input sample set $D = \{(x_k, y_k)\}_{k=1}^{m}$, learning rate η ;
4. Divide training set from D into $batch$ groups;
//sample number in each group is equal;
5. $k := 0, L_k := L_0, T_k := T_0$,
   $epoch := 0 \; Stop := 0$;
6. Randomly initialize network parameter configuration W_{ij} in the range (0, 1);
7. END
8. NET-LEARNING&TRAINING
9. BEGIN
10.   WHILE !Stop DO
11.   FOR $num := 1 \; to \; batch$ DO
12.   FOR $n := 1 \; to \; L_k$ DO
13.   Enter samples of the batch num, compute network output and function $loss_{num}$;
14.   Error backpropagation of and gradient descent method to update parameters ;
15.   Calculate output after updating W and function $loss_1$;
16.   Perturb parameters by formulas (7) and (8) $W := W + \Delta W$;
17.   Calculate output and function using the new W;
18.   IF $loss_2 > loss_1$ THEN
BEGIN
$P := exp\{(loss_1 - loss_2)/T_k\}$;
IF $P < random[0,1)$ THEN
  $W := W - \Delta W$;
//If the Metropolis criteria is not met, do not accept the new W, back to the original values
END
//This conditional statement implies that if $loss_2 \leq loss_1$, and the Metropolis criteria are met, accept the perturbed W
19.   ENDFOR (L_k)
20. $k := k + 1$;
21.   Update T_k using formula (5);
22.   Verify the stopping criterion, $top := 0 \; or \; 1$;
23.   ENDFOR $(batch)$
24. $epoch := epoch + 1$;
25.   END (WHILE);
26.   Save parameters W , and output $loss_{min}$;
27.   END

In Algorithm 3, to reduce the computational complexity, L_k can be a fixed small integer. In the next section, $L_k = L_0 = 10$. The weighting factors of setting the annealing temperature in formula (5) are $\alpha = 10$ and $\beta = 0.95 \sim 0.99$. The other parameters are given in the next section. The k records the number of temperature updating.

4 Experiments and Analysis

In this section, two commonly used datasets in the field of machine learning were selected to conduct experimental validation, performance testing, and comparative analysis on Algorithm 3. Additionally, the settings of relevant parameters in the algorithm were verified. In the comparative experiments, both the original BP algorithm and the BP algorithm based on adaptive simulated annealing used the same neural network and dataset. This means that the parameters and network configurations of the two algorithms are identical, such as the learning rate and batch division of the training set. Based on a training strategy of 70% training set and 30% validation set on each dataset, we demonstrated the accuracy of classification on the validation set. Comparative analysis was conducted based on experimental results, including optimization degree of the loss function, classification accuracy, and other indicators. Furthermore, due to the random mechanism in Algorithm 3, it underwent 30 repetitions to analyze its robustness and stability.

4.1 Fashion-MNIST Dataset

The traditional MNIST dataset is a commonly used dataset in the field of machine learning. It contains 70,000 samples consisting of 60,000 training samples and 10,000 test samples. Each sample is a grey picture of 28×28 pixel handwritten digits from 0–9, and the samples are divided into 10 categories. Recently, neural networks have been developed into deep neural networks, and the MNIST dataset is too easy for these types of networks in learning and classifying. For example, the convolutional neural network and its variants have achieve excellent results for it, their classification accuracies all exceeded 99%. In this section, we use the more complex Fashion-MNIST dataset to carry out the testing experiments to investigate learning and training effectiveness and recognition ability about the Algorithm 3.

The Fashion-MNIST dataset was provided by the Zalando research department and publicly released on GitHub in August 2017. All indicators of the dataset, such as the scale and the division of training and test sets, are exactly the same as the MNIST dataset. But, its sample images are more complex, and feature extraction is more difficult. For example, sample images include t-shirts, boots and other clothing. Therefore, the classification and recognition are more difficult than those with the MNIST dataset. Thus, the Fashion-MNIST dataset is more appropriate for examining classification and recognition capabilities of a neural network and its learning and training algorithm.

For the Fashion-MNIST dataset, the LeNet5 network is used, and trained by the BP algorithm and the adaptive simulated annealing based the BP algorithm respectively. LeNet5 consists of three convolutional layers, two pooling layers and two fully connected layers. Its structure is shown in Fig. 2 [24]. In this figure, the leftmost block named Input is the input layer, the first and third gap between different square block represents the convolution layer, the second and fourth gap between different square represents the pooling layer, the first and second bar represents the two feedforward fully connected hidden layers, and the rightmost bar represents the output layer. The numbers represent dimensions of the transformed matrix of the input samples. For example, the numbers 1 and 120 on the blue bars indicate that a vector of 120 components that was transformed

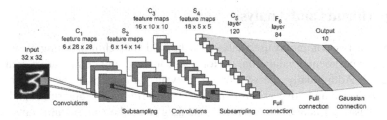

Fig. 2. Schematic diagram of the LeNet5 network structure.

by the former layer, is now input into the hidden layer. The workflow of the LeNet5 network is as follows. First, the sample is input. Then, two convolution and two pooling operations are performed alternatingly, following by a third convolution operation. Then, the output is transferred from the previous hidden layer to the latter hidden layer and output layer. Finally, the algorithm proceeds learning and training.

During learning and training processes of these two algorithms (BP algorithm and adaptive BP algorithm), the learning rate is that $\eta = 0.045$, the number of training samples in each batch is $p = 256$, that is, there are 256 images in a batch, and samples in 10 categories are roughly equally distributed to each batch. The algorithm is trained with the training set for 60 epochs. The initial temperature is that $T_0 = 100$, the cooling coefficient is $\beta = 0.992$, and the termination temperature is $T_{min} = 0.0005$.

In addition, the adaptive BP algorithm undergoes L_k times gradient descent optimization for each batch. For the fairness in comparison analysing, the BP algorithm also performs times gradient descent optimization for each batch. Furthermore, since the adaptive BP algorithm was brought randomness, the algorithm will be repeated 30 times, and the best and average results of each index are listed for comparison and analysis. Also based on the loss function value, the statistical standard deviation is computed to analyse its robustness and stability.

Having trained with the training set, then the LeNet5 network is used to identifying the test set, and several main indicators, are listed on Table 1. On the table, the accuracy rate is defined as the percentage of the number of correctly identified samples to the total number of samples in the test set. The precision rate refers to the ratio of the number of samples correctly identified in a certain category to the total number of samples in the category. The F1-score is the harmonic mean of the precision and recall. Every indicator is the average value of the indicator on 10 categories.

As shown on the Table, the best and average values of each index using the adaptive BP algorithm are higher than those using the BP algorithm. For example, the highest recognition accuracy rate is 99.14%, which is 2.23% higher than 96.91% the current best result reported on GitHub [25]. The average accuracy is also better than the current best result, which is also higher than the accuracy of the BP algorithm over 0.79%. The last column lists the standard deviation which confirms the robustness and stability of the adaptive simulated annealing based BP algorithm.

Indicators on the table can give the accurate comparative analysis, but it is short of the intuition. To obtain a more comprehensive understanding of the adaptive BP algorithm, a few scenarios are selected and displayed by graphical way to obtain some

Table 1. Comparison of indicators between BP algorithm and ASA BP algorithm (Fashion-MNIST).

BP algorithm (L_k times descent optimization)		ASA-BP algorithm(30 repetitions)			
	Recognition effect	Best recognition effect	Number of times the index is better than the BP algorithm	Average recognition effect	Standard deviation (based on loss value)
Accuracy	98.33%	99.14%	28	98.46%	1.17×10^{-5}
Precision	98.21%	99.16%	28	98.40%	
Recall	98.32%	99.14%	28	98.45%	
F1-score	98.26%	99.15%	28	98.42%	
Loss	4.60×10^{-5}	2.30×10^{-5}	28	4.35×10^{-5}	

intuitive impressions. Figure 3 shows two curves of the loss function value for the test set, which explains training results of the two algorithms. It can be observed that both curves show an overall downwards trend, but the network trained by the adaptive BP algorithm outputs smaller loss values. Therefore, it can be considered that the obtained configuration of the network parameters is better.

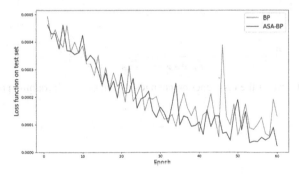

Fig. 3. Comparison curves of the loss function changes for the test set.

Separately, the downwards trend of the red curve is steep initially and then flattens slimly. Initially, fluctuations are sharp, indicating that the learning and training effects of the BP algorithm are indeed significant in the early stage. However, the progress is slow in the later stage. This reveals the gradient vanishing phenomenon in the late stage, and searching processes fall into traps of local optima, which related to the initialization parameter configuration of the network.

For the blue curve, the downwards trend is basically the same as the red curve because the perturbation is always based on the current parameter configuration of the network obtained by the BP algorithm. From the fact that the value of its loss function is mostly smaller than the value of the loss function shown by the red curve, it can be inferred that the adaptive update of the network parameters plays two roles. Firstly, it changes the starting point of the parameter configuration to the new round of gradient descent optimization, enabling the algorithm to search more optimum configuration, that is, the jump-trap is performed by the Metropolis criterion. On the second, the change in the starting point is equivalent to changing the surface of the loss function, also known as the error surface, which changes the direction of the gradient descent optimization, and gradient vanishing is avoided. Also, the adaptive tempering in the later stage, can be inferred observing the blue curve.

Figures 4 and 5 show a panoramic views of the convergence process of adaptive BP algorithm. The calculation is conducted with $p = 256$ samples per batch, then a training cycle has $batch = 235$ batches, and the total number of batches for 60 training epochs is that $batch \times epoch = 14100$. After learning and training with a batch of samples, the loss function value of the batch is calculated. Figure 4 depicts the curve of the loss function, which varies a long with the training batches, and Fig. 5 depicts the temperature curve updated with the training batches.

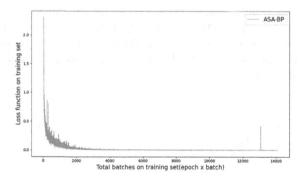

Fig. 4. Curve of the loss function changes during the batch training process.

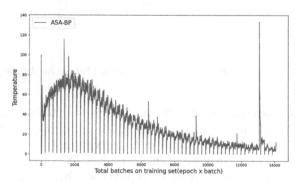

Fig. 5. Curve of temperature changes during the batch training process.

From Fig. 4, after the total training batch is 4000 (approximately 17 cycles), the loss function value is close to 0. This reflects the advantages of the fast gradient descent optimization of the BP algorithm. Reaching 6000 batches (approximately 25.5 cycles), the loss function value curve is almost flat. However there are burrs when zoomed in, indicating very small fluctuations that are clearly reflected by the temperature change curve.

As shown in Fig. 5, temperature changes amplify the small fluctuation of the loss function value. It can be observed from the figure that the temperature curve fluctuates continuously, that is, adaptive tempering occurs. However, the overall tendency is descent and tends to 0, ensuring that the algorithm eventually converges. In particular, it is also observed in this figure that there is an obvious sudden increase in temperature, which corresponds exactly to a sudden change in the value of the loss function. This phenomenon usually occurs after a complete training cycle ends. When the next training cycle begins, the loss function value will mutate due to the adaptive update of the parameter configuration, which confirms the inference by analysing Fig. 3. Further, it demonstrates that the adaptive tempering strategy and the Metropolis criterion can be combined to adjust the starting point.

However, the effect of this self-adaptive adjustment of the starting point is also closely related to the perturbation update strategy of network parameters, which needs to be designed in combination with the network structure and practical application problems. This is one point of our follow-up researches.

The total number of batches in Figs. 4 and 5 is too large to observe local parts. The loss function and temperature varying curves of the second training cycle (*epoch* = 2, 235 batches) are plotted in Figs. 6 and 7.

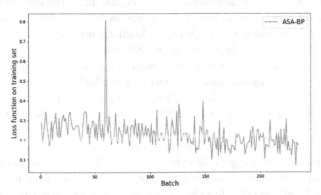

Fig. 6. Curve of the loss function changes during a batch training process in the second cycle.

These two figures better show the dependence of the temperature change with loss function value, the fluctuation state is almost the same, and the temperature change is synchronized to the loss function value. Therefore, changes of temperature also reflect the search states of the algorithm, especially at the end of the training period, the loss function and the temperature values are almost reduced to 0 at the same time.

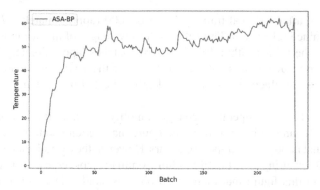

Fig. 7. Curve of temperature changes during a batch training in the second cycle.

4.2 CIFAR-10 Dataset

In the previous section, the Fashion-MNIST dataset was used to conduct a basic test experiment on the adaptive BP algorithm, and the results were compared with those of the BP algorithm. The results preliminarily show the effectiveness of the algorithm. And the optimization mechanism and search process of the algorithm are explained in detail. However, the samples of that dataset are relatively simple, and in this section, a more complex dataset, the CIFAR-10 dataset, is selected for further experiments.

The CIFAR-10 dataset consists of 10 sample classes, and of 60,000 samples totally, for a class containing 6,000 samples. There are 50,000 samples in the training set and 10,000 samples in the test set. Each sample is a 32x32 three-channel color image, and the sample categories include airplanes, birds, cats, and dogs. The color image samples obviously contain more information and features than the greyscale image samples of the Fashion-MNIST dataset. Therefore, the learning, training and classification and recognition of the CIFAR-10 dataset are more difficult.

For this dataset, the VGG-16 network is used to learning, training, classifying and recognizing the data, and its structure is shown in Fig. 8. The VGG-16 network consists of 13 convolutional layers, 5 pooling layers, 3 hidden layers of feedforward fully connected network and 1 output layer (label vector) [22]. The first and second squares both contain 2 convolutional layers and 1 pooling layer, and the third, fourth and fifth squares all contain 3 convolutional layers and 1 pooling layer. The numbers on the sides of the squares have the same meaning as those in Fig. 2 [26]. The workflow of the network is to perform the convolution and pooling operations on each block from left to right after the samples are input, then the algorithm learns and trains on the purple bars.

As in Sect. 4.1, the BP algorithm and the adaptive BP algorithm are applied to the network for experiments and comparative analysis. The learning rate is set to $\eta = 0.03$, the number of training samples for each batch is $p = 256$, in which the samples of 10 categories are roughly equally distributed, and the training set is trained for 60 epochs. The initial temperature is set to $T_0 = 100$, the cooling coefficient is $\beta = 0.992$. Also, the training epochs and the gradient descent times of the BP algorithm on each batch are the same as in Sect. 4.1. In the same way, the adaptive BP algorithm will be repeated 30

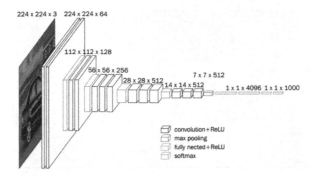

Fig. 8. Schematic diagram of the VGG-16 network structure.

times to examine its stability and robustness. Table 2 lists the recognition effects on the test set.

Table 2. Index comparison between BP algorithm and adaptive BP algorithm (CIFAR-10).

BP		ASA-BP (30 repetitions)			
	Experimental effect	Best recognition effect	Number of times the index is better than the BP algorithm	Average recognition effect	Standard deviation (based on loss)
Accuracy	91.54%	99.56%	30	99.22%	5.29×10^{-5}
Precision	91.62%	99.58%	30	99.21%	
recall	91.55%	99.51%	30	99.20%	
F1-score	91.58%	99.51%	30	99.20%	
Loss	3.94×10^{-3}	1.02×10^{-4}	30	1.90×10^{-4}	

Encouragingly, the adaptive BP algorithm are obviously superior to the BP algorithm in comparing with all indicators. For example, compared with the models not belonged to transformer, the highest classification accuracy rate reaches 99.56%, which is 0.16% higher than the current best result with this dataset and 5.2% higher than the best result [27]. The statistical analysis of the standard deviation also shows the stability and robustness of the adaptive BP algorithm with this dataset. Comparing Tables 1 and 2, the adaptive BP algorithm puts better comprehensive performance with the more difficult CIFAR-10 dataset. That means that using more complex dataset the adaptive BP algorithm will bring superiority of into full play, such as the potential global optimization ability and the generalizability of the network.

The comparison curves of the loss function value with the test set also confirm the above conclusion, as shown in Fig. 9. It can be intuitively observed from this figure that after epoch 10, further training for the BP algorithm has no effectiveness. In fact, further training actually reduces the generalizability of the network.

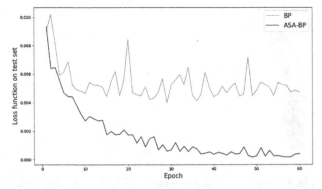

Fig. 9. Comparison curves of the loss function for the test set.

Although slight fluctuations are observed, the learning and training of the adaptive BP algorithm still searches well. Moreover, as the training period go on, the generalizability of the network is continuously improved. The comparison of the two curves shows that the BP algorithm has a fast local optimization ability, but it also easily falls into local optima and cannot escape. The adaptive simulated annealing strategy enhances the ability to escape local optima of the adaptive BP algorithm, and potentially realize global optimization.

Similar to the Sect. 4.1, Figs. 10 and 11 show panoramic views of the process in network training using the adaptive BP algorithm with the dataset. Figure 10 shows the curve of the loss function changing along with training batches, and Fig. 11 shows the variation of the temperature curve with the batches. Fluctuations can be clearly observed for both curves, indicating that the learning and training of the algorithm is ongoing to improve capabilities of the classification, identification and generalization of the network.

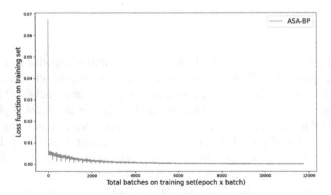

Fig. 10. Loss function curve during batch training.

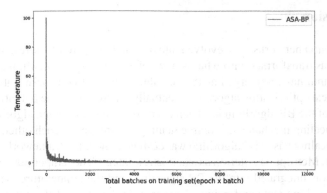

Fig. 11. Temperature curve during batch training.

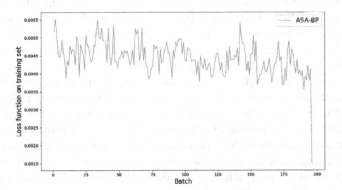

Fig. 12. Loss function curve during batch training process in the second cycle.

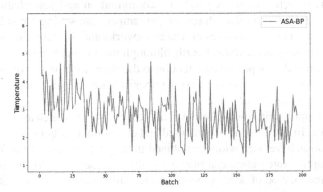

Fig. 13. Temperature curve during batch training in the second cycle.

Figures 12 and 13 show a local training situation in the second training cycle. Figure 12 shows a curve of the loss function changes, and Fig. 13 shows a curve of the temperature changes. The state is the same as that in Fig. 6 and Fig. 7 and will not be explained here.

5 Conclusion

Nowadays, neural networks have evolved into deep network architectures, and the training of them has transformed into what is now referred to as deep learning. As a result, the BP algorithm has re-emerged as the key algorithm. But considering its theoretical basis, it is a local optimization algorithm eventually. In this paper, the potential for global optimization of the BP algorithm has been enabled by integrating the algorithm into the simulated annealing mechanism. Complementing advantages of each other, an adaptive simulated annealing based BP algorithm was designed under the framework of simulated annealing with Metropolis criterion as a heuristic strategy. The temperature adaptive setting method was integrated into the BP algorithm, and the temperature synchronously reflects the searching state and guides the searching direction of the algorithm. Meanwhile being inspired by the adaptive temperature setting, the perturbation update of the connection weight and bias values of the neural network is also related to the complex of the network and the loss function adaptively, avoiding blind random perturbations. The adaptive BP algorithm was tested with the Fashion-MNIST and CIFAR-10 datasets. Results showed the high accuracy of recognition, and statistical analysis results also showed the robustness and stability of the algorithm.

The SAA is a simple single-point heuristic search algorithm, and naturally combining it with the BP algorithm will not increase the amount of computation. Regarding the promotion of researches in this paper, there are still some aspects that deserve further study. First, the additional amount of calculation owing to the adaptive simulated annealing process mainly depends on the value of L_k and the number of $batch$. Therefore, when generalizing to larger datasets, these two values can be comprehensively investigated, such as the dynamic adaptive adjustment of their ratios, $batch/L_k$. Second, for the perturbation update of the neural network parameters, we propose a strategy related to the neural network structure and loss function. However, this strategy is somewhat conservative, so more effective network parameter configuration and perturbation strategies can be comprehensively studied. Moreover, the temperature setting could be generated in combination with other structures of neural networks and their training algorithms to improve other learning algorithms. Fourth, although most of current researches on deep learning algorithms do not pay much attention to the stopping criterion of the algorithm, the common method is to set a maximum number of training cycles and then terminate the algorithm. However, as far as the integrity of the algorithm research is concerned, the setting of the stopping criterion is also a part of the algorithm research and has theoretical and significance. In addition, nature-inspired and evolutionary algorithms characterized by group evolution have been widely combined in deep learning research. We hope that the idea of adaptive integration can play a certain reference role. Meanwhile it is also our hope that the method of adaptive temperature setting can be used in designations of swarm intelligence algorithms.

Acknowledgments. This work is supported by the National Nature Science Foundation of China under Grant 62161008, in part by the Guangxi Natural Science Foundation Joint Funding Project 2020GXNSFAA159172 and 2021GXNSFBA220023,in part by the Research Basic Ability Improvement Project for Young and Middle-aged Teachers of Guangxi Universities 2021KY0604,

2022KY0606 and 2023KY0633, in part by the Special Project of Guangxi Collaborative Innovation Center of Modern Sericulture and Silk 2023GXCSSC01.

References

1. Wu, Y., Gao, R., Yang, J.: Prediction of coal and gas outburst: a method based on the BP neural network optimized by GASA. Process Saf. Environ. Prot. **133**, 64–72 (2020)
2. Zhuo, L., Zhang, J., Dong, P., et al.: An SA–GA–BP neural network-based colour correction algorithm for TCM tongue images. Neurocomputing **134**, 111–116 (2014)
3. Rere, L.M.R., Fanany, M.I., Arymurthy, A.M.: Simulated annealing algorithm for deep learning. Procedia Comput. Sci. **72**, 137–144 (2015)
4. Zhan, S., Lin, J., Zhang, Z., et al.: List-based simulated annealing algorithm for travelling salesman problem. Comput. Intell. Neurosci. **2016**, 8 (2016)
5. Fu, W.Y., Ling, C.D.: Brownian motion based simulated annealing algorithm. Chin. J. Comput. **37**(6), 1301–1308 (2014). (in Chinese)
6. Xavier-de-Souza, S., Suykens, J.A.K., Vandewalle, J., Bolle, D.: Coupled simulated annealing. IEEE Trans. Syst. Man, Cybern. **40**(2), 320–335 (2010)
7. Geng, X., Chen, Z., Yang, W., et al.: Solving the travelling salesman problem based on an adaptive simulated annealing algorithm with greedy search. Appl. Soft Comput. **11**(4), 3680–3689 (2011)
8. Wang, K., Li, X., Gao, L., et al.: A genetic simulated annealing algorithm for parallel partial disassembly line balancing problem. Appl. Soft Comput. **107**, 107404 (2021)
9. Yu, C., Heidari, A.A., Chen, H.: A quantum-behaved simulated annealing algorithm-based moth-flame optimization method. Appl. Math. Model. **87**, 1–19 (2020)
10. Liu, Y., Heidari, A.A., Cai, Z., et al.: Simulated annealing-based dynamic step shuffled frog leaping algorithm: optimal performance design and feature selection. Neurocomputing **503**, 325–362 (2022)
11. Alkhateeb, F., Abed-Alguni, B.H., Al-rousan, M.H.: Discrete hybrid cuckoo search and simulated annealing algorithm for solving the job shop scheduling problem. J. Supercomput. **78**(4), 4799–4826 (2022). https://doi.org/10.1007/s11227-021-04050-6
12. Werbos, P.: Beyond regression: new tools for prediction and analysis in the behavioural sciences. Ph. D. dissertation, Harvard University (1974)
13. Seide F., Fu, H., Droppo, J., et al.: On parallelizability of stochastic gradient descent for speech DNNs. In: 2014 IEEE International Conference on Acoustics, Speech and Signal Processing (ICASSP), pp. 235–239. IEEE (2014)
14. Rumelhart, D.E., Hinton, G.E., Williams, R.J.: Learning representations by back-propagating errors. Nature **323**(6088), 533–536 (1986). https://doi.org/10.1038/323533a0
15. Nemirovski, A., Juditsky, A., Lan, G., et al.: Robust stochastic approximation approach to stochastic programming. SIAM J. Optim. **19**(4), 1574–1609 (2009)
16. Rakhlin, A., Shamir, O., Sridharan, K.: Making gradient descent optimal for strongly convex stochastic optimization. arXiv preprint arXiv:1109.5647 (2011)
17. Javidrad, F., Nazari, M.: A new hybrid particle swarm and simulated annealing stochastic optimization method. Appl. Soft Comput. **60**, 634–654 (2017)
18. Smith, K.I., Everson, R.M., Fieldsend, J.E., et al.: Dominance-based multiobjective simulated annealing. IEEE Trans. Evol. Comput. **12**(3), 323–342 (2008)
19. Li, Y.X., Xiang, Z.L., Zhang, W.Y.: A relaxation model and time complexity analysis for simulated annealing algorithm. Chin. J. Comput. **43**(5), 796–811 (2020)

20. Dowsland, K.A., Thompson, J.M.: Simulated annealing. In: Rozenberg, G., Bäck, T., Kok, J.N. (eds.) Handbook of Natural Computing, pp. 1623–1655. Springer Berlin Heidelberg, Berlin, Heidelberg (2012)
21. Aarts, E., Korst, J.: Simulated annealing and Boltzmann Machines: A Stochastic Approach to Combinatorial Optimization and Neural Computing. John Wiley & Sons, Inc. (1989)
22. Li, Y.X., Jiang, W.C., Xiang, Z.L., Zhang, W.Y.: Relaxation model based temperature setting methods for simulated annealing algorithm. Chinese Journal of Computers. **43**(11), 2084–2100 (2020)
23. Fischetti, M., Stringher, M.: Embedded hyperparameter tuning by Simulated Annealing. arXiv preprint arXiv:1906.01504 (2019)
24. LeCun, Y., Bottou, L., Bengio, Y., et al.: Gradient-based learning applied to document recognition. Proc. IEEE **86**(11), 2278–2324 (1998)
25. Tanveer, M.S., Khan, M.U.K., Kyung, C.M.: Fine-Tuning DARTS for Image Classification. arXiv preprint arXiv:2006.09042 (2020)
26. Simonyan, K., Zisserman, A.: Very deep convolutional networks for large-scale image recognition. arXiv preprint arXiv:1409.1556 (2014)
27. Wang, L., Xie, S., Li, T., et al.: Sample-efficient neural architecture search by learning action space. arXiv preprint arXiv:1906.06832 (2019)

Research on the Important Role of Computers in the Digital Transformation of the Clothing Industry

Ping Wang and Xuming Zhang[✉]

Guangdong University of Science and Technology, Dongguan 523083, China
zhangxuming@pukyong.ac.kr

Abstract. Computers have played an important role in promoting the transformation and development of China's clothing industry, especially in having a profound impact on the digital transformation of the industry. In recent years, although more and more designers and enterprises have recognized the value of computer applications driven by the digital economy, there are still certain problems in specific applications. This article summarizes the important role of computers in the digital transformation of China's clothing industry, and proposes countermeasures for computer application from the perspective of problems.

Keywords: computer · Clothing industry · Digital transformation · effect

1 Foreword

In the process of China's economic development, the clothing industry, as a pillar industry, has played an extremely important role. In recent years, the new generation of digital technologies represented by mobile internet, cloud computing, big data, and artificial intelligence have had a disruptive impact on people's production and life. Technological innovation has also changed the business models and economic forms of traditional industries, and the clothing industry has also undergone transformation and upgrading. China issued an important strategy in 2018, clearly stating that "accelerate the promotion of digital industrialization and industry digitization", which also pointed out the direction for the digital transformation of the clothing industry.This article explores the important role of computers in the digital transformation of the clothing industry and proposes corresponding application strategies.

2 The Realistic Background of Digital Transformation in the Clothing Industry

In recent years, with the rapid development of China's digital economy, more and more industries have joined the digital queue, transforming their original production and manufacturing, management and operation models, and finding new driving forces for the development of the industry. Taking the clothing industry as an example, it has also increased its emphasis on digital transformation and development, promoting changes related to digital technology [1].

K. Li and Y. Liu (Eds.): ISICA 2023, CCIS 2147, pp. 95–102, 2024.
https://doi.org/10.1007/978-981-97-4396-4_8

On the one hand, the clothing industry fully recognizes the advantages of advanced technologies such as the internet and mobile internet, and applies advanced technological means to clothing design. Taking computer-aided technology as an example, this technology achieves online and intelligent clothing design [2]. Computer assisted technology has rich and diverse characteristics in style and pattern design, and on this basis, professional drawing software such as CorelDRAW, Painter, Freehand, etc. have emerged, further improving the accuracy and efficiency of clothing design and improving the overall level of clothing design. (see Fig. 1).

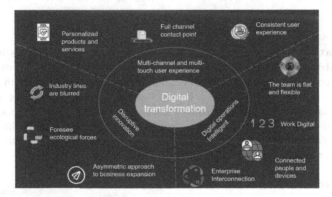

Fig. 1. Digital transformation path of clothing

On the other hand, digitization and advanced technologies are also widely used in the process of clothing production and manufacturing. The commonly used technologies include computer-aided manufacturing systems, computer-aided process design, flexible production manufacturing systems, etc. The application of these systems has changed the traditional clothing production methods and also met the diversified demands of consumers in the clothing market [3].

Although the clothing industry has achieved digital transformation in multiple aspects such as design, production, and sales, considering the actual situation, there is currently a lag in the application of advanced technologies such as computers in the clothing industry [4]. Some enterprises have not recognized the value of computer technology application and are also relatively backward in updating technological means, seriously hindering the overall level of digital transformation in the clothing industry.

3 The Important Role of Computers in the Digital Transformation of the Clothing Industry

Digital transformation is the inevitable trend of enterprise development and an important means to promote the core competitiveness of enterprises. Digital transformation enables enterprises to gain competitive advantage by improving efficiency and reducing costs while maintaining the quality of products and services. Through digital technology, enterprises can innovate in products and services and provide more personalized services to customers, thus attracting more customers. (see Fig. 2).

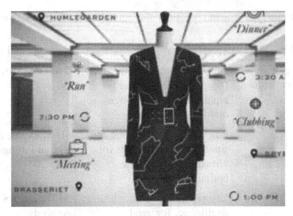

Fig. 2. Digital Management of garment design

Through digital management, enterprises can improve the utilization of internal resources, reduce the waste of resources, build intelligent production workshop, so as to provide better product quality and service for customers. Digital technology can enable enterprises to maintain competitive advantage in the fierce market competition.

Computers are an important tool for promoting the application of information technology and play a fundamental role in the digital transformation of the clothing industry. As an important representative of information technology, the application of computers in the clothing industry is to closely integrate design, production, sales and other links with science and technology, apply broader ideas to clothing, and inject more fresh blood into the industry. Based on the actual situation, computers have played an important role in the digital transformation of the clothing industry, mainly reflected in the following aspects:

Firstly, the application of computers has further improved the efficiency of clothing design and production. Computers can automate many repetitive and tedious tasks. Taking the application of CAD and CAM as examples, these two technologies can quickly generate clothing patterns, thereby accelerating the process of clothing sampling and production. Therefore, they have a positive promoting effect on improving clothing design and production efficiency [5].

In addition, the application of computers can reduce the cost investment of the clothing industry. Compared with traditional production methods, computers can be applied in various ways such as digitization and intelligence to avoid waste and losses in design and production, and to some extent, can reduce cost consumption in production. For example, in the actual production process, clothing companies cananalyze the demand and trends of the clothing market through technologies such as cloud computing and artificial intelligence, in order to reduce inventory squeezing or unsold situations (Table 1).

The application of computers can improve the quality of clothing production and manufacturing. In the process of clothing production, the application of computer and other auxiliary technologies can improve the accuracy of the production process, strengthen the overall level of control management, and thus avoid errors caused by human factors. Taking the application of 3D printing technology as an example, in

Table 1. Consumer demand questionnaire in intelligent clothing design tables.

The role of the eld	Specific role	Font size and style
Design Intelligent pattern generation Intelligent color matching	Utilizing Computer-Aided Design (CAD)software for rapid and accurate design drafting	CAD Software
	Employing 3D modeling and simulation technologies for virtual fitting and style optimization	3D Modeling Software, Virtual Fitting Technologies
	Leveraging artificial intelligence for trend prediction and design innovation	AI Design Tools, Data Analysis

clothing production and manufacturing, this technology can more accurately produce prototypes and verify the feasibility of design and production, further improving product quality and market competitiveness of clothing enterprises.

The application of computers can add more creative elements to clothing design and production. In the process of clothing production and manufacturing, the application of computer and other related technologies can allow designers to more intuitively express their ideas, see the overall effect of the design in advance, modify and optimize the details, making clothing design more innovative and presenting diverse characteristics. Taking the application of VR technology as an example, designers can conduct virtual fitting, sampling, and other work in advance to improve and optimize clothing design schemes.

Finally, the application of computers can also help fashion designers and enterprises respond more quickly to market demands. Compared to traditional market research, computer and other related technologies can better leverage the advantages of big data. By utilizing advanced technology to analyze and predict consumer market data, enterprises can timely grasp market development trends, adjust and optimize existing production strategies and product designs, and achieve sustainable development of the clothing industry [6].

4   The Application of Computers in the Digital Transformation of the Clothing Industry

Although computers have played an important role in the digital transformation of the clothing industry, their advantages and characteristics are also prominent. However, considering the actual situation, there are also obvious problems in the application of computers in the clothing industry, mainly reflected in the following aspects:

4.1   The Advantages of Computers have not Been Fully Utilized

With the innovation of computer technology, its functions and functions have been continuously strengthened, but currently the clothing industry has not fully utilized the advantages that computers should have. Taking clothing design as an example, few designers actively use computers to make changes to existing design methods, nor do they display the styles and characteristics of clothing through computers. They only use computers on the surface to carry out design activities. Essentially, they still repeat the traditional clothing design process, lacking innovation and initiative, which also hinders the innovative development of the clothing industry.

4.2   Lack of In-Depth Application of Computer Technology

The clothing industry in China has gone through a long process of development, with different process requirements from design to production. Even in the rapid development of the modern clothing industry, traditional clothing design and production processes still have a profound impact and are a resource pool for the development of the modern clothing industry. However, at present, China's clothing industry has not achieved deep integration with computer technology in design, production, and other processes. Many clothing designers have not integrated clothing design processes with the help of computers, which is not conducive to the improvement of clothing design level.

4.3   The Design of Clothing Does not Showcase the Characteristics of the Times

The aesthetic preferences for fashion design vary in different eras. Different fashion design preferences represent the aesthetic preferences of ordinary people in a specific era, serving as a recording tool for the era. To some extent, they can represent the development direction of a specific era. Fashion design that conforms to the characteristics of the era can be favored by contemporary people. Therefore, fashion designers should make full use of computer information technology to find clothing design solutions that meet the common aesthetic standards of today, and integrate the characteristics of the times into their own clothing design.

In current fashion design, many fashion designers do not pay attention to the use of computer information technology for fashion analysis, and do not pay attention to collecting fashion trends, resulting in their design works appearing single, lacking the flavor of the times, lacking hierarchy, historical depth, and artistic beauty of the times. Fashion designers who do not pay attention to the use of computer information technology will make it difficult for their designs to integrate into the context of the times. Because a large part of the clothing background information of the era is scattered in the cyberspace, without the collection and analysis of network information, there is no high-quality clothing design scheme that integrates the characteristics of the era[7]. Due to the lack of integration of fashion information, traditional fashion designers find it difficult to keep up with the trend of fashion design. Their designs lack the elements of the era's clothing in the minds of the public, which is not conducive to the improvement of the artistic quality of fashion design and the development and progress of the fashion design industry.

5 The Application Countermeasures of Computers in the Digital Transformation of the Clothing Industry

In order to promote the efficient application of computers in the clothing industry and meet the practical needs of industrial digital transformation. As participants in the clothing industry, designers, enterprises, and others need to increase their emphasis on the application of advanced technologies such as computers. In the implementation of specific work, the following aspects can be taken into consideration.

5.1 Adopting a Digital, Standardized, and Standardized Management Model

In China, the clothing industry is a traditional labor-intensive industry, and production management still follows the traditional management model. Employees in the clothing industry have low general cultural qualities and are accustomed to manual operations and management experience. Therefore, it requires a complete intelligent comprehensive management system for production, which involves managing and executing the production workshop.

Specifically, automated and intelligent production processes can optimize production planning, scheduling, and logistics, enabling rapid response to market demands and thus improving production efficiency and quality. This can not only shorten product time-to-market, but also reduce production costs, improve product quality, and increase production efficiency.

Using computers to optimize supply chain management, production process control, and warehouse management, among other aspects, can reduce production costs, inventory costs, and operating costs, and improve corporate profitability. By reducing waste and improving resource utilization, companies can maintain cost advantages in fierce market competition to reduce corporate costs. More personalized and differentiated products and services can be realized in design, development, and sales, meeting market demands and thus enhancing the innovation ability and competitiveness of enterprises. By constantly exploring new technology applications and business models, companies can seize market opportunities and achieve sustained growth. Through technologies such as intelligent logistics and supply chain management, enhancing supply chain visibility and transparency can help companies better grasp all aspects of the supply chain, manage supply chain risks and optimize supply chain costs, and improve the efficiency and reliability of the supply chain. This can not only improve customer satisfaction and service quality, but also reduce operational risks and costs.

Computer big data management, statistics and numerical analysis, presenting data and production and business situations to managers intuitively, enabling them to make timely decision-making basis, improve top-down execution ability, management ability, and feedback communication ability.

5.2 Cultivate or Recruit Versatile Talents to Participate in the Production Process

If enterprises want to develop, they must inject fresh blood, improve their vitality and innovation ability. Clothing enterprises not only need manual workers, but also advanced technical and management personnel to innovate the enterprise. Introducing advanced personnel is the introduction of advanced technology and management, which plays a crucial role in the production efficiency and market control ability of enterprises.

5.3 Representing the Advantages of Computer Information Technology in Fashion Design

With the development of computer information technology, the clothing design industry has also ushered in new development opportunities. The current concept of fashion design is constantly changing, and the perspective of fashion design is constantly expanding. Computer information technology can help fashion designers expand their design ideas, open up design ideas, and design clothing design solutions that are more in line with the requirements and trends of the times, enriching consumers' lives. Computer information technology makes it more convenient for fashion designers to access diverse clothing design materials, making it easier to obtain and track fashion design trends, and making clothing design work morestreamlined. In the future, fashion designers should rely more on computer information technology when designing clothing.

5.4 Integrating Clothing Design Processes with Computer Information Technology

The prominent feature of modern clothing design technology is the integration of traditional and modern craftsmanship. Under the traditional design model, fashion designers can only obtain traditional design materials through limited book materials and design cases. If they want to design clothing based on traditional techniques, they need to collect a large amount of clothing design content, which is neither convenient nor directly usable. Fashion designers can use computer information technology to conveniently collect traditional craft data into clothing design software, facilitating the integration of traditional and modern clothing design processes. Fashion designers can also use computer software to directly utilize traditional craftsmanship on top of modern craftsmanship, achieving free combination between different craftsmanship and achieving the complexity of design processes8. Computer information technology can also endow traditional clothing design techniques with new connotations, transforming design techniques that have fallen behind the aesthetic of the times into design works that conform to the trend of the times through improvement, promoting the improvement of clothing design level, and promoting the development of the clothing design industry.

5.5 Showcasing the Characteristics of the Times in Clothing Design

Different era characteristics will endow clothing with unique era characteristics, and different era characteristics affect the concept of clothing design. In the new era, the clothing design industry should achieve a close integration of clothing design works and computer information technology. Computer information technology can record and collect clothing design features with distinctive characteristics of the times, and designers can use these features to carry out clothing design, making clothing design have prominent characteristics of the times. Without the full application of computer information technology, fashion designers would lack convenient and fast tools to create fashion design works that are in line with the times. The ability of computer information technology to analyze and capture information is a powerful tool for achieving fashion design with the characteristics of the times. Information technology can not only collect fashion design information, but also reorganize different fashion design information, improve the effectiveness and artistry of fashion design, and achieve the diversified development of fashion design.

6 Conclusion

The emergence of information technology such as computers has had a profound impact on the development of China's clothing industry. Whether in clothing design, production manufacturing, clothing sales, and other stages, relevant entities in the clothing industry should increase their attention to computer applications, start with design optimization, strengthen digital management of production manufacturing, and further improve the overall level of computer and other information technology applications, To promote the digital development of China's clothing industry.

This paper is part of the phased research results of the education and teaching reform project "Exploration and practice of" LiuBai "Clothing Studio" of Guangdong Higher Education Association.(Guangdong teaching high letter 【2021】 No.29).

References

1. Tong, Z., Lijie, Z.: Evaluation of Digital Transformation of Listed Enterprises in China's Clothing Industry. Silk **60**(09), 1–7 (2023)
2. Xiang, L.: Application of Computer Aided Design in Textile and Clothing Modeling. Cotton Textile Technology **50**(09), 90–91 (2022)
3. Ge, L., Xiaohui, L.: Research progress on digital clothing structure design technology. Journal of Textile Science **43**(04), 203–209 (2022)
4. Sijia, N., Lei, S.: Strategic analysis of clothing brand marketing channels under digital transformation. Woolen Textile Technology **48**(04), 70–74 (2020)
5. Yuefang, F.: The Application and Development of Computer Information Technology in Fashion Design: A Review of "Fundamentals of Computer Network Technology." Chinese Science and Technology Paper **14**(07), 823 (2019)
6. Yong, G.: Performance Techniques of Computer Fashion Design.–Review of "Application of Computer Fashion Design." Printing and Dyeing Additives **35**(08), 79–80 (2018)
7. Weiming, Z., Yanghong, W.: Research on Internet plus Digital Personalized Customization Operation Mode of Clothing. Silk **55**(05), 59–64 (2018)
8. Qishu, L.: Research on the Implementation Technology of Computer Remote Assisted Clothing Design Management System. Dyeing and Finishing Technology **39**(09), 8–1 (2017)
9. Lihua, Z.: Digital technology —— the inevitable trend of the development of the garment industry in the 21st century. Shandong Textile Technology, 38–40 (2004)
10. Wanqiu, J., Liping, Z.: The role and prospect of garment CAD / CAM in modern garment enterprises. Foreign textile technology **4**, 5–8 (2000)
11. Yan, L., Zhaofeng, G.: Research on the application of intelligent technology in garment industry production. Journal of Donghua University **28**(4), 123–127 (2002)
12. Shengying, X.: On the promotion role of digital technology in the design and production of garment enterprises [D]. Shandong University of Light Industry, Shandong (2010)

Multimedia Information Retrieval Method Based on Semantic Similarity

Xuanyi Zong, Jingwen Zhao, Zhiqiang Chen, and Jinfeng He[✉]

Nantong University, Nantong 226019, China
hjf89@ntu.edu.cn

Abstract. The conventional method of multimedia information retrieval has problems such as complex operation, error in information query, and low accuracy. A multimedia information retrieval method based on semantic similarity is proposed, and during a preprocessing operation of a document, learn the methods of classical vector space models and replace dictionaries containing keyword entries with ontology library. Replace the document with the concept described in the using documents and its vectors of feature meaning, and extract and absorb the contents of the document meaningfully. To achieve efficient retrieval completion, we semantically classify documents to make preparations for outreach plans and techniques for more effective use of queried semantic vectors. When calculating the similarity of concepts and attributes, the existence relationship between each conceptual instance and the attribute force is determined, and its influencing factors are analyzed according to the conceptual characteristics to realize the whole process of completing semantic retrieval. Finally, multimedia information retrieval is completed based on semantic similarity. Through comparative experiments, the multimedia information retrieval method based on semantic similarity is compared with the traditional method and the method based on protection data analysis, and it is concluded that the method based on semantic similarity has high accuracy and applicability.

Keywords: semantic similarity · multimedia · information retrieval

1 Introduction

With the rapid development of computer technology and the internet, information including images in varies media is developing at a speed beyond human imagination, at the same time, people are facing the problem of how to effectively obtain the necessary information in the large multimedia world, rather than suffering from the lack of multimedia content in daily life. One of them is that due to the keyword-based search approach of traditional database searches [8], it is often difficult to fully describe multimedia content with just a few keywords, and the choice of image attributes as keywords is also very objective. Second, it is difficult for users to translate these informative comments in the form of some kind of code [9]. In the field of digital libraries, professional librarians organize important vocabularies in their field according to semantic hierarchy and use to

K. Li and Y. Liu (Eds.): ISICA 2023, CCIS 2147, pp. 103–112, 2024.
https://doi.org/10.1007/978-981-97-4396-4_9

find and retrieve contextual media data [1]. To improve the accuracy of searches, other semantic methods of reporting and analyzing the subject and content of a document or page are also used. However, because the current development of information retrieval is mainly through the use of interactive systems in the retrieval process to search, without considering other information [10], such as the description of conceptual features, and this conceptual semantic search method often doesn't meet the actual needs. On the basis, ontology is used to describe and characterize the semantics of user queries and documents, methods for processing concepts and functions are proposed, and basic semantic search algorithms are applied to calculate the similarity of semantic vectors found in concepts and attributes [2].

2   Design of Multimedia Information Retrieval Methods

2.1   Query and Extraction of Semantic Vectors

In the traditional multimedia information retrieval, in some areas that users query information, such as obtaining multimedia book information, specify search keywords and search to ensure that the search keywords entered by customers are complementary and limited, this effectively helps the recovery system determined the customer's search intention to a certain extent and brings more available data to the retrieval [11]. According to the role of ontology in the expression of information, keywords requested by customers can be semantically processed. Therefore, we must deeply expand the user's problem content and establish the corresponding user search semantic vector. The questions introduced by users often involve many keywords. These are the details of the user's motivation for asking questions. These questions typically include keywords, key attributes, and obtained target values. Therefore, when developing users' questions, they can be described in the ontology, and specific characteristics and attribute feature values can be extracted from the survey.

In the process of preprocessing the document, learn the methods of the classical vector space model. In the traditional vector model of documents, words, roots, or phases are first created as dictionary keywords, then each document is represented as a multidimensional vector, and finally the document is represented by various symbols such as the document's binary vector, frequency, or reverse frequency. By replacing dictionaries containing keyword entries with ontology libraries, we replace documents with meaning vectors containing concepts described in documents and their characteristics and meaningfully extract and absorb the contents of documents.

Depending on how to deal with the usage problem, each piece of information contains a topic and content, and each record also describes a system of ideas, combining statistical and semantic methods to capture keywords, and using ideas and ideas in ontology to record and summarize, which is similar to the method of user query. But from the perspective of each conceptual example, semantic presence attribute characteristics should be extracted from the document. The specific process is shown in Fig. 1:

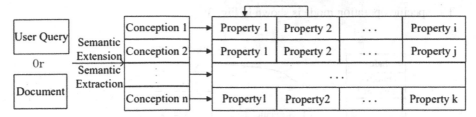

Fig. 1. Semantic vector flow for documents and user queries

2.2 Document Semantic Classification and Indexing

To achieve efficient retrieval completion, we semantically classify documents to prepare outreach plans and techniques for more effective use of queried semantic vectors. According to the results of recording semantic vector records, information is obtained from semantic vectors based on private ontology trees. This is shown in Fig. 2:

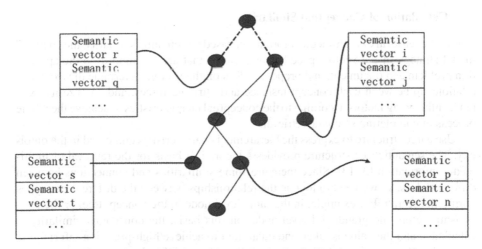

Fig. 2. Semantic classification of documents under the Ontology conceptual tree

In general, the semantic feature vector of a text involves each concept ontology and its characteristics, locates the appearance of the text on the category corresponding to each concept, and logically establishes a hierarchical control structure corresponding to the text library itself, thus laying the cornerstone for the semantic indexing function of the text. To facilitate retrieval operations, the classified documents are first semantically indexed. The first step is to insert the ontology concept into the index file, arrange it in dictionary order, and then create an ordered linked list, which first refers to the document semantic vector characteristics under the concept, and then creates a pointer to the link in the index file, and designs the list on the pointer, so as to facilitate the loading object to be queried through the pointer during the buffering process, without scrolling through the entire document set, so as to reduce the time required for retrieval. Then, the semantic property vector of the document is linked to the corresponding document.

The specific operation mode is shown in Fig. 3:

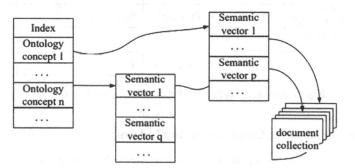

Fig. 3. Document semantic index

2.3 Calculation of Conceptual Similarity

Before the retrieval of multimedia information based on semantic similarity, the "partial" model strategy of the vector space borrowed tools that are meaningful to the application problem and document, and proposed ideas on the vector. Determine the existence relationship between each concept instance and attribute forces, and analyze its existence influencing factors according to the conceptual characteristics to achieve the whole process of completing semantic retrieval.

Use a tree structure to express the hierarchical characteristics embodied in the ontology. The semantic tree structure provides a theoretical basis for the retrieval of multimedia information [3, 13]. Since there are some similarities and connections between different ideas, so we need to look at the relationships between the different tree types during the search. For example, in the same level node in the concept tree, or the relationship before the grandchild level node, on this basis, the conceptual similarity is calculated, and the value is taken and quantified to achieve high-precision information retrieval. Before making calculations, the following are defined:

If in the ontological concept of a tree structure, there is such a hierarchical relationship between Concepts A and B ——Concepts A and B are parent-child levels to each other, and A and B have the same branch conceptual relationship. The distance between concept A and concept B is calculated by the following equation: d(A,B)

$$d(A,B) = dep(B) - dep(A) \tag{1}$$

where dep(A) and dep(B) are depth values for concept A and concept B in the hierarchy respectively. The structure diagram is shown in Fig. 4:

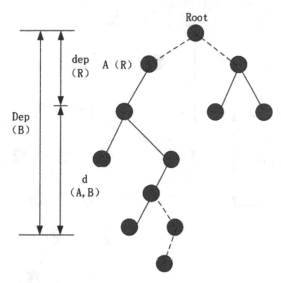

Fig. 4. A and B are the structural diagrams of the concept of the same branch

If concepts A and B are not ancestral in the ontological concepts of the tree hierarchy. Then this relationship is generally called heterobranched state. And in this case, there is no other formal connection between concept A and concept B before. Suppose that concept R is the common ancestral hierarchy of concepts A and B in the tree hierarchy where there is no correlation, and that they have the longest distance. In general, concept R is the conceptual minimum of concept A and concept B, represented as R(A,B), the size of the distance between them is calculated by the following equation:

$$d(A,B)=d(A,R)+d(B,R) \qquad (2)$$

Therefore, there are only three possibilities for obtaining any two conceptual relationships in the ontology concept tree, namely the same branch concept, the hetero branched concept, and the same concept.

When calculating the conceptual similarity, from the perspective of the same branch concept or hetero branch concept [12], if the relationship between concept A and concept B gradually becomes estranged as the distance between the two concept values increases, the conceptual similarity between A and B will also decrease on this basis. It is expressed as a functional relationship that the similarity y of the dependent variable gradually decreases as the distance x of the independent variable increases [4]. From the concept of homobranch or heterobranch, if the depth dep value between concept A and concept B is larger, the higher the value of the attribute that A and B existed before. It is expressed as a function relationship that the similarity y of the dependent variable gradually increases as the depth x of the independent variable increases [5].

In addition, there is a certain relationship with the number A, B of concepts itself and the number of semantic concepts associated with it. If A and B it is the same branch conceptual structure as shown in Fig. 4, the A(R) subconcept consists of two parts, of which the two parts are B relevance and A,B semantic conceptual relevance.

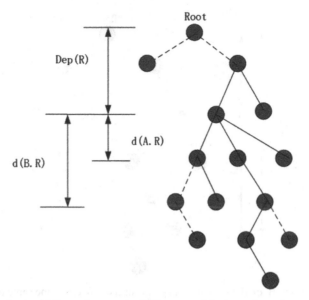

Fig. 5. A and B are the structural diagrams of the heterobranch concept

On this basis, when the A, B weight value of the semantic correlation concept is the minimum value, the correlation value A,B is the largest. In another case, when A, B the heterobranched conceptual structure shown in Fig. 5 is structured, the R subconcept consists of A3 parts: the subconcept, B the subconcept, and A,B the related semantic concept, and if the proportion of the latter is low, the A,B correlation is higher. It is calculated that if the semantic similarity value is 1, there is no relationship between the related concepts and their depth, and the distance between the two concepts is zero. The A and B conceptual similarity between defined concepts is as follows:

$$
Sim() = \begin{cases} \left(1 - \frac{\alpha}{dep(R(A,B))+1} \times \frac{\beta}{d(A,B)} \times \frac{son(B)}{son(A)}\right), d(A, B) \neq 0 \\ \left(1 - \frac{\alpha}{dep(R(A,B))+1} \times \frac{\beta}{d(A,B)} \times \frac{son(A)+son(B)}{son(R)}\right), d(A, B) \neq 0 \\ 1, d(A, B) = 0 \end{cases} \quad (3)
$$

In Eq. (3), dep(R(A,B)) is the concept A and B the concept's nearest root conceptual depth; d(A,B) is the size A of the concept and B the distance between concepts; son(A) and son(B) are the number of all nodes of the root-based subtree A,B in the concept tree, respectively; a and β adjusts the weight value of (R(A,B))and d(A,B) separately, and sei the value range to $0 \leq sim(A, B) \leq 1$.

Any ontology concept will have instances in various cases, and the criterion for distinguishing them is the difference in the value of the attribute [6, 14]. The same attributes may exist in instances with different concepts, and on this basis, the conceptual similarity between instances is calculated to complete the implementation of information retrieval.

After calculating the conceptual similarity of semantic vectors and the similarity of conceptual case features, the complete semantic similarity between semantic vectors can be obtained [7, 15]. When semantic extension technology is applied to information retrieval, the similarity is mainly manifested in the ability to extend the concept to meet the query needs of users.

Let the semantic vector V1 be A1[p1],A2[p2],. . . ,Am[pm], and the semantic vector V2 be B1[Q1],B2[Q2],. . . ,Bm[Qm].

When performing similarity calculation, each conceptual instance output by the user when querying the semantic vector and all conceptual instances in the retrieved document feature semantic vector are analyzed and verified one by one to determine if the content is relevant. On this basis, the similarity of the concept and the semantic vector of document features is compared, and the maximum value is set to Max.

$$Sim_v(V_1, V_2) = \frac{1}{m} \sum_{i=1}^{m} Max\big(w \cdot Sim_c(A_i, B_j) + (1 - \omega) \cdot Sim_p(P_i, Q_j)\big) \qquad (4)$$

In Eq. (4) V1 is the semantic vector value in the user query process, V2 is a semantic vector value in a document feature, w is the weight value relationship between conceptual similarity and attribute similarity, simv(V1, V2) is the similarity value between V1 and V2 , simc(Ai, Bj) is the semantic vector i between concept A and concept B and the closest similarity C of the semantic vector j, m is the total quantity.

3 Comparative Experiments

3.1 Experimental Description

After this paper designs and realizes the retrieval of multimedia information based on semantic similarity, In order to verify the feasibility of the proposed method, the multimedia information retrieval method based on semantic similarity is compared with the traditional method and the multimedia information retrieval method based on data analysis.

3.2 Experimental Preparation

First, select the test environment in the experiment from the network documents on the homepage of www.ustc.edu.cn, www.pku.edu.cn and other universities as Windows PC. The parameters are Genuine Intel® CPU T2080@2.67GHz CPU, 16GB memory, 1TB hard disk. The operating system is Windows 10 and the software is known to have three problems when it comes to performance. Before the experiment, debug the various parameters to ensure that they can be used normally.

3.3 Experimental Results

In the experimental process, the number of information after multimedia information retrieval by three methods was used as a comparison index, and the weight value of conceptual similarity was taken as a variable. The experimental results are as follows: (It is known that in this experiment, the multimedia information retrieval should be 1200 by full search)

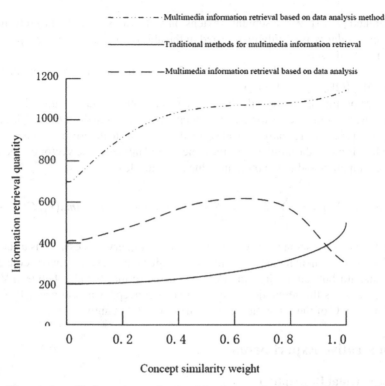

Fig. 6. The number of information retrievals of the three methods under different weight values

The data obtained from the above experiments can be seen that when the traditional method of multimedia information retrieval is used, when the weight value is 0, the number of information retrieval is 200, which is the minimum value between the three. When the weight value is 1.0, the number of information retrieved is 600, which is closest to the standard value compared with other data, but the accuracy rate is only 50%, and when the weight value is between 0-0.8, the number of information retrieved is between 200-400. In the multimedia information retrieval based on the data analysis method, when the weight value is 0.6-0,7, the number of information retrieval is the highest, 800, its accuracy rate is 67%, but when it is greater than 0.6, the number of information retrieval decreases, and when the weight value is 1.0, the number of information retrieval is only 200, overall, the use of multimedia information retrieval based on data analysis, the results will fluctuate greatly, there is volatility. The multimedia information retrieval method based on semantic similarity designed in this paper is greater than the other two methods under any one weight value, and the number of information retrieval is between 750-1200, and the accuracy rate can reach up to 100%. It can be seen that the method designed in this paper can maintain a high accuracy rate when retrieving information, and has more advantages and usability (Fig. 6).

4   Conclusion

In this article, we will introduce how to search semantic information based on ontology. Based on Ontology, an appropriate user query semantic extension method, a method for extracting semantic information from documents, and a standard for calculating semantic vector similarity are proposed. Experimental results show that the search effect of the text design method is high compared with the other two methods.

References

1. Kokula Krishna Hari, J.S.K.: The Proceedings of the International Conference on Information Engineering, Management and Security 2014: ICIEMS 2014. Association of Scientists, Developers and Faculties (2014)
2. Chang, S.-F.: Content based multimedia retrieval: lessons learned from two decades of research, In: Proceedings of the 19th ACM international conference on Multimedia, in MM '11, pp. 1–2. New York, NY, USA, Association for Computing Machinery, (2011) https://doi. org/10.1145/2072298.2072300
3. Yu, B.: Research on information retrieval model based on ontology. EURASIP J. Wirel. Commun. Netw. **2019**(1), 30 (2019). https://doi.org/10.1186/s13638-019-1354-z
4. Zhang, J., Wang, X., Zhang, H., Sun, H., Liu, X.: Retrieval-based neural source code summarization, In: Proceedings of the ACM/IEEE 42nd International Conference on Software Engineering, in ICSE '20, pp. 1385–1397. New York, NY, USA, Association for Computing Machinery, (2020) https://doi.org/10.1145/3377811.3380383
5. Harispe, S., Ranwez, S., Janaqi, S., Montmain, J.: Semantic Similarity from Natural Language and Ontology Analysis. In:Synthesis Lectures on Human Language Technologies. Cham, Springer International Publishing (2015) https://doi.org/10.1007/978-3-031-02156-5
6. Sánchez, D., Batet, M.: A semantic similarity method based on information content exploiting multiple ontologies. Expert Syst. Appl. **40**(4), 1393–1399 (2013). https://doi.org/10.1016/j. eswa.2012.08.049
7. Yin, M.: Personalized advertisement push method based on semantic similarity and data mining, in 2021 Third International Conference on Inventive Research in Computing Applications (ICIRCA), pp. 1476–1479. (2021) https://doi.org/10.1109/ICIRCA51532.2021.9544770
8. Hliaoutakis, A., Varelas, G., Voutsakis, E., Petrakis, E.G.M., Milios, E.: Information Retrieval by Semantic Similarity. IJSWIS **2**(3), 55–73 (2006). https://doi.org/10.4018/jswis.200607 0104
9. Lew, M.S., Sebe, N., Djeraba, C., Jain, R.: Content-based multimedia information retrieval: State of the art and challenges, ACM Trans. Multimedia Comput. Commun. Appl. **2**(1), 1–19 (2006), https://doi.org/10.1145/1126004.1126005
10. Meghini, C., Sebastiani, F., Straccia, U.: A model of multimedia information retrieval. J. ACM **48**(5), 909–970 (2001). https://doi.org/10.1145/502102.502103
11. Rinaldi, A.M., Russo, C., Tommasino, C.: A knowledge-driven multimedia retrieval system based on semantics and deep features, Future Internet. **12**(11), Art. no. 11 (2020) https://doi. org/10.3390/fi12110183
12. Varelas, G., Voutsakis, E., Raftopoulou, P., Petrakis, E.G.M., Milios, E.E.: Semantic similarity methods in wordNet and their application to information retrieval on the web, In: Proceedings of the 7th annual ACM international workshop on Web information and data management, in WIDM '05, pp. 10–16. New York, NY, USA, Association for Computing Machinery (2005) https://doi.org/10.1145/1097047.1097051

13. Kambau, R.A., Hasibuan, Z.A.: Unified concept-based multimedia information retrieval technique, in 2017 4th International Conference on Electrical Engineering, Computer Science and Informatics (EECSI), pp. 1–8 (2017). https://doi.org/10.1109/EECSI.2017.8239086
14. Ping, C.A.I., Zhiqiang, W., Xianghua, F.: Semantic-based cross-media information retrieval technology. Microelectron. Comp. **27**(3), 102-105 (2010)
15. Shuang, L., Liang, B., Tianyuan, Y., Yuhua, J.: A cross-media semantic similarity measurement method based on bidirectional learning ordering. Comp. Sci. **44**(S1), 84-87 (2017)

Iterative Learning Control for Encoding-Decoding Method with Data Dropout at Both Measurement and Actuator Sides

Yongxian Chen[1] and Yunshan Wei[1,2(✉)]

[1] School of Electronics and Communication Engineering, Guangzhou University, Guangzhou 510000, Guangdong, China
`weiys@gzhu.edu.cn`
[2] Key Laboratory of On-Chip Communication and Sensor Chip of Guangdong Higher Education Institutes, Guangzhou 510000, Guangdong, China

Abstract. This paper investigates the problem of iterative learning control in networked structures, with a specific focus on addressing random data dropout at both measurement and actuator sides to achieve zero-error tracking performance. An encoding and decoding mechanism is introduced into the system. Initially, the system output undergoes encoding, quantization, and subsequent transmission to the controller. When the data are received, they are decoded and applied to generate the input for the next iteration. Subsequently, the generated input undergoes the same process as the output transmission, including encoding, quantization, transmission, and decoding. The proposed approach's convergence is proven under random data dropout, and the effectiveness of the scheme is demonstrated through convergence analysis and illustrative examples.

Keywords: Iterative learning control · Encoding and decoding · Data dropout

1 Introduction

Iterative Learning Control (ILC) is an intelligent control strategy used to handle repetitive tasks within a specified time interval. ILC utilizes previous tracking and input information to update the input information for the current iteration, aiming to improve tracking performance. ILC was initially proposed by Arimoto et al., and has made substantial advancements in both theory and practical applications over the past few decades [1, 2].

With the rapid advancement of communication and network technologies, it has become feasible to operate systems and learning controllers remotely through networks. However, control faces challenges such as the inability to transmit actual signals with lossless accuracy and the significant communication burden involved in network data transmission. In [3–5], researchers addressed the issue of continuous data dropout in systems. Specifically, they compensated for the continuously lost output data in the latest iteration by successfully estimating the predicted information with the same time

K. Li and Y. Liu (Eds.): ISICA 2023, CCIS 2147, pp. 113–126, 2024.
https://doi.org/10.1007/978-981-97-4396-4_10

stamps from the previous iteration using a multi-step prediction model. In this regard, many researchers have made important contributions, including the introduction of quantization mechanisms. In their studies, [6] introduced logarithmic quantizer and generated quantized output signals using infinite-level quantizer. Subsequent research focused on quantized ILC [7, 8]. [7] introduced changes to the quantized object while retaining the logarithmic quantizer. This change is evident in the fact of transmitting the quantization error instead of the quantized output. [8] discussed three cases: input quantization, output quantization, and error quantization. For the case of output quantization, it was found that convergence is bounded and zero-error convergence cannot be achieved.

Our research motivation is to address an important problem: how to achieve zero-error tracking performance using a simple quantizer, such as uniform quantizer. In [9], an encoding-decoding mechanism was proposed to enhance tracking performance, employing a uniform quantizer as the quantization method. In [10–13], encoding-decoding mechanisms using the uniform quantizer were considered in the case of data dropout, but the encoding-decoding mechanisms were only applied to either the measurement or actuator side. The encoding-decoding mechanism in [14] was applied to both measurement and actuator sides, demonstrating zero-error convergence under both infinite-level and finite-level quantizer. However, [14] did not consider random data dropout. [15] proposed the concept of random data dropout occurring simultaneously at both measurement and actuator sides. Can we incorporate random data dropout into the encoding-decoding mechanism at both measurement and actuator sides?

The main contribution of this paper is to address the encoding and decoding mechanisms at both measurement and actuator sides facing random data dropout. Specifically, it models data dropout using a Bernoulli distribution and takes into account the differences in initial states to achieve asymptotic convergence.

This paper is organized as follows: Sect. 2 presents the problem formulation and provides a comprehensive description of the encoder and decoder. Section 3 explains the P-type learning algorithm and offers a detailed analysis of the convergence of the proposed scheme. In Sect. 4, simulation results are presented, demonstrating the effectiveness of the proposed approach using a permanent magnet linear motor. Finally, Sect. 5 concludes the paper.

2 Problem Formulation

Consider the following linear discrete-time time-varying system:

$$\begin{cases} x_k(t+1) = A(t)x_k(t) + B(t)u_k(t) \\ y_k(t) = C(t)x_k(t) \end{cases} \tag{1}$$

where $x_k(t) \in R^n$, $u_k(t) \in R^l$ and $y_k(t) \in R^p$ are the state, input and output, respectively. $k = 1, 2, \ldots$ Denotes the iteration number, and $t = 0, 1, \ldots, T_d$ where T_d is the iteration length. $A(t)$, $B(t)$ and $C(t)$ are suitable time-varying matrices with appropriate dimensions. We assume that $C(t)B(t) \neq 0$ corresponds to the case where the relative degree of the system is one.

In order to facilitate further analysis, it is necessary to make the following assumptions.

Assumption 1: The desired reference $y_d(t)$ is realizable in the sense that there exist $x_d(0)$ and $u_d(t)$ such that.

$$\begin{cases} x_d(t+1) = A(t)x_d(t) + B(t)u_d(t) \\ y_d(t) = C(t)x_d(t) \end{cases} \tag{2}$$

Assumption 2: The initial state values can be precisely reset asymptotically in the sense that $x_k(0) \to x_d(0)$ as $k \to \infty$. Then

$$\lim_{k \to \infty} E\{\|x_d(0) - x_k(0)\|_\infty\} = 0. \tag{3}$$

Remark 1: Assumption 1 suggests the presence of desired input capable of generating the desired signal. In Assumption 2, it is necessary for the initial state $x_k(0)$ can be reset gradually. This condition aimed at allowing space for the design of appropriate initial value to achieve this asymptotic reset condition, as demonstrated in methods proposed in [3]. Compared to existing encoding and decoding mechanism based ILC schemes, Assumption 2 relaxes the initial condition.

For configuration, communication between the plant and the controller is restricted capability, and the transmitted information is quantized to reduce communication burden. However, the quantization process introduces quantization errors that affect the system output. The aim of this paper is to develop an appropriate learning algorithm to address the issues of quantization errors and data dropouts, such that the system input signal achieves convergence.

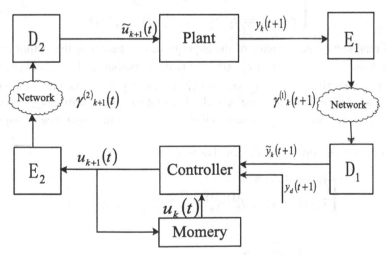

Fig. 1. Block diagram of ILC with encoding-decoding mechanism and data dropouts.

The quantization at both measurement and actuator sides is integrated with a refined encoding-decoding mechanism that exists data dropout, as illustrated in Fig. 1, where E_1, D_1, E_2, and D_2 denote the output encoder, output decoder, input encoder, and input decoder, respectively. The system output is first encoded by E_1 and transmitted, then the encoded data are decoded by D_1 for controller updating. After that, the generated input is encoded by E_2 and decoded by D_2 similarly.

For convenience, the unit uniform quantizer was used,

$$Q(n) = \begin{cases} 0, & \text{if } -\frac{1}{2} \leq n \leq \frac{1}{2} \\ j, & \text{if } \frac{2j-1}{2} < n < \frac{2j+1}{2} \\ -Q(-n) & \text{if } n < -\frac{1}{2} \end{cases} \tag{4}$$

where $j = 1, 2, \ldots$ is the output of quantizer, and n denote an arbitrary value. For a vector $n = [n_1, \cdots, n_m]^T$, the quantizer is formulated as $Q(n) = [Q(n_1), \cdots, Q(n_m)]^T$.

The encoder E_1 and E_2 are designed as follows:

$$\begin{cases} \varsigma_0^{(1)}(t) = 0 \\ s_{k+1}^{(1)}(t) = Q\left(\dfrac{y_{k+1}(t+1) - \varsigma_k^{(1)}(t)}{d_k^{(1)}}\right) \\ \varsigma_{k+1}^{(1)}(t) = d_k^{(1)} \gamma_{k+1}^{(1)}(t+1) s_{k+1}^{(1)}(t) + \varsigma_k^{(1)}(t) \end{cases} \tag{5}$$

and

$$\begin{cases} \varsigma_0^{(2)}(t) = 0 \\ s_{k+1}^{(2)}(t) = Q\left(\dfrac{u_{k+1}(t) - \varsigma_k^{(2)}(t)}{d_k^{(2)}}\right) \\ \varsigma_{k+1}^{(2)}(t) = d_k^{(2)} \gamma_{k+1}^{(2)}(t) s_{k+1}^{(2)}(t) + \varsigma_k^{(2)}(t) \end{cases} \tag{6}$$

where 0 represents a zero vector of the appropriate dimension as the output or input of the system. $y_k(t)$, $s_k^{(1)}(t)$, and $\varsigma_k^{(1)}(t)$ are the input, output, and internal state of the encoder E_1, respectively. $u_k(t)$, $s_k^{(2)}(t)$, and $\varsigma_k^{(2)}(t)$ correspond to the input, output, and internal state of the encoder E_2, respectively. The unit uniform quantizer $Q(\cdot)$ is defined in (4). $d_k^{(i)}$, $i = 1, 2$ are scaling sequences that are to modify the difference in magnitude between the internal states and the signals of the encoders.

The decoder D_1 and D_2 are designed as follows:

$$\begin{cases} \tilde{y}_0(t+1) = 0 \\ \tilde{y}_{k+1}(t+1) = d_k^{(1)} \gamma_{k+1}^{(1)}(t+1) s_{k+1}^{(1)}(t) + \tilde{y}_k(t+1) \end{cases} \tag{7}$$

and

$$\begin{cases} \tilde{u}_0(t) = 0 \\ \tilde{u}_{k+1}(t) = d_k^{(2)} \gamma_{k+1}^{(2)}(t) s_{k+1}^{(2)}(t) + \tilde{u}_k(t) \end{cases} \tag{8}$$

where $\tilde{y}_k(t)$ (estimated output) is the output of decoder D_1, which is the estimate $y_k(t)$. Moreover, $\tilde{u}_k(t)$ (system input) is the output of decoder D_2, which is the estimate of the $u_k(t)$.

According to the framework, the system input is $\tilde{u}_k(t)$. $u_k(t)$ and $\tilde{u}_k(t)$ are expressed the generated input and system input, respectively. Then, the system formulation becomes

$$\begin{cases} x_k(t+1) = A(t)x_k(t) + B(t)\tilde{u}_k(t) \\ y_k(t) = C(t)x_k(t) \end{cases} \qquad (9)$$

Remark 2: We choose to utilize the unit uniform quantizer due to its convenience, cost-effectiveness, and commonly used in practical applications. Our objective is to demonstrate that the proposed encoding-decoding scheme can attain asymptotic zero-error tracking performance even when working with quantized data. It is important to note that random data dropouts might occur during the transmission between the encoder and decoder. The encoding-decoding mechanism maintains the output value from the previous iteration when data dropout occurs to ensure that there is no zero output.

In order to describe the encoding-decoding mechanism with random data dropouts, define $\gamma_k^{(i)}(t), (l = 1, 2),\ t \in \{0, 1, \cdots, T_d\}$ to be a stochastic variable following Bernoulli distribution and take binary values of 0 and 1. Namely, both $\gamma_k^{(i)}(t),\ (i = 1, 2)$ are equal to 1 if encoding-decoding mechanism data is successfully transmitted, and 0 otherwise. Furthermore, $P\left[\gamma_k^{(i)}(t) = 1\right] = \gamma_i(t), (i = 1, 2)$ where $0 < \gamma_i(t) < 1, (i = 1, 2)$. Note that both encoding-decoding mechanism work individually, assuming that $\gamma_i(t), (i = 1, 2)$ are independent.

The characteristics of the encoding-decoding mechanism in the presence of random data dropouts are first clarified. The two crucial relationships between system output $y_k(t)$ and estimated output $\tilde{y}_k(t)$, as well as the association between generated input $u_k(t)$ and system input $\tilde{u}_k(t)$, will be verified. By replacing the expression of $s_k^{(1)}(t)$ form the encoder E_1 into the estimated output $\tilde{y}_k(t+1)$ in (7), we can get

$$\begin{aligned} \tilde{y}_{k+1}(t+1) &= d_k^{(1)}\gamma_{k+1}^{(1)}(t+1)s_{k+1}^{(1)} + \tilde{y}_k(t+1) \\ &= d_k^{(1)}\gamma_{k+1}^{(1)}(t+1)Q\left(\frac{y_{k+1}(t+1) - \varsigma_k^{(1)}(t)}{b_k^{(1)}}\right) + \tilde{y}_k(t+1) \\ &\quad d_k^{(1)}\gamma_{k+1}^{(1)}(t+1)\left[\frac{y_{k+1}(t+1) - \varsigma_k^{(1)}(t)}{d_k^{(1)}} + \vartheta_{k+1}^{(1)}(t+1)\right] + \tilde{y}_k(t+1) \\ &= \gamma_{k+1}^{(1)}(t+1)\left[y_{k+1}(t+1) - \varsigma_k^{(1)}(t) + d_k^{(1)}\vartheta_{k+1}^{(1)}(t+1)\right] + \tilde{y}_k(t+1), \end{aligned} \qquad (10)$$

where $\vartheta_{k+1}^{(1)}(t+1) = Q\left(\frac{y_{k+1}(t+1)-\varsigma_k^{(1)}(t)}{d_k^{(1)}}\right) - \frac{y_{k+1}(t+1)-\varsigma_k^{(1)}(t)}{d_k^{(1)}}$ is the output quantization error. Apparently, each dimension of $\vartheta_k^{(1)}(t)$ is limited to 1/2 owing to the inherent characteristic of a unit uniform quantizer.

If $\gamma_{k+1}^{(1)}(t+1) = 1$, we can get

$$\tilde{y}_{k+1}(t+1) = y_{k+1}(t+1) + d_k^{(1)}\vartheta_{k+1}^{(1)}(t+1) + \tilde{y}_k(t+1) - \varsigma_k^{(1)}(t). \qquad (11)$$

For $\tilde{y}_k(t+1) - \varsigma_k^{(1)}(t)$. By mathematical inductive method, $\tilde{y}_k(t+1) - \varsigma_k^{(1)}(t) = 0$ is always valid. Then, we can obtain

$$\tilde{y}_{k+1}(t+1) = y_{k+1}(t+1) + d_k^{(1)}\vartheta_{k+1}^{(1)}(t+1). \tag{12}$$

If $\gamma_{k+1}^{(1)}(t+1) = 0$, we can get $\tilde{y}_{k+1}(t+1) = \tilde{y}_k(t+1)$, then (10) can be rewritten in the following from:

$$\tilde{y}_{k+1}(t+1) = \begin{cases} y_{k+1}(t+1) + d_k^{(1)}\vartheta_{k+1}^{(1)}(t+1), & \gamma_{k+1}^{(1)}(t+1) = 1 \\ \tilde{y}_k(t+1), & \gamma_{k+1}^{(1)}(t+1) = 0 \end{cases} \tag{13}$$

Similarly, the following equation can be obtained for the generated input $u_k(t)$ and system input $\tilde{u}_k(t)$:

$$\tilde{u}_{k+1}(t) = \begin{cases} u_{k+1}(t) + d_k^{(2)}\vartheta_{k+1}^{(2)}(t), & \gamma_{k+1}^{(2)}(t) = 1 \\ \tilde{u}_k(t). & \gamma_{k+1}^{(2)}(t) = 0 \end{cases} \tag{14}$$

From (13) and (14), it can be observed that the difference between the original values before encoding and decoded values after the transmission is a product of a scaling sequence and a limited quantization error when there is no data dropout. The estimated output and system input remain the same as those of the previous iteration when there is data dropout. Therefore, by choosing the scaling sequences $d_k^{(i)}(i = 1, 2)$ converging to zero, it becomes feasible to achieve asymptotically accurate estimation, resulting in zero-error tracking performance.

3 Main Results

In this section, we use the estimated system output and the information whether data dropout occurs into the learning algorithm. We employ the conventional P-type update law as following:

$$u_{k+1}(t) = u_k(t) + \gamma_{k+1}^{(1)}(t+1)L[y_d(t+1) - \tilde{y}_k(t+1)], \tag{15}$$

where L denote the learning gain matrix. Moreover, the actual input signal utilized for the plant is provided

$$\tilde{u}_{k+1}(t) = \gamma_{k+1}^{(2)}(t)\tilde{u}_{k+1}(t) + \left[1 - \gamma_{k+1}^{(2)}(t)\right] \cdot \tilde{u}_k(t). \tag{16}$$

We first present the supporting lemmas. Denote $\Delta u_k(t) = u_d(t) - u_k(t)$, and $\Delta \tilde{u}_k(t) = u_d(t) - \tilde{u}_k(t)$ as the error of the generated input and system input, respectively. Define the augmented input error

$$\Delta u_k^e(t) = \left[(\Delta u_k(t))^T, (\Delta \tilde{u}_k(t))^T\right]^T. \tag{17}$$

The augmented input error can be characterized as follows. Denote $\Delta x_k(t) = x_d(t) - x_k(t)$ as the state error.

Lemma 1. The regression presented below holds for the augmented input error given in (17),

$$
\Delta u_{k+1}^e(t) = P_k(t)\Delta u_k^e(t) + W_k(t)\Delta x_k(t) + G_k(t)d_{k-1}^{(1)}\vartheta_k^{(1)}(t+1)
$$
$$
+ H_k(t)d_{k-1}^{(2)}\vartheta_k^{(2)}(t+1) - \gamma_{k+1}^{(2)}(t)Id_k^{(2)}\vartheta_{k+1}^{(2)}(t),
\tag{18}
$$

where $P_k(t) = \begin{bmatrix} I - \gamma_k^{(1)}(t+1)LC(t+1)B(t) & 0 \\ \gamma_{k+1}^{(2)}(t)\left[I - \gamma_k^{(1)}(t+1)LC(t+1)B(t)\right] & \left[1 - \gamma_{k+1}^{(2)}(t)\right]I \end{bmatrix}$, $W_k(t) =$

$\begin{bmatrix} -\gamma_k^{(1)}(t+1)LC(t+1)A(t) \\ -\gamma_{k+1}^{(2)}(t)\gamma_k^{(1)}(t+1)LC(t+1)A(t) \end{bmatrix}$, $G_k(t) = \begin{bmatrix} \gamma_k^{(1)}(t+1)L \\ \gamma_{k+1}^{(2)}(t)\gamma_k^{(1)}(t+1)L \end{bmatrix}$, and $H_k(t) =$

$\begin{bmatrix} \gamma_k^{(1)}(t+1)LC(t+1)B(t) \\ \gamma_{k+1}^{(2)}(t)\gamma_k^{(1)}(t+1)LC(t+1)B(t) \end{bmatrix}$.

This lemma describes the random asynchronization between the generated input and system input, which is proven using the random matrix. The contraction mapping property of the matrix $P_k(t)$ is important for the convergence analysis of the regression model (18). This property is confirmed in the Lemma 2.

Proof of Lemma 1:

Based on (15), (13), (12), (9), and (2), we can get

$$
\Delta u_{k+1}(t) = \Delta u_k(t) - L\gamma_k^{(1)}(t+1)\left[y_d(t+1) - \tilde{y}_k(t+1)\right]
$$
$$
= \Delta u_k(t) - L\gamma_k^{(1)}(t+1)\left[y_d(t+1) - y_k(t+1) - d_{k-1}^{(1)}\vartheta_k^{(1)}(t+1)\right]
$$
$$
= \Delta u_k(t) - L\gamma_k^{(1)}(t+1)[C(t+1)A(t)\Delta x_k(t) + C(t+1)B(t)\Delta u_k(t)
$$
$$
- C(t+1)B(t)d_{k-1}^{(2)}\vartheta_k^{(2)}(t) - d_{k-1}^{(1)}\vartheta_k^{(1)}(t+1)]
$$
$$
= \left[I - \gamma_k^{(1)}(t+1)LC(t+1)B(t)\right]\Delta u_k(t) - \gamma_k^{(1)}(t+1)LC(t+1)A(t)\Delta x_k(t)
$$
$$
+ \gamma_k^{(1)}(t+1)LC(t+1)B(t)d_{k-1}^{(2)}\vartheta_k^{(2)}(t) + \gamma_k^{(1)}(t+1)Ld_{k-1}^{(1)}\vartheta_k^{(1)}(t+1).
\tag{19}
$$

Substituting (19) into $\Delta \tilde{u}_{k+1}(t) = \gamma_{k+1}^{(2)}(t)\Delta \tilde{u}_{k+1}(t) + \left(1 - \gamma_{k+1}^{(2)}(t)\right)\Delta \tilde{u}_k(t)$ leads

to

$$
\Delta \tilde{u}_{k+1}(t) = \gamma_{k+1}^{(2)}(t)\Delta \tilde{u}_{k+1}(t) + \left(1 - \gamma_{k+1}^{(2)}(t)\right)\Delta \tilde{u}_k(t)
$$
$$
\Delta \tilde{u}_{k+1}(t) = \gamma_{k+1}^{(2)}(t)\left[\Delta u_{k+1}(t) - d_k^{(2)}\vartheta_{k+1}^{(2)}(t)\right] + \left(1 - \gamma_{k+1}^{(2)}(t)\right)\Delta \tilde{u}_k(t)
$$
$$
= \gamma_{k+1}^{(2)}(t)\Delta u_{k+1}(t) - \gamma_{k+1}^{(2)}(t)d_k^{(2)}\vartheta_{k+1}^{(2)}(t) + \left(1 - \gamma_{k+1}^{(2)}(t)\right)\Delta \tilde{u}_k(t)
$$
$$
= \gamma_{k+1}^{(2)}(t)\left[I - \gamma_k^{(1)}(t+1)LC(t+1)B(t)\right]\Delta u_k(t) - \gamma_{k+1}^{(2)}(t)\gamma_k^{(1)}(t+1)LC(t+1)A(t)\Delta x_k(t)
$$
$$
+ \gamma_{k+1}^{(2)}(t)\gamma_k^{(1)}(t+1)LC(t+1)B(t)d_{k-1}^{(2)}\vartheta_k^{(2)}(t) + \gamma_{k+1}^{(2)}(t)\gamma_k^{(1)}(t+1)Ld_{k-1}^{(1)}\vartheta_k^{(1)}(t+1)
$$
$$
- \gamma_{k+1}^{(2)}(t)d_k^{(2)}\vartheta_{k+1}^{(2)}(t) + \left(1 - \gamma_{k+1}^{(2)}(t)\right)\Delta \tilde{u}_k(t).
\tag{20}
$$

Based on (19) and (20), taking into account the augmented input error $\Delta u_k^e(t)$, and the correlation matrices $P_k(t), W_k(t), G_k(t)$, and $H_k(t)$, it is evident that the regression model (18) holds. The proof is complete.

Lemma 2. if the learning gain matrix L in (18) satisfies $\|I - LC(t+1)B(t)\|_\infty < 1$, then we have.

$$\sup_t E\{\|P_k(t)\|_\infty\} < 1. \tag{21}$$

The proof of Lemma 2 is provided in [15].

The primary theorem is given as follows.

Theorem 1: Consider system (9) with Assumptions 1 and 2. Apply the learning algorithm (15) and (16) with encoding-decoding mechanism (5), (6), (7) and (8) in the random data dropouts. If the learning gain matrix L in (15) and (16) satisfies.

$$\|I - LC(t+1)B(t)\|_\infty < 1 \tag{22}$$

If we can choose the scaling sequences $d_k^{(i)}$ $(i = 1, 2)$ converging to zero, the system output achieves asymptotic zero-error tracking performance along the iteration axis. That is $\lim_{k\to\infty} E\{\|y_d(t) - y_k(t)\|_\infty\} = 0$.

Proof: By applying $\|\cdot\|_\infty$ to both sides of the regression for the augmented input error (18). We can get

$$\left\|\Delta u_{k+1}^e(t)\right\|_\infty \le \|P_k(t)\|_\infty\left\|\Delta u_k^e(t)\right\|_\infty + \|W_k(t)\|_\infty\|\Delta x_k(t)\|_\infty + \|G_k(t)\|_\infty\left\|d_{k-1}^{(1)}\vartheta_k^{(1)}(t+1)\right\|_\infty$$
$$+\|H_k(t)\|_\infty\left\|d_{k-1}^{(2)}\vartheta_k^{(2)}(t+1)\right\|_\infty + \gamma_k^{(2)}(t)\left\|d_k^{(2)}\vartheta_{k+1}^{(2)}(t)\right\|_\infty \tag{23}$$

Taking mathematical expectation to (23), we can obtain

$$E\left\{\left\|\Delta u_{k+1}^e(t)\right\|_\infty\right\} \le E\{\|P_k(t)\|_\infty\} \cdot E\left\{\left\|\Delta u_k^e(t)\right\|_\infty\right\} + E\{\|W_k(t)\|_\infty\} \cdot E\{\|\Delta x_k(t)\|_\infty\} + E\{\|G_k(t)\|_\infty\}$$
$$\times E\left\{\left\|d_{k-1}^{(1)}\vartheta_k^{(1)}(t+1)\right\|_\infty\right\} + E\{\|H_k(t)\|_\infty\} \cdot E\left\{\left\|d_{k-1}^{(2)}\vartheta_k^{(2)}(t+1)\right\|_\infty\right\}$$
$$+E\left\{\gamma_k^{(2)}(t)\right\} \cdot E\left\{\left\|d_k^{(2)}\vartheta_{k+1}^{(2)}(t)\right\|_\infty\right\}. \tag{24}$$

Noticing system (9) and the desired reference model (2), we have

$$\Delta x_k(t) = A(t-1)\Delta x_k(t-1) + B(t-1)\Delta \tilde{u}_k(t-1). \tag{25}$$

Then taking $\|\cdot\|_\infty$ to both sides of (25), we have

$$\|\Delta x_k(t)\|_\infty \le \|A(t-1)\|_\infty\|\Delta x_k(t-1)\|_\infty + \|B(t-1)\|_\infty\|\Delta \tilde{u}_k(t-1)\|_\infty$$
$$\le \cdots\cdots$$
$$\le \sum_{i=0}^{t-1}\prod_{j=i+1}^{t-1}\|A(j)\|_\infty\|B(i)\|_\infty\|\Delta \tilde{u}_k(i)\|_\infty + \prod_{j=0}^{t-1}\|A(j)\|_\infty\|\Delta x_k(0)\|_\infty. \tag{26}$$

We further take mathematical expectation to (26),

$$
\begin{aligned}
E\{\|\Delta x_k(t)\|_\infty\} &\leq \sum_{i=0}^{t-1}\prod_{j=i+1}^{t-1} E\{\|A(j)\|_\infty\}\cdot E\{\|B(i)\|_\infty\}\cdot E\{\|\Delta\tilde{u}_k(i)\|_\infty\} \\
&\quad + \prod_{j=0}^{t-1}E\{\|A(j)\|_\infty\}\cdot E\{\|\Delta x_k(0)\|_\infty\} \\
&\leq n_B\sum_{i=0}^{t-1} n_A^{t-i-1}E\{\|\Delta\tilde{u}_k(i)\|_\infty\} + n_A^t E\{\|\Delta x_k(0)\|_\infty\} \\
&\leq n_B\sum_{i=0}^{t-1} n_A^{t-i-1}E\{\|\Delta u_k^e(i)\|_\infty\} + n_A^t E\{\|\Delta x_k(0)\|_\infty\}
\end{aligned}
\tag{27}
$$

where $n_A=\max_t\|A(t)\|_\infty$ and $n_B=\max_t\|B(t)\|_\infty$. $\Delta\tilde{u}_k(t)$ is part of $\Delta u_k^e(t)$, we have $\|\Delta\tilde{u}_k(t)\|_\infty \leq \|\Delta u_k^e(t)\|_\infty$ for all t.

Now substituting (27) into (23) lead to

$$
\begin{aligned}
E\{\|\Delta u_{k+1}^e(t)\|_\infty\} &\leq E\{\|P_k(t)\|_\infty\}\cdot E\{\|\Delta u_k^e(t)\|_\infty\} + E\{\|W_k(t)\|_\infty\}\cdot n_B\sum_{i=0}^{t-1}n_A^{t-i-1}E\{\|\Delta u_k^e(i)\|_\infty\} \\
&\quad + E\{\|G_k(t)\|_\infty\}\cdot E\{\|d_{k-1}^{(1)}\vartheta_k^{(1)}(t+1)\|_\infty\} + E\{\|H_k(t)\|_\infty\}\cdot E\{\|d_{k-1}^{(2)}\vartheta_k^{(2)}(t+1)\|_\infty\} \\
&\quad + E\{\gamma_k^{(2)}(t)\}\cdot E\{\|d_k^{(2)}\vartheta_{k+1}^{(2)}(t)\|_\infty\} + n_A^t E\{\|\Delta x_k(0)\|_\infty\}\cdot E\{\|W_k(t)\|_\infty\}
\end{aligned}
\tag{28}
$$

Now, the classical $\|\cdot\|_\lambda$ can be employed. In particular, by multiplying both sides of last inequality with $\alpha^{-\lambda t}$ where $\alpha>1$ and $\lambda>1$ are defined, and then taking supremum over all time instances t,

$$
\begin{aligned}
\sup_t\left(\alpha^{-\lambda t}E\{\|\Delta u_{k+1}^e(t)\|_\infty\}\right) &\leq \sup_t E\{\|P_k(t)\|_\infty\}\cdot\sup_t\left(\alpha^{-\lambda t}E\{\|\Delta u_k^e(t)\|_\infty\}\right) + \sup_t E\{\|W_k(t)\|_\infty\} \\
&\quad \times n_B\sup_t\alpha^{-\lambda t}\left(\sum_{i=0}^{t-1}n_A^{t-i-1}E\{\|\Delta u_k^e(i)\|_\infty\}\right) + \sup_t\alpha^{-\lambda t}\left(E\{\|G_k(t)\|_\infty\}\cdot E\{\|d_{k-1}^{(1)}\vartheta_k^{(1)}(t+1)\|_\infty\}\right) \\
&\quad + \sup_t\alpha^{-\lambda t}\left(E\{\|H_k(t)\|_\infty\}E\{\|d_{k-1}^{(2)}\vartheta_k^{(2)}(t+1)\|_\infty\}\right) + \sup_t\alpha^{-\lambda t}\left(E\{\gamma_k^{(2)}(t)\}\cdot E\{\|d_k^{(2)}\vartheta_{k+1}^{(2)}(t)\|_\infty\}\right) \\
&\quad + \sup_t\alpha^{-\lambda t}n_A^t E\{\|\Delta x_k(0)\|_\infty\}\cdot E\{\|W_k(t)\|_\infty\} \\
&\leq n_P\sup_t\left(\alpha^{-\lambda t}E\{\|\Delta u_k^e(t)\|_\infty\}\right) + n_W n_B\sup_t\alpha^{-\lambda t}\left(\sum_{i=0}^{t-1}n_A^{t-i-1}E\{\|\Delta u_k^e(i)\|_\infty\}\right) \\
&\quad + \frac{1}{2}(n_G+n_H+\gamma_2)\bar{d}_{k-1} + n_A n_W E\{\|\Delta x_k(0)\|_\infty\}
\end{aligned}
\tag{29}
$$

where $\sup_t\alpha^{-\lambda t}\leq 1$, $n_P=\sup_t E\{\|P_k(t)\|_\infty\}$, $n_W=\sup_t E\{\|W_k(t)\|_\infty\}$, $\gamma_2=E\{\gamma_k^{(2)}(t)\}$, $n_G=\sup_t E\{\|G_k(t)\|_\infty\}$, and $n_H=\sup_t E\{\|H_k(t)\|_\infty\}$. Because each dimension of

$\vartheta_k^{(1)}(t)$ is limited by 1/2, then $\sup_t E\left\{\left\|\vartheta_k^{(i)}(t)\right\|_\infty\right\} \leq \frac{1}{2}, i = 1, 2$. Denote $\overline{d}_{k-1} = \max\left\{d_{k-1}^{(1)}, d_{k-1}^{(2)}, d_k^{(2)}\right\}$.

Let $\alpha > n_A$, then it is observed that

$$\sup_t \alpha^{-\lambda t}\left(\sum_{i=0}^{t-1} n_A^{t-i-1} E\{\|\Delta u_k^e(i)\|_\infty\}\right) \leq \sup_t\left(\alpha^{-\lambda t}\sum_{i=0}^{t-1} \alpha^{t-i-1} E\{\|\Delta u_k^e(i)\|_\infty\}\right).$$
$$\leq \sup_t\left(\alpha^{-\lambda i} E\{\|\Delta u_k^e(i)\|_\infty\}\right)\frac{1 - \alpha^{-(\lambda-1)t}}{\alpha^\lambda - \alpha}$$

(30)

Define a new $\|\cdot\|_\lambda$ of $\Delta u_k^e(t)$ as $\left\|\Delta u_k^e(t)\right\|_\lambda = \sup_t\left(\alpha^{-\lambda t} E\{\|\Delta u_k^e(t)\|_\infty\}\right)$.

$$\left\|\Delta u_{k+1}^e(t)\right\|_\lambda \leq \left(n_P + n_W n_B \frac{1 - \alpha^{-(\lambda-1)t}}{\alpha^\lambda - \alpha}\right) \cdot \left\|\Delta u_k^e(t)\right\|_\lambda$$
$$+ \frac{1}{2}(n_G + n_H + \gamma_1)\overline{d}_{k-1} + n_A n_W E\{\|\Delta x_k(0)\|_\infty\}$$

(31)

From Lemma 2, we find $n_P = \sup_t E\{\|P_k(t)\|_\infty\} < 1$. Let $\alpha > \max\{1, n_A\}$, then there always exists a sufficiently large λ such that $0 < n_P + n_W n_B \frac{1-\alpha^{-(\lambda-1)t}}{\alpha^\lambda - \alpha} < 1$. Obviously, if we select the adjustment sequence $d_k^{(i)}(i = 1, 2)$ satisfying the condition that $d_k^{(i)} \to 0(i = 1, 2)$ as $k \to \infty$, then $\lim_{k\to\infty} \overline{d}_{k-1} \to 0$. Thus, from Assumption 2, we have $\lim_{k\to\infty}\left\|\Delta u_k^e(t)\right\|_\lambda = 0$, then $\lim_{k\to\infty}\left\|\Delta u_k^e(t)\right\|_\infty = 0$ for all instances t. It is apparent that $\lim_{k\to\infty}\|\Delta\tilde{u}_k(t)\|_\infty = 0$. Furthermore, we know $\lim_{k\to\infty}\|\Delta x_k(t)\|_\infty = 0$ and then $\lim_{k\to\infty}\|y_d(t) - y_k(t)\|_\infty = 0$ for all instances t. This completes the proof.

4   Illustrative Simulations

To validate the proposed results, the following permanent magnet linear motor (PMLM) model is used [16]

$$\begin{cases} x(t+1) = x(t) + v(t)\Delta, \\ v(t+1) = v(t) - \Delta\dfrac{k_1 k_2 \psi_f^2}{Rm}v(t) + \Delta\dfrac{k_2\psi_f}{Rm}u(t), \\ y(t) = v(t), \end{cases}$$

where v and x are the rotor velocity and motor position, respectively. Δ, R, m, and ψ_f are the sampling period, the resistance of stator, the rotor mass and the flux linkage with $\Delta = 0.01s$, $R = 8.6$, $m = 1.635kg$, $\psi_f = 0.35Wb$. $k_1 = \pi/\tau$ and $k_2 = 1.5\pi/\tau$ with $\tau = 0.031m$ are the pole pitches. The desired trajectory is given by $y_d(t) = 0.5 - \sin(0.05\pi t) - 0.5\cos(0.02t), 0 \leq t \leq 0.5$.

In Assumption 2, the initial state is selected as $x_d(0) = 0$ and $x_k(0) = 2/k$, such that $\lim_{k\to\infty} x_k(0) = x_d(0)$, and the initial input is set as $u_0(t) = 0$. Then, we choose the learning gain L as 7, such that $\|I - LC(t+1)B(t)\|_\infty < 1$. The scaling sequence $d_k^{(i)}, i = 1, 2$ are set as $0.9 \cdot (2/3)^k$ and $0.1 \cdot (2/3)^k$, respectively. For the model of data dropout, different data dropout rates will influence convergence speeds. All in all, for arbitrary time instant, the number of successfully transmitted information tends to infinity, as $k \to \infty$.

The system output y_k is shown in Fig. 2 that demonstrates a gradual improvement in tracking performance as the iteration number increases. For the model of data dropout, the simulation is simply set to the data dropout rate of 20% for $\gamma_k^{(i)}, i = 1, 2$.

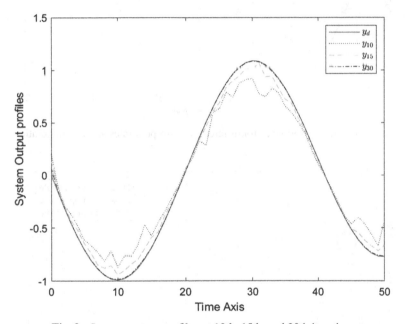

Fig. 2. System output profiles at 10th, 15th, and 30th iteration.

A higher convergence speed is observed with a smaller data dropout rate. To provide a more distinct assessment of the convergence accuracy at the same number of iterations, Fig. 3 illustrates the output tracking profiles for different data dropout rates at the 30th iteration.

To demonstrate the zero-error convergence property intuitionally, we plot the maximum tracking error $\max_t |e_k(t)|$ along the iteration axis in Fig. 4, where $e_k(t) = y_d(t) - y_k(t)$. To demonstrate the convergence speed for different data dropout rates, we have chosen the data dropout rates of 20%, 50% and 70% to observe their influence.

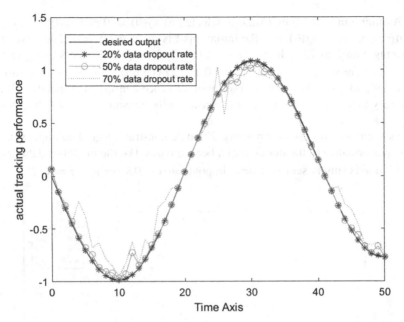

Fig. 3. Tracking performance for different data dropout rates at the 30th iteration.

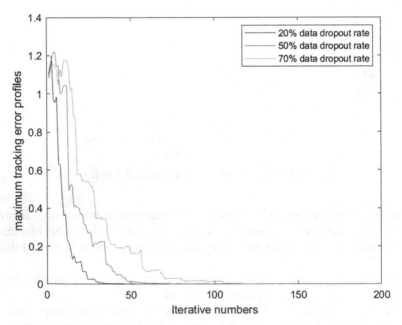

Fig. 4. Maximal tracking error profiles under different data dropout rates.

5 Conclusion

This paper investigates the ILC problem in the presence of stochastic data dropout in the encoding-decoding mechanism. The encoding-decoding mechanism employs a simple compensation mechanism to ensure stable execution of the controller, thereby improving tracking performance under limited information. As the overall number of iterations rises, the frequency of synchronous updates between the generated input and the actual input increases, leading to asymptotic convergence. The effectiveness of the proposed algorithm is validated through illustrative simulations. However, this paper only considers the case of infinite-level quantizer and does not address the scenario of finite-level quantizer. In our future studies, we will further investigate the influence of the encoding and decoding mechanism on the tracking performance.

Acknowledgement. This work was supported by the Key Laboratory of On-Chip Communication and Sensor Chip of Guangdong Higher Education Institutes [2023KSYS002].

References

1. Ahn, H.-S., Chen, Y., Moore, K.L.: Iterative learning control: brief survey and categorization. IEEE Trans. Syst., Man, Cybern. C **37**(6), 1099–1121 (2007). https://doi.org/10.1109/TSMCC.2007.905759
2. Shen, D.: Iterative learning control with incomplete information: a survey. IEEE/CAA J. Autom. Sinica **5**(5), 885–901 (2018)
3. Shen, D., Zhang, C., Xu, Y.: Intermittent and successive ILC for stochastic nonlinear systems with random data dropouts. Asian J. Control **20**(3), 1102–1114 (2018)
4. Zhang, Z., Li, Z., Guo, S., et al.: A novel successive updating scheme of iterative learning control for networked control system with output data dropouts. In: IEEE 12th Data Driven Control and Learning Systems Conference (DDCLS), pp. 1702–1707. IEEE (2023)
5. Huang, L., Sun, L., Wang, T., et al.: Optimal input filtering for networked iterative learning control systems with packet dropouts and channel noises in both sides. Int. J. Robust Nonlinear Control **32**(9), 5086–5104 (2022)
6. Bu, X., Wang, T., Hou, Z., et al.: Iterative learning control for discrete-time sys-tems with quantised measurements. IET Control Theory Appl. **9**(9), 1455–1460 (2015)
7. Xu, Y., Shen, D., Bu, X.: Zero-error convergence of iterative learning control using quantized error information. IMA J. Math. Control. Inf. **34**(3), 1061–1077 (2017)
8. Bu, X., Hou, Z., Cui, L., et al.: Stability analysis of quantized iterative learning control systems using lifting representation. Int. J. Adapt. Control Signal Process. **31**(9), 1327–1336 (2017)
9. Zhang, C., Shen, D.: Zero-error convergence of iterative learning control using uniform quantizer with encoding and decoding method. In: 2017 36th Chinese Control Conference (CCC), pp. 3473–3478. IEEE (2017)
10. Huo, N., Shen, D.: Encoding–decoding mechanism-based finite-level quantized iterative learning control with random data dropouts. IEEE Trans. Autom. Sci. Eng. **17**(3), 1343–1360 (2019)
11. Tao, Y., Huang, Y., Tao, H., et al.: Gradient-based iterative learning control for signal quantization with encoding-decoding mechanism. In: 2023 IEEE 12th Data Driven Control and Learning Systems Conference (DDCLS), pp. 184–189. IEEE (2023)

12. Zhang, H., Chi, R., Hou, Z., et al.: Data-driven iterative learning control using a uniform quantizer with an encoding–decoding mechanism. Int. J. Robust Nonlinear Control **32**(7), 4336–4354 (2022)
13. Huang, Y., Tao, H., Chen, Y., et al.: Point-to-point iterative learning control with quantised input signal and actuator faults. Int. J. Control **97**(6), 1361–1376 (2023)
14. Shen, D., Zhang, C.: Zero-error tracking control under unified quantized iterative learning framework via encoding–decoding method. IEEE Trans. Cybern. **52**(4), 1979–1991 (2022)
15. Jin, Y., Shen, D.: Iterative learning control for nonlinear systems with data dropouts at both measurement and actuator sides. Asian J. Control **20**(4), 1624–1636 (2018)
16. Huo, N., Zhang, C., Shen, D.: Uniformly quantized ilc with encoding and decoding mechanism under random data dropouts. In: 2019 Chinese Control And Decision Conference (CCDC), pp. 3459–3464. IEEE (2019)

A Domain Adaptive Segmentation Label Generation Algorithm for Autonomous Driving Scenarios

Kangshun Li[✉] and Tian Feng

College of Mathematics and Informatics, South China Agricultural University,
Guangzhou 510642, Guangdong, China
Likangshun632@sina.com

Abstract. Semantic segmentation algorithm is a cornerstone algorithm in the field of autonomous driving. The complex and variable data in the production environment makes the data domain in the production environment seriously offset, which causes significant performance degradation of the network model in the development environment data. In this paper, we propose a pseudo-label generation algorithm (EOPL) with smooth assignment of label expansion and erosion based on the adversarial learning algorithm and self-supervised learning strategy in domain adaptive learning. In this paper, experiments are conducted on GTAV and Cityscapes datasets, and the algorithm is validated using multiple network models. The experimental results show that under the condition that the amount of source domain data is reduced by half and the number of domain classification training iterations is shortened by half, the mIoU evaluation index of the FADA model with EOPL improves by 1.5% overall and 10 percentage points higher for individual class IoU evaluation index; while the ADVENT model with EOPL improves by 5% for most classes and 30% for individual classes, which not only shortens the running time, but also improves the efficiency of image segmentation.

Keywords: Semantic Segmentation · Domain Adaptive Learning · Self-supervised Learning · Soft Label Smoothing

1 Introduction

It is in this context that the domain adaptive algorithm is proposed, which focuses on the assumption of independent homogeneous distribution and aims to train the model to learn new unlabeled data that has never been labeled in the production environment, so that the network model is able to learn the domain invariant features, such as shape, style, size, etc., between different datasets [1–3, 10, 11]. Unsupervised training of the model and improvement of model performance is achieved by reducing the distance between the feature space and data distribution between the trained and untrained datasets of the network model. That is to say, the data distribution of the neural network model fitted in the laboratory environment is "close" to the data distribution in the generative environment, in order to facilitate the subsequent representation, the trained dataset with

K. Li and Y. Liu (Eds.): ISICA 2023, CCIS 2147, pp. 127–134, 2024.
https://doi.org/10.1007/978-981-97-4396-4_11

real-value labels is referred to as the source domain dataset; the untrained dataset without real-value labels is referred to as the target domain dataset. Compared with other algorithms, the domain adaptive algorithm proposes a new domain classifier component to identify the domain from which the current batch of data is trained, through which the domain adaptive algorithm is applied to solve the problem of domain bias between the source and target domains of the network model due to the differences in the characteristics and distribution of the data. To put it simply, the algorithm takes a "fragile" model trained in a laboratory environment and adapts it to the data in a production environment to make it more "robust", so that the feature extraction and classification effects of the model can be improved. The general flowchart of the domain adaptive algorithm is shown in Fig. 1.

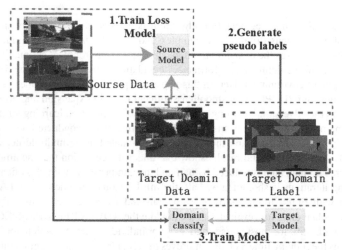

Fig. 1. Flowchart of Domain Adaptive Algorithm The source domain dataset is labeled and can be used to train the source domain model; the target domain dataset is unlabeled, so it is necessary to pass the target domain data to the source domain model, and use the output of the source domain model as the labels, which can be provided to the target domain model for training; and the domain classifier module enhances the performance of the network.

2 Related Work

2.1 Domain Adaptation at Training Time

In domain adaptive algorithms [12, 13], a model trained with the source domain data is used to detect the target domain data, and the higher confidence detection results from the source domain model are used as "labels" to help train the target domain network model. Therefore, this labeling of the output results from the source domain network model is called pseudo-labeling. The mathematical expression can be found in Eq. 1

$$\mathcal{L}_{ce} = -\sum_{i=1}^{H \times W} \sum_{c=1}^{C} \hat{y}_t^{(i,k)} \log\left(p_t^{(i,k)}\right) \tag{1}$$

$\mathcal{L}_{ce}$ represents the cross-entropy loss function, H, W are the height and width values of the input images, $\hat{y}_t^{(i,k)}$ represents the pseudo-label obtained by inputting the target domain data from the model, and $p_t^{(i,k)}$ is the label output value of the target domain model. c represents the first few classes. This cross-entropy function is a typical loss function used for classification.

As a domain discriminator for binary classification, if $\hat{y}_t^{(i,c)}$ is used as the binary classification source domain as well as the target domain label, $p_t^{(i,c')}$ represented in Eq. (8) is the output of the domain classification result for each pixel after the softmax function, the task of the domain classifier is to discriminate each pixel for the binary classification domain, and the equation is expressed as

$$\hat{y}_t^{(i,k)} = \begin{cases} 1, \text{ if } k = argmax_{k'} p_t^{(i,k')} \\ 0, \text{ otherwise} \end{cases} \tag{2}$$

This is not the case for ADVENT network [14], which is called Adversarial entropy, because the authors propose a method for converting probabilistic maps into moisture maps. This entropy mapping is different from probabilistic mapping, but they both have the goal of minimization. For the ADVENT network it is entropy minimization, which the proposers of the network refer to as a successful method for semi-supervised learning. Like the proponents of the FADA network, the proponents of the ADVENT network usually overfit the models trained on the source domain data, and this overfitting meets the definition of low entropy: i.e., a large amount of feeding energy (training set) is required, while the models trained corresponding to the lack of labels of the target domain data are high entropy models. From the entropy point of view, the methods that the authors of the ADVENT network have in mind are: 1 Using the entropy value as a proxy as a training loss as well as a counteracting loss 2 Global information matching on the weighted self-information of the batch data (source data as well as target data). The unique formula of ADVENT network is how to convert the probability map into an entropy map, which is listed here, while other formulas are more common:

$$\mathcal{L}_{prob\_2\_entropy} = -\frac{p_t^{(i,k)} \times \log_2^{p_t(i,k)+1 \times e^{-30}}}{\log_2^c} \tag{3}$$

$p_t^{(i,k)}$ represents the probability map, c represents the channel of the probability map.

2.2 Domain Adaptation at Testing Time

Traditional domain adaptive algorithms have an assumption of "having access to source domain data". That is, the source domain data is utilized for adversarial learning against the target domain data as explained in Sect. 1.2. Often source domain datasets are not accessible when in a production environment. Test-time adaptation (TTA) is proposed in this context: during testing, a model with only source domain data as well as source domain parameters must adapt itself to generalize to the data under the target domain [4], a methodology proposed by Dequan's team at the University of Brooklyn. A series of other excellent algorithms have been derived from their work, such as the Sachin team that proposed conjugate pseudo-labeling [5, 6]. The cornerstone of the TTA algorithm is the pseudo-labeling generation algorithm, and thus the pseudo-labeling generation algorithm will also have a place.

$$\mathcal{L}_{trainloss} = \mathcal{L}_{ce}(x^s, y^s) + \mathcal{L}_{ce}(x^s)$$
$$\mathcal{L}_{testloss} = \mathcal{L}_{ce}(x^t) \tag{4}$$

2.3 Self-supervised Learning

When faced with a situation where there is no hand-constructed labeled data, supervised learning methods are unable to compute the error through the loss function or update the weights of the neural network through backpropagation. Self-supervised learning methods solve this problem, i.e., where does the "knowledge" come from, and previous work such as BDL networks [7], CAG networks [8], and CBST networks [9, 10] start from this approach. In addition, there are many new works in the field of re-self-supervised segmentation algorithms in recent years, such as the Transformer-based InternImage proposed by the Tsinghua University team [15, 16]; and a kind of deep difference network proposed by Zhang Jin's team to process by encoding meta-knowledge [17]. Typically, the addition of a self-supervised learning module will be accompanied by the emergence of the problem that arises with that module, namely how to ensure that the pseudo-labels submitted to the network are strictly classified and must be correct. It has been the practice to use the model trained on the source domain data to perform a network forward pass on the target domain data, and then hard pruning regularization is performed directly on each prediction confidence in the data to prevent overfitting of the target domain model. Hard pruning regularization involves deflating the confidence of pixels whose confidence exceeds a threshold, but even after "many experiments", carefully set threshold pruning still produces noisy pseudo-labels during self-supervised training.

3 Method

In order to prevent the source domain network model from overfitting on the source domain dataset, the authors of the FADA network model set a confidence threshold pruning, which is effective when applied to the source domain network model; however, when the method is also applied to the target domain network model, the classification ability of the network will be lacking because the target domain network model has not undergone supervised training of the data, which results in poor or even counterproductive effects of pruning on the target domain network model. Model is ineffective or even counterproductive. The authors of the ADVENT network model utilize the traditional adversarial learning strategy, and direct self-supervised learning will not be able to effectively learn the relevant knowledge of the dataset. So they all have the defect of pseudo-label generation. The algorithm improvement work done in this paper is based on the defect described above: "The unsupervised training of the target domain model classifier has insufficient classification ability, so the traditional pseudo-labeling generation strategy is flawed". The traditional pseudo-label generation strategy is to input the classification probability result P_t into the classifier after the feature map is classified, whereas the inflated corrosive label and smooth pseudo-label generation algorithm is different, after the target domain data is classified by the classifier to obtain the classification result P_t, instead of inputting the classification result P_t directly into the domain classification module D, P_t is firstly subjected to a threshold binarization in order to generate the masks. The classification result P_t will be a tensor of the same shape as P_t after threshold binarization. If P_t is graphed, certain object categories can be seen as outlines, erosion for categories that can be classified better to regularize; expansion for categories that are poorly classified to aid in training; and closure operations for categories that are prone to noise in large areas such as roads. Obviously, after the binarization of the classification results, the expansion location is filled with 1, and the corrosion is filled with 0. The opening and closing operations are the combination of the expansion and corrosion operations before and after.

Depending on the value r_t of the pixel position $Position_{(x,y)}$, there will be different operations, $r_t = 1$ will assign the average value P_{cavg} of the current class M = 1, P_{cavg} if it is not a pixel position if it is not the maximum value of the current pixel position $Position_{(x,y)}$, a certain degree of reduction of P_{cavg}, such as self-subtraction 0.02; $r_t = -1$, the classification probability is the same as above, with a certain degree of reduction, such as self-subtraction 0.02; $r_t = 0$ will not be processed, and finally the generated pseudo-profile will be obtained. Such as self-subtraction of 0.02; $r_t = -1$, the classification probability is the same as above, a certain degree of reduction, such as self-subtraction of 0.02; $r_t = 0$ is not processed, and finally the generated pseudo-label $\hat{y}$ is obtained, and the formula expression can be seen below:

Algorithm 1: EOPL

Input: classification probability $P_t = [P_1, P_2, ..., P_N]^T$, of the output image after inputting the target domain dataset into the source domain model, mask map $M_t = [M_1, M_2, ..., M_N]^T$ after binarization of P_t, data batch size of one training session B, a certain class in the category C, the training tensor is （B,C,W,H）, W,H are the width and height of P_t

Output: A pseudo-labeled map $\hat{y}$ with a different probability and range than y_t

$B_{best} \leftarrow$ findBestBinaryparame() // Parameter thresholds for which binarization work

$M_{best} \leftarrow$ expand_rode_label(M) // Masks after an inflationary erosion operation on the mask M

$B_{best}, M_{best} \leftarrow$ Paremeter_Initial() // Initialization parameters

for i = 0; i ≤ B; i ← i + 1 **do**

  for j = 0; j ≤ C; j ← j + 1 **do**

    compute M_{best} //Equation 11

    if M_{best} -M=1 **then**

    Updated pseudo probability P   //Equation 12

    if M_{best} -M=-1 **then**

    Updated pseudo probability P   // Equation 13

    Smooth Pseudo Label $\hat{y}$ // Equation 14

  end

 end

end

---

4 Experiment

The implementation strategies of domain adaptive experiments are usually as follows: first, train the network on the source domain model, and then load the source domain model weight file into the target domain model for domain classification confrontation training. The abscissa is the number of iterations selected by the model, and the ordinate is the evaluation index mIoU. Only source domain data participate in training, and mIoU is 0.3388. According to Fig. 2, the soft label smoothing strategy can increase by 10–20% in individual categories (the only-src used as a reference) Fig. 3.

In this experiment, the best result of 40,000 iterations is 46.32%. After using the soft label smoothing strategy, the best result of FFADA network in the same equipment and the same variable is 47.88%, which is 0.9% higher than the result claimed by the FADA network, and nearly 1.5% higher than the FADA model. In terms of speed, when the number of iterations is set to 24,000, then FADA-EOPL network can achieve the highest accuracy of the model, which is almost twice the speed of 40,000 iterations designed

by FADA network. According to the Fig. 2, the oscillation problem of mIoU in FADA network during the training coincides with the oscillation problem that described in TransDA model, and FADA-EOPL can effectively solve the oscillation problem Table 1
.

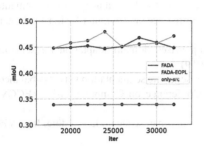

Fig. 2. FADA network mIoU evaluation metrics

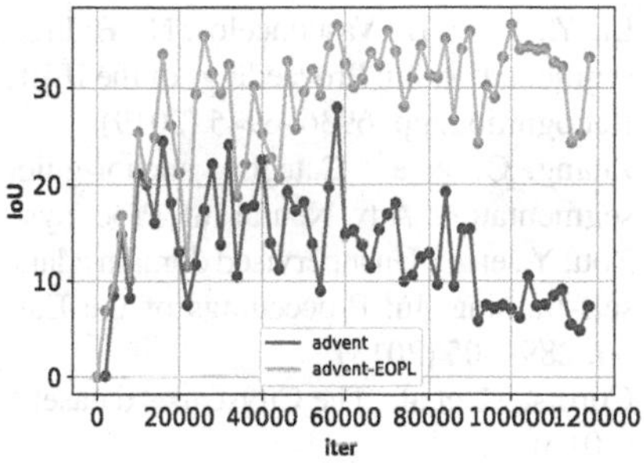

Fig. 3. IoU metrics for ADVENTG bicycle class

Table 1. SOTA Model Comparison

Model	mIoU	iter	bicycle	train	rider	sign	person
FADA	46.32	40000	28.33	23.80	33.80	17.96	58.57
Expand-smooth-FADA	47.88	24000	30.62	26.03	34.25	18.68	58.91
gain	**+1.56**	**−16000**	**+2.29**	**+2.23**	**+0.45**	**+0.72**	**+0.46**
ADVENT	41.04	70000	27.98	10.4	25.63	12.83	57.98
Expand-smooth-ADVENT	40.29	50000	36.52	13.03	29.86	18.16	59.07
gain	**−0.75**	**−20000**	**+8.54**	**+2.63**	**+4.23**	**+5.33**	**+1.09**

Acknowledgements. This work was supported by Natural Science Foundation of Guangdong Province with No. 2020A1515010784 and with the Key Field Special Project of Guangdong Provincial Department of Education with No.2021ZDZX1029.

References

1. Wang, R., et al.: Medical image segmentation using deep learning: A survey. IET Image Proc. (2020)
2. Pan, S.J., Yang, Q.: A survey on transfer learning. IEEE Trans. Knowl. Data Eng. **22**(10), 1345–1359 (2009)
3. Mirzadeh, S.I., et al.: Improved knowledge distillation via teacher assistant. Proc. AAAI Conf. Artif. Intell. **34**(04), 5191–5198 (2020)

4. Wang, D., et al.: Tent: fully test-time adaptation by entropy minimization. In: International Conference on Learning Representations (2021)

5. Goyal, S., et al.: Test-time adaptation via conjugate pseudo-labels. arXiv preprint arXiv:2207. 09640 (2022)

6. Chen, Y., et al.: Domain adaptive faster R-CNN for object detection in the wild. In: Proceedings of the IEEE Conference on Computer Vision and Pattern Recognition, pp. 3339–3348 (2018)

7. Li, Y., Yuan, L., Vasconcelos, N.: Bidirectional learning for domain adaptation of semantic segmentation. In: Proceedings of the IEEE/CVF Conference on Computer Vision and Pattern Recognition, pp. 6936–6945 (2019)

8. Zhang, Q., et al.: Category anchor-guided unsupervised domain adaptation for semantic segmentation. Adv. Neural Inf. Proc. Syst. **32** (2019)

9. Zou, Y., et al.: Unsupervised domain adaptation for semantic segmentation via class-balanced self-training. In: Proceedings of the European Conference on Computer Vision (ECCV), pp. 289–305 (2018)

10. Cordts, M. et al.: The Cityscapes dataset for semantic urban scene understanding. In: CVPR (2016)

11. Richter, S.R., Vineet, V., Roth, S., Koltun, V.: Playing for data: ground truth from computer games. In: Leibe, B., Matas, J., Sebe, N., Welling, M. (eds.) European Conference on Computer Vision (ECCV). LNCS, vol. 9906, pp. 102–118. Springer International Publishing (2016). https://doi.org/10.1007/978-3-319-46475-6_7

12. Wang, H., et al.: Classes matter: a fine-grained adversarial approach to cross-domain semantic segmentation. In: European Conference on Computer Vision, pp. 642-659. Springer, Cham (2020).https://doi.org/10.1007/978-3-030-58568-6_38

13. Ben-David, S., Blitzer, J., Crammer, K., et al.: A theory of learning from different domains. Mach. Learn. **79**(1), 151–175 (2010)

14. Vu, T.H., Jain, H., Bucher, M., Cord, M., Pˊerez, P.: Advent: adversarial entropy minimization for domain adaptation in semantic segmentation. In: CVPR (2019)

15. Liu Z, et al.: Swin transformer: hierarchical vision transformer using shifted windows. In: Proceedings of the IEEE/CVF International Conference on Computer Vision, pp. 10012–10022 (2021)

16. Wang, W., et al.: Internimage: exploring large-scale vision foundation models with deformable convolutions. In: Proceedings of the IEEE/CVF Conference on Computer Vision and Pattern Recognition, pp. 14408–14419 (2023)

17. Zhang, J., Zhang, X., Zhang, Y., et al.: Meta-knowledge learning and domain adaptation for unseen background subtraction. IEEE Trans. Image Process.Process. **30**, 9058–9068 (2021)

Visualization Analysis of Convolutional Neural Network Processes

Hui Wang, Tie Cai, Yong Wei[✉], and Zeming Chen

Shenzhen Institute of Information and Technology, Shenzhen 518172, China
46536895@qq.com

Abstract. Convolutional neural network visualization can help deep learning researchers better understand the concepts and principles of convolution. However, the existed analysis methods have some shortness, such as input and internal features of the model. Through visualization, researchers can have a clearer view of each step in convolution operations, including input, convolution kernel, convolution operation, and output, thereby better understanding the essence and function of convolution is needed. So, we propose a visualization analysis methods based on the random meaning perturbation (VARMP). The proposed method adds the random meaning perturbation to input image, then we can find the minimum deleting mask. From the experiment results, we can find that the visualization analysis results show the real train process of convolutional neural network.

Keywords: Convolutional visualization · optimizing convolutional neural networks · convolution kernel

1 Introduction

The performance of neural network models is becoming increasingly powerful and widely used to solve various computer related tasks, demonstrating excellent capabilities. However, humans do not fully understand the operating mechanism of neural network models. In recent years, artificial intelligence (AI) has become one of the most important scientific research fields with enormous social influence. AI technology has been widely applied in various fields [1, 2]. With the development of scalable high-performance infrastructure, AI systems have become Indispensable tools in many fields and even surpassed human levels in completing more and more complex tasks [3, 4]. However, the excellent performance of AI systems in prediction, recommendation, and decision support is often achieved through the use of complex neural network models that hide the logic of internal processes, which are commonly referred to as black box models [5–7] Neural network models approximate the relationships between variables in the dataset through nonlinear, non monotonic, and non polynomial functions, which makes the internal operating principles highly opaque Neural network models often obtain correct prediction results in the training set due to errors, resulting in excellent performance in training but poor performance in practice [8–11] Therefore, the black box nature of neural networks makes it difficult for humans to fully trust the decisions of neural network models.

© The Author(s), under exclusive license to Springer Nature Singapore Pte Ltd. 2024
K. Li and Y. Liu (Eds.): ISICA 2023, CCIS 2147, pp. 135–141, 2024.
https://doi.org/10.1007/978-981-97-4396-4_12

The current interpretable research methods are complex and diverse. Sorting out different interpretable methods and classifying them is an essential task. Many literature have proposed classification methods for neural network interpretability problems from different perspectives. Reference summarizes the dimensions of defining interpretive algorithms: global or local interpretability: the model may be completely interpretable or only a single decision may be interpretable; Time limit: The time that users are free or allowed to spend understanding and explaining; The nature of user expertise: Users who use the model may have different background knowledge and experience. The classification method proposed in reference for the interpretation method of black box models is based on characteristics such as the type of problem to be solved, the type of interpretation method, the type of black box model, and the type of input data. In reference, interpretation methods are divided into feature-oriented interpretation methods, global feature-based interpretation methods, conceptual model interpretation methods, surrogate model interpretation methods, local pixel based interpretation methods, and human centered interpretation methods. Reference [1] proposes a suggested classification based on the type of explanation returned by the explanation method and the format of the data being analyzed and using the fidelity, stability, robustness, and running time of the explanatory model as evaluation indicators, a portion of the explanatory methods were selected for quantitative comparison. Reference focuses on the purpose of creating explanatory methods and the ways to achieve this goal, and summarizes interpretable methods into four categories: Methods for explaining complex black box models, methods for creating white box models, methods for promoting fairness and limiting discrimination, and methods for analyzing the sensitivity of model predictions. Reference also divides explanatory algorithms into three categories based on their purpose: analog data processing is used to establish connections between the input and output of the system; Used to explain the representation of internal network data; Used to explain the generation network.

Human beings have a willingness to further understand neural network models For models with poorer decision-making abilities than humans, it is hoped that after a deep understanding of the model, problems can be identified and solved, thereby helping the model improve performance For models with similar decision-making abilities to humans, it is hoped that the decision results can be explained, so that humans can trust the model and apply it For models with better decision-making abilities than humans, it is hoped that their decision-making mechanisms can be analyzed to help humans better and more deeply understand the problems that need to be solved. Explainable AI (XAI) [12] research aims to interpret artificial intelligence models in a way that is understandable to humans [13], enabling humans to understand the internal operational logic and decision results of models, providing convenience for model troubleshooting and widespread use The research on visualizing and interpreting neural network models has attracted increasing attention In 2018, the European Parliament introduced provisions on automated decision-making in the General Data Protection Regulation (GDPR), stipulating that data subjects have the right to access relevant explanatory information involved in automated decision-making In addition, in 2019, the Advanced Expert Group on Artificial Intelligence proposed ethical guidelines for trustworthy artificial intelligence Although there are different opinions on these provisions in the law [14, 15], there is a general agreement on the necessity and urgency of implementing such a principle The National Institute of Standards and Technology (NIST) of the United States released four principles on XAI in August 2020 [12]: provability (explanatory results can be proven by evidence), usability (explanatory results can be understood by the model's users and meaningful to them), accuracy (explanatory results must accurately reflect the model's operating mechanism) Limitations (explaining the results can identify situations that are not suitable for its own operation). So, we propose a visualization analysis methods based on the random meaning perturbation. The proposed method adds the random meaning perturbation to input image, then we can find the minimum deleting mask.

2 Visualization Method Based on the Random Meaning Perturbation

With the development of interpretability research, there have been many literature discussions and summaries on the interpretability of neural networks Some literature provides detailed discussions on key issues in interpretability research reference focusing on explaining the issue of data representation within networks explored the reasons why specific inputs lead to specific outputs, the information contained within the network itself, and proposed two approaches to explain the neural network process: approximating the original model by creating a surrogate model, or highlighting the most relevant small portion of calculations by creating a saliency map Reference focuses on distinguishing different definitions of interpretability conceptually, abstracting from the exact neural network structure and application fields, and outlines the techniques for explaining deep neural network models reference outlines different types of explanatory methods and comments on their practicality in practice, starting from the differences in explanatory models in explaining content, audience, and purpose. Reference discusses a series of key technical challenges faced in interpretable machine learning from different perspectives, including a series of classic problems in interpretable machine learning, such as the challenge of constructing sparse models for tabular data, the challenge of additive models. The challenges of case-based reasoning, supervised and unsupervised problem-solving, dimensionality reduction, interpretable reinforcement learning, etc., and provide analysis of interpretable techniques corresponding to different challenges reference focuses on the issue of explaining the vulnerability and robustness of neural network models in adversarial attacks. By utilizing perturbations in the input, the loss results are visualized for qualitative analysis. At the same time, quantitative indicators are proposed to evaluate the inherent robustness of the interpretive definition model, and the model is quantitatively analyzed.

The most typical method of interpreting based on ideal samples is to find a representative input sample by maximizing the activation value of a neuron, channel, or layer. This method is called activation maximization (AM). The activation maximization method is an optimization method originally used in unsupervised networks. Reference first applied this method to convolutional neural networks to solve deep network visualization problems in image classification tasks in reference, a method for visualizing a model to specify a class is introduced. Given a convolutional neural network and a class, a sample of the selected class with the most interest is generated in the input space The specific method is to target the neurons representing the specified class c in the fully connected classification layer of the convolutional neural network, with an activation value of $S_c(I)$ for class c. A random input sample I is used to activate the neurons. During the backpropagation process, the network weight is kept unchanged, and the input sample I is iteratively optimized to obtain the input sample that maximizes the activation value Sc (I) for the specified class of neurons, Thus obtaining the features learned by the specified category in the convolutional neural network In formula (1), λ Represents a regular term parameter. This visualization result can output the features learned by the network in different categories:

$$\arg\max_I S_c(I) - \lambda \|I\|_2^2 \tag{1}$$

The activation maximization method can be generalized to a more general approach, thereby obtaining ideal samples of any neuron in the network. For any neuron i in the network, find an optimal input sample x^* to maximize the activation function $a_i(x^*)$ of the neuron, that is:

$$x^* = \arg\max_x (a_i(x) - R_\theta(x)) \tag{2}$$

3 Experiment Results Analysis

An advantage of the proposed method is that the generated visualizations are clearly interpretable. For example, the deletion game produces a minimal mask that prevents the network from recognizing the object. We used the Citrus Huanglongbing data sets as experimental data set.

It provides another optimization approach for ideal samples, which applies image blur operators and image deblurring operators to generate ideal samples. These two types of operators are achieved by using Gaussian low-pass filters for convolution and deconvolution operations. Image blur operation is mainly used to filter out high-frequency noise, while image deblurring operation is mainly used to offset the blur caused by image blur operation. This algorithm can better extract detailed information from the background and foreground. Reference applied this algorithm to generate images of interest for convolutional layer filters in different networks, proving that the optimized generated images are more interpretable. The method results are shown in Fig. 1. For generating images of interest using different filters in different convolutional layers in the VGG network, it can be seen that the features extracted by different filters are different. In order to obtain a generated image that is closer to the original image, reference added an image generator network for synthesizing images. By continuously optimizing the generator input to maximize the activation value of specified neurons in the network, the output image of the generator, which is the image of interest to the specified neuron, is obtained.

The commonality of interpretation methods based on ideal samples is that these methods focus on showcasing the features learned by network units in the model, that is, generating ideal samples that activate the network units the most. The visualization effect varies for different interpretation methods based on ideal samples. This type of

Fig. 1. The analysis results of VARMP.

research typically aims to achieve better visualization results, with research directions including reducing sample noise and generating samples with semantic information.

The advantage of the interpretation method based on ideal samples is that the basic principle is simple, it can display the features learned by the model network units, to a certain extent, it restores the operating principle of the network, and is similar to the way humans perceive neural networks, so the entire process is easy to understand. The drawbacks of this type of method are that it is difficult to construct the objective function for a specific network, the iterative optimization process is not easy, and the backprop-agation optimization process has information loss issues. Moreover, the ideal samples obtained through such methods often lack clear semantic information and are difficult to match with the way humans perceive things, resulting in weaker interpretability of the generated samples. This type of method is used for understanding and studying network units, and usually has a good explanatory effect on relatively simple network structures.

4 Conclusion

This article provides a detailed discussion on the interpretability of CNN, including the definition and necessity of research on model interpretability, representative research on model interpretability and related classification algorithms, classification methods of new interpretable. The proposed method adds the random meaning perturbation to input image, then we can find the minimum deleting mask. The current research on models that can explain this topic is still in a relatively early stage, and the willingness of humans to understand neural network models is still high. We hope to achieve more intelligent, understandable, and transparent interpretable algorithms in future research.

Acknowledgement. This paper is This work was supported by by Guangdong Basic and Applied Basic Research Foundation under Grant No. 2022A1515011447, National Natural Science Foundation Youth Fund Project of China under Grant No. 62203310, the Shenzhen Fun damental Research fund under No. Grant JCYJ20190808100203577, the Shenzhen Fun damental Research fund under No. Grant 0220820010535001 and Shenzhen Institute of Information Technology Key Laboratory Project under Grant No. SZIIT2023KJ005.

References

1. Maaten, L., Postma, E., Herik, J.: Dimensionality reduction: a comparative review. Review. Lit. Arts Am. **10**(1) (2009)
2. Bodria, F., Giannotti, F., Guidotti, R., Naretto, F., Pedreschi, D., Rinzivillo, S.: Benchmarking and survey of explanation methods for black box models. arXiv preprint arXiv:2102.13076, (2021)
3. Kong, X.W., Tang, X.Z., Wang, Z.M.: A survey of explainable artificial intelligence decision. Syst. Eng. Theor. Pract. **41**(2), 524–536 (2021) (in Chinese). https://doi.org/10.12011/SET P2020-1536]
4. Goyal, Y., Wu, Z.Y., Ernst, J., Batra, D., Parikh, D., Lee, S.: Counterfactual visual explanations. In: Proceedings of the 36th International Conference on Machine Learning, pp. 2376–2384. Long Beach: PMLR (2019)
5. Wang, Y.L., Su, H., Zhang, B., Hu, X.L.: Interpret neural networks by identifying critical data routing paths. In: Proceedings of the 2018 IEEE/CVF Conference on Computer Vision and Pattern Recognition, Salt Lake City, pp. 8906–8914. IEEE(2018). https://doi.org/10.1109/CVPR.2018.00928]
6. Pasquale, F.: The black box society: the secret algorithms that control money and information. Bus. Ethics Q. **26**(4), 568–571 (2016). https://doi.org/10.1017/beq.2016.50
7. Rudin, C.: Stop explaining black box machine learning models for high stakes decisions and use interpretable models instead. NatureMachine Intell. **1**(5), 206–215 (2019). https://doi.org/10.1038/s42256-019-0048-x
8. Su, J.M., Liu, H.F., Xiang, F.T., Wu, J.Z., Yuan, X.S.: Survey of interpretation methods for deep neural networks. Comput. Eng. **46**(9), 1–15 (2020) (in Chinese). https://doi.org/10.19678/j.issn.1000-3428.0057951]

9. Schramowski, P., et al.: Making deep neural networks right for the right scientific reasons by interacting with their explanations. Nat. Mach. Intell. **2**(8), 476–486 (2020). https://doi.org/10.1038/s42256-020-0212-3

10. Zech, J.R., Badgeley, M.A., Liu, M., Costa, A.B., Titano, J.J., Oermann, E.K.: Variable generalization performance of a deep learning model todetect pneumonia in chest radiographs: a cross-sectional study. PLoS Med. **15**(11), e1002683 (2018). https://doi.org/10.1371/journal.pmed.1002683

11. Badgeley, M.A., et al.: Deeplearning predicts hip fracture using confounding patient and healthcare variables. NPJ Digit. Med. **2**(1), 31 (2019). https://doi.org/10.1038/s41746-019-0105-1

12. Hamamoto, R.: Application of artificialintelligence technology in oncology: towards the establishment of precision medicine. Cancers **12**(12), 3532 (2020). https://doi.org/10.3390/cancers12123532]

13. Comandè, G.: Regulating Algorithms' Regulation? First Ethico-Legal Principles, Problems, and Opportunities of Algorithms. In: Cerquitelli, T., Quercia, D., Pasquale, F. (eds.) Transparent Data Mining for Big and Small Data, pp. 169–206. Springer International Publishing, Cham (2017). https://doi.org/10.1007/978-3-319-54024-5_8

14. Wachter, S., Mittelstadt, B., Floridi, L.: Why a right to explanation of automated decision-making does not exist in the general dataprotection regulation. Int. Data Priv. Law **7**(2), 76–99 (2017). https://doi.org/10.1093/idpl/ipx005

15. Lipton, Z.C.: The mythos of model interpretability: in machine learning, the concept of interpretability is both important and slippery. Queue **16**(3), 31–57 (2018). https://doi.org/10.1145/3236386.3241340

Packet Performance Predictor Based on Graph Isomorphism Network for Neural Architecture Search

Yue Liu$^{(\boxtimes)}$, Jiawang Li, Zitu Liu, and Wenjie Tian

School of Computer Engineering, Science, Shanghai University, Shanghai, China
yueliu@shu.edu.cn

Abstract. Architecture performance predictor is an important way to evaluate an intermediate neural architecture to improve the efficiency of Neural Architecture Search (NAS). However, it is difficult to learn cluster distribution and isomorphism of architecture representation of the neural network which is very important for the architecture performance prediction. This paper proposes a method named GIN-P^3 (Graph Isomorphism Network-based Packet Performance Predictor) to accurately predict the performance of intermediate architectures in NAS. Firstly, we design a clustering-based architecture grouping strategy, which clusters architectures with different characteristics to different groups by using the latent space representation of the architectures based on the variational autoencoder. Secondly, the isomorphism characteristics of each architecture group are learned through Graph Isomorphism Network. Following this, the architecture performance predictor is further constructed and embedded in the NAS based on reinforcement learning through the direct prediction of the architectures sampling by LSTM to improve the efficiency of architecture search. The experiments conducted on the NAS-Bench-101and NAS-Bench-201 show that compared with baseline methods including support vector regression, peephole, gradient-boosted decision trees, and RNAS, GIN-P^3 can obtain a higher prediction correlation in performance prediction; further, it can find relatively high-performance neural architecture more quickly.

Keywords: Neural Architecture Search · Performance Predictor · Graph Isomorphism Network

1 Introduction

Designing high-performance neural network architectures for specific datasets remains a challenging and time-consuming task, requiring trial and error experiments and domain knowledge. Neural Architecture Search (NAS) [4] has emerged as a promising approach to automate this process and has been successfully applied in tasks like image classification [1], object detection [2], and language modeling [3]. However, NAS algorithms are computationally expensive due to the evaluation of deep neural networks (DNNs) during the search. To address this problem, researchers propose performance predictors to estimate DNN fitness without training, reducing computational costs. Constructing performance predictors faces two main challenges: how to effectively determine the performance of neural network architecture and how to represent the relationship between architecture and performance in performance predictors.

To effectively determine the performance of neural network architectures, some researchers have attempted to convert network architectures into corresponding representations, for input into the performance predictors. However, discrete representations lead to excessively high dimensionality of the search space, thereby increasing architectural redundancy and optimization difficulty. Researchers are trying to map discrete representations of network architectures to continuous spaces [8, 16, 18]. In [8], an autoencoder is applied for encoding neural network architectures into continuous representations; meanwhile, the gradient ascent is used to maximize predictors and discover new high-performance architectures. The architecture representation can be thought of as a string of coding, which are regarded as the DNA of the neural architecture. However, due to the uniqueness of architecturally encoded representations, there inevitably exists architectures with the same adjacency matrix or same operations matrix but different representations in the search space. For an accurate prediction of the neural network performance, it is necessary to consider the Isomorphism characteristics of the architecture in the building of the performance predictor.

To represent the relationship between the architecture and performance, some researchers [5, 7] built performance predictors to estimate architecture performance from an encoding of the architecture, thereby predicting the final performance of the neural network architecture directly. The ability of neural predictors to accurately predict the performance of neural architectures is critical to search strategies using neural predictors [24, 25]. DARTS [6] employs continuous relaxation to achieve gradient-based optimization, where the graph is represented by a convex combination of the possibilities of all edges. Existing performance predictor construction methods only build a single-agent model to predict the performance of neural network architectures. The effectiveness of the predictors can be further improved if performance predictors are built with different architectures for different neural networks to learn different correlations.

Therefore, in this paper, we propose a method named GIN-P^3 (Graph Isomorphism Network-based Packet Performance Predictor) that investigates the isomorphism and data distribution of neural architecture in performance predictor construction. GIN-P^3 can accurately predict the performance of intermediate architectures in NAS. The main contributions of this work are as follows:

- To distinguish the relationship between different architectures and performance, we propose a Clustering based Architecture Packet Grouping Strategy, which improves the performance prediction accuracy by grouping neural network architectures with different characteristics.
- To solve the isomorphism of architecture representation in neural network performance prediction, we propose a performance predictor construction method based on graph isomorphism network, which learns the isomorphism features of the architecture.

- Several experiments are executed on NAS-banch-101 and NAS-banch-201. The experimental results show that the proposed method outperforms the state-of-the-art models on NAS benchmarks and more quickly finds relatively high-performance neural architecture.

The rest of the paper is structured as follows: Sect. 2 introduces related works in the field of NAS. The details of GIN-P^3 are presented in Sect. 3. Subsequently, Sect. 4 discusses the details of our experimental setup and the results of our experimentation. Finally, in Sect. 5, we provide our conclusions.

2 Related Works

The architecture representation effectively reduces the time consumption in NAS. Recently, researchers have attempted to map discrete representations of network architectures to continuous spaces. For example, Li et al. [17] used a variational graph autoencoder to learn a continuous representation space for the architecture and build a performance predictor in this space. Yan et al. [19] employed a variational autoencoder to learn unsupervised representations of neural network architectures. Cheng et al. [26] preserved graph correlation information by adding similarity loss to capture the graph topology information. These different representation methods encoded the architectures of neural networks into the latent spaces. However, similarity loss ignores the relationship between representation and architecture, resulting in the same adjacency matrix or same operations matrix but different representations in the search space. Thus, we propose a new Architecture Packet Grouping Strategy to identify architecture structures and more efficiently predict architecture performance.

Based on the architecture presentation, performance predictors are typically developed to predict the performance of DNNs by avoiding the time-consuming training process. At present, existing performance predictors can be generally classified into three categories: shallow training strategy-based performance predictors [2, 9], learning curves–based performance predictors [10, 11], and end-to-end performance predictors [12−14, 23, 27]. The representation of architecture is fed into these performance predictors, which then forecast the performance of the architecture. Deng et al. [17] employed a multilayer perceptron as end-to-end performance predictors to predict the performance of the architecture. Xu et al. [16] proposed an RNAS performance predictor to rank different neural network architectures to discover promising architectures. However, performance predictors predict performance according to different architectural representations, but different representations may be produced by isomorphism neural network architectures. To distinguish the relationship between different architectures and performance, this work uses different performance predictors to predict the performance of the architecture groups with similar characteristics.

3 Methods

This study investigates the isomorphism and data distribution of neural architecture in performance predictor construction and proposes a method named GIN-P^3, which is CAPGS (Clustering based Architecture Packet Grouping Strategy) and GIN-PP (Graph

Isomorphism Network–based Performance Predictor). Then, it further applies reinforcement learning (RL) based NAS to design high-performance neural architectures effectively. The process of GIN-P^3 is shown in Fig. 1 to use the cluster distribution of neural architectures, CAPGS employs variational autoencoders (VAE) to learn the latent space representation of architectures, then groups the architectures in the latent space based on the K-Means method (Sect. 3.1). Moreover, GIN-PP learns the isomorphism characteristics of each group based on the graph isomorphism network (GIN) and constructs the performance predictor by minimizing the difference between the predicted and the real performance (Sect. 3.2). Finally, to efficiently design high-performance neural architecture, GIN-P^3 is embedded in RL-based NAS (Sect. 3.3). It uses LSTM as the controller to sample sub-architecture in the search space and predicts the performance using GIN-P^3 as feedback to RL to update the LSTM parameters, thereby improving the search efficiency and effectiveness of NAS.

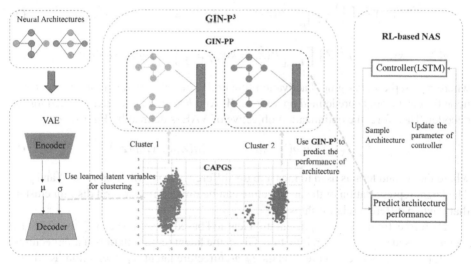

Fig. 1. The process of GIN-P3.

3.1 Clustering Based Architecture Packet Grouping Strategy

The neural architecture search space can be viewed as a directed acyclic graph consisting of N nodes and E edges, where each node represents an operation in a neural network (e.g., convolution and pooling). Therefore, the architecture can be viewed as an upper triangular adjacency matrix $A \in R^{N \times N}$ representing the connection relationship between nodes and a one-hot operation matrix $X \in R^{N \times O}$ represent the type of node operation, where O is the number of operation types that a node can choose from.

To reduce the redundancy of discrete architecture representation and to learn the intrinsic properties of architecture, VAE is first employed to learn the latent space representation of architecture. The GIN-P^3 is based on the unsupervised representation learning method proposed by Arch2vec [19]. It learns the latent representation by an encoder, which is defined by Eq. (1).

$$q(Z|X,\tilde{A}) = \prod_{i=1}^{n} q(z_i|X,\tilde{A}) \tag{1}$$

$$q(z_i|X,\tilde{A}) = \mathcal{N}\left(z_i|\mu_i, diag\left(\sigma_i^2\right)\right) \tag{2}$$

where μ and σ represent the mean and variance variables in the latent space learned by the VAE. z is obtained from the reparameterization technique proposed by Kingma et al. [20]. $\tilde{A} = A + A^T$ represents the transformation of the originally directed neural network into undirected. The goal of the decoder is to reconstruct the adjacency matrix $\tilde{A}$ and node operation matrix based on latent variables Z, which are defined in Eq. (3) and Eq. (4).

$$P(\tilde{A}|Z) = \prod_{i=1}^{N} \prod_{j=1}^{N} P(\tilde{A}_{ij}|z_i, z_j), P(\tilde{A}_{ij} = 1|z_i, z_j) = \sigma\left(z_i^T z_j\right) \tag{3}$$

$$p\left(\tilde{X} = [k_1,\dots,k_O]^T|Z\right) \prod_{i=1}^{N} P(\tilde{X}_i = k_i|z_i) = \prod_{i=1}^{N} softmax(WZ + b)_{i,k_i} \tag{4}$$

where $\sigma(.)$ represents sigmoid activation function. The softmax function represents the node to select from O predefined operations. W And b are the parameters trained for the decoder. Therefore, the optimization objective of VAE is defined by Eq. (5).

$$E_{q(Z|X,\tilde{A})}\left[\log p(X,\tilde{A}|Z)\right] - KL(q(Z|X,\tilde{A})||p(Z)) \tag{5}$$

where the second term is used to measure the difference in distribution for $q(.)$ and $p(.)$. The latent representation of the architecture can be divided into several disjoint clusters that contain different relationships.

Therefore, for the latent representation learned based on arch2vec [19], the K-Means method is used to group the architecture. Given the latent representation of architectures, $Z = \{z_1 \dots, z_n\}$, where z_i represents the i_{th} architecture latent representation. K-Means distinguishes the cluster center with the minimum distance from l cluster centers, and the distance measurement is defined by Eq. (6).

$$d\left(z_i, c_j\right) = \frac{\sum_{k=1}^{m} z_{ik} \cdot c_{jk}}{\sqrt{\sum_{k=1}^{m} z_{ik}^2} \cdot \sqrt{\sum_{k=1}^{m} c_{jk}^2}} \tag{6}$$

where $d\left(z_i, c_j\right)$ represents the distance between i_{th} architecture and j_{th} architecture of the cluster center. m represents the dimension of the latent space learned by VAE. When architectures have been clustered by K-Means, based on latent representation, the architectural grouping for each cluster covers different architecture characteristics. Hence, GIN [15] is employed to construct performance predictors for architectures in each cluster, thereby improving the effectiveness of the performance predictor.

3.2 Performance Predictor Based on Graph Isomorphism Network

The goal of a graph neural network is to learn the representation of graph structure data and node features. Here, the operation matrix of architecture is used to represent the node feature of the graph neural network, whereas the adjacency matrix is used to represent the connection of adjacent nodes. Graph neural network based on neighborhood aggregation can be divided into three parts: aggregate (aggregating first-order neighborhood features), combine (combining the features of neighborhood aggregation with the current features to update the current node), and readout (converting the features of all nodes into the graph features for graph classification or regression tasks). Therefore, a layer graph neural network can be represented with Eq. (7) and Eq. (8).

$$x_v^{(k)} = AGGREGATE^{(k)}\left(\left\{X_u^{(k-1)} : u \in A(v)\right\}\right), X_v^{(1)} = X_v \qquad (7)$$

$$X_v^{(k)} = COMBINE^{(k)}\left(X_v^{(k-1)}, x_v^{(k)}\right) \qquad (8)$$

where $x_v^{(k)}$ represents the neighborhood feature of the architecture nodes after the aggregation of k_{th} layer neural network. $A(v)$ represents the set of nodes adjacent to the node v, $X_v^{(k)}$ represents the feature of the current node v of k_{th} layer, and X_v represents the operation matrix of the node v. Xu et al. [15] proposed a GIN to solve the problem of graph isomorphism where the aggregate function and the readout function are both injective functions. Therefore, aggregate is set as the sum function, and combine is set as $1 + \varepsilon$. .. It also introduces a multilayer perceptron (MLP) to guarantee injective property. The GIN framework based on MLP and SUM is defined in Eq. (9).

$$X_v^{(k)} = MLP^{(k)}\left(\left(1 + \varepsilon^{(k)}\right) \cdot X_v^{(k-1)} + \sum_{u \in A(v)} X_u^{(k-1)}\right) \qquad (9)$$

where ε is the learnable parameter in GIN. Because the performance prediction aims to complete the graph-level regression task, it is necessary to use the readout function to convert the node features into the graph features. It sums up all nodes features in each iteration for graph features and then connects them. Therefore, the graph features can be defined by Eq. (10).

$$X_G = CONCAT\left(\left\{\sum_{v \in G} X_v^{(k)} | k = 0, ..., K\right\}\right) \qquad (10)$$

where K represents the number of layers of GIN and G represents all nodes in the neural architecture.

To solve the problem of isomorphism in neural architecture, a performance predictor is constructed based on GIN. It minimizes the mean square error between the predicted performance and the real performance, and its objective function can be defined by Eq. (11).

$$\min_{W_P} \frac{1}{n} \sum_{i=1}^n \left[PP(A_i, X_i) - y_i\right]^2 \qquad (11)$$

where PP represents the performance predictor. When the feature vector X_G of neural architecture has been learned by GIN, MLP is used to learn the relationship between architecture and performance, that is, $PP(A, X) = MLP(X_G)$. W_P is the learned parameter of the performance predictor. y_i represents the real performance of neural architecture (A_i, X_i), and n is the number of training data for performance prediction.

3.3 Neural Architecture Search Jointly by GIN-P^3 and Reinforcement Learning

To further improve the efficiency of the RL-based NAS, we use the performance predictor constructed by GIN-P^3 into neural architecture search as a performance evaluator of neural networks. Specifically, based on the search space representation of Arch2vec [19], an architecture search is conducted in the learned latent space. It uses LSTM as a controller to generate variables in latent space and reconstruct them into neural architecture through a decoder. Moreover, reinforcement learning is adopted as a search strategy to update the parameters of the controller. The objective function is defined by Eq. (12).

$$\nabla_\theta J(\theta) = \nabla_\theta \frac{1}{T} \sum\nolimits_{t=1}^{T} \left[\log \pi_\theta((A_t, X_t), z_t) \left(\sum\nolimits_{k=t+1}^{T} \gamma^{k-t-1} PP(A_k, X_k) - b \right) \right]$$

(12)

where θ represents the parameters of LSTM. T represents the number of architectures sampled in one iteration of RL. z_t represents the latent variable of architecture t. $\pi_\theta((A_t, X_t), z_t)$ is the probability of sampling from the multivariate normal distribution with z_t as the mean vector and unit matrix as the covariance matrix. $PP(A_t, X_t)$ represents the predicted performance of GIN-P^3 for architecture t. Here, to reduce the variance of the strategy gradient and ensure the unbiasedness of gradient direction, a constant b is introduced as the baseline. Additionally, $\gamma \in [0, 1]$ is the discount factor, which indicates that the rewards should be related to previous performance predictions.

For the neural network performance evaluation, GIN-P^3 is adapted to directly predict the sampled neural architectures. It calculates the distance between the current architecture and each cluster center based on Eq. (6). Subsequently, it uses the corresponding predictor to predict the performance of architectures per the nearest cluster. This is shown in Eq. (13).

$$\lambda = argmin_{i \in \{1,...,k\}} d(z, c_i)$$

(13)

where λ represents the cluster of the current architectures. Here, through GIN-P^3, the performance feedback can be obtained directly without architecture training, thereby improving the efficiency and effectiveness of the NAS to a certain extent while designing better neural network architecture for the actual task.

4 Experiment

4.1 Experimental Datasets

Experiments are conducted on the NAS-Bench-101 [22] and NAS-Bench-201 [28] benchmark to evaluate the effectiveness of GIN-P^3. The NAS-Bench-101 is the first proposed architecture dataset for NAS research. It contains 423,624 different convolutional neural network architectures, which are both trained on CIFAR-10 for image classification from scratch to full convergence to get validation and test accuracy. NAS-Bench-201, with different search spaces, results from CIFAR-100 dataset, and more diagnostic information, contains a total of 15625 different convolutional neural network architectures. In the performance prediction experiment, 0.025% (105), 0.1% (423), 1% (4,236), and 10% (42,362) architectures with performance annotation are sampled from the NAS-Bench-101. And 1% (156), 5% (781), 10% (1,562), and 20% (3,125) architectures with performance annotation are sampled from the NAS-Bench-201. Hold-out validation is used to evaluate the effectiveness of the performance predictor, and 10% of architectures are selected as a test set, which is disjoint with the architectures with annotation.

4.2 Experimental Setting

Evaluation Metrics. Three metrics were employed to measure the prediction results: Kendall's Tau (KTau), Pearson's correlation coefficient (P), and root mean square Error (RMSE). KTau measures the ranking correlation between predicted performance values and actual labels. RMSE measures the deviation of predictions from actual values directly, and P measures the linear correlation between predictions and actuals. The higher values of KTau, P, and the lower value of RMSE indicate a better result of GIN-P^3.

Parameter Setting. To find the optimal number of clusters, we used the CH index to evaluate the effect of clustering. During the experiment, under different amounts of architecture-performance data, the number of clusters was set from 2 to 20 for K-Means clustering. Due to the data distribution of different proportions is different, we set the number of clusters with the largest CH index to train predictors.

4.3 The Effectiveness of Performance Predictors

Tables 1–4 report that the performance of GIN-P^3 is competitive with the compared baselines. Specifically, on the NAS-Bench-101 dataset, the advantage of GNI-P^3 is less obvious when the proportion of architectural performance data is small (0.025% and 0.1%). SVR and GBDT only learn sequential features of layer connections and operation types, so they have advantages in small datasets. Nevertheless, GNI-P^3 has comparable performance to SVR and GBDT when the architecture-performance ratio is 0.1% of the data. As architectural performance data increases (1% and 10%), thereby increasing architectural redundancy and optimization difficulty. Since GIN-P^3 can learn architectural features of isomorphic neural network architectures, its performance is superior. Peephole considers only the size of the convolutional kernel and the number of channels when determining the architecture of a neural network, ignoring the type and connections of layers. As a result, Peephole is unable to find the architecture's intrinsic properties, thereby having a negative effect. Compared with Peehole, RNAS can better learn spatial structure information, but this representation will bring a certain amount of architectural redundancy, and the isomorphism of the architecture cannot be learned. Therefore, the KTau, P, and RMSE of RNAS are all weaker than those of GIN-P^3. In addition, Table 3 and Table 4 show the experimental results of GIN-P^3 on NAS-Bench-201. When the size of architectural performance data (1% and 5%) is small, SVR and GBDT are still competitive method. However, at the 5% ratio, the KTau value of GINP^3 was 0.0598 and 0.0218 higher than that of SVR and GBDT, respectively. Moreover, the P and RMSE of GIN-P3 were 0.8246 and 0.1172 respectively, second only to GBDT. When the size of architectural performance data is 10% and 20%, the KTau values of GIN-P^3 are 0.7660 and 0.8261 respectively, and the performance P and RMSE have achieved obvious advantages compared with other methods. Therefore, it can be concluded that GIN-P^3 has higher effectiveness than the com-pared methods, which can more accurately predict the performance of neural net-work architectures.

Table 1. Comparison results on prediction performance for 0.025%% and 0.1% architectures on the NAS-Bench-101.

Method	0.025% (105)			0.1% (423)		
	KTau	P	RMSE	KTau	P	RMSE
SVR	0.2880 ± 0.0026	0.2549 ± 0.0058	0.0763 ± 2.030e-04	0.4045 ± 0.0031	0.3748 ± 1.381e-03	0.0753 ± 8.774e-06
Peephole [17]	0.1384 ± 0.0424	0.0154 ± 0.0002	0.0687 ± 0.0005	0.2186 ± 0.0394	0.1796 ± 0.0007	0.0574 ± 0.0003
GBDT	**0.4406 ± 2.430e-03**	**0.2675 ± 8.410e-03**	0.0669 ± 1.044e-04	0.4692 ± 5.989e-03	**0.4243 ± 3.766e-04**	**0.0534 ± 1.825e-06**
RNAS [16]	0.2827 ± 0.0017	0.2113 ± 0.0011	0.0655 ± 2.006e-04	0.3392 ± 0.0340	0.2296 ± 0.0170	0.2489 ± 0.0012
GCN [21]	0.1162 ± 0.0052	0.1392 ± 0.0074	**0.0635 ± 0.0002**	0.3201 ± 0.0678	0.3871 ± 0.0881	0.0578 ± 0.0018
GIN-PP	0.1225 ± 0.0734	0.1298 ± 0.0028	0.1515 ± 0.0085	0.3524 ± 0.0868	0.2430 ± 0.0318	0.0748 ± 0.0023
GIN-P^3	0.1444 ± 0.0598	0.1541 ± 0.0007	0.1230 ± 0.0035	**0.5166 ± 0.0396**	0.3889 ± 0.0006	0.0612 ± 0.0044

4.4 The Validation of Neural Architecture Search

To validate the effectiveness of GIN-P^3 in discovering high-performance architectures in the NAS, a comparison with RNAS and GBDT was performed. We selected only 0.025%, 0.1%, and 1% of the architecture-performance data in NAS-Bench-101 to train different performance predictors. The final performance is the result of the optimal architecture determined by RL and performance predictor on the test set. Simultaneously, we gave a wall-clock time budget of 1×10^6 s to perform the architecture search and repeated 10 experiments with different initialization architectures.

Table 2. Comparison results on prediction performance for 1% and 10% architectures on the NAS-Bench-101.

Method	1% (4236)			10% (42362)		
	KTau	P	RMSE	KTau	P	RMSE
SVR	0.5441 ± 0.0004	0.4968 ± 5.748e-05	0.0638 ± 3.517e-06	0.5489 ± 0.0001	0.5469 ± 3.699e-05	0.0573 ± 3.514e-06
Peephole [17]	0.3101 ± 0.0829	0.3326 ± 0.0003	0.0564 ± 0.0006	0.3358 ± 3.195e-3	0.3402 ± 3.544e-4	0.0556 ± 4.528e-5
GBDT	0.6421 ± 1.171e-04	0.5372 ± 8.222e-05	0.0498 ± 3.079e-06	0.6678 ± 1.617e-05	0.5531 ± 2.247e-05	0.0498 ± 2.175e-06
RNAS [16]	0.6120 ± 0.0005	0.4993 ± 0.0004	0.0520 ± 2.350e-06	0.6523 ± 0.0758	0.7032 ± 0.0038	0.0431 ± 2.410e-06
GCN [21]	0.3106 ± 0.0462	0.3464 ± 0.0009	0.0579 ± 0.0001	0.3583 ± 0.0256	0.3643 ± 0.0016	0.0573 ± 6.454e-5
GIN-PP	0.5611 ± 0.0455	0.4682 ± 0.0048	0.0537 ± 4.824e-4	0.6372 ± 0.0407	0.3816 ± 0.0065	0.0451 ± 3.589e-4
GIN-P^3	**0.6526 ± 0.0761**	**0.5912 ± 8.652e-4**	**0.0482 ± 3.524e-4**	**0.6786 ± 0.0325**	**0.7303 ± 0.0064**	**0.0414 ± 2.456e-4**

Table 3. Comparison results on prediction performance for 1% and 5% architectures on the NAS-Bench-201.

Method	1% (156)			5% (781)		
	KTau	P	RMSE	KTau	P	RMSE
SVR	**0.6025 ± 5.674e-04**	0.6972 ± 2.119e-04	0.1447 ± 6.657e-05	0.6982 ± 2.729e-05	0.8031 ± 1.159e-04	0.1207 ± 3.886e-05
Peephole [17]	0.2780 ± 0.0074	0.4056 ± 0.0091	0.1867 ± 1.556e-04	0.4519 ± 0.0001	0.6244 ± 0.0010	0.1569 ± 5.402e-05
GBDT	0.5993 ± 7.459e-04	**0.7436 ± 0.0008**	**0.1350 ± 7.432e-05**	0.6862 ± 1.281e-04	**0.8348 ± 0.0002**	**0.1134 ± 3.468e-05**
RNAS [16]	0.2383 ± 7.690e-04	0.4022 ± 3.452e-03	0.1862 ± 3.736e-05	0.4019 ± 4.344e-04	0.6376 ± 1.274e-03	0.1566 ± 1.359e-05
GCN [21]	0.1259 ± 1.524e-03	0.1685 ± 2.211e-04	0.2045 ± 2.371e-06	0.1865 ± 2.639e-05	0.2159 ± 2.730e-05	0.2014 ± 2.369e-05
GIN-PP	0.4082 ± 0.0007	0.5327 ± 0.0031	0.2000 ± 1.2283e-04	0.5525 ± 0.0005	0.7013 ± 0.0011	0.1540 ± 8.102e-05
GIN-P^3	0.4221 ± 0.0048	0.5429 ± 9.04e-03	0.1960 ± 2.832e-04	**0.7080 ± 0.0012**	0.8246 ± 1.024e-03	0.1172 ± 1.038e-04

Table 4. Comparison results on prediction performance for 10% and 20% architectures on the NAS-Bench-201.

Method	10% (1562)			20% (3125)		
	KTau	P	RMSE	KTau	P	RMSE
SVR	0.7310 ± 4.064e-05	0.8467 ± 9.863e-05	0.1088 ± 1.415e-05	0.7548 ± 4.668e-05	0.8822 ± 4.458e-05	0.0978 ± 3.331e-05
Peephole [17]	0.4743 ± 0.0001	0.6639 ± 0.0004	0.1520 ± 2.720e-05	0.6762 ± 0.0001	0.4881 ± 0.0005	0.1495 ± 2.549e-05
GBDT	0.7176 ± 6.128e-05	0.8576 ± 0.0001	0.1069 ± 5.873e-05	0.7372 ± 7.584e-05	0.8744 ± 0.0001	0.1034 ± 1.893e-05
RNAS [16]	0.6242 ± 2.881e-04	0.8505 ± 3.406e-04	0.1113 ± 8.261e-05	0.7342 ± 7.308e-05	0.9388 ± 2.711e-05	0.0705 ± 1.221e-05
GCN [21]	0.1928 ± 6.092e-04	0.2091 ± 7.152e-04	0.2009 ± 1.090e-06	0.1706 ± 1.973e-03	0.1899 ± 2.491e-03	0.2016 ± 2.891e-06
GIN-PP	0.5696 ± 0.0012	0.7308 ± 0.0017	0.1446 ± 1.204e-04	0.6762 ± 0.0008	0.8508 ± 0.0008	0.1166 ± 1.989e-04
GIN-P^3	**0.7660 ± 0.0020**	**0.9347 ± 3.672e-04**	**0.0749 ± 1.167e-04**	**0.8261 ± 0.0012**	**0.9748 ± 6.432e-05**	**0.0497 ± 9.395e-05**

The experimental results are shown in Table 5, which show the effectiveness of GIN-P^3 in embedding NASs to discover well-performing neural network architectures. The first column represents the performance difference between the search architecture and the optimal network architecture in the current search space and the second column represents the ranking of the network architectures in all search space. For the test regret and rank indicator on different numbers of architectures, GIN-P^3 again wins the highest value. On 0.025% architecture-performance data, the rank in the NAS-Bench-101 search space has improved significantly, from 11.231% (47,577) and 1.171% (5,066) to 0.128% (541), respectively. Compared with the other two methods, our method improves the accuracy of performance prediction by solving the problem of isomorphism of architecture and exploiting the characteristics of class cluster distribution, thus providing more accurate performance feedback for architecture search.

To verify the efficiency of the proposed method in the architecture search process, we compare the time-dependent trends of the architecture performance searched using the three methods on three different amounts of architecture-performance data. In Fig. 2, the search process of RNAS is unstable at three different quantities, indicating that RNAS is biased in the performance prediction of some neural network architectures, resulting in inaccurate performance evaluation of network architectures. Additionally, our proposed method can have a relatively high search performance over time. Therefore, the experimental results show that our method can search well-performing neural network architectures more efficiently than the other two methods, thus further demonstrating the effectiveness of our method.

Table 5. Comparison of different performance predictors for reinforcement learning neural architecture search.

Methods	0.025%(105)		0.1%(423)		1%(4236)	
	Test regret	Rank (%)	Test regret	Rank (%)	Test regret	Rank (%)
RNAS [16]	0.0162 ± 0.0043	11.231	0.0143 ± 0.0030	7.942	0.0137 ± 0.0039	7.332
GBDT	0.0084 ± 0.0019	1.171	0.0046 ± 0.0018	0.066	0.0053 ± 0.0017	0.128
Ours	0.0053 ± 0.0017	0.128	0.0044 ± 0.0012	0.055	0.0041 ± 0.0024	0.038

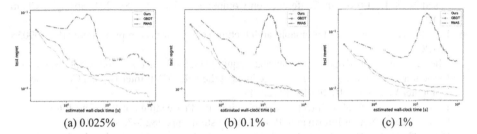

(a) 0.025%   (b) 0.1%   (c) 1%

Fig. 2. Comparison of predictor discovery of architectural performance over search time.

5 Conclusions

To efficiently design high-performance architecture, it is necessary to understand the isomorphism and cluster distribution of a neural architecture to build performance predictors. Therefore, we propose a novel method called GIN-P^3 to accurately predict the performance of intermediate architectures in NAS. Firstly, GIN-P^3 employs K-Means to cluster the architecture latent representation based on the variational autoencoder for classifying the architectures with different characteristics. And then, the isomorphism characteristics of each architecture group are learned through a GIN to further construct the architecture performance predictor. Furthermore, GIN-P^3 is embedded in RL based NAS through the direct prediction of architecture sampling by LSTM. Several experiments conducted on the NAS-Bench-101 and NAS-Bench-201 benchmarks show that GIN-P^3 can obtain comparable, even lower test regret compared with the traditional NAS method. We discovered that building an architectural performance predictor in a divide-and-conquer manner for different neural networks to learn different associations can further improve the effectiveness of the predictor. For future work, the causal approach can be introduced to explain the relationship between architecture and performance and further improve the interpretability of the NAS.

Acknowledgments. This work is supported by the National Key Research and Development Program of China (Grant No. 2021YFB3802101), the Programs of National Natural Science Foundation of China (No. 52073169 and No. 92270124).

References

1. Niyas, S., Pawan, S.J., Kumar, M.A., et al.: Medical image segmentation with 3D convolutional neural networks: a survey. Neurocomputing **493**(7), 397–413 (2022)
2. Mi, J.X., Wang, X.D., Zhou, L.F., et al.: Adversarial examples based on object detection tasks: a survey. Neurocomputing **519**(28), 114–126 (2023)
3. Li, J., et al.: Accelerating neural architecture search for natural language processing with knowledge distillation and earth mover's distance. In: Proceedings of the 44th International ACM SIGIR Conference on Research and Development in Information Retrieval, pp. 2091–2095 (2021)
4. Liu, Y., et al.: A survey on evolutionary neural architecture search. IEEE trans. Neural Netw. Learn. syst. **34**(2), 550–570 (2021)
5. Ning, X., et al.: A generic graph-based neural architecture encoding scheme for predictor-based NAS. In: European Conference on Computer Vision, pp. 189–204. Springer, Cham (2020)
6. Liu, H., et al.: Darts: differentiable architecture search. arXiv preprint arXiv:1806.09055 (2018)
7. Chen, Z., et al.: Not all operations contribute equally: hierarchical operation-adaptive predictor for neural architecture search. In: Proceedings of the IEEE/CVF International Conference on Computer Vision, pp. 10508–10517 (2021)
8. Luo, R., et al.: Neural architecture optimization. In: Proceedings of the 32nd International Conference on Neural Information Processing Systems, pp. 7827–7838 (2018)
9. Pham, H., et al.: Efficient neural architecture search via parameters sharing. In: International Conference on Machine Learning, pp. 4095–4104. PMLR (2018)
10. Yoo, J., et al.: Photorealistic style transfer via wavelet transforms. In: Proceedings of the IEEE/CVF International Conference on Computer Vision, pp. 9036–9045 (2019)
11. Krizhevsky, A., et al.: Imagenet classification with deep convolutional neural networks. Commun. ACM **60**(6), 84–90 (2017)
12. Srivastava, R.K., et al.: Training very deep networks. Advances in neural information processing systems. In: Advances in Neural Information Processing Systems, vol. 28 (2015)
13. Huang, G., et al.: Deep networks with stochastic depth. In: European Conference on Computer Vision, pp. 646–661. Springer, Cham (2016)
14. Iandola, F.N., et al.: SqueezeNet: AlexNet-level accuracy with 50x fewer parameters and< 0.5 MB model size. arXiv preprint arXiv:1602.07360 (2016)
15. Xu, K., et al.: How powerful are graph neural networks? In: International Conference on Learning Representations (2018)
16. Xu, Y., et al.: RNAS: architecture ranking for powerful networks. arXiv preprint arXiv:1910.01523 (2019)
17. Deng, B., et al.: Peephole: predicting network performance before training. arXiv preprint arXiv:1712.03351 (2017)
18. Li, J., et al.: Neural architecture optimization with graph vae. arXiv preprint arXiv:2006.10310 (2020)
19. Yan, S., et al.: Does unsupervised architecture representation learning help neural architecture search? In: Advances in Neural Information Processing Systems, vol. 33, pp. 12486–12498 (2020)
20. Kingma, D.P., Welling, M.: Auto-encoding variational bayes. arXiv preprint arXiv:1312.6114 (2013)
21. Liu, C., et al.: Progressive neural architecture search. In: Proceedings of the European Conference on Computer Vision, pp. 19–34 (2018)

22. Ying, C., et al.: NAS-bench-101: towards reproducible neural architecture search. In: International Conference on Machine Learning, vol. 97, pp. 7105–7114. PMLR (2019)
23. Zoph, B., Le, Q.V.: Neural architecture search with reinforcement learning. arXiv preprint arXiv:1611.01578 (2016)
24. Luo, R., et al.: Semi-supervised neural architecture search. In: Advances in Neural Information Processing Systems, vol. 33 (2020)
25. Dudziak, L., et al.: BRP-NAS: Prediction-based NAS using GCNS. In: Advances in Neural Information Processing Systems, vol. 33, pp. 10480–10490 (2020)
26. Cheng, H. P., et al.: NASGEM: neural architecture search via graph embedding method. In: Proceedings of the AAAI Conference on Artificial Intelligence, vol. 35, pp. 7090–7098 (2021)
27. Huang, M., et al.: Arch-Graph: acyclic architecture relation predictor for task-transferable neural architecture search. In: Proceedings of the IEEE/CVF Conference on Computer Vision and Pattern Recognition, pp.11881–11891 (2022)
28. Dong, X., Yang, Y.: NAS-bench-201: Extending the scope of reproducible neural architecture search. arXiv preprint arXiv:2001.00326 (2020)

Big Data Analysis and Information Security

Reversible Data Hiding Algorithm Based on Adaptive Predictor and Non-uniform Payload Allocation

Dan He[✉]

Dongguan City University, Dongguan 523000, Guangdong, China
hdhzhd@126.com

Abstract. Reversible data hiding is a technique that enables the secure embedding and complete extraction of data without reducing the quality of the carrier image. It has significant application value in fields such as medical images, military images, and digital forensics. However, existing reversible data hiding methods often need clarification on embedding capacity, image quality, and the trade-off between computational complexity and robustness. This paper proposes a reversible data hiding algorithm based on adaptive predictor and non-uniform payload allocation. The algorithm first uses an adaptive predictor to predict the image and then dynamically allocates different embedding bits according to the size and distribution of the prediction error, thus achieving non-uniform payload allocation. The algorithm only changes the low bits of the prediction error when embedding data, thus ensuring the high fidelity of the image quality. The algorithm can fully recover the original image when extracting data, thus achieving reversibility. The paper conducts experiments on various types of images, and the results show that the algorithm outperforms existing reversible data hiding methods in terms of embedding capacity and image quality while having lower computational complexity and stronger robustness.

Keywords: Reversible Data Hiding · Adaptive Predictor · Payload Allocation

1 Introduction

In the network era, the security and integrity of data are under unprecedented threat, and information security has become a focal issue of social concern. Data hiding [1], encompassing techniques such as steganography and digital watermarking [2], involves embedding and transmitting confidential information within multimedia carriers like images, audio, and videos. Steganography aims to conceal sensitive information within a carrier to avoid detection by third parties. At the same time, digital watermarking seeks to hide identifying information within a carrier for purposes such as copyright protection, authentication, and traceability. Traditional data hiding methods often introduce certain distortions to carriers, which are intolerable in critical applications such as medical images, court evidence, and military intelligence. A technique known as reversible data hiding (RDH) has been developed to tackle this challenge. RDH can completely restore

K. Li and Y. Liu (Eds.): ISICA 2023, CCIS 2147, pp. 159–169, 2024.
https://doi.org/10.1007/978-981-97-4396-4_14

the original carrier, extract confidential data, and achieve lossless data hiding effects. However, RDH's core challenge lies in designing efficient embedding and extraction methods while balancing embedding capacity and distortion rate. Embedding capacity refers to the maximum length of confidential information that can be embedded, while distortion rate indicates the degree of difference between the carrier after embedding and the original carrier. These two metrics are mutually constrained, typically demonstrating a trade-off where increasing embedding capacity leads to higher distortion rates and vice versa.

Consequently, the quest for RDH methods with the lowest distortion rates under a given embedding capacity has become a central focus and challenge in this field. In the current academic research landscape, optimizing this balance and enhancing the performance of RDH systems are crucial issues propelling continuous development in this domain. Against this backdrop, this paper explores and proposes RDH algorithms with high embedding capacity and low distortion rates, addressing data security and image quality requirements.

Various algorithms have been proposed domestically and internationally to address the issue of RDH, categorized into four main types: lossless compression [3], histogram shifting (HS) [4], prediction error expansion (PEE) [5, 6], and integer transform [7, 8]. The earliest lossless compression algorithm was introduced by Celik et al. [3] in 2005. The fundamental concept involves leveraging the redundancy in image data to achieve lossless compression, thereby creating space for embedding secret data. These algorithms exhibit the advantage of providing a higher embedding capacity without introducing perceptible distortions, preserving the integrity and quality of the original image. However, their drawback lies in the requirement for a high compression rate and increased computational complexity during the compression process. While lossless compression algorithms demonstrate favorable performance in terms of embedding capacity and image quality, their effectiveness may be compromised when attempting to conceal substantial amounts of data due to limitations imposed by the image's compression rate.

HS is a technique that utilizes the histogram characteristics of an image to achieve RDH. It embeds secret data into the peak or valley of the histogram. Then, it shifts the other parts of the histogram, thus ensuring that the image can be completely restored to its original state after extracting the secret data. Ni et al. [4] proposed the first histogram shifting algorithm in 2006, which has a high embedding capacity and a low visual distortion while keeping the file size of the cover image unchanged. However, the limitation of this algorithm is that it requires extra transmission of some information, such as the peak position, to enable the extraction of data and the recovery of the image, which increases the communication burden and security risk. To overcome these problems, some researchers have subsequently proposed many improved HS algorithms. For example, Wang et al. [9] proposed a general framework for data embedding based on multiple histogram shifting, which can significantly improve the embedding capacity while reducing image quality loss.

PEE is a technique that leverages the spatial correlation of an image for RDH. It embeds secret data into the predictive errors of the image. Subsequently, it expands the range of predictive errors, ensuring the complete recovery of the image to its original state after secret data extraction. Thodi and Rodriguez [5] initially proposed the predictive error expansion algorithm in 2007, demonstrating high embedding capacity and low distortion rates. However, it notably depends on the prediction methods employed for images and necessitates transmitting additional auxiliary information. Several improved PEE algorithms have been introduced to address these challenges. For instance, Sachnev et al. [10] proposed an RDH algorithm based on rhombus prediction. This algorithm partitions the image into two non-overlapping sets resembling a checkerboard pattern, with one set used for prediction and the other for embedding data, enabling dual-layer embedding and enhancing the algorithm's embedding capacity. Kumar et al. [11] discussed a review of the working predictors being used in PEE-based RDH and introduced a predictor using extreme gradient boosting.

This study proposes an innovative RDH algorithm based on an adaptive predictor and non-uniform payload allocation techniques. Firstly, the algorithm enhances the accuracy and capacity of data embedding by employing an adaptive predictor for precise prediction of image pixel values. Subsequently, the algorithm utilizes the extension of prediction errors for data embedding. Following this, a non-uniform payload allocation strategy is implemented based on the distribution characteristics of prediction error histograms during the payload allocation phase. This strategy optimizes the utilization of image pixel distribution, reducing the number of ineffective shifted pixel points and effectively minimizing image distortion. Experimental results demonstrate significant improvements in embedding capacity and visual distortion compared to existing techniques.

2   Related Work

This section will delve into two core concepts closely related to our study: the rhombus predictor [10] and prediction error expansion (PEE) [5]. As an efficient image prediction technique, the rhombus predictor analyzes and predicts pixel values through a distinctive geometric shape, providing a reliable foundation for data hiding. Simultaneously, the PEE technique increases the capacity for hidden data by exploiting the expansion of prediction errors to maximize image quality preservation. In the related work section, we will retrospectively review the development of these technologies, assess their applications in the field of RDH, and engage in an in-depth discussion on how they theoretically support and technologically inspire our study.

2.1   Rhombus Predictor

The rhombus predictor [10] is an efficient image prediction technique that focuses on enhancing the accuracy of pixel value estimation. By considering the correlation between adjacent pixels, this method leverages the four pixels surrounding the central pixel in a rhombus-shaped cell to achieve high-performance prediction of the central pixel. The rhombus predictor fully exploits the spatial distribution characteristics of pixels in the image, imparting adaptability and contextual awareness to pixel prediction. The structure of the rhombus predictor is illustrated in Fig. 1.

For a pixel $p_{i,j}$, its four neighboring pixels (including $p_{i,j-1}, p_{i+1,j}, p_{i,j+1}$, and $p_{i-1,j}$) are employed to predict the pixel value of $p_{i,j}$. These five pixels form a cell to conceal one bit of data. Specifically, all pixels in the image are categorized into "Cross" and "Dot" sets (see Fig. 1). During the first data embedding round, the "Cross" set is employed for data embedding, while the "Dot" set is utilized for pixel prediction. In the second data embedding round, the roles of the two sets are interchanged: the "Cross" set is used for pixel prediction, and the "Dot" set is employed for data embedding. The central pixel $p_{i,j}$ is predicted based on its four adjacent pixels: $p_{i,j-1}$, $p_{i+1,j}$, $p_{i,j+1}$, and $p_{i-1,j}$, and the predicted pixel value $p'_{i,j}$ is expressed as follows:

$$p'_{i,j} = \frac{p_{i,j-1} + p_{i+1,j} + p_{i,j+1} + p_{i-1,j}}{4} \tag{1}$$

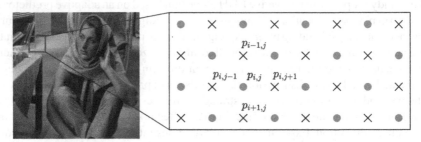

Fig. 1. Rhombus predictor. Predicting the pixel value $p_{i,j}$ within the Dot set involves utilizing the four neighboring pixel values from the Cross set, with subsequent expansion for concealing a single bit of data.

Based on the predicted pixel $p'_{i,j}$ and the original pixel $p_{i,j}$, the prediction error $e_{i,j}$ is calculated as follows:

$$e_{i,j} = p_{i,j} - p'_{i,j} \tag{2}$$

2.2 Prediction Error Expansion (PEE)

Prediction error expansion (PEE) [5] is a technique that embeds secret data by expanding the prediction error. Before performing the prediction error expansion, the pixel value $p_{i,j}$ is predicted to obtain the predicted pixel value $p'_{i,j}$, and the prediction error $e_{i,j}$ is calculated. We introduce the algorithms for data embedding and extraction separately to ensure the reversibility of data embedding and extraction. At the data embedding end, each prediction error $e_{i,j}$ is expanded or shifted, with the specific method as follows:

$$\tilde{e}_{i,j} = \begin{cases} e_{i,j} + m, & if e_{i,j} = 0 \\ e_{i,j} - m, & if e_{i,j} = -1 \\ e_{i,j} + 1, & if e_{i,j} > 0 \\ e_{i,j} - 1, & if e_{i,j} < -1 \end{cases} \tag{3}$$

where $m \in \{0, 1\}$ represents the data bit pending embedding. Bins -1 and 0 are expanded for data embedding, while the remaining bins are shifted to create vacancies, ensuring reversibility. Ultimately, the cover pixel $p_{i,j}$ undergoes modification to $\tilde{p}_{i,j} = p'_{i,j} + \tilde{e}_{i,j}$ to produce the marked pixel.

At the data extraction end, the marked pixel $\tilde{p}_{i,j}$ is predicted using the same predictor, yielding $p'_{i,j}$. Subsequently, the marked prediction error is computed as $\tilde{e}_{i,j} = \tilde{p}_{i,j} - p'_{i,j}$. Then, for each prediction error $\tilde{e}_{i,j}$, the original prediction error is restored as follows:

$$
e_{i,j} = \begin{cases} \tilde{e}_{i,j}, & \text{if } \tilde{e}_{i,j} \in \{0, -1\} \\ \tilde{e}_{i,j} - 1, & \text{if } \tilde{e}_{i,j} > 0 \\ \tilde{e}_{i,j} + 1, & \text{if } \tilde{e}_{i,j} < -1 \end{cases} \tag{4}
$$

Simultaneously, the embedded data can be extracted as $m = 0$ if $\tilde{e}_{i,j}$ belongs to $\{-1, 0\}$, or $m = 1$ if $\tilde{e}_{i,j}$ belongs to $\{-2, 1\}$. Finally, the cover pixel is restored as $p_{i,j} = p'_{i,j} + e_{i,j}$.

3 Methodology

This section will elaborate on the methodology adopted in our research. Initially, we will discuss the design and implementation of the adaptive predictor, a mechanism capable of dynamically adjusting prediction strategies based on the varying content of images to enhance prediction accuracy. Subsequently, we will expound on the non-uniform payload allocation strategy, which optimizes the performance and security of data hiding by judiciously distributing the embedding payload. The synergy of these two approaches provides robust support for our image processing techniques, enabling them to excel in various application scenarios.

3.1 Adaptive Predictor

For a grayscale image of size $M \times N$, we divide it into four non-overlapping subsets and perform adaptive pixel value prediction. As shown in Fig. 2, the details of the image segmentation are represented by different colors: red, green, blue, and yellow correspond to subsets S_1, S_2, S_3, and S_4, respectively. Except for the boundary pixels, the pixel prediction method within each subset is consistent. This paper will focus on subset S_1 to elaborate on the pixel value prediction method. Subset S_1 is used for data embedding during the first round of data embedding, while the other three subsets are used to predict S_1. Let the current pixel be denoted as $p_{i,j}$, where i and j represent the row and column indices of the pixel, respectively. The predictive context for this pixel is defined as $PC(i, j)$, initially set as $PC(i, j) = \{p_{i-1,j}, p_{i,j+1}, p_{i+1,j}, p_{i,j-1}\}$. The maximum and minimum values in the set $PC(i, j)$ are denoted as PC_{max} and PC_{min}, i.e., $PC_{max} = max(PC(i, j))$, $PC_{min} = min(PC(i, j))$. To reflect the variation more accurately in pixel values around the current pixel $p_{i,j}$, we set a threshold T to quantify this change. Depending on the variation in pixel values around $p_{i,j}$, we adopt an adaptive method to predict the value of $p_{i,j}$.

When $PC_{max} - PC_{min} \leq T$, indicating minor changes in surrounding pixels, we use the four neighboring pixel values of $p_{i,j}$ for prediction. The predicted value $p'_{i,j}$ is expressed as:

$$
p'_{i,j} = \frac{p_{i-1,j} + p_{i,j+1} + p_{i+1,j} + p_{i,j-1}}{4} \tag{5}
$$

When $PC_{max} - PC_{min} > T$ indicates significant changes in surrounding pixels, we use the eight neighboring pixel values of $p_{i,j}$ for prediction. The predicted value $p'_{i,j}$ is expressed as:

$$p'_{i,j} = \frac{p_{i-1,j} + p_{i,j+1} + p_{i+1,j} + p_{i,j-1} + p_{i-1,j+1} + p_{i+1,j+1} + p_{i+1,j-1} + p_{i-1,j-1}}{8}$$

(6)

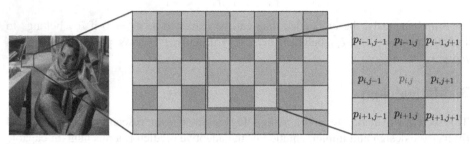

Fig. 2. Adaptive predictor: predicting the pixel value $p_{i,j}$ based on the four or eight neighboring pixel values. (Color figure online)

After obtaining the predicted value $p'_{i,j}$, we calculate the prediction error $e_{i,j} = p_{i,j} - p'_{i,j}$. Subsequently, we will employ the PEE algorithm for data embedding.

3.2 Non-uniform Payload Allocation Strategy

We conduct a detailed statistical analysis of the distribution of prediction errors by computing the predicted values and prediction errors for all pixels in the cover image. The statistical results indicate a relatively high proportion of prediction errors, taking values of 0 and –1. Consequently, we propose a non-uniform payload allocation strategy to achieve effective and distributed embedding of secret data. This algorithm integrates dynamic prediction and weight adjustment strategies, maximizing the overall embedding capacity while preserving the image's visual quality.

Initially, for an image of size $M \times N$, we tallied the number of pixels with prediction errors equal to –1 and 0, denoted as $Total(-1, 0)$. Subsequently, for each subset S_t, we computed the quantities of pixels with prediction errors equal to –1 and 0, represented as $count^t_{-1}$ and $count^t_0$, where t denotes the subset index with values in $\{1, 2, 3, 4\}$. Next, we calculated the proportions of prediction errors equal to –1 and 0 within each subset relative to the entire image, denoted as W_t. The specific calculation method for W_t is as follows:

$$W_t = \frac{count^t_{-1} + count^t_0}{Total(-1, 0)}$$

(7)

where W_t retains two decimal places.

Subsequently, based on the computed weights, we dynamically adjusted the embedding capacity for each subset. Specifically, if a subset's weight is larger, more payload is allocated to that subset; conversely, less payload is allocated if a subset's weight is smaller. Assuming the total payload is denoted as L, the payload allocated to subset S_t is given by $L_t = L \times W_t$.

Ultimately, through four rounds of data embedding operations, we completed the embedding of the entire payload. This process ensures the effective embedding of secret data and a delicate balance in image quality.

3.3 Auxiliary Information Embedding

It is imperative to embed auxiliary information into the cover image during the embedding process to ensure the reversibility of data embedding and extraction. Specifically, 'this auxiliary information comprises a location map LM, the total payload L, and payload allocation weights W_t for four subsets. The purpose of the location map is to prevent pixel value overflow and underflow during data embedding. To achieve this objective, we record the positions in the cover image where pixel values are 0 and 255 before pixel prediction. The pixels at these positions will not be used for data embedding. Utilizing an arithmetic compression algorithm, we compressed the location map and embedded the compressed map as part of the payload into the cover image.

Furthermore, the total payload L occupies 20 bits, while each payload weight W_t takes up 8 bits. The chosen bit lengths are designed to ensure accurate data embedding and extraction while minimizing adverse effects on the quality of the cover image. Our design addresses the requirement for reversibility and integrates preventive measures against potential issues such as overflow and underflow during the data embedding process. This comprehensive consideration ensures our algorithm's robustness and the cover image's protection, meeting the high standards demanded by the scientific research community.

4 Experimental Results and Analysis

In this study, we propose a novel RDH algorithm based on an adaptive predictor and a non-uniform payload allocation strategy. We conduct a series of experiments to assess the algorithm's effectiveness thoroughly. This involves evaluating its performance based on two key aspects: the distribution characteristics of prediction errors and the processed images' peak signal-to-noise ratio (PSNR) values. The analysis of the prediction error histogram distribution provides an intuitive basis for validating the effectiveness of the adaptive prediction algorithm. At the same time, the computation of PSNR values serves as a vital metric to evaluate the preservation of visual quality in the images. We select 1338 color images from the UCID dataset as experimental samples, converting them into corresponding grayscale images for further analysis. The dimensions of these images are uniformly set at 512×384 or 384×512. Figure 3 illustrates examples of images used in the experiments. The secret data is a random bit sequence generated by a random function.

All experiments were conducted on a PC with the Windows 10 operating system, an Intel Core i5-10400F CPU, and 16 GB of RAM, utilizing the MATLAB R2020a environment. In the subsequent sections, we will explain the experimental design and result analysis, demonstrating our algorithm's superior performance and application potential.

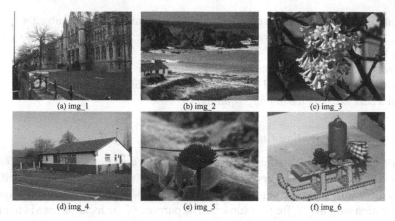

Fig. 3. Six experimental image examples.

4.1 Prediction Errors Distribution

To comprehensively evaluate the performance of the adaptive predictor, we conduct a thorough analysis of the histogram distribution of the prediction errors. In the UCID dataset, we randomly select six representative grayscale images for the experiment, as shown in Fig. 3. These images cover various image features, from simple textures to complex details. We calculate the prediction errors for each image and plot the corresponding histograms. A comparative analysis is performed to demonstrate the superiority of the proposed adaptive predictor over the traditional diamond predictor [10] in terms of pixel prediction accuracy, as depicted in Table 1.

Table 1. Prediction error distribution percentage.

Prediction error	[−255, −2]	−1	0	[1, 255]
Diamond predictor	40.35%	7.26%	7.78%	44.61%
Proposed predictor	39.85%	8.25%	8.77%	43.13%

The experimental results indicate that, compared with the diamond predictor, the adaptive predictor exhibits a more concentrated distribution of prediction errors around 0 and −1 for most test images. Notably, the proportion of the prediction error distribution between −1 and 0 is significantly higher for the adaptive predictor, aligning closely with our designed non-uniform load distribution strategy. This distribution characteristic not only underscores the robust adaptability of the adaptive predictor across diverse image content but also provides an ideal environment for the efficient embedding of secret data.

Furthermore, we observe that the adaptive predictor can dynamically adjust the prediction model based on local image features, achieving more precise predictions in regions with complex textures. This capability significantly enhances the concealment and security of the secret data embedding process. In summary, our algorithm demonstrates innovation in theory and exhibits outstanding performance and broad applicability in practical applications.

4.2 Image Fidelity

In this study, we comprehensively assess the impact of data hiding technology on the visual quality of images. We adopt PSNR as an objective metric to quantify the fidelity of images before and after data embedding. The experimental subjects are selected from the UCID dataset, and thorough testing is conducted under varying payload conditions to record changes in PSNR values. Specifically, we choose six grayscale images with distinct texture and brightness features, as illustrated in Fig. 3. The data embedding ranges from 5,000 to 15,000 bits, with increments of 1,000 bits. We compare the performance of our proposed method with three existing approaches (referred to as Thodi [5], Sachnev [10], and Wang [9]) in terms of PSNR values, and the experimental results are presented in Fig. 4.

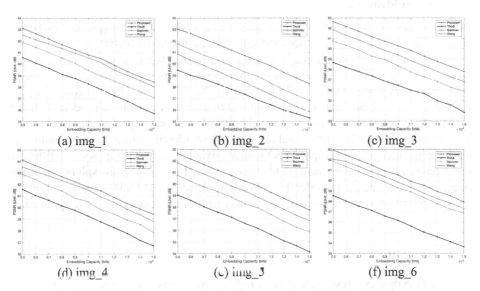

Fig. 4. PSNR comparison under the same amount of data embedding.

Subsequently, to further validate the universality and stability of our approach, we randomly select 20 images from the UCID dataset, constructing a more extensive experimental dataset. The PSNR values for these images are computed under an data embedding of 10,000 bits. All experimental results are meticulously documented in Table 2, facilitating direct comparison and analysis.

The experimental outcomes demonstrate that even at higher data embedding levels, our algorithm maintains elevated PSNR values, substantiating the exceptional performance of our data hiding technology in preserving the visual quality of images. Additionally, compared to alternative methods, our algorithm exhibits superior performance

Table 2. PSNR results for 20 images.

Image	img_1	img_2	img_3	img_4	img_5	img_6	img_7
PSNR	60.79	60.71	60.27	61.79	60.24	60.54	59.86
Image	img_8	img_9	img_10	img_11	img_12	img_13	img_14
PSNR	59.89	63.43	60.45	60.98	61.65	61.22	59.98
Image	img_15	img_16	img_17	img_18	img_19	img_20	
PSNR	57.89	60.43	61.21	60.54	60.78	60.34	

across various testing conditions, particularly in scenarios of high payload, where the protection of image quality is notably pronounced. These experimental findings validate our approach's efficacy and provide robust support for subsequent research and applications.

4.3 Image Fidelity Analysis

This study aims to validate the effectiveness of a reversible data hiding algorithm based on an adaptive predictor and a non-uniform payload distribution strategy. We conduct a comprehensive evaluation through a series of rigorous experiments. The experimental results unequivocally demonstrate that our algorithm maintains high visual quality across different payload levels, as evidenced by the stability of PSNR values. Its performance is particularly noteworthy compared to alternative methods, especially under high payload conditions, illustrating the algorithm's efficacy in preserving image quality even when embedding substantial amounts of confidential information. In summary, our experimental analysis emphasizes the dual advantages of the proposed algorithm in secret data embedding and image quality preservation. Future research explores the algorithm's performance across different embedding capacities and image types and investigates optimization strategies to adapt the algorithm to a broader range of application scenarios.

5 Conclusion

This paper proposes a reversible data hiding algorithm based on an adaptive predictor and a non-uniform effective payload allocation strategy. The adaptive predictor dynamically adjusts the prediction model based on the local features of the image, facilitating more accurate predictions and reducing prediction errors. The non-uniform payload allocation strategy boosts embedding efficiency by assigning distinct embedding capacities to four-pixel sets based on the distribution of prediction errors. We conduct a series of experiments to evaluate the performance of the proposed algorithm in terms of embedding capacity and image quality. The experimental results validate the dual advantages of the proposed algorithm in secret data embedding and image quality preservation. Future research can further explore the algorithm's performance across different embedding capacities and various types of images and investigate optimization strategies to adapt the algorithm to a broader range of application scenarios.

Acknowledgments. This work is supported by 2023 Dongguan Social Development Science and Technology Project General Project with the Grant No. 20231800903852, 2023 Guangdong Province General Universities Youth Innovative Talents Projects with the Grant No. 2023KQNCX140, and 2021 Guangdong Province University Scientific Research Projects with the Grant No. 2021ZDZX1029.

References

1. Yin, Z., Xiang, Y., Zhang, X.: Reversible data hiding in encrypted images based on multi-MSB prediction and Huffman coding. IEEE Trans. Multimedia **22**(4), 874–884 (2020)
2. He, W., Cai, Z., Wang, Y.: High-fidelity reversible image watermarking based on effective prediction error-pairs modification. IEEE Trans. Multimedia **23**, 52–63 (2021)
3. Celik, M., Sharma, G., Tekalp, A., Saber, E.: Lossless generalized-LSB data embedding. IEEE Trans. Image Process. **12**(2), 253–266 (2005)
4. Ni, Z., Shi, Y., Ansari, N., Su, W.: Reversible data hiding. IEEE Trans. Circuits Syst. Video Technol. **16**(3), 354–362 (2006)
5. Thodi, D., Rodriguez, J.: Expansion embedding techniques for reversible watermarking. IEEE Trans. Image Process. **16**(3), 721–730 (2007)
6. He, W., Xiong, G., Weng, S., Cai, Z., Wang, Y.: Reversible data hiding using multi-pass pixel-value-ordering and pairwise prediction-error expansion. Inf. Sci. **467**, 784–799 (2018)
7. Weng, S., Zhao, Y., Pan, J., Ni, R.: Reversible watermarking based on invariability and adjustment on pixel pairs. IEEE Signal Process. Lett. **15**, 721–724 (2008)
8. Peng, F., Li, X., Yang, B.: Adaptive reversible data hiding scheme based on integer transform. Signal Process. **92**, 54–62 (2021)
9. Wang, J., Chen, X., Ni, J., Mao, N., Shi, Y.: Multiple histograms-based reversible data hiding: framework and realization. IEEE Trans. Circuits Syst. Video Technol. **30**(8), 2313–2328 (2020)
10. Sachnev, V., Kim, H., Nam, J., Suresh, S., Shi, Y.: Reversible watermarking algorithm using sorting and prediction. IEEE Trans. Circuits Syst. Video Technol. **19**(7), 989–999 (2009)
11. Kumar, R., Sharma, D., Dua, A., Jung, K.: A review of different prediction methods for reversible data hiding. J. Inf. Secur. Appl. **78**, 103572 (2023)

Research on Bayberry Traceability Platform Based on Blockchain

Hongyu Xiao[1], Zihang Gao[1,2], Xiaojun Cui[1(✉)], and Nannan Zhao[1]

[1] Wenzhou Vocational College of Science and Technology, Wenzhou 325006, China
cxjxhy@163.com
[2] Cangnan Industrial Research Institute of Modern Agriculture, Cangnan,
Shenzhen 325899, China

Abstract. Aiming at the actual business process of the bayberry industry chain, we developed a bayberry whole industry chain information traceability system based on blockchain technology. At the same time, a database and blockchain dual-mode data storage mechanism is designed to optimize the operating efficiency of blockchain computing nodes in the industry chain. Finally, according to user feedback and test results, the system is optimized twice. In this work, we can provide auxiliary means for upgrading the competition level of bayberry industry.

Keywords: Modern Agriculture · bayberry · traceability Platform · blockchain

1 Introduction

The current traceability methods in the agricultural industry are relatively simple, and they face many problems such as traditional traceability trust issues, user information leakage issues, information island issues, and information system centralization issues. Traditional traceability methods are easy to be tampered with in the middle, making anti-counterfeit traceability useless [1]. Both consumers and manufacturers hope to use effective anti-counterfeiting and traceability methods to decentralize and increase the subject of trustworthy endorsement, which can not only prevent serious leakage of user personal information, but also improve the reliability and authenticity of product information [2]. Blockchain technology just enough to meet the requirements.

According to the analysis, experts and scholars have carried out a certain degree of research on blockchain technology to realize the information traceability system in the agricultural field. However, for the bayberry industry, the existing blockchain systems (such as grain and oil food systems, miscellaneous grain product systems, and livestock asset authentication systems) may not be applicable. The reason is that in a specific application environment, if the data storage and query efficiency of the blockchain have not been optimized, and the native system is directly applied to the scene application, the throughput and delay of the information traceability system of the entire industry chain will be reduced. It has become the bottleneck of the operation efficiency of the traceability system, and ultimately cannot guarantee the data security and authenticity of the traceability information between the production, processing and circulation nodes of the agricultural product industry chain.

K. Li and Y. Liu (Eds.): ISICA 2023, CCIS 2147, pp. 170–178, 2024.
https://doi.org/10.1007/978-981-97-4396-4_15

2 Relate Work

According to research, in recent years, researchers have successively explored the application framework of blockchain technology in the agricultural field [3]. For example, Ronaghi [1] proposed a model for evaluating the maturity of blockchain technology in the agricultural supply chain, and discussed the application of the blockchain maturity model in supply chain management. Khan et al. designed a metaheuristic-based genetic algorithm [2], which is used to receive production details and processes in the agricultural blockchain, and analyze predictive pricing through real-time scheduling, management, and monitoring records. Bera et al. [3] used private blockchain technology to realize the intelligent precision agricultural authentication function and key management scheme, which greatly optimized the computing cost and performance security of data transmission under the agricultural Internet of Things.

3 Research Framework

3.1 Research Content

The blockchain technology research for the information traceability of the whole crop industry chain is one of the hot spots in recent years, facing the challenges of the information traceability of the bayberry industry chain. At present, there are more than dozens of agricultural information traceability systems based on blockchain technology, but none of them can be directly applied to the bayberry industry. Bayberry as an important industry in the agricultural field, urgently needs to develop a special information traceability system for this industry chain. The paper will develop an information traceability system for the entire industry chain applicable to bayberry, build and optimize the corresponding data storage mechanism and the performance of related blockchain computing nodes. The platform framework of the research is shown in Fig. 1. The specific research contents are as follows.

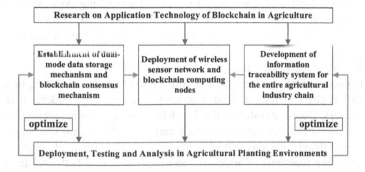

Fig. 1. The platform framework of the research.

This paper is oriented to the application scenario of the whole industry chain of bayberry, and uses the existing open source framework (such as Hyperledger Fabric) as the carrier to design a whole industry chain information of bayberry consisting of the display layer, user layer, business layer, blockchain layer and data security layer Traceability system. The system mainly includes the following three parts.

Development and performance optimization of information traceability system. For the data collected at each business stage of the whole industry chain of bayberry, by invoking the smart contract verification deployed on the blockchain network, the data is broadcast in the blockchain network, so that each node reaches a consensus on the uploaded data, and then pack the data summaries generated by hash algorithm conversion into blocks and save them in the distributed ledger on the blockchain network. At the same time, the complete data information of the supply chain and the mapping relationship between the blocks and data returned by the smart contract, and finally complete the information traceability of the entire industry chain of bayberry, ensuring that the traceability data cannot be falsified.

Design and implement a dual-mode data storage mechanism of the database and blockchain, optimize the operation efficiency of blockchain computing nodes in the industry chain, and effectively reduce the pressure of data storage on each node in the blockchain network. Low query efficiency and other issues. At the same time, the Kafka consensus mechanism is optimized to effectively sort the transactions of multiple participants on the chain, providing high data throughput and low latency processing capabilities.

Combining with the national food safety law to write smart contracts, so that the data of the red bayberry industry chain can be standardized and regulated, to verify the contract data to be uploaded to the chain in real time, monitor the transaction information on the blockchain network, and trigger the verification mechanism. Legal information is stored to ensure the reliability of bayberry data and the credibility of the traceability system.

Bayberry data collection and blockchain computing node deployment. By deploying wireless sensor networks in bayberry planting bases, automatic data collection of bayberry physiological information during the planting process (such as flowering stage, fruit setting stage, hard shell stage, maturity stage, and picking stage, etc.) is realized. At the same time, according to the growth cycle of bayberry, the specific date and time period will remind the operator to enter the business operation information of the corresponding stage. For blockchain computing nodes, the paper chooses two solutions to test and compare at the same time:

In the bayberry planting base, divide according to the node computing function, and configure different blockchain dedicated servers. These include Orderer node servers, Peer node servers, Comitter node servers, block browser application servers, and system application servers. This application solution can make the overall system reliable and efficient, but the disadvantage is that the deployment cost is relatively high.

The computing functions of nodes are no longer distinguished, and the above five types of servers are uniformly deployed on remote cloud platforms (such as Tencent Cloud, Alibaba Cloud, Huawei Cloud, etc.). This solution can greatly reduce the operating cost of the underlying technology of the blockchain, simplify the construction and operation and maintenance of the blockchain, and at the same time face the information traceability of the whole industry chain of bayberry, and has stronger system robustness, but the disadvantage lies in the later data Migration will be more difficult.

Carry out large-scale application and promotion. Large-scale application and promotion of blockchain technology in the form of on-site meetings or training courses in the bayberry planting base. Through the demonstration and deployment of the information traceability system of the whole industry chain of bayberry, the transformation of blockchain technology in the field of bayberry and the implementation of technology will be accelerated. At the same time, it provides two different construction schemes for easy selection, adapting to local conditions, and meeting the management needs of different users. Through the means of information technology, build the bayberry brand, enhance consumers' awareness of bayberry agricultural products, increase the added value of the brand, so as to achieve agricultural efficiency, increase farmers' income, and promote the rapid and sustainable development of agriculture.

3.2 Key Technologies

The key technologies of this paper mainly include the following four parts.

Identification technology. Identification is one of the keys throughout the entire supply chain. In-depth research on the current mainstream two-dimensional code, RFID and other identification technologies, select the appropriate technology to build a traceable quality safety and supervision information system. For the source identification of various bayberry varieties in the bayberry planting base and the untrustworthy data source of the supervision system, the identity authentication scheme provides effective support for data confidentiality and equipment identity reliability, and realizes fine-grained identity identification of data on the blockchain. The effective supervision of bayberry products is realized based on the information traceability system of the whole industry chain of bayberry, and the quality status of bayberry in various channels is monitored in real time through sensor equipment, in order to realize the dynamic evaluation and audit of bayberry and prevent the risk of counterfeiting and shoddy products.

Data description method. In view of the complex and diverse data types of the entire supply chain of bayberry-related agricultural products, significant time-space coupling, strong randomness, and high discreteness, this paper intends to analyze the supply of bayberry-related agricultural products from the aspects of time, space, and source Chain data, combined with object-oriented thinking to cluster agricultural product supply chain data, on this basis, construct metadata standards for agricultural product supply chain related to bayberry to meet the need for consistent expression of agricultural product supply chain data.

Dual-mode data storage mechanism. Bayberry supply chain has the characteristics of many nodes, long supply chain and wide coverage. Each link on the chain contains a large amount of data. If it is uploaded to the blockchain network at one time, the upload speed will be slow, the load of the blockchain is heavy, the operating cost is high, and the requirements for the hardware facilities of each node in the blockchain network will be very high. In this paper, the on-chain and off-chain dual-mode storage mechanism, through the smart contract verification of the detailed data information of each link in the bayberry supply chain, and finally stored in the relational database together with the location information of the blockchain where the data is located. The data stored in the blockchain network is the hashed data and the unique identification code corresponding to the data. In this way, the operational efficiency of the blockchain is improved while ensuring the security and credibility of the data.

Blockchain consensus mechanism. The Kafka consensus mechanism sorts all transaction information through Kafka, supports separate sorting of multiple channels in the system, and is a distributed streaming information processing platform that supports multi-channel partitioning. It has a crash fault tolerance mechanism, but it cannot provide protection against malicious attacks in the network. Therefore, by adding information encryption algorithms in the Kafka consensus mechanism, such as symmetric encryption algorithms, asymmetric encryption algorithms, and hash algorithms, data integrity, privacy, and valid transaction credentials are guaranteed, and digital signatures are used to ensure transaction security, especially elliptic curve encryption. The algorithm generates public-private key pairs and the elliptic curve digital signature algorithm guarantees the non-repudiation of transactions as a representative, and realizes the security verification of data through zero-knowledge proof and multi-party security calculation, completes the consensus on the data on the chain, and conducts a check on the nodes in each upload process Identity verification, if the node certificate is unreliable, the consensus will be re-consensus, and finally the operation of storing the node data on the chain will be completed.

4   Implementation Plan

4.1   System Construction

In this paper, Fabric is used as the carrier framework of the information traceability system of the whole industry chain of bayberry, and a wireless sensor network is built to monitor various types of bayberry physiological information, and an observation and analysis model of bayberry is established to realize data collection related functions. Based on the blockchain technology, the bayberry traceability blockchain (alliance chain) is abbreviated as: the bayberry chain, which makes the planting information, detection information, and picking process records transparent, traceable and unalterable, and realizes the growth, detection, picking, and logistics processes of bayberry. The whole process is tracked, and each piece of information is recorded on the blockchain. The tamper-proof property of the blockchain is used to ensure the authenticity of the data, and the time stamp technology is used to ensure that each process can be checked. This will greatly enhance the credibility of the system and increase the participation enthusiasm of the public, various planting cooperatives, government departments, and logistics companies. The architecture is divided into two parts, the upper layer and the lower layer, with a total of 5 layers. The bottom part includes the storage layer and the blockchain service layer, on which the transactions and data records of the entire network are carried out. The upper part includes an interface layer that provides external and internal interfaces, the on-chain code layer that provides contract-related services written according to specific requirements, and an application layer that provides information collection and query services to users. The internal interface is used for internal communication between nodes, such as broadcasting blocks. The external interface is for external users, such as querying and accepting new transactions. The code layer on the chain has the functions of transaction sending, status checking, and status storage. According to the requirements of the bayberry chain traceability service system, the architecture of the goods tracing system is shown in Fig. 2.

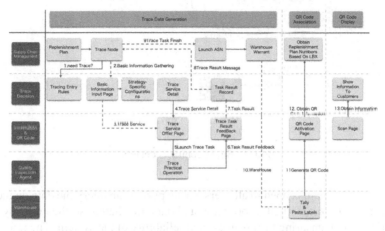

Fig. 2. The architecture of the goods tracing system.

4.2 Smart Contracts Design

The smart contracts can then be used to drive the blockchain system. First of all, all participants must negotiate together, and after reaching an agreement, the results of the negotiation will be deployed on the blockchain in the form of smart contracts. A number of states and transition rules, conditions that trigger contract execution, etc. are pre-specified in the smart contract. The blockchain can monitor the status of smart contracts in real time, and activate and execute smart contracts by checking external data sources and confirming that the trigger conditions are met. The life cycle information of bayberry seasons is written into the blockchain, and the characteristics of the blockchain are immutable, trustless, transparent, distributed storage, and strong damage resistance to prevent false information and transparent production processes. The operating principle of the smart contract is shown in Fig. 3.

There are mainly four types of roles in this system, namely planting enterprises (cooperatives), supervision and management units, testing institutions, and the public. Each type of role corresponds to a type of node in the blockchain. Each user joining the system is an independent node in the network and has a unique blockchain address. This address is used to uniquely identify the identity of the node, and serves as an interface for reading information about the node from the blockchain.

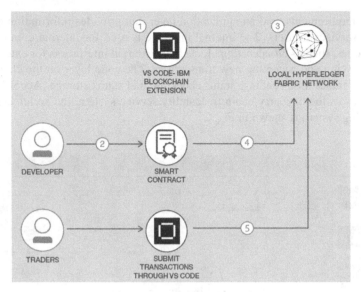

Fig. 3. Smart contracts design.

The bayberry chain is open to all users, so the public can query the data on the chain through the open port, and all information and data are shared, so the entire system is highly transparent, ensuring the openness and reliability of the system. The information of every link in the whole process of bayberry growth is completely stored on the blockchain and cannot be tampered with. Any fraud will be discovered.

Through the above design scheme, the construction of the bayberry industry blockchain system can be initially realized, and then the data collected by the wireless sensor network at the flowering stage, fruit setting stage, hard shell stage, ripening stage, and picking stage can be manually entered into the business of the corresponding stage The operation information is uploaded to the system, and finally connected to the blockchain computing node network to complete the overall system construction. The information traceability system of bayberry whole industry chain to be developed adopts a layered system architecture, which is divided into: data security layer, blockchain/back-end technology layer, business layer and user layer from bottom to top.

4.3 Blockchain Mechanism Optimization

Since the members of this research group all work within a region and belong to the same work unit, the effective organization of the research personnel can be guaranteed. The project leader is responsible for coordinating the division of labor and cooperation of the project, and adopts the principle of "selecting a topic, setting a time, assigning a person to be responsible, being responsible to the end, and division of labor and cooperation". The specific operation is to discuss and decide the topic and implementation time after the research group consults with experts, and the members of the research group are responsible for the division of labor to the end, submit the research data within a given period, and the research group analyzes and discusses the summary data to form the preliminary results. On the basis of the above achievements, we consulted experts and corrected opinions to form mid-term results. At the same time, we fed back the mid-term results to experts, extracted opinions, corrected countermeasures, and formed later results.

The entire industrial chain of agricultural products has the characteristics of many participating main nodes, long industrial chain, wide scope, large amount of data and multi-source heterogeneity. All the data of each node is uploaded to the blockchain network. Not only is the upload speed slow, but it is also easy to cause network congestion, which leads to high pressure on data storage of each node in the blockchain network, low query efficiency, and great hidden dangers of data integrity. It also affects the data storage system. Higher requirements are put forward for equipment performance and input cost, which affects the implementation of the blockchain-based traceability system. There are five main pressures on on-chain storage, namely consensus, computing, network, storage, and access.

For this reason, this project proposes the on-chain and off-chain dual-mode storage mechanism as a storage strategy for the traceability and collaborative management of bayberry agricultural product information. The data output by each node of the industrial chain is standardized and normalized. Secondly, smart contracts are used to verify the standardized detailed data of each node, and most of the verified bayberry agricultural product industrial chain data and blockchain location information are stored locally or in the cloud. In relational and non-relational databases on the server.

Then use MD5 to calculate the key traceability information of bayberry agricultural products on the chain together with the partial data (images, videos, etc.) and the signature of the holder, and build an index off the chain, and only perform accurate reading and writing of Key-Value on the chain. At the same time, in order to ensure the privacy of the smart contract, the smart contract can also be stored under the chain if necessary, using the calculation node to record the calculation of the contract, and the consensus node to record the state record of the contract. Finally, for the storage of traceability data under the chain It should be as detailed as possible, the data calculated by the hash algorithm on the chain should be as streamlined as possible, and the data on the chain must go through consensus, so using this dual-mode storage mechanism as a collaborative management storage strategy can be flexibly dealt with The impact of network

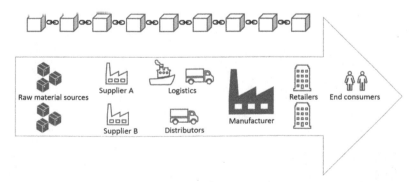

Fig. 4. The designed storage model of information traceability collaborative management process.

congestion, transmission delay, etc. The fast query of data on the chain achieves a balance of efficiency, cost and privacy security. The designed storage model of information traceability collaborative management process is shown in Fig. 4.

5 Conclusion

In this paper, we apply blockchain technology to the bayberry industry for the first time, and develops a set of information traceability system suitable for the entire industry chain of bayberry, and finally completes the information traceability of the entire bayberry industry chain, ensuring that the traceability data cannot be falsified. By optimizing the dual-mode data storage mechanism and the blockchain consensus mechanism in the system, the overall performance of the system can be improved, the processing capacity of high data throughput and low delay can be improved, and the reliability of bayberry data and the credibility of the traceability system can be ensured.

Acknowledgments. This work was supported by the Basic Agricultural Science and Technology Project of Wenzhou under Grant N20220003 and National Social Science Found of China "Reaserch on Virtual Reality Media Narrative" (Grant No.21&ZD326).

References

1. Ronaghi, M.H.: A blockchain maturity model in agricultural supply chain. Inf. Proc. Agric. **8**(3), 398–408 (2021)
2. Khan, A.A., Shaikh, Z.A., Belinskaja, L., et al.: A blockchain and metaheuristic-enabled distributed architecture for smart agricultural analysis and ledger preservation solution: a collaborative approach. Appl. Sci. **12**(3), 1487 (2022)
3. Bera, B., Vangala, A., Das, A.K., et al.: Private blockchain-envisioned drones-assisted authentication scheme in IoT-enabled agricultural environment. Comput. Stan. Interfaces **80**, 103567 (2022)

Research on Satellite Navigation and Positioning Based on Laser Point Cloud Data

Yuming Sun and Hua Wang[✉]

Guangdong University of Science and Technology, Dongguan 523000, China
460864372@qq.com

Abstract. The existing satellite navigation and positioning technology can't meet the requirement of target positioning accuracy, and the results are not obtained in time. In order to improve the accuracy and real-time of satellite navigation and positioning, a satellite navigation and positioning method based on laser point cloud data is proposed. In order to reduce the influence of interference factors, the mean filter method is used to denoise the laser point cloud data, and the Euclidean threshold region segmentation method is used to segment discrete data points. Aiming at extracting geometric features of objects, corner features of point cloud data are extracted based on geometric features. On the basis of the preprocessing results, the absolute position information, angular velocity and driving speed of satellite navigation are acquired by laser point cloud data, and the navigation datum line and characteristic points in the navigation path are acquired by image processing in machine vision. Satellite navigation and positioning based on laser point cloud data are completed. The simulation results show that the proposed method can effectively improve the positioning accuracy and operation efficiency, and can meet the practical application requirements.

Keywords: Laser point cloud data · Satellite navigation · Positioning · Region Segmentation

1 Introduction

Satellite navigation and positioning technology is mainly used to locate and monitor the target through the position, speed, time and other information provided by the global positioning and navigation system, so as to obtain accurate information of the target [1, 2]. Satellite navigation and positioning technology has the following advantages: (1) the whole space-time; (2) All weather; (3) Continuous real-time navigation and so on. These advantages promote satellite navigation and positioning technology to become a more widely used positioning technology at present. Therefore, once the satellite navigation and positioning system appeared, it was applied in various military and security fields, which prompted great changes in aviation, navigation and other technologies. Meanwhile, it also increased information services in various fields and promoted the development of emerging industries [3].

Because of its high research value, many researchers have carried out related research and achieved some research results. Literature [4] proposed a direct acquisition algorithm

© The Author(s), under exclusive license to Springer Nature Singapore Pte Ltd. 2024
K. Li and Y. Liu (Eds.): ISICA 2023, CCIS 2147, pp. 179–186, 2024.
https://doi.org/10.1007/978-981-97-4396-4_16

for satellite navigation positioning P-code. XFAST was used to process local sequences, and on this basis, the search scope was expanded. Meanwhile, the mean value processing method was used to process local sequences and received sequences, reducing the amount of computation and thus achieving the purpose of improving the speed of target acquisition. The experimental results show that this method is effective and can realize the positioning of satellite navigation, and the positioning speed is fast, but the positioning accuracy is not high. Literature [5] designed and tested the satellite laser reflector. The method of literature [5] designed the laser reflector under the premise of fully considering the far-field diffraction and other conditions. According to the far-field diffraction energy distribution theory of the Angle reflector, the size of the reflector is optimized, and the velocity difference compensation Angle is designed by Angle compensation method to optimize the parameters of the satellite laser reflector. The analysis of the test results shows that the parameter design of the method is reasonable, the ranging accuracy and range can meet the requirements of the practical application, but there is a problem of low efficiency in the application.

Due to the current satellite navigation and positioning technology is not very mature, this paper proposes a satellite navigation and positioning method based on laser point cloud data, and has obtained a relatively ideal research effect.

2   Methods

2.1   Preprocessing of Laser Point Cloud Data

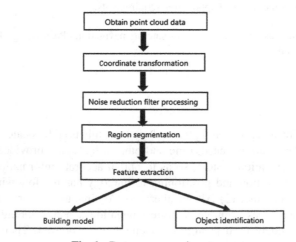

Fig. 1. Data preprocessing steps

The coordinates obtained by laser scanning tree are polar coordinates, but the data preprocessing requires cartesian coordinates, so coordinate transformation is needed. The polar coordinates of any data points in the liDAR are set as (r, θ), and the index coordinates of the above data points are given as follows (x, y) (Fig. 1):

$$\begin{cases} x = r \cos \theta \\ y = r \sin \theta \end{cases} \tag{1}$$

r represents the distance value, and θ represents the scanning Angle.

In the process of data preprocessing, it will be interfered by various external factors, and the instrument itself also has measurement errors and other influencing factors, which lead to tracking errors in the process of Lidar measurement [6, 7]. If the noise is not smooched in time, the accuracy of feature point extraction and the quality of model reconstruction will be negatively affected, and even the reconstructed curve will not be smooth [8]. Therefore, in the process of point cloud data extraction, it needs to be processed accordingly. The above processing process is essentially the related data noise reduction processing. In this paper, the mean filtering method is used to denoise the relevant data. This method can simplify the complex problem and the denoising effect is very ideal.

On the basis of denoising, this paper selects the Euclidean threshold region segmentation method for region segmentation. The discrete data points collected within the set period are divided [9, 10], the main purpose of which is to divide the above data into multiple disconnected areas.

Suppose that any region contains n point, then:

$$A\{X_i, Y_i\} = |i = 1, 2, \cdots, n| \tag{2}$$

where, A represents the set of all points in any region; $\{X_i, Y_i\}$ represents the number of all discrete points in the coordinate system [11].

The specific steps of region segmentation are as follows:

(1) The following formula is used to calculate the distance between two adjacent points:

$$D_j = \sqrt{\left(X_j - X_{j+1}\right)^2 + \left(Y_j - Y_{j+1}\right)^2} \tag{3}$$

D_j represents the distance between two adjacent points; (X, Y) represents the cartesian coordinates of discrete points in the set of points

(2) The Douglas-Peucker algorithm is used to obtain the specific distance relationship between two points, which T_D represents the threshold value. If $D_j > T_D$, it indicates that (x_i, y_i) is one of the segmentation point in the above region [12, 13]. The above segmentation points can divide A into two different regions, and the above process is repeated until the data in the study area has been cycled, and we can get the final N segments $\{A_1, A_2, \cdots, A_N\}$.

(3) Calculate the total number of discrete points in each region.

(4) Count the total number of data points in different areas. If the total number of data points in a certain area is less than or equal to 3, it indicates that the area is a noisy area, and the area with noise needs to be deleted.

Based on the denoising results, corner features of point cloud data were extracted. Corner point represents the discontinuity between each line segment, and it is the point where two line segments intersect. Generally, a corner point corresponds to a corner object [14]. Constraints to be met for corner point extraction are given in detail as follows:

(1) The intersection between two line segments must be a corner point.
(2) The end points of a complete line segment must be corner points.

In this paper, the corner feature extraction method based on geometric features is used to extract the corner feature of point cloud data [15], and the above segmsegmed area is recursively used for corner detection. If the starting point and the ending point in this area are on the same line, it indicates that the point farthest away from this line can be calculated by the following formula.

To sum up, it is necessary to determine whether the calculated distance is greater than the preset threshold. If it is greater than the given threshold, it indicates that the point is a corner point, and its feature extraction is carried out [16]. On the basis of the above, two-dimensional map is described through a straight line:

$$d = |(y(i) - y(1)) \cdot \cos \gamma + (x(i) - x(l)) \cdot \sin \gamma| \tag{4}$$

where, $(x(i), y(i))$ represents the cartesian coordinates of the I-th scan point, and γ represents the line segment formed by the connection of the starting point and the ending point, the Angle between the X axes, d is the straight-line distance between the points, $|\cdot|$ is the notation for the absolute value.

Set the first and the last points of region b_i as $p_m(X_m, Y_m)$ and $p_n(X_n, Y_n)$, respectively. The distance between the two points is set to be L. The distance between point M_k and L in region b_i is denoted as d_k, and set the distance threshold between this point and L as d_T, suppose $d_k > d_T$, Then point P_k can be used to divide the above segmentation points, Otherwise, the segmentation ends. On the basis of the above, the least square method is selected to fit the data in b_i, and the above operation process is continued until all the features are extracted. The details are shown in the following figure (Fig. 2):

The value of threshold d_T has different degrees of influence on the feature extraction results. If the value of d_T is large, the feature points will be omitted. And If the value of d_T is small, the segmentation will be too dense, Both of them will have negative effects on line segment extraction. On the basis of the above, the dynamic threshold method was improved, the proportion parameter was set, and the length of the line segment was dynamically adjusted by analyzing the different values of the proportion parameter, which could effectively increase the precision of muscle segment extraction and reduce the error.

2.2 Satellite Navigation and Positioning Based on Laser-Point Cloud Data

Based on the laser point cloud data preprocessing results obtained in Sect. 2.1, the overall flow chart of the proposed method is shown as follows:

The GPS part in Fig. 3 is mainly responsible for providing the absolute coordinate position information of the satellite, etc. The machine vision part mainly preprocesses the collected information to obtain the relative position coordinates of the known nodes in the navigation path. After the processing of the above two parts, the two groups of

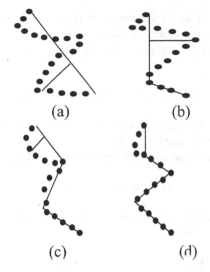

Fig. 2. Specific steps of line feature extraction

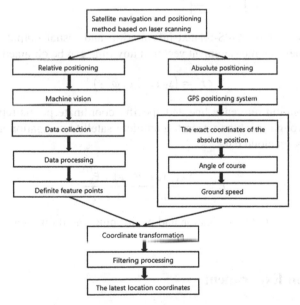

Fig. 3. Principle of satellite navigation and positioning based on laser-point cloud data

information are unified into the same coordinates, and the corresponding filtering is carried out to obtain the latest position information [17]. Machine vision positioning mainly obtains the navigation reference line in the image and determines the feature points in the navigation path. Through the selected candidate points, the Hough transform method based on the known candidate points is used for linear fitting, and the distribution center of the candidate points is obtained. After obtaining the specific navigation path, select any point and set it as the feature point representing the path, and record the specific position of the image in the coordinate system in detail.

The specific coordinate position of the target point can be obtained through the following formula:

$$x_p = x_v + x_p^{vision} \cos \varphi_v + y_p^{vision} \sin \varphi_v \tag{5}$$

φ_v represents the specific heading Angle of satellite navigation [18].

In order to effectively reduce the estimation error in the filtering estimation of state variables, the following nonlinear models should be considered:

$$x_k = A_{k-1} x_{k-1} + w_{k-1} \tag{6}$$

On the basis of the above, the operating state of the system and the covariance of filtering error are predicted, then:

$$\begin{cases} \xi_k^{(i)} = A_{k-1} \xi_{k-1}^{(i)}, i = 1, 2, \cdots, 2n \\ \widehat{x}_k = \sum_{i=0}^{2n} \omega_i^m \xi_k^{(i)} \end{cases} \tag{7}$$

By setting the above GPS-related information and visual output information as observation values, the measurement vector at any time can be obtained:

$$\zeta K = \left(x_{v,k}, y_{v,k}, \varphi_{v,k} \right) \tag{8}$$

Obtain navigation reference line and specific coordinate points representing crop row features through machine vision to complete satellite navigation and positioning based on laser point cloud data:

$$E = \frac{\left(x_{v,k}, y_{v,k}, \varphi_{v,k} \right) * \xi_k^{(i)}}{A_{k-1} x_{k-1} + w_{k-1}} \tag{9}$$

To sum up, the satellite navigation and positioning based on laser-point cloud data is completed.

3 Simulation Experiment

In order to verify the comprehensive effectiveness of the proposed satellite navigation and positioning method based on laser-point cloud data, City A is selected as the research object. Experimental environment: Cpu: Core4300, OS: Windows xp professional SP3, memory: 3G, experimental platform: Matlab2010a. The positioning accuracy and positioning efficiency were used as experimental indexes to compare and analyze the methods in literature [4], literature [5] and the proposed methods.

(1) Positioning accuracy (%):. In order to verify the positioning effect of the proposed method, the positioning accuracy of the method in reference [4], the method in reference

[5] and the proposed method is compared. The specific comparison results are shown in Fig. 4:

It can be seen from the analysis of Fig. 4 that the positioning accuracy of the proposed method has been significantly improved compared with the other two methods, indicating that the proposed method can accurately carry out satellite navigation and positioning. This is because in the process of data preprocessing before satellite navigation and positioning, the proposed method smooths the noise in order to reduce the interference of various external factors and the instrument itself, thus improving the accuracy of positioning.

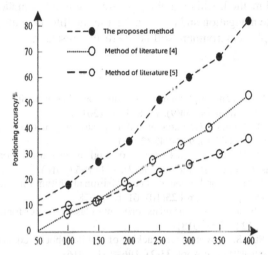

Fig. 4. Comparison results of positioning accuracy of different methods

Table 1. Comparison of operation efficiency

Number of simulation experiments	The proposed method	Method of literature [4]	Method of literature [5]
5	98	90	94
10	99	91	90
15	99	93	87
20	96	90	85
25	98	87	82
30	99	85	80

According to the analysis of Table 1, the maximum operating efficiency of the proposed method is 99%, which is the highest among the three methods. The operation efficiency of the method in literature [4] is the second. The operation efficiency of the method in reference [5] is the lowest among the three methods. The above data show that the proposed method can obtain the satellite navigation and positioning information in time, and reduce the positioning time, so that the practical application value of the method is higher.

4   Conclusion

Aiming at the problems existing in traditional satellite navigation and positioning methods, a satellite navigation and positioning method based on laser-point cloud data is proposed. The simulation results show that the proposed method has higher positioning accuracy, and the overall operation efficiency is significantly improved compared with the traditional method. Although the proposed method has made some progress in the study of satellite navigation and positioning, the satellite navigation and positioning technology in complex environment still needs further research.

References

1. Li, W., Fung, J.: Design and implementation of distributed laser positioning system based on TDOF. Chin. J. Sens. Actuators **30**(9), 1438–1446 (2017)
2. Song, K., Wang, B., Tang, C.: Research on indoor positioning method based on laser ranging scanning. Laser Infrared **46**(8), 938–942 (2016)
3. Chen, H., Yang, Z., Guo, P., et al.: Research on high-precision location algorithm of laser spot center. Trans. Beijing Inst. Technol. **36**(2), 181–185 (2016)
4. Xu, H., Zeng, F.: Research on P-code direct acquisition algorithm for satellite navigation and positioning. Electro-optics Control **25**(10), 61–65 (2018)
5. Lv, H., Chen, N., Zhong, Z.: Design and experiment of laser reflector for test satellite. Infrared Laser Eng. **005**(9), 122–130 (2018)
6. Yung, J., Liu, G., Sun, S.: Moving target tracking of mobile robot based on laser and monocular vision fusion. Control Theory Appl. **33**(2), 196–204 (2016)
7. Wu, B., Zhang, F., Qu, X., et al.: Design of monocular vision point tracking system for LiDAR. Nanotechnol. Precis. Eng. **16**(1), 93–100 (2018)
8. Zheng, S., Fang, L., Xu, Z.: Calibration and attitude correction of monocular vision-laser ranging positioning system. Mech. Sci. Technol. Aerospace Eng. **36**(12), 1926–1934 (2017)
9. Wang, X., Wang, X., He, M.: Background weighted mean shift target tracking algorithm based on histogram ratio. High Power Laser Part. Beams **28**(5), 13–17 (2016)
10. Huang, J., Li, Y., Hu, L.: Research on Lidar target tracking based on data association and improved statistical model. Automot. Eng. **40**(3), 356–362 (2018)
11. Chen, H., Sun, Y., Wang, Y., et al.: Chin. J. Lasers **45**(1), 160–167 (2018)
12. Li, Y., Wu, J., Zhao, S., et al.: Secondary synchronous switching method for low - and medium-orbit satellite cross-layer laser link. Acta Electronica Sinica **45**(3), 762–768 (2017)
13. Lu, F.Y., Hua, Z., Ran, L., et al.: Fast dynamic target location based on RFID phase and laser. Comput. Eng. **491**(08), 314–320 (2018)
14. Qian, F., Yang, M., Zhang, X.: Opt. Precis. Eng. **24**(11), 2880–2888 (2016)
15. Dong, W., Wang, X., Wu, N., et al.: Implementation of indoor visible light positioning system based on LED light intensity. Opt. Commun. Technol. **41**(3), 12–14 (2017)
16. Yang, X., Zhang, Z., Zhang, C., et al.: New optocoupler motor positioning technology based on STM32. Appl. Electron. Technique **479**(5):66–68+72 (2018)
17. Feng, X., Guo, Q., Han, C., et al.: Method for determining the position of zero optical path difference in interferogram. J. Infrared Millimeter Waves **36**(6), 795–798 (2017)
18. Zhao, B., Ma, G.: Research on center subpixel location of line laser fringe. Electron. Des. Eng. **25**(24), 184–188 (2017)

Research and Application of System with Bayberry Blockchain Based on Hyperledger Fabric

Hongyu Xiao[1], Zihang Gao[1,2], Xiaojun Cui[1(✉)], and Nannan Zhao[1,2]

[1] Wenzhou Vocational College of Science and Technology, Wenzhou 325006, China
cxjxhy@163.com
[2] Cangnan Industrial Research Institute of Modern Agriculture, Cangnan 325899, China

Abstract. This paper takes the performance analysis of bayberry blockchain system as the research topic. First, the bayberry blockchain system is applied and deployed based on the Hyperledger Fabric framework to realize the sharing of production, circulation, market, consumption and other data in the supply chain among relevant participants. Then the performance indicators such as query throughput, consensus throughput, consistency throughput, average latency, and fail rate of the bayberry blockchain system were analyzed.

Keywords: Hyperledger Fabric · Bayberry · Blockchain

1 Introduction

With the improvement of standard of living, consumption has put forward higher requirements for traceable high quality agricultural products. The timeliness and richness of information of the traditional post traceability system of agricultural products cannot meet the needs of consumers at all [1]. The blockchain technology can monitor the origin of bayberry and so assist construct reliable chains of bayberry supply and increase customer confidence [2].

Blockchain is a chained data structure that combines data blocks in chronological order and is cryptographically guaranteed to be an unchangeable and unforgeable distributed ledger. Bitcoin, Ethereum, and Hyperledger are all typical blockchain systems. Among them, Hyperledger Fabric is the most popular enterprise-level blockchain framework. Fabric adopts a loosely coupled design to modularize components such as consensus mechanism and identity verification, so that it can be easily selected according to the application scenario during the application process. Module. We designed the bayberry blockchain system is developed and applied using Hyperledger Fabric. The performance of Fabric is one of the issues that users are most concerned about. However, there is currently no authoritative and neutral organization that conducts performance tests on Fabric and gives test reports based on recognized rules. This is probably due to the following reasons. The first reason, Fabric is still under rapid development and has not yet provided detailed, neutral and recognized test rules. The second reason, Fabric

K. Li and Y. Liu (Eds.): ISICA 2023, CCIS 2147, pp. 187–195, 2024.
https://doi.org/10.1007/978-981-97-4396-4_17

network structure (network bandwidth, disk IO, computing resources, etc.), configuration parameters (such as block size, endorsement strategy, number of channels, status database, etc.), consensus algorithm (solo, kafka, pbft, etc.) will all affect the evaluation. As a result, it is difficult to build test models that reflect the full picture of the fabric. The third reason, Fabric transaction process is complex and has many differences from traditional databases, and it is not suitable for traditional testing solutions and tools.

This paper focuses on the bayberry blockchain system of Hyperledger Fabric, builds a test model, conducts actual tests on community-native Fabric and cloud blockchain (based on Fabric), identifies the performance bottlenecks of community-native Fabric, and attempts to use the dynamic scalability and rapid performance provided by cloud Blockchain. The PBFT algorithm is tuned to improve several key evaluation indicators.

2 Relate Work

According to research, in recent years, researchers have successively explored the application and deployment of blockchain technology in the agricultural products [3]. For example, Sajja [4] discusses blockchain technology applications in food supply chains, agricultural insurance, smart agriculture and agricultural goods transactions. Torky clearly discussed the main functions and strengths of the common blockchain platforms used in managing various sub-sectors in precision agriculture such as crops, livestock grazing, and food supply chain [5]. Hang [6] proofs the concept that integrates a legacy fish farm system with the Hyperledger Fabric blockchain is implemented on top of the proposed architecture. The efficiency and usability of the proposed platform are demonstrated through a series of experiments using various metrics.

3 Bayberry Blockchain System

3.1 System Framework

We take the application of information transparency in the entire supply chain of bayberry related agricultural products as the research object, and the ultimate goal is to solve the high quality and low price of bayberry related agricultural products and low consumer trust. The research work on the current information technology and related research progress, and in-depth investigation and research on the whole process of existing bayberry related agricultural product supply chain, and on this basis, put forward a research plan and research plan on the transparent supply chain of bayberry related agricultural products, and analyze the feasibility of the research plan. The system framework of the research is shown in Fig. 1.

Hardware and Infrastructure Layer. The blockchain infrastructure layer is the various hardware that helps run the network. The infrastructure support system includes professional sensors, intelligent monitoring equipment, 4G network, etc. Through 4G mobile communication network and other transmission methods to realize information interaction and transmission, sensor network nodes and video monitoring nodes transmit data to the data acquisition server. The computer hardware layer is the carrier of the

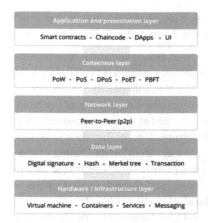

Fig. 1. The platform framework of the research

bayberry blockchain system. We rely on cloud services, it is equipped with the software and hardware environment required by the platform to meet public network conditions, including physical environments such as network systems, database systems, and web server systems.

Data Layer. The bayberry blockchain system data center is used to store and manage data, responsible for data storage logic rules, data reading and writing, data backup, etc. The data center of the bayberry traceability system mainly stores the data collected by the system, such as bayberry orchard data, bayberry batch data, bayberry product data, bayberry growth traceability data, user tracking data, blockchain data, etc. At the same time, the data center also includes a hard disk storage center, which mainly includes the collected picture data, the picture data uploaded by the system, and the anti-counterfeiting and traceable QR code picture data generated by the system.

Network Layer. The network layer in the blockchain performs a data acquisition and processing function. The data acquisition and processing program mainly includes meteorological data acquisition software, sensor data acquisition software, and monitoring picture acquisition software. Use the general data interface to collect the data of the hardware equipment deployed in the bayberry garden to the data center of the bayberry traceability system. The data collection is automatic, once an hour, and the data is stored in the database. The monitoring pictures are stored in the hard disk with the date as the file name, and the collected data is used as the data source of the anti-counterfeiting traceability data.

Consensus Layer. The consensus layer is an important layer in the operation of the blockchain. This layer is an important foundation in creating a decentralized network. So, without a consensus layer the blockchain network cannot verify transactions and no blocks can be validated. The main task of the consensus layer is to validate and sequence blocks and ensure all nodes reach an agreement on which blocks should be added. If several blocks are created simultaneously due to network congestion, the consensus layer ensures that only one block is added to the network. Blockchain protocols set the criteria for reaching a consensus. Every blockchain has different requirements for this. Additionally, this layer is also often called the consensus mechanism.

Application Layer. The application layer in the blockchain contains various protocols and technologies that users directly interact with. The application layer in the blockchain contains various protocols and technologies that users directly interact with blockchain management program. The blockchain management program first extracts data from the collected data as traceable data, and at the same time calls the data formatting module to format the data into JSON format. Traceability data is divided into two parts: automatic traceability data and manual traceability data. Automatic traceability data consists of weather data, sensor data, and picture data. The image data includes two parts: the hard disk link address of the image, and the hash value of the image. Since the picture is relatively large, the overhead of putting it into the blockchain is relatively high, so the data of the picture address and hash value are used. The image address can effectively find the corresponding image, and the hash value of the image is used to prevent tampering. If the hard disk image is tampered with later, the hash value will also change, and the comparison with the data in the blockchain will fail to pass the verification. Therefore, this method can reduce the block chain and network overhead while ensuring the security of the picture, effectively ensuring that the picture is the original picture collected by monitoring.

3.2 Key Technologies

The key technologies of this paper mainly include the following five parts.

Identification Technology. Identification is one of the keys throughout the supply chain. In-depth research on current mainstream one-dimensional codes, two-dimensional codes, RFID and other identification technologies, and select appropriate technologies to build a traceable quality safety and regulatory information platform.

Data Description Method. In view of the complex and diverse data types of the entire supply chain of alpine bayberry-related agricultural products, significant time-space coupling, strong randomness, and high discreteness, the project team intends to analyze the supply chain data of alpine bayberry-related agricultural products from the aspects of time, space, and source. The object idea clusters the agricultural product supply chain data, and on this basis, constructs the metadata standard of the bayberry-related agricultural product supply chain to meet the need for consistent expression of the agricultural product supply chain data.

Frontier Technologies of Information Science such as Blockchain, Big Data Storage and Utilization. The decentralization of the blockchain reduces the difficulty of trust endorsement, and can organize more authoritative supply chain participants to supplement and jointly maintain more commodity data and improve the trust of users. Blockchain technology cannot be tampered with, Features such as time stamps allow transactions to be traced back, and when supply chain data is tampered with, effective accountability can be achieved. On the basis of anonymity, the blockchain also uses various technical means such as encryption and verification methods to effectively protect.

We study the data structure composition of bayberry-related agricultural product supply chain, design the logical organization and storage strategy of seamless massive spatio-temporal data sets, and realize the efficient management of supply chain data. Combine the data sharing requirements of different granularities in practical applications, and design a data parallel sharing framework, to meet the needs of multiple users and multiple scenarios to concurrently access agricultural product supply chain data. Using cloud computing technology and big data technology to analyze and process the stored massive big data, it provides assistance for enterprises to make production decisions and the government to formulate relevant support policies. With the help of blockchain technology, farms, farmers, certification agencies, governments, sales companies, logistics and warehousing companies, etc. arc added to the alliance chain nodes, and the information on the commodity raw material circulation process, production process, commodity circulation process, and marketing process is integrated and consolidated. Write into the blockchain, and use the anti-tampering property of the blockchain to realize the whole process traceability of one object and one code.

Internet of Things Technology. In the era of big data, using the Internet of Things technology, any person, machine, or object connected to the network can be used as a means of information collection, such as various sensors (cameras, monitoring points) connected to the network, various machines (IP address) can collect information, and individuals can also collect location, social public opinion and social event information through mobile terminals. Ubiquitous information collection provides a richer source of data. When building the platform, make full use of monitoring cameras, sensors and other IoT devices for data collection.

HTML5 and CSS3 Cross-Platform Website Building and Mobile Screen Adaptive Technology. The HTML5 is the development trend of future websites. It not only allows the website to have a better visual user experience, but also can be adapted to multiple terminals on PC, mobile, tablet, and WeChat, which greatly reduces costs. Using HTML5 and CSS3 technology, a cross-platform website can be developed at one time, and PC mobile screen have a good browsing experience. Mobile display is superior, and it is convenient for social sharing and dissemination. Browsers like Firefox, Chrome, and Apple's Safari support direct playback of multimedia elements, and web pages can have more cool dynamic effects. It can be indexed by search engines, which is conducive to SEO promotion. Eliminate APP installation, easy to distribute and update, and solve the pain points of many APP distribution updates.

The adaptive mobile screen technology means that the page layout can be adjusted adaptively for different screen resolutions and sizes of the mobile terminal. a specific version. This concept was born to solve mobile internet browsers. Adaptive technology can provide users of different terminals with a more comfortable interface and better user experience, and with the popularity of large-screen mobile devices, more and more websites adopt this technology and innovate. This paper will use the mobile terminal screen adaptive technology, which can not only improve the user experience, but also greatly shorten the development time, and adapt to the current situation of various sizes of mobile devices in the mainstream market.

3.3 System Deployment

For the bayberry blockchain system consensus mechanism, the bottom layer of the system uses the alliance chain framework Hyperledger Fabric, and uses the object-oriented language Go language to develop smart contracts. The designed underlying architecture of the blockchain traceability system implements a blockchain system that uploads data through smart contracts by applying the container engine technology Docker to download the images of each module in the architecture. Each key control point is equivalent to an organization. The certificate authority server generates user certificates. Each organization configures the key-value database CouchDB to back up blockchain data to achieve distributed storage. The consensus module consists of the sorting node Kafka, the management node Zookeeper and the consensus node Orderer. The sorting node Kafka acts as a message queue in the alliance chain framework Hyperledger Fabric, sorting transactions sent by the consensus node Orderer.

The management node Zookeeper maintains the election process of the leader and follower of the sorting node Kafka in the consensus to ensure the normal operation of the consensus module; the consensus node Orderer is responsible for receiving the information incoming from the node and broadcasting the information sorted by the sorting node Kafka. This system deploys 6 hosts as servers. While realizing distributed multi-hosts, different hosts can synchronize information and share data, it also has high stability and reduces failures and even crashes. The network topology of bayberry blockchain system is shown in Fig. 2.

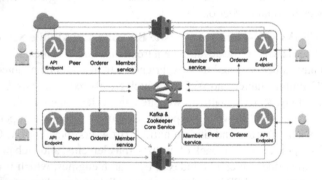

Fig. 2. The network topology of bayberry blockchain system

Divide nodes according to their computing functions and configure dedicated servers for different blockchains. These include Orderer node server, Peer node server, Comitter node server, Block browser application server, and System application server. This application solution can make the overall system reliable and efficient, but the disadvantage is that the deployment cost is high. The unified deployment of the above 5 types of servers on the remote cloud platform can greatly reduce the operating costs of the underlying technology of the blockchain, simplify the construction and operation and maintenance of the blockchain, and at the same time, face the information traceability of the entire industry chain of bayberry, with stronger capabilities. The system is robust, but the disadvantage is that later data migration will be more difficult.

In the system testing process involved in this project, two solutions were selected for simultaneous deployment to better reflect the impact of hardware limitations on the operating performance of the blockchain system under different needs. Most blockchain network topologies currently assume fixed parameters such as network bandwidth, packet loss rate, and RTT (Round-Trip Time). However, such a stable network status usually does not occur in actual application scenarios, especially in rural environments. Based on the above reasons, a series of network sniffing tools can be used to collect data packets sent and received in network scenarios within a certain period of time. After subsequent processing, the real-time network status can be reproduced through software routing, and finally network performance is optimized for different solutions.

4 System Performance Testing

4.1 Transaction Process Analysis

In the Fabric transaction process, different roles are involved, each role assumes different functions. Peer can be subdivided into Endorser node and submission Committer node, and consensus is completed by the Orderer.

The application client initiates a transaction proposal to the blockchain network through the SDK. The transaction proposal sends the contract identification, contract method and parameter information to be called for this transaction, as well as client signature and other information to Endorser. The proposal process of bayberry blockchain system is shown in Fig. 3.

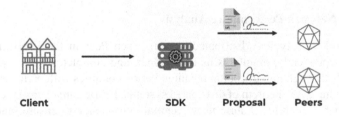

Fig. 3. The proposal process of bayberry blockchain system

After receiving the transaction proposal, the Endorser node verifies the signature and determines whether the submitter has the right to perform the operation. After passing the verification, the smart contract is executed, and the result is signed and returned to the application client.

After receiving the information returned by the Endorser node, the application client determines whether the proposal results are consistent and whether it is executed with reference to the specified endorsement policy. If there is not enough endorsement, the application client terminates the processing; otherwise, the application client. The client packages the data together to form a transaction, signs it, and sends it to Orderers.

Orderers perform consensus sorting on the received transactions, then package a batch of transactions together according to the block generation strategy, generate a new block, and send it to the Committer node. The transaction submission process of bayberry blockchain system is shown in Fig. 4.

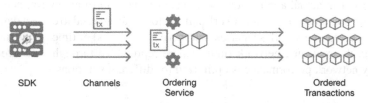

SDK   Channels   Ordering   Ordered
                               Service   Transactions

Fig. 4. The transaction submission process of bayberry blockchain system

After receiving the block, the Committer node will verify each transaction in the block, check whether the input and output the transaction depends on are consistent with the status of the current blockchain, and append the block to local blockchain and modify the world state.

To complete a transaction through Fabric, the client must perceive three steps (collecting endorsements, submitting sorting, and confirming the results), while reading and writing from a traditional database only requires initiating a request and waiting for confirmation. If you use classic testing tools such as JMeter, you need to wrap the fabric SDK with the RESTFul interface, which increases the complexity of the evaluation. The evaluation results in this article are all tested and generated by the Caliper tool.

4.2   Fabric Network Performance Analysis

Fabric network is a typical distributed system. Each Peer in the Fabric network is deployed independently, maintains its own ledger, and completes status synchronization internally through gossip communication. Fabric complies with partition tolerance. According to the CAP theorem of distributed systems, Fabric cannot ensure consistency while ensuring availability. Fabric uses eventual consistency to ensure that all nodes finally reach an agreement on the world state. This process is the process of Orderer consensus and Peer verification and confirmation. Therefore, in our test model, we mainly examine the indicators such as query throughput, consensus throughput, consistency throughput, average latency, and fail rate of the bayberry blockchain system.

Each Peer node appends the block to the channel's chain, and for each valid transaction, the write set is submitted to the current state database. The bayberry blockchain system will send out an event to notify the SDK that this transaction has been immutably attached to the chain, and will also notify the transaction verification result whether it is valid or invalid.

Fabric network query performance is actually an endorsement request. The peer side mainly contains three processes. The three processes are to verify proposal signature, check whether the channel ACL is met, and simulate the execution of transactions and sign the results. It can be seen that the read performance of a single node (8 vCPU, 16G) is around 1800 TPS (Transaction Per Second). Observing the monitoring indicators, we found that the CPU usage is around 75%, which is close to full load, while the memory usage is only around 20%. It is not difficult to understand that the endorsement process involves a lot of verification and signature work, which are both computationally intensive. Type operation. According to the partition tolerance of the blockchain in compliance with the CAP theorem, we can horizontally expand the peers within the organization to improve performance.

5 Conclusion

This paper has completed the application and performance analysis of bayberry blockchain system. The system includes two parts, the hardware deployment and the software construction. The hardware mainly includes video surveillance equipment and sensor equipment. The video surveillance and sensors need to be deployed in the target bayberry garden as a data source for data collection. The software includes the bayberry blockchain system running and performance analysis. Our system performance testing work mainly consists of two parts. The first is to extract the evaluation index according to the business characteristics, and the second is to establish a stable and testable business model.

Acknowledgments. This work was supported by the Basic Agricultural Science and Technology Project of Wenzhou under Grant N20220003 and National Social Science Found of China "Reaserch on Virtual Reality Media Narrative" (Grant No. 21&ZD326).

References

1. Dasaklis, T.K., Voutsinas, T.G., Tsoulfas, G.T., et al.: A systematic literature review of blockchain-enabled supply chain traceability implementations. Sustainability **14**(4), 2439 (2022)
2. Xiong, H., Dalhaus, T., Wang, P., et al.: Blockchain technology for agriculture: applications and rationale. Front. Blockchain **3**, 7 (2020)
3. Demestichas, K., Peppes, N., Alexakis, T., et al.: Blockchain in agriculture traceability systems. a review. Appl. Sci. **10**(12), 4113 (2020)
4. Sajja, G.S., Rane, K.P., Phasinam, K., et al.: Towards applicability of blockchain in agriculture sector. Mater. Today: Proc. **80**, 3705–3708 (2023)
5. Torky, M., Hassanein, A.E.: Integrating blockchain and the internet of things in precision agriculture: Analysis, opportunities, and challenges. Comput. Electron. Agric. **178**, 105476 (2020)
6. Hang, L., Ullah, I., Kim, D.H.: A secure fish farm platform based on blockchain for agriculture data integrity. Comput. Electron. Agric. **170**, 105251 (2020)

A Multiparty Reversible Data Hiding Scheme in Encrypted Domain Based on Hybrid Encryption

Bing Chen, Lu Chai, Yong Wang[✉], Jingkun Yu, and Wanhan Fang

School of Cyber Security, Guangdong Polytechnic Normal University, Guangzhou 510665, China

26613213@qq.com

Abstract. The existing multiparty reversible data hiding in encrypted domain (MRDHED) is unstable since its embedding capacity depends on the original image distribution or the number of generated encrypted images. In this paper, a stable MRDHED method by hybrid encryption is proposed. The advanced encryption standard (AES) algorithm is adopted to encrypt an original image to generate an AES-based image. The AES-based image is divided into multiple non-overlapping sub-images by image segmentation. The sub-images are reassembled to generate multiple reassembled images in a loop. A (2, 2) Chinese remainder theorem-based (CRT-based) secret sharing is used to encrypt the reassembled images to obtain the corresponding encrypted images. With the homomorphism of the CRT-based secret sharing, the secret message can be concealed into each encrypted image. Experimental results exhibit that the proposed method has stable embedding capacity regardless of the original image distribution and the number of generated encrypted images, while improving the computational overhead.

Keywords: Reversible Data Hiding · Multiparty Data Hiding · Hybrid Encryption · Secret Sharing

1 Introduction

When it comes to media security techniques, data hiding is a popular technique by embedding secret message into digital media [1, 2]. As a branch of data hiding, reversible data hiding in encrypted domain (RDHED) conceals secret message into an encrypted image to generate a marked encrypted image, and reconstructs the concealed secret message and the original image from the marked encrypted image [3, 4]. Due to the reversibility of RDHED, it is mainly used for tampering detection and integrity authentication in military, judicial and medical fields.

In recent years, many significant results have been achieved for RDHED. In [5], Puteaux et al. proposed two high-capacity schemes by predicting the most significant bit (MSB), namely, the prediction error correction based method and the embedded prediction error method. To make full use of the redundancy of original image, Yin et al. [6] adopted multi-MSB prediction and Huffman coding to reserve embedding room for

© The Author(s), under exclusive license to Springer Nature Singapore Pte Ltd. 2024
K. Li and Y. Liu (Eds.): ISICA 2023, CCIS 2147, pp. 196–207, 2024.
https://doi.org/10.1007/978-981-97-4396-4_18

data hiding. In [7], Yi et al. proposed that the small pixel blocks of images are labeled via adopting a parametric binary tree (PBT), and block permutation and modulation are introduced to preserve the correlations of the small pixel blocks for data hiding. Inspired by the method of [7], Wu et al. [8] treated the whole original image as a block and labeled the block using a PBT algorithm, which realizes improved embedding capacity. In [9], chunk encryption with exclusive-or operation is employed to encrypt the original image to preserve the correlation of the image. The redundancy matrix is introduced to represent the chunks in encrypted image to create embedding room. Similarly, block encryption with same random value is employ to preserve the correlation of the image [10], and thus the prediction difference of the encrypted image is close to that of the original image. In [11], Fu et al. used adaptive coding and stream cipher to preserve the correlation of the image and adaptively compressed part of the image to create embedding room. In order to make the encrypted images generated by symmetric encryption easy for data hiding, homomorphic encryption is employed to reconstruct the encrypted images in [12], which provides comprehensive merits in data expansion, computational overhead, embedding capacity, etc. In addition, Qiu et al. [13] proposed a general RDHED framework to address the issue of different embedding capacities caused by different reserving room techniques.

Different from the RDHED scheme, multiparty reversible data hiding in encrypted domain (MRDHED) encrypts the image into different encrypted images, and then assigns these encrypted images to different data hiders for data embedding [14, 15]. MRDHED scheme ensures that even if some data hiders are damaged by attacks, the receiver can still obtain a sufficient number of marked encrypted images from unaffected data hider for data extraction and image reconstruction, which further improves the security of the original image. In [16], a secret sharing-based MRDHED model with four cases is given, and a separable case with high embedding capacity is derived. In this method, the overflow pixels in the original image are shrunk by labeling the location information and embedding the location information into the original image. The generated encrypted images are easily adopted for data hiding by replacing least significant bits with the bits of secret message. In [17], another MRDHED scheme with improved secret sharing is proposed, in which multiple pixels are encrypted at a time, and thus the generated encrypted images are smaller than the original image. In addition, the embedding capacity is improved when more encrypted images are generated compared with the method in [16].

However, the MRDHED scheme cannot achieve stable performance, that is, the embedding capacity varies with the original image distribution or the number of generated encrypted images. To this end, this paper proposes a stable MRDHED scheme based on hybrid encryption. In the proposed scheme, a hybrid encryption method combining advanced encryption standard (AES) and Chinese remainder theorem-based (CRT-based) secret sharing is used to encrypt the original image to generate multiple encrypted images. The secret message is concealed into each encrypted image to generate corresponding marked encrypted image by using the homomorphism of the CRT-based secret sharing. The receiver collects enough marked encrypted images to reconstruct the original image and concealed secret messages. The main contributions of this paper are given as follows:

1) Stable embedding capacity. The embedding capacity is not affected by the original image distribution and the number of generated encrypted images since the embedding room is created by the image encryption algorithm.

2) Reduced computational overhead. The proposed scheme directly performs hybrid encryption operation on the original image without additional operations, which effectively reduces the computational overhead in the image encryption stage.

The remainder of this paper is organized as follows. Section 2 briefly introduces a CRT-based secret sharing. In Sect. 3, the details of the proposed MRDHED scheme based on hybrid encryption are presented, while Sect. 4 provides the experimental results and analysis. Finally, the conclusion is given in Sect. 5.

2 Preliminary

In this paper, a (t, n) threshold CRT-based secret sharing [18] is used for image encryption, where $2 \leq t \leq n$. It splits a secret into n shares and distributes them to multiple participants. Any t or more shares can reconstruct the secret, but any $t - 1$ or less shares give no clue of the secret. The details are described as follows.

To split a secret s, a prime q and n strictly increasing modulus $k_1, k_2, \cdots k_n$ are selected, and satisfied.

1) $q > s$,
2) $gcd(k_i, k_j) = 1, i \neq j. gcd(\cdot)$ is the greatest common divisor,
3) $gcd(q, k_i) = 1$,
4) $M' = \prod_{l=1}^{t} k_l$ and $N' = q \prod_{l=1}^{t-1} k_{n-l+1}$ with $M' > N'$,

where $i = 1 \cdots n, j = 1 \cdots n$. For each selected k_i, a corresponding share y_i is calculated by the congruence formula, that is,

$$y_i \equiv y(mod \, k_i), \tag{1}$$

where $y = s + Aq$, A is a random integer with $0 \leq A < \frac{M'}{q} - 1$, and $\lfloor \cdot \rfloor$ is the floor function. The generated share y_i is distributed to i th participant.

On the other hand, the secret s can be reconstructed by collecting the shares of any t participants. Without loss of generality, it is assumed that any t shares $y_{i_1}, y_{i_2}, \cdots, y_{i_t}$ are collected. With modulus $k_{i_1}, k_{i_2}, \cdots, k_{i_t}$, a congruence equation is constructed by

$$\begin{cases} y \equiv y_{i_1} (mod \, k_{i_1}) \\ y \equiv y_{i_2} (mod \, k_{i_2}) \\ \cdots \\ y \equiv y_{i_t} (mod \, k_{i_t}) \end{cases} \tag{2}$$

According to the Chinese remainder theorem, it follows from Eq. (2) that

$$y \equiv \sum_{l=1}^{t} y_{i_l} N_l N_l^{-1} (mod \, N), \tag{3}$$

where $N = \prod_{l=1}^{t} k_{i_l}$, $N_l = N/k_{i_l}$, and $N_l N_l^{-1} = 1 \bmod k_{i_l}$. Thus, the secret s can be calculated by $y \bmod q$.

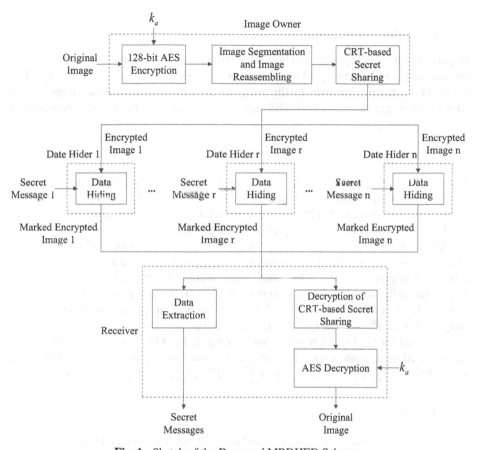

Fig. 1. Sketch of the Proposed MRDHED Scheme.

3 The Proposed Scheme

In this section, a MRDHED method based on hybrid encryption is proposed, in which AES algorithm and CRT-based secret sharing are used for hybrid. The framework of the proposed scheme is shown in Fig. 1, which involves image encryption stage, data hiding stage, and data extraction and image reconstruction stage. In the image encryption stage, the image owner encrypts the original image to generate multiple encrypted images using a hybrid encryption algorithm. The hybrid encryption combines AES algorithm and CRT-based secret sharing. The generated encrypted images are sent to different data hiders for data hiding. In the data hiding stage, each data hider first encrypts the message into the secret message, and then conceals the secret message into the encrypted image to generate a marked encrypted image. By collecting sufficient marked encrypted images from undamaged data hiders, the receiver can either extract the concealed secret message or reconstruct the original image.

3.1 Image Encryption Stage

To perform image encryption, a hybrid encryption algorithm is introduced, in which the original image is first encrypted by AES algorithm and then further encrypted by CRT-based secret sharing. Specifically, the image owner encrypts an original image into an AES-based image using a 128-bit AES algorithm with an encryption key, as represented as

$$I = EA_{k_a}(C), \tag{4}$$

where C is the original image, $EA_{k_a}(\cdot)$ is the 128-bit AES algorithm with encryption key k_a, and I is the generated AES-based image. In fact, the AES algorithm is a block cipher, i.e., one block with 128 bits is encrypted at a time. The original image is thus divided into blocks and then encrypted according to Eq. (4).

After that, the image owner converts the AES-based image into multiple encrypted images by using a CRT-based secret sharing. Firstly, the AES-based image is divided into n non-overlapping sub-images by image segmentation. Let the sub-images be I_u, $1 \leq u \leq n$. Secondly, the n sub-images are reassembled in a loop to generate n reassembled images, denoted as R^u, $1 \leq u \leq n$. Each reassembled image is composed of r sub-images, that is, $R^u = (I_{u_1}, I_{u_2}, \cdots, I_{u_r})$, $u_v = (u_{v-1} + 1) mod n$, $1 \leq u_1 \leq n$, $2 \leq r \leq n$. Finally, a (2, 2) CRT-based secret sharing by Eq. (1) is used to converted the n reassembled images into n encrypted images. Let the n encrypted images be $[P^{u1}, P^{u2}]$, where $P^{u1} = (P_{u_1}^{u1}, P_{u_2}^{u1}, \cdots, P_{u_r}^{u1})$ and $P^{u2} = (P_{u_1}^{u2}, P_{u_2}^{u2}, \cdots, P_{u_r}^{u2})$, respectively.

3.2 Data Hiding Stage

When receiving the encrypted image, each data hider can conceal secret message into it to generate a marked encrypted image, where the secret message is obtained by encrypting the original message. To generate a marked encrypted image, the data hider modifies a portion of encrypted image according to the secret message. Denote a bit of secret message as $b \in \{0, 1\}$. For simplicity, a pair of encrypted elements (P_1^{11}, P_1^{12}) is taken as an illustration. Given a threshold T, if the encrypted element $P_1^{11} < T$ and $b = 0$, or the encrypted element $P_1^{11} \geq T$ and $b = 1$, the encrypted element P_1^{11} (resp. P_1^{12}) is directly set to the marked encrypted element, denoted as PM_1^{11} (resp. PM_1^{12}). On the contrary, the marked encrypted element is calculated by

$$PM_1^{11} = P_1^{11} + U_1^{11}(mod\ k_1) \tag{5}$$

and

$$PM_1^{12} = P_1^{12} + U_1^{12}(mod\ k_2), \tag{6}$$

where U_1^{11} and U_1^{12} are calculated by Eq. (1) with $s = 0$ and different random integer A, respectively. Similarly, the encrypted elements in different data hider can also be used to conceal secret message. With the corresponding secret messages, the data hiding is implemented by modifying the encrypted images. Accordingly, the marked encrypted images are generated.

3.3 Data Extraction and Image Reconstruction Stage

By collecting sufficient marked encrypted image, the original image can be reconstructed, as well as the concealed secret message. According to the hybrid encryption algorithm, the receiver needs to obtain any $n - r + 1$ marked encrypted images to achieve image reconstruction. Meanwhile, the concealed secret message is extracted from the corresponding $n - r + 1$ marked encrypted image.

For image reconstruction, the receiver first performs the decryption of the CRT-based secret sharing and then the decryption of the AES algorithm. In the CRT based secret sharing, the reassembled image is obtained by directly decrypting the marked encrypted images. For a pair of marked encrypted elements (PM_1^{11}, PM_1^{12}), the associated reassembled unit is calculated by

$$R_1^1 \equiv PM_1^{11}k_2k_2^{-1} + PM_1^{12}k_1k_1^{-1}(mod\ k_1k_2). \tag{7}$$

In this way, all the reassembled elements can be calculated, and the reassembled image is reconstructed. Subsequently, the AES-based image I is obtained from the restored reassembled image. Finally, the original image C can be reconstructed by decrypting the AES-based image, that is,

$$C = DA_{k_a}(I), \tag{8}$$

where $DA_{k_a}(\cdot)$ is the decryption of AES algorithm with encryption key k_a.

For the data extraction, the receiver first extracts the concealed secret message, and then decrypts the secret message to get the original message. The concealed secret message is extracted by judging the relationship between encrypted element and threshold T. Specifically, the bit of secret message is extracted by

$$b = \begin{cases} 0, & if\ PM_1^{11} < T \\ 1, & if\ PM_1^{12} \geq T \end{cases} \tag{9}$$

When all the concealed bits are extracted from a marked encrypted image, a secret message is obtained. When all the secret messages are extracted from the $n - r + 1$ marked encrypted images, the data extraction is accomplished.

4 Experimental Results and Analysis

Six grayscale images with different features selected from the USC-SIPI database [19] are used for experiments, all of which are sized by 512×512, as shown in Fig. 2. For color images, we convert them into grayscale images using the weighting method. The message to be concealed is randomly generated through a pseudorandom number generator.

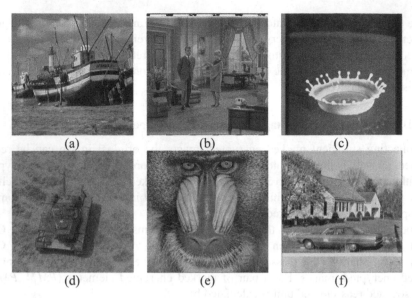

Fig. 2. Six test images. (a) Boat, (b) Couple, (c) Splash, (d) Tank, (e) Baboon, and (f) House.

4.1 Feasibility

The feasibility experiment of the proposed method is given in Fig. 3, in which the AES-based image is divided into 4 non-overlapping sub-images, and a reassembled image consists of 2 sub-images. The image Boat is encrypted by using the 128-bit AES algorithm, and an AES-based image Boat is generated, as shown in Fig. 3(f). The AES-based image Boat is reassembled, and four reassembled images are generated, as shown in Fig. 3(b)–(e). With a (2, 2) CRT-based secret sharing, the four reassembled images are converted into four encrypted images, as shown in Fig. 3(g)–(j). For each encrypted image, a bit is concealed into a pair of encrypted elements to generate a marked encrypted element. Figure 3(k)–(n) shows the corresponding four marked encrypted images. By decrypting any three marked encrypted images, a reconstructed image can be obtained. Figure 3(o)–(r) shows three reconstructed images, all of which are the same as the original image Boat.

4.2 Embedding Capacity

Generally, embedding capacity is measured by embedding rate (bit per pixel, bpp), as calculated as

$$Embedding\ rate = \frac{Total\ bits\ embedded\ in\ each\ encrypted\ image}{Total\ pixels\ of\ each\ encrypted\ image}.$$

In the proposed method, a bit is concealed into a pair of encrypted elements using homomorphism of CRT-based secret sharing. In [16], bit replacement is used to conceal a bit into a bit-plane of a partially encrypted pixel. Both of the method in [16] and the proposed method vacate embedding room after image encryption by the data hiders.

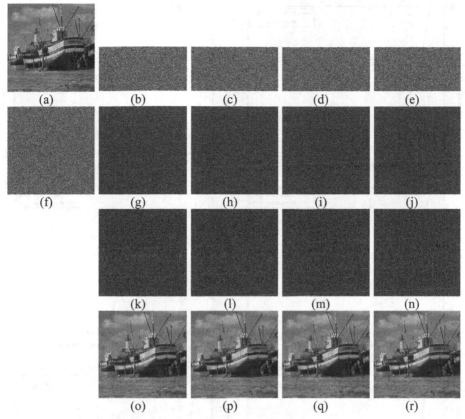

Fig. 3. Experimental results of the proposed method in terms of feasibility. (a) The original image, (b)–(e) the four reassembled images generated from the AES-based image, (f) the AES-based image, (g)–(j) the four encrypted images, (k)–(n) the four marked encrypted images, (o)–(r) the four reconstructed image from any three marked encrypted images.

Instead, the method in [17] vacates embedding room before image encryption by the content owner. The method can obtain a high embedding rate since the embedding room is reserved from the original image for data hiding. Of course, the embedding rate of the method is determined by the distribution of the original image. Smoothly distributed images can achieve a high embedding rate, while complexly distributed images can only achieve a lower embedding rate. However, content owner is only willing to encrypt the original image due to the limitations of the computing source. Therefore, the proposed method is compared with the method in [16].

The result of the comparison is shown in Fig. 4, where the number of generated encrypted images is considered. It is shown that the method in [16] achieves a higher embedding rate when the number of generated encrypted images is small, and the embedding rate decreases with the increase of the number of generated encrypted images. The embedding rate of the proposed method does not change with the change of the number

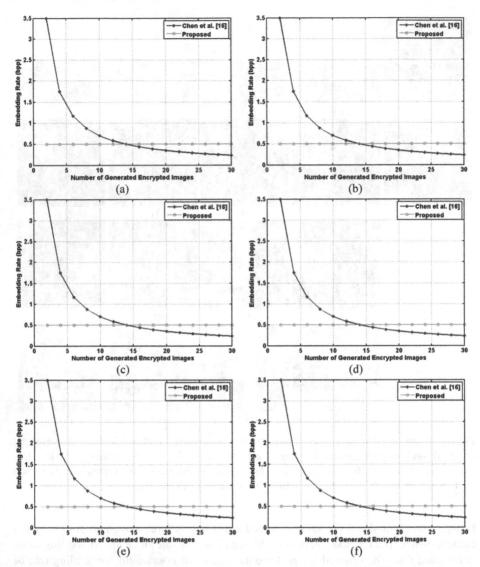

Fig. 4. Comparison of the embedding rate for different test images with different numbers of encrypted images. (a) Boat, (b) Couple, (c) Splash, (d) Tank, (e) Baboon, and (f) House.

of generated encrypted images. It means that the proposed method can achieve a stable embedding rate, that is, the distribution of the images and the number of generated encrypted images will not affect the embedding rate.

4.3 Computational Overhead

This experiment aims to evaluate the computational overhead of the proposed method and the method in [17]. All the programs are developed by Matlab 2010b, and run on 64-bit Windows 10 Professional with Intel Core i7-11700 CPU @2.50 GHz, 16 GB RAM. In the proposed method, the original image is encrypted into encrypted images by hybrid encryption. In the method of [17], to encrypt the original image into encrypted images, the proportion of correctly predicted pixels is calculated using median edge detector under a given dataset, and the optimal embedding layer is obtained based on the calculated proportion. With the optimal embedding layer, the multiple MSBs of pixels are predicted to obtain prediction error and the predictable pixels and unpredictable pixels are generated with the prediction error. Finally, the reference information is adopted to handle overflow pixels. So far, the handled pixels could be used for encryption. Obviously, there are a lot of additional operations on the original image, which increases computational overhead in the image encryption stage. The comparison of the computational overhead of the image encryption stage is illustrated in Table 1, in which $n = 4$ and $n = 8$ are considered, respectively. That is, four encrypted images and eight encrypted images are generated in the experiment, respectively. Besides, $r = 2$ is adopted in the proposed method, and $t = 3$ and $t = 7$ is used in [17] for $n = 4$ and $n = 8$, respectively. It can be seen that the proposed method owns lower computational overhead in the image encryption stage.

Table 1. Computational overhead comparison (in second) under different encryption strategies for different test images.

Image	Hua et al. [17]		Proposed	
	$n = 4$	$n = 8$	$n = 4$	$n = 8$
Boat	172.49	104.41	17.38	16.76
Couple	144.04	96.74	17.35	16.98
Splash	175.16	106.52	17.49	16.74
Tank	170.82	103.26	17.39	16.70
Baboon	160.35	100.31	17.39	16.83
House	198.90	115.68	17.28	16.58

5 Conclusion

This paper presents a MRDHED method based on hybrid encryption, in which the stable embedding rate and improved computational overhead are achieved. In the image encryption, an original image is firstly encrypted into an AES-based image by the 128-bit AES algorithm, and then, the AES-based image is divided into n non-overlapping sub-images by image segmentation. The sub-images are reassembled into n reassembled images with r sub-images in a loop. Finally, the n reassembled images is encrypted into n encrypted images by a (2, 2) CRT-based secret sharing. To achieve data hiding, the encrypted images are modified to marked encrypted images by the homomorphism of the CRT-based secret sharing. On the receiver side, with any $n - r + 1$ marked encrypted images, the reassembled images is primarily reconstructed by the decryption of the CRT-based secret sharing, and then the original image is reconstructed by the inverse operation of the AES algorithm. Experimental results and analysis are also given to illustrate the effectiveness of the presented method.

Acknowledgment. This work is supported by the National Natural Science Foundation of China (No. 62102101) and the Doctoral Scientific Research Foundation of Guangdong Polytechnic Normal University (No. 2021SDKYA101). Bing Chen and Lu Chai contributed equally to this work, and they are co-first authors of this work.

References

1. Li, G., Feng, B., He, M., Weng, J., Lu, W.: High-capacity coverless image steganographic scheme based on image synthesis. Signal Process.: Image Commun. **111**, 116894 (2023)
2. Chen, B., Wu, X., Lu, W., Ren, H.: Reversible data hiding in encrypted images with additive and multiplicative public-key homomorphism. Signal Process. **164**, 48–57 (2019)
3. Wu, H.T., Cheung, Y.M., Zhuang, Z., Xu, L., Hu, J.: Lossless data hiding in encrypted images compatible with homomorphic processing. IEEE Trans. Cybern. **53**(6), 3688–3701 (2023)
4. Qin, C., Qian, X., Hong, W., Zhang, X.: An efficient coding scheme for reversible data hiding in encrypted image with redundancy transfer. Inf. Sci. **487**, 176–192 (2019)
5. Puteaux, P., Puech, W.: An efficient MSB prediction-based method for high-capacity reversible data hiding in encrypted images. IEEE Trans. Inf. Forensics Secur. **13**(7), 1670–1681 (2018)
6. Yin, Z., Xiang, Y., Zhang, X.: Reversible data hiding in encrypted images based on multi-MSB prediction and Huffman coding. IEEE Trans. Multimedia **22**(4), 874–884 (2020)
7. Yi, S., Zhou, Y.: Separable and reversible data hiding in encrypted images using parametric binary tree labeling. IEEE Trans. Multimedia **21**(1), 51–64 (2019)
8. Wu, Y., Xiang, Y., Guo, Y., Tang, J., Yin, Z.: An improved reversible data hiding in encrypted images using parametric binary tree labeling. IEEE Trans. Multimedia **22**(8), 1929–1938 (2020)
9. Liu, Z.L., Pan, C.M.: Reversible data hiding in encrypted images using chunk encryption and redundancy matrix representation. IEEE Trans. Dependable Secure Comput. **19**(2), 1382–1394 (2022)
10. Xu, D., Su, S.: Reversible data hiding in encrypted images with separability and high embedding capacity. Signal Process.: Image Commun. **95**, 116274 (2021)
11. Fu, Y., Kong, P., Yao, H., Tang, Z., Qin, C.: Effective reversible data hiding in encrypted image with adaptive encoding strategy. Inf. Sci. **494**, 21–36 (2019)
12. Chen, B., Yin, X., Lu, W., Ren, H.: Reversible data hiding in encrypted domain by signal reconstruction. Multimed. Tools Appl. **82**, 1203–1222 (2023)
13. Qiu, Y., Ying, Q., Yang, Y., Zeng, H., Li, S., Qian, Z.: High-capacity framework for reversible data hiding in encrypted image using pixel prediction and entropy encoding. IEEE Trans. Circuits Syst. Video Technol. **32**(9), 5874–5887 (2022)

14. Hua, Z., Wang, Y., Yi, S., Zheng, Y., Liu, X., Chen, Y., Zhang, X.: Matrix-based secret sharing for reversible data hiding in encrypted images. IEEE Trans. Dependable Secure Comput. **20**(5), 3669–3686 (2023). https://doi.org/10.1109/TDSC.2022.3218570

15. Xiong, L., Han, X., Yang, C.N., Shi, Y.Q.: Robust reversible watermarking in encrypted image with secure multi-party based on lightweight cryptography. IEEE Trans. Circuits Syst. Video Technol. **32**(1), 75–90 (2022)

16. Chen, B., Lu, W., Huang, J., Weng, J., Zhou, Y.: Secret sharing based reversible data hiding in encrypted images with multiple data-hiders. IEEE Trans. Dependable Secure Comput. **19**(2), 978–991 (2022)

17. Hua, Z., Wang, Y., Yi, S., Zhou, Y., Jia, X.: Reversible data hiding in encrypted images using cipher-feedback secret sharing. IEEE Trans. Circuits Syst. Video Technol. **32**(8), 4968–4982 (2022)

18. Asmuth, C., Bloom, J.: A modular approach to key safeguarding. IEEE Trans. Inf. Theory **29**(2), 208–210 (1983)

19. The USC-SIPI image database Homepage. http://sipi.usc.edu/database/. Last accessed 16 Aug 2023

Research on Smart Agriculture Big Data System Based on Spark and Blockchain

Yuming Sun and Hua Wang[✉]

Guangdong University of Science and Technology, Dongguan 523000, China
460864372@qq.com

Abstract. In order to improve the intellectualization of the agricultural industry, this paper designs and implements a secure agricultural big data ecosystem on Spark big data platform, which allows farmers and wholesalers to know the sales status of the agricultural market in real time. All sales data are distributed in different agricultural institutions, and some are even decentralized registration, to ensure the independent storage of these data. Blockchain, as a distributed ledger technology for multiple parties to maintain and back up information security, is a good breakthrough for agricultural data sharing innovation. This system implements the agricultural big data center on the Spark big data platform. Through the data synchronization module and independent data acquisition system, the original distributed data is centrally stored and analyzed. The agricultural products sales information system uses the advantages of Spark big data platform to provide personalized agricultural products services for farmers and wholesalers.

Keywords: Big data · Blockchain · Sales of Agricultural Products · Spark

1 Introduction

The development pattern of agriculture can be divided into three stages, namely, traditional agriculture with manual labor as its main feature, modern agriculture with machinery and equipment as its main feature, and smart agriculture with the help of emerging technologies. Agriculture 3.0 is a smart agriculture that uses emerging technologies to boost agricultural production. Today, with continuous technological innovation and development, emerging technologies such as cloud computing, Internet of Things and big data have been adopted in the whole life cycle of agricultural production, processing, transportation and e-commerce sales, making the agricultural production process more intelligent, agricultural sensor data more diverse, and agricultural monitoring and early warning more accurate. The birth of smart agriculture will greatly improve the reliability and efficiency of agricultural production, ensure the sustainable development of agriculture at the same time, and effectively advance agricultural information from traditional to digital to intelligent. Combined with big data technology, accurate perception, efficient circulation and dynamic update of agricultural data can be realized. Through the construction of agricultural big data resource pool, the transformation of "data-driven" thinking of smart agricultural decision-making and management can be established, and a

© The Author(s), under exclusive license to Springer Nature Singapore Pte Ltd. 2024
K. Li and Y. Liu (Eds.): ISICA 2023, CCIS 2147, pp. 208–213, 2024.
https://doi.org/10.1007/978-981-97-4396-4_19

new digital agricultural ecology of perception, decision-making, control and assessment with data can be formed.

The smart agriculture big data architecture is composed of the base layer, the perception layer, the support layer, the application and display layer, the unified security operation and maintenance guarantee system and the standard specification system. Unified security guarantee, standards and operation and maintenance guarantee are the basis and premise of smart agriculture big data architecture, which can effectively constrain the overall construction process and guarantee the safety, operation and maintenance of the overall agricultural information construction. The infrastructure layer is a basic platform that provides computing, storage, network and security resources for the construction of smart agriculture. It can provide security guarantee for all kinds of data resources, prevent the loss and leakage of data information, and realize unified management and scheduling of data resources for different businesses. After the basic environment is guaranteed, it is the establishment of the perception system of the Internet of Things. Unlike other industries, agriculture needs various sensors and video monitoring devices to carry out real-time monitoring of the whole life cycle of crops, livestock and poultry. Therefore, intelligent agriculture needs to deploy video monitoring, agricultural conditions monitoring equipment, temperature and humidity sensors, light sensors and other devices to ensure the source of basic agricultural data. In addition to sensing and monitoring data, the overall construction of smart agriculture also needs to collect agricultural production environment, marketing, management and control data. Through the agricultural big data resource center, various data sources are collected and stored, so as to form basic database, theme database and thematic database to support various smart agricultural application systems built in the application and display layer.

The system uses the Spark big data platform to store massive data. Convenient scalability greatly reduces the difficulty of storage and system upgrade. Based on the above considerations, Spark platform, a personalized safety-oriented big data ecosystem, is established. The system uses the Spark big data platform for massive data storage and scalability, greatly reducing storage and maintenance costs. The Spark Big Data platform implements the agricultural big data center, which uses the data synchronization module and an independent data acquisition system to centrally store and analyze the original distributed data. As the data scale of users and agricultural products continues to expand, the computing capability of the Spark cluster can solve the problem of large-scale concurrent access during traffic peaks, ensuring the robustness of the system pairs.

2 Related Works

Scholars mainly conducted research on the integration and application of blockchain with finance, education, finance and accounting, e-commerce trading, real economy, etc. For example, Li Fengyang et al. designed a blockchain order trading platform for the mutual trust of agricultural e-commerce trading. As for the countermeasures of the agricultural application mechanism of blockchain, Liu Hongchao et al. analyzed the mechanism of blockchain and the safe production of agricultural products from the perspectives of information asymmetry, game theory and externality, and then proposed the control mechanism. If the former applies the existing theories of other disciplines to the mechanism of agricultural blockchain, then Zhu Sizhu et al. analyze the application mechanism of blockchain in agricultural product quality and safety traceability, the modernization level of industrial chain and supply chain, loans and financing in rural areas, insurance and insurance claims payment through the form of cases, and put forward countermeasures and suggestions on research and development, infrastructure, talents, scenarios and coordination mechanism. In terms of the agricultural field and prospect of blockchain application, Liu Ruyi et al. believe that the current realized circulation of agricultural products includes cross-border transaction alliance chain, logistics alliance chain, traceability alliance chain and financing blockchain.

Lu believes that agricultural blockchain is currently mainly applied in the high-end agricultural field of self-certification and anti-counterfeiting, and the application has not been widely rolled out. Shang Jie et al. believe that agricultural blockchain has many advantages such as policy, technology and internal environmental opportunities, so the prospect is promising. Liang Xiaohe et al. conducted bibliometric analysis through the Dewinter Patent Index database (DII) and Dewinter Patent Innovation Platform (DI) databases, and also confirmed the above conclusions about the application of agricultural blockchain. Blockchain deepens information technology integration and technological upgrading. New technologies will integrate and improve the quality of traditional industries in more advanced fields, and give rise to new agricultural business forms and models. First, blockchain decentralized distributed ledger technology is the original innovation breakthrough of technology, and has become the main potential for change, such as evidence preservation and traceability, which is the most important link in the safety chain of agricultural products. Second, the integrated innovation of blockchain, as a combination of several technologies, is progressive rather than subversive. For example, encryption technology ensures the security of information storage, and time stamp technology ensures the continuous record of time, which is the basis of the control of the whole cycle time of agriculture from production to consumption, and records whether the product quality is expired and whether the intermediate link is delayed. Third, blockchain as a general technology to support other technology platforms, such as electricity, the Internet as an infrastructure to serve the broad scenarios of human society operation, including rural areas benefit the majority of the population.

3   Research Methodology

With the deepening of the rural revitalization strategy, in order to accelerate the implementation of rural revitalization and development, promote the adjustment of industrial structure, and deepen the innovation of management mechanism, it is particularly necessary to accelerate the reform of management mode and promote the implementation of precision marketing strategy under the existing development mode of characteristic product industry.

Since the agricultural sector includes agricultural production, as well as the management of agricultural market supply and modern logistics, the application of any technology is not separate; They must be comprehensive and systematic. Agricultural product safety, pesticides, agricultural product sellers and wholesalers will be included in the scope of technical monitoring through RFID technology.

Apache Spark is a fast, versatile computing engine designed for large-scale data processing. Spark is an open-source Hadoop MapReduce-like general parallel framework created by UC Berkeley AMP lab (AMP Lab of University of California, Berkeley). Spark has the advantages of Hadoop MapReduce. However, unlike MapReduce, the intermediate output results of a Job can be stored in the memory, so that HDFS data is no longer read or written. Therefore, Spark is better applicable to the iterative MapReduce algorithm, such as data mining and machine learning. The Spark ecosystem is shown in Table 1:

Table 1. Spark ecosystem

Application type	Time span
Complex batch data processing	Minutes to hours
Interactive query based on historical data	Tens of seconds to minutes
Data processing based on real-time data stream	Hundreds of milliseconds to seconds

In the past, batch query required Hadoop, interactive query Apache and stream processing query required Storm. If different scenario requirements were used to solve user requirements between different scenarios, data input and output between different scenarios could not be shared seamlessly, and system maintenance cost was high. Three different teams are required to maintain the software. Spark can solve problems such as seamless sharing, high cost, and insufficient resource utilization. One Spark software stack can meet different user requirements.

Spark has strong adaptability. It can read and write native data from HDFS, Cassandra, HBase, S3, and Techyon at the persistence layer. It can use Mesos, YARN, and the Standalone resource manager to schedule jobs to compute Spark applications.

On HDFS, large files are divided into equal parts by default. In the HDFS introduction document, the default value is 64 MB. Spark stores intermediate data in memory. Spark provides two types of operations on data sets, namely Transformations and Actions.

4 Experiments and Analysis

4.1 Spark Architecture

Usually, when the amount of data to be processed exceeds the scale of a single machine (for example, our computer has 4GB memory, but we need to process more than 100GB of data), we can choose spark cluster for calculation. Sometimes, the amount of data to be processed is not large, but the calculation is very complicated and requires a lot of time. At this time, we can also choose to make use of the powerful computing resources of spark cluster for parallel computing. Its architecture diagram use graphics instead (see Figs. 1, 2 and 3).

Based on Spark Core, different user requirements can be derived. It can be deployed in a unified manner with Hadoop clusters, and resources can be scheduled and managed using YARN. YARN is the Hadoop cluster manager. Technical personnel only need to master one technology.

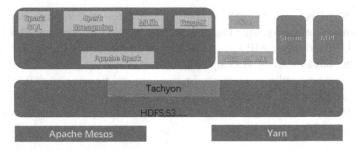

Fig. 1. Spark infrastructure

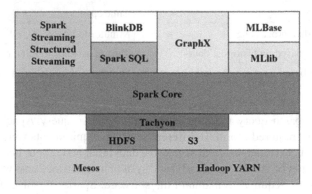

Fig. 2. Hadoop cluster manager architecture

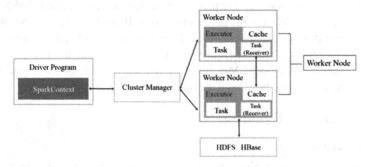

Fig. 3. Worker nodes manager architecture

4.2 Build Smart Agriculture Big Data Ecosystem Based on Spark

The Hadoop cluster is deployed in a single layer network topology, mainly to demonstrate the feasibility of the system, which will reduce many unnecessary problems in development. Of course, in a real production environment, this is obviously not appropriate; The most appropriate still requires the typical two-tier network topology. The Hadoop cluster of this system adopts the master/slave architecture. From a scalability and performance perspective, in a large-scale Hadoop cluster configuration, assign different component roles to different machines to avoid individual failures and entire cluster failures. This system cluster consists of 8 hosts, using Ubuntu 2.4STD operating system, using Gigabit LAN to ensure data transmission. The cluster uses unified installation directories hadoop:/usr/local/hadoop, Hbase:/ust local/Hbase, and Zookeeper: /usr/local/zookeeper.

5 Conclusion

With the deepening of agricultural information, the eastern region, with its advantages in economy, technology and talents, has made high level and high starting point investment to consolidate and lead the new technological revolution. The western region, including Guizhou Province, has taken blockchain as another strategic choice to catch up with the late in rural revitalization after the first trial of new technologies such as big data and artificial intelligence. The agricultural block chain is not only the selected answer favored by the developed areas to keep ahead development, but also the required answer which is urgently changing its own situation. It is of great significance to our country, especially the western regions, to realize high quality economic and social development in the new stage. Based on Spark big data platform, this paper designs and implements a smart agriculture big data ecosystem. It is designed against the backdrop of the growing trend of smart agriculture big data ecosystem. The system mainly designed every link of agricultural products from production (arable land, seeding, fertilization, plant protection, harvest), storage to sales, from tillage, seeding, planting, plant protection, harvest to grain drying storage, from crop growth state, planting management, crop processing and sales to peripheral services, integrated information technology and agricultural production, relying on the Internet of Things and big data to achieve integration and mutual Connect the module. And the use of blockchain technology to ensure the safety and reliability of data.

References

1. Binance.2020.Binance APIs. https://docs.binance.org/api-reference/dex-api/paths.html. Accessed 1 January 2020
2. Blockchain.2020.Blockchain Explorer. https://www.blockchain.com/api. Accessed 1 January 2020
3. BlockCypher.2020. API documentation. https://www.blockcypher.com/dev/bitcoin/#introduction. Accessed 1 January 2020
4. Bloomberg.2020. Hackers Steal $40 Million Worth of Bitcoin From Binance Exchange. https://www.bloomberg.com/news/articles/2019-05-08/crypto-exchange-giant-binance-reports-a-hack-of-7-000-bitcoin. Accessed 1 January 2020
5. CCN.2020. Someone Tried to Hack Etherscan. https://www.ccn.com/someone-tried-to-hack-etherscan-using-the-comment-section. Accessed 1 January 2020
6. Coindesk.2020.Binance Customer Data Has Leaked. https://www.coindesk.com/binance-kyc-issue. Accessed 1 January 2020
7. Etherscan.2020. Etherscan APIs. https://etherscan.io/apis. Accessed 1 January 2020
8. Wolfgang Gräther, Sabine Kolvenbach, Rudolf Ruland, Julian Schütte, Christof Torres, and Florian Wendland. 2018. Blockchain for Education: Lifelong Learning Passport. In Proceedings of 1st ERCIM Blockchain Workshop. -, -, 0
9. State of the DApps. 2022. DApp Statistics. https://www.stateofthedapps.com/stats
10. Oraclize.2020.API documentation. https://docs.provable.xyz/#home. Accessed 1 January 2020

Efficient Public Key Encryption Equality Test with Lightweight Authorization on Outsourced Encrypted Datasets

Chengyu Jiang, Sha Ma[✉], and Hao Wang

College of Mathematics and Informatics, South China Agricultural University,
Guangzhou 510642, China
martin_deng@163.com

Abstract. Cloud computing changes our way of life by offering advantages in convenience and cost savings. However, the security and privacy issues associated with this technology should not be taken lightly. In various cloud computing applications such as shared contacts, categorization of users on social applications, and unpublished papers plagiarism detection, public key encryption with equality test (PKEET) plays a significant role in their privacy protection. In the existing PKEET schemes, the authorization algorithms can only authorize an encrypted data or all encrypted data, but cannot authorize any encrypted data collections. In addition, the majority of these schemes use computationally expensive pairing operations. Based on this, we propose a new pairing free scheme that can authorize specified sets of encrypted data thereby reducing the number of authorization trapdoors. Additionally, our scheme demonstrates high efficiency in simulation experiment compared to related works.

Keywords: Cloud computing · equality test · privacy protection

1 Introduction

Cloud storage plays a crucial role in cloud computing, providing convenient options for data backup, sharing, and prevention of loss. As cloud storage technology advances, users must prioritize the security of their stored data. The data uploaded by users may contain sensitive information such as contacts and unpublished papers. If these data are leaked due to a network attack or malicious sale by cloud servers, it could result in serious consequences. An effective solution to address these concerns is to encrypt the data before uploading it. But encrypting data can cause it to lose its original information, and without decryption, it is difficult to perform meaningful operations on it. Many techniques, such as data mining [1] and near-duplicate detection [13], typically involve performing set intersection operations. By utilizing set intersection operations, we can obtain common features and labels among different users, thereby providing

© The Author(s), under exclusive license to Springer Nature Singapore Pte Ltd. 2024
K. Li and Y. Liu (Eds.): ISICA 2023, CCIS 2147, pp. 214–224, 2024.
https://doi.org/10.1007/978-981-97-4396-4_20

a foundation for subsequent calculations. Matching encrypted data becomes a highly crucial technical issue in this process, necessitating the search for effective solutions.

The public key encryption with equality test [14] (PKEET), which compares whether two ciphertexts encrypted with different public keys belong to the same plaintext, is considered a promising approach to address this problem. PKEET has been widely used for privacy protection in scenarios such as fingerprint identification [2,12], friend matching [3], and discovering the same illness patients [5].

1.1 Related Work

Yang et al. [14] proposed the first groundbreaking PKEET scheme in 2010, and later Tang et al. introduced the authorization mechanism into PKEET [10,11]. Ma et al. proposed public key encryption with delegated equality test [7] and efficient public key encryption with equality test supporting flexible authorization [6]. Huang et al. [4] introduced the concepts of user-level authorization and ciphertext-level authorization in PKEET. Wang et al. [12] also conceived the scheme that provides four different types of authorization. In 2020, Willy et al. [9] introduced a novel scheme called public-key encryption with multi-ciphertext equality test (PKE-MET). This scheme allows the tester to verify whether multiple ciphertexts contain the same message, while preserving the confidentiality of each individual ciphertext. Inspired by PKE-MET, Zhang et al. [16] put forwoard to an efficient public key encryption with set equality test (PKE-SET) in 2023. At the same year, Yang et al. [15] proposed revocable public key encryption with equality test without pairing (RPKEET). However, the aforementioned schemes do not involve the concept of ciphertext set authorization. Ciphertext set authorization refers to the data owner granting authorization for a specified set of ciphertexts, with the result obtained by performing test algorithm once. For example, if there are n ciphertexts to be tested, the aforementioned scheme requires generating n ciphertext-level authorizations and performing the test algorithm n times. But in our scheme, only one authorization for the n ciphertexts needs to be generated and the test algorithm only be performed once.

1.2 Our Contribution

Our proposed scheme makes the following contributions:

1. We propose the public key encryption equality test with lightweight authorization (PKEET-LA) concept and formalize the definition of PKEET-LA, which provides three different types of authorization: user-level, single-ciphertext level, and ciphertext set level. Then we construct the scheme of PKEET-LA.
2. We propose, for the first time, the concept of ciphertext set authorization and implement it. It allows for authorizing a specified set of n ciphertexts as a whole, which significantly reduces the number of authorizations and saves storage space.

3. Our test algorithm does not involve exponentiation and pairing operations, which is highly favorable for devices with limited computational resources. Additionally, it also saves computational resources for server.

1.3   Organization

In Sect. 2, we formalize the framework flow, definition and security model of PKEET-LA, while the scheme is given in Sect. 3. In Sect. 4, we provide corresponding security analysis. The comparison of our scheme and related works is given in Sect. 5. Finally we summarize in Sect. 6.

2   Public Key Encryption Equality Test with Lightweight Authorization

2.1   Framework Flow

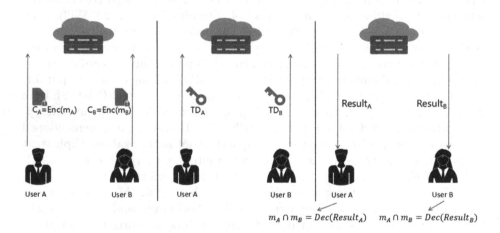

Fig. 1. Framework flow.

As shown in Fig. 1, initially, users in the system generate the public parameters collaboratively. Then, each user generates their own public and private keys. Subsequently, anyone can encrypt their data using the intended recipient's public key and upload the ciphertext to the cloud server for storage. To compute the intersection of their encrypted data with another user's data, a user sends the trapdoor to the cloud server. Upon trapdoor reception, the cloud server computes the intersection of their data and returns the result to the user.

2.2   Definition

A PKEET-LA consists of the following eleven algorithms.

1) Setup(λ): Given a security parameter λ, the algorithm outputs the public parameter pp.
2) KeyGen(pp): Given pp, the algorithm generates a pair of public and private keys (PK, SK).
3) Encrypt(pp, PK, m): Given pp, PK and data $m \in \mathbb{M}$, the algorithm computes the corresponding ciphertext $C \in \mathbb{C}$, where $\mathbb{M}$ is plaintext space and $\mathbb{C}$ is ciphertext space.
4) Decrypt(pp, SK, C): Given pp, SK and C, the algorithm returns m.
5) Aut$_u$(SK): Given SK, the algorithm outputs user-level trapdoor TD^1.
6) Aut$_c$(SK, C): Given SK and C, the algorithm outputs single-ciphertext level trapdoor TD^2.
7) Aut$_s$($SK, \{C^1, ..., C^n\}$): Given SK and a ciphertext set $\{C^1, ..., C^n\}$, the algorithm outputs ciphertext set level trapdoor TD^0.
8) Test$_1$(C_A, TD_A^1, C_B, TD_B^1): Given C_A, TD_A^1, C_B, and TD_B^1, where A and B are different users, the algorithm outputs 1 if C_A and C_B are generated on the same data, or 0 otherwise.
9) Test$_2$(C_A, TD_A^2, C_B, TD_B^2): Given C_A, TD_A^2, C_B, and TD_B^2, where A and B are different users, the algorithm outputs 1 if C_A and C_B are generated on the same data, or 0 otherwise.
10) Test$_3$(C_A, TD_A^1, C_B, TD_B^2): Given C_A, TD_A^1, C_B, and TD_B^2, where A and B are different users, the algorithm outputs 1 if C_A and C_B are generated on the same data, or 0 otherwise.
11) Test$_4$($\{C_A^1, ..., C_A^n\}, TD_A^3, \{C_B^1, ..., C_B^n\}, TD_B^3$):   Given $\{C_A^1, ..., C_A^n\}, TD_A^3$, $\{C_B^1, ..., C_B^n\}$ and TD_B^3, where A and B are different users, the algorithm outputs 1 if $\{C_A^1, ..., C_A^n\}$ and $\{C_B^1, ..., C_B^n\}$ are generated on the same data, or 0 otherwise.

Consistency. For the consistency property, two conditions must be satisfied:

1. Let PK and SK be public key and private key. The equation must holds: $\forall m = $ Decrypt(pp, SK, C), where $C = $ Encrypt(pp, PK, m) and $m \in \mathbb{M}$.
2. Let $C = $ Encrypt(pp, PK, m), $TD^1 = $ Aut$_u$(SK), $TD^2 = $ Aut$_c$(SK, C) and $TD^3 = $ Aut$_s$($SK, \{C^1, ..., C^n\}$). The equations must hold:
   a) If m_A = m_B,   Test$_1$(C_A, TD_A^1, C_B, TD_B^1)   =   1, Test$_2$(C_A, TD_A^2, C_B, TD_B^2) = 1 and Test$_3$ (C_A, TD_A^1, C_B, TD_B^2) = 1.
   b) If $\{m_A^1, ..., m_A^n\} = \{m_B^1, ..., m_B^n\}$, Test$_4$($\{C_A^1, ..., C_A^n\}, TD_A^3, \{C_B^1, ..., C_B^n\}$, TD_B^3) = 1.

2.3   Security Model

The security of our scheme is built upon the Computational Diffie-Hellman (CDH) assumption. The following is the definition of CDH:

Let $\mathbb{G} =< g >$ be a group of prime order p. We say that the CDH problem is hard, if given a tuple $(g, g^a, g^b) \in \mathbb{G}^3$ for random $g \in \mathbb{G}$, and $a, b \in \mathbb{Z}_p$, any probabilistic polynomial-time adversary $\mathcal{A}$ computes g^{ab} with negligible advantage ε,

$$\text{Adv}_{\mathcal{A}}^{CDH} = \Pr[\mathcal{A}(g, g^a, g^b) = g^{ab}] \leq \varepsilon.$$

In order to define the security of PKEET-LA, we consider the following two types of adversaries in our scheme.

Outside adversary. These adversaries are similar to malicious users, who do not possess the target user's private key and authorization trapdoor.

Inside adversary. These adversaries are typically played by malicious cloud server, who possesses the target user's authorization but not their private key.

To streamline security analysis, we introduce the following games for evaluating security properties.

Game 1: Let $\mathcal{A}_1$ be an outside adversary. The game between $\mathcal{A}_1$ and challenger $\mathcal{C}$ is as follow.

1. **Setup** : Upon receiving a security parameter λ, $\mathcal{C}$ generates the public parameter pp and the public/private key pair (PK, SK) by running Setup(pp) and KeyGen(pp) algorithms. Then $\mathcal{C}$ sends PK to $\mathcal{A}_1$.
2. **Phase 1** : $\mathcal{A}_1$ is permitted to make polynomially many queries to the following oracle.
   - $O^{Decrypt}$: Upon receiving a ciphertext C, if C is valid, $\mathcal{C}$ returns m; otherwise returns $\perp$.
3. **Challenge** : $\mathcal{A}_1$ picks two equal-length data m_0^*, m_1^* which are submitted to $\mathcal{C}$ for challenge. $\mathcal{C}$ picks $\rho \in \{0, 1\}$ randomly, and then computes $C_\rho^* \leftarrow$ Encrypt(pp, PK^*, m_ρ^*) as the challenge ciphertext to return.
4. **Phase 2** : $\mathcal{A}_1$ makes queries as same as in **Phase 1**. In addition, C_ρ^* cannot be queried to $O^{Decrypt}$.
5. **Guess**: $\mathcal{A}_1$ outputs its guess. If $\rho = \rho'$, $\mathcal{A}_1$ wins this game. The advantage of $\mathcal{A}_1$ is defined as $Adv_{PKEET\text{-}LA, \mathcal{A}_1}^{IND\text{-}CCA}(\lambda) = |\Pr[\rho = \rho'] - \frac{1}{2}|$.

Definition 1: Our PKEET-LA is indistinguishability under chosenciphertext attacks (IND-CCA) secure if for any probabilistic polynomial time (PPT) $\mathcal{A}_1$ whose advantage $Adv_{PKEET\text{-}LA, \mathcal{A}_1}^{IND\text{-}CCA}(\lambda)$ is negligible in the security parameter λ.

Game 2: Let $\mathcal{A}_2$ be an inside adversary. The game between $\mathcal{A}_2$ and challenger $\mathcal{C}$ is as follow.

1. **Setup** : Upon receiving a security parameter λ, challenger $\mathcal{C}$ generates the public parameter pp and the public/private key pair $(PK, SK) = ((pk_1, pk_2), (sk_1, sk_2))$ by running Setup(pp) and KeyGen(pp) algorithms. Then $\mathcal{C}$ sends PK and sk_2 to $\mathcal{A}_2$.
2. **Phase 1** : $\mathcal{A}_2$ is permitted to make polynomially many queries to the following oracle.
   - $O^{Decrypt}$: This oracle is as same as in game 1.
   - O^{Aut_u}: On input a user-level trapdoor query, it returns TD_1.

- O^{Aut_c}: On input a single-ciphertext level trapdoor query, it returns TD_2.
- O^{Aut_s}: On input a ciphertext set level trapdoor query, it returns TD_3.

3. **Challenge** : $\mathcal{C}$ randomly selects m^* and then computes $C^* \leftarrow$ Encrypt(pp, PK^*, m^*) as the challenge ciphertext to return.

4. **Phase 2** : $\mathcal{A}_2$ makes queries as same as in **Phase 1**. In addition, C^* cannot be queried to $O^{Decrypt}$.

5. **Guess:** $\mathcal{A}_2$ outputs its guess. If $m' = m^*$, $\mathcal{A}_2$ wins this game. The advantage of $\mathcal{A}_2$ is defined as $Adv_{PKEET\text{-}LA,\mathcal{A}_2}^{OW\text{-}CCA}(\lambda) = |\Pr[m' = m^*]|$.

Definition 2: Our PKEET-LA is one-wayness under chosen ciphertext attacks (OW-CCA) secure if for any probabilistic polynomial time (PPT) $\mathcal{A}_2$ whose advantage $Adv_{PKEET\text{-}LA,\mathcal{A}_2}^{OW\text{-}CCA}(\lambda)$ is negligible in the security parameter λ.

3 PKEET-LA Structure

3.1 Scheme

- Setup(λ): On input security parameter λ, the algorithm outputs the public parameters $pp = (\mathbb{G}, p, g, H_1, H_2, H_3)$, where $\mathbb{G}$ is a group with prime order p, g is the generator $g \in \mathbb{G}$. H_1, H_2 and H_3 are three collision-resistant hash function: $H_1 : \mathbb{G} \rightarrow \mathbb{G}, H_2 : \{0,1\}^\lambda \rightarrow \mathbb{G}, H_3 : \mathbb{G} \times \mathbb{G} \rightarrow \{0,1\}^{\lambda+l}$, where l is bit-length of the elements in $\mathbb{Z}_p^*$.
- KeyGen(pp): On input pp, the algorithm randomly selects $sk_1 \in \mathbb{Z}_p^*, sk_2 \in \mathbb{Z}_p^*$ and outputs key pair $(SK, PK) = ((sk_1, sk_2), (pk_1 = g^{sk_1}, pk_2 = g^{sk_2}))$.
- Encrypt(pp, PK, m): On input pp, PK and data m, the algorithm selects $r \in \mathbb{Z}_p^*$ randomly and computes ciphertext $c_1 = g^r, c_2 = H_1(pk_2^r) \cdot H_2(m), c_3 = H_3(pk_1^r, c_2) \oplus (m||r)$. Finally, it returns $C = (c_1, c_2, c_3)$.
- Decrypt(pp, sk, C): On input pp, SK and C, the algorithm first recovers $(m'||r')$ by computing $H_3(c_1^{sk_1}, c_2) \oplus c_3$. If $c_1 = g^{r'}$ and $c_2 = H_1(c_1^{sk_2}) \cdot H_2(m')$ hold, it returns m'.
- Aut$_u$(SK): With SK, the algorithm returns user-level trapdoor $TD^1 = sk_2$.
- Aut$_c$(SK, C): On input SK and C, the algorithm returns single-ciphertext level trapdoor $TD^2 = c_1^{sk_2}$.
- Aut$_s$($SK, \{C^1, ..., C^n\}$): On input SK and ciphertext set $\{C^1, ..., C^n\}$, the algorithm outputs ciphertext set level trapdoor $TD^3 = \prod_{i=1}^{n} H_1(pk_2^{r_i})$ as return, where n is the size of ciphertext set.
- Test$_1$(C_A, TD_A^1, C_B, TD_B^1): On input C_A, TD_A^1, C_B, and TD_B^1, the algorithm computes $\frac{c_{A,2}}{c_{A,1}^{TD_A^1}} = H_2(m_A)$ and $\frac{c_{B,2}}{c_{B,1}^{TD_B^1}} = H_2(m_B)$. If $H_2(m_A) = H_2(m_B)$, it returns 1; otherwise, it returns 0.
- Test$_2$(C_A, TD_A^2, C_B, TD_B^2): On input C_A, TD_A^2, C_B and TD_B^2, the algorithm computes $\frac{c_{A,2}}{TD_A^2} = H_2(m_A)$ and $\frac{c_{B,2}}{TD_B^2} = H_2(m_B)$. If $H_2(m_A) = H_2(m_B)$, it returns 1; otherwise, it returns 0.
- Test$_3$(C_A, TD_A^1, C_B, TD_B^2): On input C_A, TD_A^1, C_B and TD_B^2, the algorithm computes $\frac{c_{A,2}}{c_{A,1}^{TD_A^1}} = H_2(m_A)$ and $\frac{c_{B,2}}{TD_{B,2}} = H_2(m_B)$, If $H_2(m_A) = H_2(m_B)$, it returns 1; otherwise, it returns 0.

- $\mathsf{Test}_4(\{C_A^1, ..., C_A^n\}, TD_A^3, \{C_B^1, ..., C_B^n\}, TD_B^3)$: On input $\{C_A^1, ..., C_A^n\}$, TD_A^3, $\{C_B^1, ..., C_B^n\}$ and TD_B^3, the algorithm computes $\frac{\prod_{i=1}^n c_{A,2}^i}{TD_A^3} = \prod_{i=1}^n H_2(m_A^i)$ and $\frac{\prod_{i=1}^n c_{B,2}^i}{TD_B^3} \prod_{i=1}^n H_2(m_B^i)$. If $\prod_{i=1}^n H_2(m_A^i) = \prod_{i=1}^n H_2(m_B^i)$, it returns 1; otherwise, it returns 0.

3.2 Consistency of the Scheme

Theorem 1. *Our PKEET-LA scheme satisfies the consistency property.*

Proof. Our analysis is as below:

1. Because $c_1^{sk_1} = g^{sk_1 r} = pk_1^r$, $H_3(c_1^{sk_1}, c_2) \oplus c_3 = H_3(c_1^{sk_1}, c_2) \oplus H_3(pk_1^r, c_2) \oplus (m\|r) = (m\|r)$.
2. From the above scheme, we have the following equations that hold:

$$\frac{c_{A,2}}{c_{A,1}^{TD_A^1}} = \frac{H_2(m_A) \cdot H_1(pk_{A_2}^r)}{H_1(g^{rsk_{A,2}})} = \frac{H_2(m_A) \cdot H_1(g^{rsk_{A,2}})}{H_1(g^{rsk_{A,2}})} = H_2(m_A)$$

$$\frac{c_{A,2}}{TD_A^2} = \frac{H_2(m_A) \cdot H_1(g^{rsk_{A,2}})}{H_1(g^{rsk_{A,2}})} = H_2(m_A)$$

$$\frac{\prod_{i=1}^n c_A^i}{TD_A^3} = \prod_{i=1}^n \frac{H_2(m_A^i) \cdot H_1(pk_{A,2}^{r_i})}{H_1(pk_{A,2}^{r_i})} = \prod_{i=1}^n H_2(m_A^i)$$

If $m_A = m_B$, we can get:

$$\frac{c_{A,2}}{c_{A,1}^{TD_A^1}} = \frac{c_{B,2}}{c_{B,1}^{TD_B^1}}, \frac{c_{A,2}}{TD_A^2} = \frac{c_{B,2}}{TD_B^2}, \frac{c_{A,2}}{c_{A,1}^{TD_A^1}} = \frac{c_{B,2}}{TD_B^2}$$

So $\mathsf{Test}_1(C_A, TD_A^1, C_B, TD_B^1) = 1$, $\mathsf{Test}_2(C_A, TD_A^2, C_B, TD_B^2) = 1$ and Test_3 $(C_A, TD_{A_1}, C_B, TD_{B_2}) = 1$ hold.
If $\{m_A^1, ..., m_A^n\} = \{m_B^1, ..., m_B^n\}$, we can get:

$$\frac{\prod_{i=1}^n c_A^i}{TD_A^3} = \frac{\prod_{i=1}^n c_B^i}{TD_B^3}$$

So $\mathsf{Test}_4(\{C_A^1, ..., C_A^n\}, TD_A^3, \{C_B^1, ..., C_B^n\}) = 1$ holds.

4   Security Analysis

Theorem 2. *Our PKEET-LA scheme satisfies the IND-CCA security against outside adversary A_1 under the CDH assumption in the random oracle model.*

Proof. Given a CDH problem instance $(\mathbb{G}, p, g, g^a, g^b)$, where $g \in \mathbb{G}$ and unknown $a, b \in \mathbb{Z}_p^*$. The challenger $\mathcal{C}$ controls simulator $\mathcal{B}$ to play with the adversary A_1 with ϵ advantage to compute g^{ab} in the following game 1.

Setup : $\mathcal{B}$ sets the public parameter $pp = (\mathbb{G}, p, g, H_1, H_2, H_3)$, where $H_1 \sim H_3$ are three random oracles. $\mathcal{B}$ randomly selects $PK = (pk_1, pk_2) = (g^a, g^{au})$ as PK to return, where u is randomly selected.

Phase1: $\mathcal{B}$ responds $\mathcal{A}_1$'s queries as follows:

H_1-*query* : It maintains a initially empty list $L_{H_1}(x, X)$. With input x, if x has been queried, it returns the X from L_{H_1} ; otherwise, it picks $X \in \mathbb{G}$ randomly to store and return.

H_2-*query* : It maintains a initially empty list $L_{H_2}(y, Y)$. With input y, if y has been queried, it returns the Y from L_{H_2} ; otherwise, it picks $Y \in \mathbb{G}$ randomly to store and return.

H_3-*query* : It maintains a initially empty list $L_{H_3}(z_1, z_2, Z)$. With input z, if Z has been queried, it returns the Z from L_{H_3} ; otherwise, it picks $Z \in \{0,1\}^{\lambda+l}$ randomly to store and return.

$O^{Decrypt}$: On input a ciphertext $C = (c_1, c_2, c_3)$, it queries Z from H_3 to recover $(m||r)$, then queries Y from H_2 by m. If $c_1 = g^r$ and there exists $X \in L_{H_1}$ such that $c_2 = X \cdot Y$ holds, it returns m; otherwise $\perp$.

Challenge : $\mathcal{A}_1$ submits two equal-length messages m_0^*, $m_1^* \in \{0,1\}^\lambda$. $\mathcal{B}$ picks $\rho \in \{0,1\}$ randomly to generate ciphertext $C_\rho^* = (c_1^* = g^b, c_2^* = H_1(g^{aub}) \cdot H_2(m_\rho^*), c_3^* = H_3(g^{ab}, H_1(g^{aub}) \cdot H_2(m_\rho^*)) \oplus (m_\rho^*||b)$. Finally, $\mathcal{B}$ returns C_ρ^* to $\mathcal{A}_1$.

Phase2 : Same as in **Phase1**, but $\mathcal{A}_1$ cannot query $O^{Decrypt}$ on C_ρ^*.

Guess . $\mathcal{A}_1$ outputs its guess ρ^*. If $\rho^* = \rho$, $\mathcal{A}_1$ wins this game and $\mathcal{B}$ can compute g^{ab} as the solution for BDH problem instance.

Because the game does not abort and $\mathcal{A}_1$ has ϵ advantage to win. The advantage that $\mathcal{B}$ can choose the correct element in L_{H_3} to solve BDH problem is ϵ/q_{H_3}, where q_{H_3} is the number of times that $\mathcal{A}_1$ asks H_3.

Theorem 3. *Our PKEET-LA scheme satisfies the OW-CCA security against outside adversary A_2 under the CDH assumption in the random oracle model.*

Proof. Given a CDH problem instance $(\mathbb{G}, p, g, g^a, g^b)$, where $g \in \mathbb{G}$ and unknown $a, b \in \mathbb{Z}_p^*$, the challenger $\mathcal{C}$ controls simulator $\mathcal{B}$ to play with the adversary $\mathcal{A}_2$ with ϵ advantage to compute g^{ab} in the following game 2.

Setup : $\mathcal{B}$ sets the public parameter $pp = (\mathbb{G}, p, g, H_1, H_2, H_3)$, where $H_1 \sim H_3$ are three random oracles. $\mathcal{B}$ randomly selects $PK = (pk_1, pk_2) = (g^a, g^{sk_2})$, where sk_2 is random, and returns PK and sk_2.

Phase1: $\mathcal{B}$ responds as same as in game 1, except O^{Aut_u}, O^{Aut_c} and O^{Aut_s}.

O^{Aut_u} : On input a user-level trapdoor query, it returns TD_1.

O^{Aut_c} : On input a ciphertext C, it returns TD_2.

O^{Aut_c} : On input a ciphertext set $C = C^1, ..., C^n$, it returns TD_3.

Challenge : $\mathcal{B}$ randomly selects m^*, and generates ciphertext $C^* = (c_1^* = g^b, c_2^* = H_1(g^{sk_2 b}) \cdot H_2(m^*), c_3^* = H_3(g^{ab}, H_1(g^{sk_2 b}) \cdot H_2(m^*)) \oplus (m^*||b)$. Finally, $\mathcal{B}$ returns C^* to $\mathcal{A}_1$.

Phase2 : Same as in **Phase1**, but $\mathcal{A}_2$ cannot query $O^{Decrypt}$ on C^*.

Guess : $\mathcal{A}_2$ outputs its guess m'. If $m^* = m'$, $\mathcal{A}_2$ wins this game and $\mathcal{B}$ can compute g^{ab} as the solution for BDH problem instance.

Due to the page limit, we omit the process of probability analysis. The advantage that $\mathcal{B}$ solves BDH problem is $\frac{\epsilon - \frac{1}{2\lambda}}{q_{H_3}}$, where q_{H_3} is the number of times that $\mathcal{A}_2$ asks H_3.

5   Performance Assessment

In this section, we make a comparison of PKEET-LA with related pairing-free PKEET [8,15,16]. We list the primary computational cost and storage space in Table 1. Compared to [8,15,16], our scheme is significantly more efficient in Enc and TEST algorithms, and it is also moderately efficient in Dec and Aut$_u$ algorithms. In addition, only our scheme has the AUT$_s$ algorithm. In terms of storage, our ciphertext exhibits a shorter length compared to [8,15] as well.

To intuitively show the computational comparison, we implement the four schemes mentioned above using the C++ language on a computer equipped with an AMD R5 3.6 GHz CPU, 8 GB memory and the Windows 10 operation system. In Fig. 2, the X-axis represents the number of ciphertexts produced by executing the algorithms, while the Y-axis represents the execution time of the algorithms. In Fig. 2(a), compared to schemes [15,16], and [8], the time cost of our encryption algorithm is reduced by about 90%, 44%, and 58% respectively. In Fig. 2(b), the time cost of our decryption algorithm is similar to schemes [8,16], but it is reduced by approximately 45% compared to [15]. According to Fig. 2(c), our test algorithm has a much lower time cost compared to the other

Table 1. Comparison of computational and storage cost.

Scheme	Enc	Dec	Aut$_s$	Test	$|C|$
PKE-SET [16]	$3exp$	$2exp$	×	$2exp$	$|\mathbb{G}| + 3|\mathbb{Z}_p^*| + 2\lambda$
RPKEET [15]	$6exp$	$5exp$	×	$2exp$	$2|\mathbb{G}| + 6|\mathbb{Z}_p^*| + \lambda$
GPKEET/BP [8]	$6exp$	$3exp$	×	$2exp$	$4|\mathbb{G}| + 3|\mathbb{Z}_p^*| + \lambda$
PKEET-LA	$3exp$	$3exp$	0	0	$2|\mathbb{G}| + |\mathbb{Z}_p^*| + \lambda$

exp: The exponentiation's computational cost.
$|\mathbb{G}|, |\mathbb{Z}_p^*|$: The element's bit length of $\mathbb{G}$ and $\mathbb{Z}_p^*$ respectively.

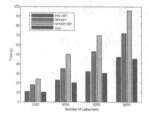 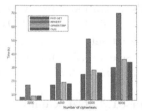 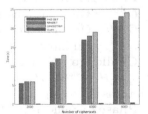

(a) Running time of Enc.   (b) Running time of Dec.   (c) Running time of Test.

Fig. 2. Experiment comparison.

three schemes, which is one of our advantages. Another advantage is that our scheme enables ciphertext set authorization. Instead of generating n trapdoors for testing n ciphertexts, our scheme only requires generating one trapdoor.

6 Conclusion

To address the issue of lacking ciphertext set authorization in PKEET, we propose the concept of public key encryption equality test with lightweight authorization (PKEET-LA) and construct a concrete scheme. The scheme reduces the number of trapdoors and saves storage space by implementing ciphertext set authorization. We define its security model and provide security proofs. Finally, by simulation experiments, our scheme shows excellent efficiency, indicating its suitability for devices with limited computational resources and storage space.

Acknowledgments. This work is supported by the Guangdong Basic and Applied Basic Research Foundation (2024A1515012666) and the National Natural Science Foundation of China (61872409).

References

1. Aggarwal, C.C., Yu, P.S.: A general survey of privacy-preserving data mining models and algorithms. In: Privacy-preserving data mining: models and algorithms, pp. 11–52 (2008)
2. Blanton, M., Gasti, P.: Secure and efficient protocols for iris and fingerprint identification. In: Computer Security–ESORICS 2011: 16th European Symposium on Research in Computer Security, Leuven, Belgium, September 12-14 (2011)
3. Du, J., Ma, S., Yang, T., Huang, Q.: Authenticated identity-based encryption scheme with equality test for cloud-based social network. In: 2023 7th International Conference on Cryptography, Security and Privacy (CSP), Tianjin, China, April 21-23, 2023, pp. 27–34. IEEE (2023)
4. Huang, K., Tso, R., Chen, Y.C., Rahman, S.M.M., Almogren, A., Alamri, A.: Pke-aet: public key encryption with authorized equality test. Comput. J. **58**(10), 2686–2697 (2015)
5. Lu, J., Li, H., Huang, J., Ma, S., Au, M.H.A., Huang, Q.: An Identity-based encryption with equality test scheme for healthcare social apps. Comput. Stand. Interfaces **87**, 103759 (2024)
6. Ma, S., Huang, Q., Zhang, M., Yang, B.: Efficient public key encryption with equality test supporting flexible authorization. IEEE Trans. Inf. Forensics Secur. **10**(3), 458–470 (2014)
7. Ma, S., Zhang, M., Huang, Q., Yang, B.: Public key encryption with delegated equality test in a multi-user setting. Comput. J. **58**(4), 986–1002 (2015)
8. Shen, X., Wang, B., Wang, L., Duan, P., Zhang, B.: Group public key encryption supporting equality test without bilinear pairings. Inf. Sci. **605**, 202–224 (2022)
9. Susilo, W., Guo, F., Zhao, Z., Wu, G.: Pke-met: public-key encryption with multiciphertext equality test in cloud computing. IEEE Trans. Cloud Comput. **10**(2), 1476–1488 (2020)

10. Tang, Q.: Towards public key encryption scheme supporting equality test with fine-grained authorization. In: Parampalli, U., Hawkes, P. (eds.) Information Security and Privacy, pp. 389–406. Springer Berlin Heidelberg, Berlin, Heidelberg (2011). https://doi.org/10.1007/978-3-642-22497-3_25
11. Tang, Q.: Public key encryption supporting plaintext equality test and user-specified authorization. Secur. Commun. Networks **5**(12), 1351–1362 (2012)
12. Wang, Y., Huang, Q., Li, H., Xiao, M., Ma, S., Susilo, W.: Private set intersection with authorization over outsourced encrypted datasets. IEEE Trans. Inf. Forensics Secur. **16**, 4050–4062 (2021)
13. Xiao, C., Wang, W., Lin, X., Yu, J.X., Wang, G.: Efficient similarity joins for near-duplicate detection. ACM Trans. Database Syst. **36**(3), 1–41 (2011)
14. Yang, G., Tan, C.H., Huang, Q., Wong, D.S.: Probabilistic public key encryption with equality test. In: Pieprzyk, J. (ed.) CT-RSA 2010. LNCS, vol. 5985, pp. 119–131. Springer, Heidelberg (2010). https://doi.org/10.1007/978-3-642-11925-5_9
15. Yang, T., Ma, S., Du, J., Jiang, C., Huang, Q.: Revocable public key encryption with equality test without pairing in cloud storage. The Computer Journal p. bxad006 (2023)
16. Zhang, X., Ma, S., Jiang, C., Zhou, P.: An efficient public key encryption with set equality test. In: 2023 7th International Conference on Cryptography, Security and Privacy (CSP), Tianjin, China, April 21-23, 2023, pp. 163–169. IEEE (2023)

Fake News Detection Model Incorporating News Text and User Propagation

Shuxin Yang, Jiahao Li, and Weidong Huang[✉]

Jiangxi University of Science and Technology, Ganzhou 341000, China
18279766975@139.com

Abstract. Fake news detection aims to detect the authenticity of news from different perspectives to maximize the performance of detecting fake news. In recent years, scholars have been engaged in the research of fake news detection, with their studies primary focusing on supervised learning. However, these studies require strenuous time and efforts to be spent on labeling datasets. To this end, this paper proposes a fake news detection model incorporating news text and user propagation named NT-UP by using unsupervised learning. NT-UP mainly consists of three components: text representation learning module, text feature extraction module and user propagation module. Specifically, the text representation learning module encodes news text and user tweets into a news vector and tweet vectors, respectively. The text feature extraction module is designed to extract news text features and in the user propagation module, the news propagation network is constructed by using news vector and tweet vectors, and then the graph feature is extracted from the news propagation network by using the graph contrastive learning based on unsupervised learning. The performance of the model is tested on real-world datasets, and experimental results show that the performance of NT-UP is better than that of other models.

Keyword: Fake News Detection · Unsupervised Learning · Channel Attention Mechanism · Feature Fusion

1 Introduction

With the development of mobile Internet, our quality of life has been greatly enhanced, and the number of netizens has also grown rapidly. Netizens use their mobile phones to search and read their preferred content on social platforms in their spare time. This phenomenon not only promotes the vigorous development of the self-media industry, but also injects new vitality into Internet content. However, some in self-media with ill motives seeks popularity and attention by spreading rumors, network violence and other harmful behaviors [1], thus causing serious interference and damage to the Internet public opinion ecology. Therefore, strengthening the study on fake news detection is crucial to the governance of the network.

In recent years, scholars have devoted themselves to developing effective methods to identify fake news and promoted the advancement of fake news detection technology.

K. Li and Y. Liu (Eds.): ISICA 2023, CCIS 2147, pp. 225–239, 2024.
https://doi.org/10.1007/978-981-97-4396-4_21

One technical route of existing news detection methods determines true and fake news by extracting features from news content such as news text, pictures, or videos, and another technical route determines true and fake news by extracting graph structure features from news propagation graphs. With the advancement of fake news detection work, some scholars have implemented fake news detection by combining multiple features, which has significantly improved the performance of fake news detection [2]. However, these detection methods based on supervised learning have a common limitation: the time-consuming effort of labeling datasets. Thus prediction performance of the model for fake news needs to be further improved.

In light of the issues noted above, this paper proposes a novel fake news detection model incorporating news text and user propagation named NT-UP. To be specific, NT-UP mainly consists of three components: text representation learning module, text feature extraction module and user propagation module. The main contributions of this paper are as follows:

(1) To improve the performance of the model in extracting text features, the text feature extraction module is designed to extract text features of the news by combining the pooling layer and the channel attention mechanism.
(2) To enhance the model's ability to extract graph features of news propagation networks, the graph data augmentation method is employed to highlight the structure of the news propagation network. In addition, we use graph contrastive learning to extract the graph structure features.
(3) Due to the introduction of unsupervised learning, this model realizes the extraction of text features and graph features on unlabeled datasets, and enables fake news detection after feature fusion. This paper uses unlabeled real-world datasets to verify the performance of the model, and the experimental results show that the model works well.

2   Related Work

Existing works on fake news detection usually have three ways: extracting features from the content of the news, extracting graph features from the propagation path of the news, or extracting features from the news content and propagation path respectively and, fusing them into news embeddings.

2.1   Content-Based Fake News Detection

News content generally includes text, audio, video, and pictures. Different methods are used to extract features in news content to detect fake news. Scholars usually use deep learning to extract text features.

For example, Chen et al. [3] proposed a method based on recurrent neural networks, which has good performance when process text and other content. On this basis, Smith optimized and proposed a method for fake news detection using Bi-directional Recurrent Neural Network (Bi-RNN). Bi-directional RNN is used to capture context information in both forward and backward directions. Context information strengthens the contextual features in news texts and improves the model's performance in detecting fake news. Shu et al. [4] started from the perspective of news texts and social context, used a recurrent neural network (RNN) to extract text features, and used the LSTM (Long Short-Term Memory) to extract temporal features in the context, and realized the detection of fake news by fusing text features and temporal features. Ni et al. [5] proposed a multi-view attention model based on attention mechanism to extract multiple features such as text, images, and social network structures from news, and stitch the features into news features to predict fake news. Qian et al. [6] constructed a fake news detection model based on a hierarchical multimodal contextual attention network to model the multimodal information and multilevel semantic relationships of news. Zhang et al. [7] addressed the issue that most existing models focus only on the content published by authors, ignoring the social emotion expressions reflected in news comments. Therefore, they extracted emotional features from news text and news comments separately, and input the extracted features into a false news detection classifier to achieve false news identification. Mansouri et al. [8] trained a fake news detection method based on semi-supervised learning method, which extracts features by correlating the relationship between users participating in the news and the corresponding posts, and the method was used to perform well in the task of identifying fake news. Li et al. [9] trained the model with labeled news data, and used the confidence function to improve the quality of the model's labeling of unlabeled data, and realized a semi-supervised fake news detection method. Wang et al. [10] provided pseudo labels for users' comments, selected a data annotator to mark pseudo labels for users' comments, and used reinforcement learning technology to improve the quality of pseudo labels labeled by users, Finally, Wang used a fake news discriminator to train the model with real label and pseudo label data to improve the accuracy of the model in identifying fake news. Singhal et al. [11] adopted a transfer learning approach and built a pre-trained model capable of extracting text, image, and audio features from large-scale annotated datasets, With the help of transfer learning technology, the model also showed good fake news detection performance in small or newer datasets. Lin et al. [12] used data distillation technology to train the teacher model through supervised learning method, annotate the unlabeled Chinese data set with pseudo labels, and then train the student model to realize unsupervised learning to identify fake news in Chinese news.

However, fake news detection based on neural networks have limitations in extracting text features, Additionally, the training time and the complexity of the model will also be affected by the length of the news. Due to this, other solution needs to be designed to improve the fake news detection task.

2.2 Graph-Based Fake News Detection

As study on the detection of fake news continues to advance, scholars have discovered that the propagation pathways of fake news on social media differ from those of genuine news.

Therefore, scholars have been attempting to use Graph Neural Networks to extract the graph features of news propagation networks for the purpose of detecting fake news. For example, Ren et al. [13] implemented a fake news detection method based on heterogeneous networks, which uses two attention mechanisms to learn the complex relationships among nodes in news data and make predictions about fake news. Wang et al. [14]

proposed a fake news detection via knowledge-driven multimodal graph convolutional networks. They first modeled the graph data transcribed from text and visual information as a heterogeneous graph, and then realized the detection of fake news by learning the latent semantics in the heterogeneous graph. The above works focus on modeling the news itself, and the impact of propagation between nodes is not considered. Therefore, Yuan et al. [15] carried out the detection of fake news by considering the local and global relationships between nodes and edges in the graph. Sun et al. [16] discovered that the early propagation structures of fake news differ from those of real news. Thus, they identified fake news in the early stages of news propagation by capturing global preference modules and local context learning modules from users' common propagation behaviors. Bian et al. [17] focused on the spread and dispersion of rumors. They extracted rumor spread features from top-down propagation and dispersion features from bottom-up propagation. They also used root node enhancement methods to highlight text features in the root node, achieving the identification of fake news. Konkobo et al. [18] utilize news content information, public evaluation of news, and author credibility information to respectively evaluate the opinions expressed by users as well as assess the credibility of users. They establish a user-based small network structure for news dissemination, and use these findings as input for a false news classifier to achieve false news detection.

In summary, fake news detection is mainly studied in two directions: content and graph. Scholars typically combine various features of news to improve the robustness and accuracy of models. Moreover, most of these studies are based on deep learning or supervised learning, indicating that there is still significant room for improvement in fake news detection.

3   Problem Definition

Definition 1 (News Texts): News authors typically write news texts in various forms such as reports, analyses, comments, and editorials with the aim of conveying the content, development trends, and potential impacts of current events to the public. Since the model cannot directly learn text features, preprocessing is required for news texts. The collection of news texts is represented as $P = \{P_1, P_2, \ldots, P_m\}$, where m represents the number of news, the news texts are encoded into news vectors $N = \{N_1, N_2, \ldots, N_m\}$.

For the set of retweets written by users who retweeted news article $P_i (1 \leq i \leq m)$ denoted as $C^{P_i} = \left\{ c_1^{P_i}, c_2^{P_i}, \ldots, c_o^{P_i} \right\}$, an encoder is used to encode the set of tweets into tweet vectors $U^{P_i} = \left\{ u_1^{P_i}, u_2^{P_i}, \ldots, u_o^{P_i} \right\}$, where o is the number of retweets involved in retweeting news P_i.

Definition 2 (User Propagation Network): With the continuous growth in the number of users retweeting news, constructing a news propagation network with news vector and user nodes is beneficial for extracting the graph structural features of User Propagation Network. Therefore, we consider the tweet vectors set $U^{P_i} = \left\{ u_1^{P_i}, u_2^{P_i}, \ldots, u_o^{P_i} \right\}$ as user nodes. We construct the user propagation network based on the news text vector and user nodes according to user retweet information. For example, in the news propagation network formed when news text P_i is retweeted, its root node is set as the news vector

N_i, and the child nodes are the user nodes $U^{P_i} = \left\{ u_1^{P_i}, u_2^{P_i}, \ldots, u_o^{P_i} \right\}$. If tweet $c_1^{P_i}$ is written by a user after reading and retweeting the source news P_i then the tweet vector $u_1^{N_i}$ becomes a child node of N_i, If tweet $c_2^{N_i}$ is retweeted and written by a user after reading tweet $c_1^{N_i}$, then $u_2^{N_i}$ becomes a child nodes of $u_1^{N_i}$.

4 Method

This paper proposes a fake news detection model named NT-UP based on unsupervised learning. As shown in Fig. 1, NT-UP mainly consists of three components: text representation module, text feature extraction module and user propagation module.

4.1 Text Representation Module

Since users on social platforms are not limited in the number of times they can retweet news and write tweets, this paper encodes the collected news texts and the first tweets written when users retweet news into news vector and tweet vectors, respectively. Tweets vectors are regarded as user nodes, providing data support for subsequent modules. For the news text collections $P = \{P_1, P_2, \ldots, P_m\}$, each news text undergoes stop-word removal and tokenization.

For example, for one news text P_i, it is encoded into a text sequence set $S^{P_i} = \left\{ s_1^{P_i}, s_2^{P_i}, \ldots, s_n^{P_i} \right\}$, where n represents the number of words in the text sequence. An encoder is used to encode the news text sequence into word vectors for the news text. Afterward, the generated word vectors are averaged to create a news vector N_i that represents the overall semantics of news P_i.

To obtain user node, let's take news P_i as an example. The collection of all user tweets that retweeted this news is denoted as $C^{P_i} = \left\{ c_1^{P_i}, c_2^{P_i}, \ldots, c_0^{P_i} \right\}$. For instance, for $c_j^{P_i} (1 \leq j \leq 0)$, stop-word removal and tokenization are performed to obtain the text sequence of the tweet set $Q_j^{P_i} = \left\{ q_{j_1}^{P_i}, q_{j_2}^{P_i}, \ldots, q_{j_h}^{P_i} \right\}$, where h represents the number of words in the tweet text. An encoder is used to encode the text sequence into word vectors for the tweet, followed by averaging to obtain the tweet vector $u_j^{P_i}$. The tweets of other users undergo the same process to obtain the tweet vectors set $U^{P_i} = \left\{ u_1^{P_i}, u_2^{P_i}, \ldots, u_o^{P_i} \right\}$. These tweet vectors are denoted as user nodes.

4.2 Text Feature Extraction Module

This module consists of five parts: fully connected layer, max-pooling layer, average-pooling layer, shared network, and channel attention mechanism. The module is designed to extract text features from news vector.

Taking the news vector N_i of news P_i as an example, word embeddings is used to encode the word vectors, positional encoding, and grammatical features in N_i, which

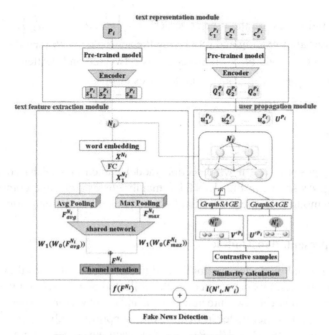

Fig. 1. The overview of the NT-UP framework

are then merged into a text matrix X^{N_i}. Next, we use fully connected layer to calculate a higher-dimensional representation $X_1^{N_i}$.

While the pooling layers focuses on enhancing the ability to extract and recognize text features in text feature extraction tasks, the Max Pooling layer (MP) focuses more on local features and their correlations, making it advantageous for extracting local features from news. However, as it may lose sequential information in the text during process. The Avg Pooling layer (AP) helps to preserve it and is less affected by factors such as word length and syntax in text feature extraction. As such, both MP and AP are used to aggregate the text matrix $X_1^{N_i}$, resulting in the extraction of local context features $F_{max}^{N_i}$, and global semantic features $F_{avg}^{N_i}$.

To reduce the computational resources and time required for channel attention calculation, two MLP and hidden layer are constructed as Shared Network. The pooling layer features $F_{max}^{N_i}$ and $F_{avg}^{N_i}$ are forwarded to the shared network for dimension reduction in channel map $F^{N_i} \in R^{C \times 1 \times 1}$. After the shared network performs channel spaces dimension transformation on the pooling layer features, channel spaces $W_1\left(W_0\left(F_{max}^{N_i}\right)\right)$ and $W_1\left(W_0\left(F_{avg}^{N_i}\right)\right)$ for Max Pooling features and Avg Pooling features are obtained. These two sets of channel spaces are weighted to yield a channel space F^{N_i} that represents richer text features in news:

$$F^{N_i} = \sigma\left(W_1\left(W_0\left(F_{max}^{N_i}\right)\right) + W_l\left(W_0\left(F_{avg}^{N_i}\right)\right)\right) \tag{1}$$

where σ represents the sigmoid function, $W_0 \in R^{C \times \frac{C}{r}}$, $W_1 \in R^{\frac{C}{r} \times C}$, where r indicates the ratio of feature dimension reduction, and C represents channel.

Then calculate F^{N_i} using the formula (2) to get the text feature $f(F^{N_i})$:

$$f(F^{v_i}) = 1/(1 + exp(-F^{v_i})) \tag{2}$$

4.3 User Propagation Module

In this module, a graph-based unsupervised learning approach called graph contrastive learning is used to extract the graph features of user propagation networks. Continuing with the example of news P_i and tweets generated by users when retweeting the news, the user propagation network G^{N_i} is constructed using news vector P_i and user nodes set $U^{P_i} = \left\{ u_1^{P_i}, u_2^{P_i}, \ldots, u_0^{P_i} \right\}$.

In this network, the root node is the news vector N_i, and the child nodes are user nodes U^{P_i}, connected based on user retweeting information in the construction of the news propagation network. Next, the graph contrastive learning is applied to extract graph features from the user propagation network G^{N_i}. To diversify the node distribution and mitigate over-fitting during model training, a graph data augmentation method T is used to delete edges in G^{N_i} based on the size of edge features, transforming the structure of the user propagation network G^{N_i} into a subgraph $\widetilde{G^N} - T(G^{N_i})$, where $\widetilde{G^N}$ has a root node $\tilde{N}$ and child nodes representing user nodes set $V^{P_i} = \left\{ v_1^{P_i}, v_2^{P_i}, \ldots, v_0^{P_i} \right\}$.

To further enhance the model's performance, GraphSAGE (Graph SAmple and aggreGatE) is employed to aggregate node representations in the graph. GraphSAGE is an effective graph neural network designed to capture similarities between nodes and local structures, making it suitable for tasks involving node aggregation. It is used to encode nodes in both G^{N_i} and $\widetilde{G^N}$. Specifically, GraphSAGE samples and aggregates neighbor nodes for nodes in G^{N_i}, and the updated graph G'^{N_i} contains a root node N_i' and user nodes set $U'^{P_i} = \left\{ u_1'^{P_i}, u_2'^{P_i}, \ldots, u_0'^{P_i} \right\}$. The same operation is performed on $\widetilde{G^N}$, resulting in the updated subgraph $\widetilde{G'^N}$ with a root node N_i'' and user nodes set $V'^{P_i} = \left\{ v_1'^{P_i}, v_2'^{P_i}, \ldots, v_0'^{P_i} \right\}$. The node representations obtained through the aggregation operations above can better capture the structure and features of vertices in the subgraph. The aggregation and update operation in this process is shown as follows:

$$x' = \sigma(mean(x + Neighbor(x))) \tag{3}$$

This formula represents the aggregation and update of current node by aggregating neighboring nodes during nodes forward propagation in the subgraph, where $Neighbor(x)$ represents the neighboring nodes of node x in the subgraph.

After aggregating and updating nodes representations in the graph, a graph contrastive learning approach is used to compare the structures of G'^{N_i} and $\widetilde{G'^N}$. First, the root nodes N_i' and N_i'' of G'^{N_i} and $\widetilde{G''^N}$ are defined as root node samples (N_i', N_i''). Then, the root node N_i' from G'^{N_i} and user nodes U'^{P_i} are defined as intra-group contrastive samples set $(N_i', U'^{P_i}) = \left\{ \left(N_i', u_1'^{P_i} \right), \left(N_I', u_2'^{P_i} \right), \ldots, \left(N_I' u_0'^{P_i} \right) \right\}$. Similarly, the root node N_i' from G'^{N_I} and user nodes V'^{P_i} from $\widetilde{G'^N}$ are defined as inter-group contrastive samples set $(N_i', V'^{P_i}) = \{(N_i', v_1'^P), (N_i', v_2'^P), \ldots, (N_i', v_o'^P)\}$. The contrastive loss $l(N_i', N_i'')$ between G'^{N_i} and $\widetilde{G'^N}$ is computed using the following graph contrastive learning formula:

$$l(N_i', N_i'') = \log \frac{exp^{\theta(N_i', N'')/k}}{exp^{\theta(N_i', N'')/k} + \sum_{j=1}^0 exp^{\theta\left(N_i', u_j'^{P_i}\right)/k} + \sum_{j=1}^0 exp^{\theta\left(N_i', v_j'^{P_i}\right)/k}} \tag{4}$$

In this formula, k is the temperature parameter that determines the focus on samples in graph contrastive learning. $\theta(y, z) = s(p(y, z))$ computes the cosine similarity between node y in G'^{N_i} and node z in $\widetilde{G'^N}$, where $p(y, z)$ represents the nonlinear projection of node y into node z.

4.4 News Prediction Module

The text features computed by the text feature extraction module as $f(F^{N_i})$ and the contrast loss calculated by the user propagation module as $l(N_i', N_i'')$ are fused into news embeddings. These news embeddings are then used in the fake news discriminator to calculate the binary cross-entropy loss using the following formula, enabling the binary classification of fake news P_i in the fake news detection:

$$D = softmax\left(l(N_i', N_i'') + f\left(F^{N_i}\right)\right) \tag{5}$$

5 Experimental Settings and Results

5.1 Dataset and Data Preprocessing

To validate the effectiveness of the NT-UP model, FakeNewsNet [19] is used here. The FakeNewsNet dataset encompasses two subsets, POL and GOS, which contain a substantial number of real and fake news samples. These samples span a diverse range of topics and time periods. The dataset provides a wealth of textual data to train the NT-UP model, thereby enabling a more precise evaluation of the NT-UP model's performance in analyzing text data within social media platforms. The dataset also includes tweet written by users participating in news propagation on Twitter. The POL contains 314 news and 40,740 user tweets. Among these news texts, there are 157 true news texts and 157 fake news texts. In the GOS, there are 5,464 news texts and 308,798 user tweets with 2,732 true news texts and 2,732 fake news texts.

For encoding news texts and user tweets, the pre-trained spaCy model is used for tokenization and stop-word removal. Word2Vec embeddings are applied to encode news texts and user tweets into word embeddings, and the BERT is employed to further encode text sequences. Although the BERT model excels in text sequence encoding, it has limitations regarding input length, especially when handling large volumes of text. To alleviate training difficulties, the BERT-Large model with a larger maximum sequence length (512 tokens) is selected for encoding news texts. This model demonstrates better performance and shorter processing times. After encoding the news texts with this model, Avg Pooling is performed to generate news vector representing the overall semantics of the news text. In the case of user tweets about shared news texts, due to their typically shorter length and the large number of users, using BERT for training user tweet requires significant computational resources. Thus, a token-based model with a shorter maximum sequence length is employed to encode each user's tweets. During experimentation, setting the maximum sequence length to 16 tokens proved to be the most efficient. After encoding each user's tweets into tweet vectors, an Avg Pooling operation is applied to obtain a collection of tweet vectors, which serves as the user nodes.

The news vector and user nodes are used to construct user propagation networks based on user sharing information. For the POL, the network is constructed using 314 news texts and 41,054 user tweets, resulting in an average of 131 user nodes per news texts. In the case of the GOS, the network consists of 5,464 news texts and 314,262 user nodes, averaging 58 user nodes per news. The details are summarized in Table 1.

Table 1. Constructed Graph Data.

Data	POL	GOS
Total News	314	5,464
Total Nodes	41,054	314,262
Total Edges	40,740	308,798
Avg. Nodes per News	131	58

5.2 Comparative Models

To validate the effectiveness of the NT-UP model, we compared it with other six representative fake news detection models as follows:

(1) CSI [20]: An LSTM-based model for extracting features from news text to predict fake news.
(2) BERT | MLP. A model utilizing BERT and MLP to extract features from news texts for predicting fake news.
(3) GCNFN [21]: The initial model employs graph convolutional network to extract features from the news propagation graph. It merges the extraction of graph features from user propagation and text features from user comments.
(4) GNN-CL [22]: A model that integrates graph neural network and diffpool to extract graph features from the news propagation network.
(5) GLAN [15]: This model constructs a heterogeneous graph by representing Twitter text, user tweets, and user relationships. It employs this graph to extract connections between news and users.
(6) PPC [23]: Leveraging recurrent neural networks and convolutional neural networks, this model extracts user node attributes and news propagation paths to tackle the binary classification issue of identifying fake news.

(7) UPFD [2]: This model employs graph neural networks to capitalize news texts features and user preference features.

5.3   Comparative Experiments

The experimental platform server is configured as follows: Intel(R) Xeon E5–2678 processors, 30GB of RAM, and a NVIDIA RTX A2000 12GB GPU. All models run in the following environment: Ubuntu 18.04 operating system, Python 3.8 programming environment, PyTorch 1.11.0 deep learning framework, CUDA 11 general-purpose computing toolkit, cuDNN 8 computational acceleration package, Python-Geometric 2.3.1 deep learning library, and the Adam optimizer. The default parameter settings are as follows: hidden layer dimension is 128, learning rate is 0.05, dropout ratio is 0.6, batch size is 128, feature dimension reduction rate r is 16, and temperature parameter k is 0.1. The model's performance is tested on two datasets. After conducting 15 separate tests on each dataset, the experimental results are averaged to determine the model's final performance. The experimental results indicate that the NT-UP model achieves its best performance on the GOS after training for 100 epochs, while it requires 150 epochs to reach optimal performance on the POL.

Fake news detection is a binary classification task, and accuracy (ACC) and F1-score (F1) are adopted as evaluation metrics. The experimental results are presented in Table II (the best for each evaluation metric are indicated in bold).As shown in Table 2, NT-UP exhibits higher advantages compared to other models. In the POL, NT-UP demonstrates the ACC of 87.93% and the F1 score of 87.96%, surpassing the second-ranked UPFD by 3.62% and 3.31%. In the GOS, NT-UP achieves the ACC of 97.62% and the F1 score of 97.59%, showing a performance advantage of 0.53% and 0.52% over the second-ranked UPFD.

Table 2. Comparative Experimental Performance

Performance Comparison (%)				
Dataset	POL		GOS	
Evaluation	ACC	F1	ACC	F1
CSI	75.33	83.36	78.20	82.34
BERT + MLP	71.04	71.03	85.76	85.75
GCNFN	82.35	86.85	96.38	95.59
GNNCL	62.90	62.25	95.11	95.09
UPFD	84.31	84.65	97.09	97.07
PPC	62.21	64.56	90.38	90.33
NT-UP	**87.93**	**87.96**	**97.62**	**97.59**

5.4   Ablation Experiment

As shown in Table 3, NT-UP has higher advantages than the comparative models. In the POL, NT-UP showed 87.93% ACC and 87.96% F1, and its performance exceeded the second-ranked UPFD by 3.62% and 3.31%. In the performance of the GOS, NT-UP performed with the ACC of 97.62% and the F1 of 97.59%, its performance also has 0.53% and 0.52% advantages over the second-ranked UPFD.

Table 3. Comparative Experimental Performance

Performance Comparison (%)

Dataset	POL		GOS	
Evaluation	ACC	F1	ACC	F1
NT-UP	87.93	87.96	97.62	97.59
w/o CL	85.78	86.18	97.54	97.53
w/o At	86.88	86.69	97.31	97.35
w/o CL & At	84.31	84.65	97.09	97.07

(1) "w/o CL" indicates that the graph features are extracted from the news propagation network without using the graph contrastive learning method in the user propagation module of the model. The performance test results of the model from this group show that after deleting the graph contrastive learning method, the ACC and F1 performance data of the model in identifying fake news from the GOS dataset are decreased by 0.08% and 0.06% respectively. The performance figures of ACC and F1 for identifying fake news in POL dataset are reduced by 2.15% and 1.78% respectively.
(2) "w/o At" indicates the removal of the channel attention mechanism from the context semantic encoder, while retaining the fully connected layer to extract the contextual semantic features of news text. From the experimental results, it can be observed that without utilizing the channel attention mechanism, the performance of NT-UP in the GOS dataset decreases by 1.05% and 1.27% in terms of the ACC and F1 metrics for detecting fake news, respectively. In the POL dataset, the performance in detecting fake news decreases by 0.31% and 0.24% in the ACC and F1 metrics, respectively.
(3) "w/o CL& At" means that neither graph contrastive learning method nor channel attention mechanism is used in the news propagation module and the context semantic encoder. From the experimental results, it can be seen that the model's performance in the GOS dataset decreases by 3.62% and 3.31% in the ACC and F1 metrics, respectively. In the POL dataset, the ACC and F1 metrics decrease by 0.53% and 0.52%, respectively.

According to the above analysis, it can be reasonably concluded that in the NT-UP model obtained by optimization and improvement, the user propagation module implemented by using graph contrast learning method can make this module more effective in the task of extracting graph features, and the text feature extraction module composed of channel attention mechanism can make this module have a more satisfactory performance. The news embedding obtained by fusing the graph features and text features extracted by these two modules can make the model show better accuracy in the task of identifying fake news, which proves that the introduction of these two groups of methods into the model has a good improvement in the performance of the model.

5.5 Analysis of Hyperparameters

The size of parameter r affects the ability of the text feature extraction module to extract news text features, and the size of parameter k is related to the degree of attention paid by the graph comparison learning method in the user propagation module to the news propagation network. In order to explore the influence of different parameter values r and k on the detection of fake news by the NT-UP model, we focused on setting the reduction ratio r of the text feature extraction module and the temperature parameter k of the user propagation module to test the ability of NT-UP model to detect fake news in the POL and GOS datasets. In the experiments, we keep one parameter constant in the experiment to test the NT-UP performance under different parameters setting. The following charts (Figs. 2 and 3) show the experimental results.

As shown in Fig. 2, the experimental results presented by adjusting parameter r show that a smaller value of r may lead to the underfitting of NT-UP, and more training wheels are needed to obtain the best performance. On the contrary, a larger value of r may lead to premature convergence of NT-UP, resulting in overfitting and failing to give full play to the model's performance in detecting fake news. Therefore, when r is set to 16, the model will not be able to detect fake news. The model showed better performance when the epochs of training was set to 150. On the other hand, the experimental results obtained by using parameter k of different sizes to measure the degree of attention paid to the news propagation network are shown in Fig. 3. When a larger parameter k is set, the model pays more attention to the news propagation network, resulting in the appearance of overfitting of the model and reducing the overall performance of the model in detecting fake news. However, more training epochs need to be trained when the parameter size is set to 0.1. When the number of training epochs of the model is set to 150, the model has better fake news detection performance compared with other parameter Settings. When the parameter k is set lower, the module pays less attention to the news propagation network and fails to perform the underfitting phenomenon of model performance. Based on the above experiments to test the model performance by adjusting the size of the hyperparameter, we can conclude that when the parameter r is set to 16, k is set to 0.1, and the training epochs is 150 rounds, the NT-UP model can achieve the best performance in detecting fake news.

In conclusion, by setting the two hyperparameters of the NT-UP model to the appropriate size, that is, the reduction ratio r and the temperature parameter k, the performance of the model in detecting fake news can be maximized. Too large a parameter value may result in overfitting and premature convergence, while too small a parameter value may require more training epochs to achieve satisfactory results.

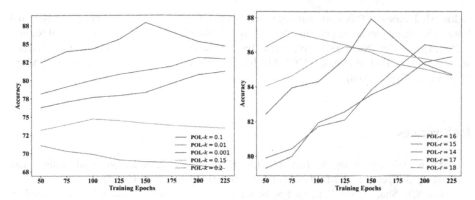

Fig. 2. Performance results of NT-UP under different parameters in POL.

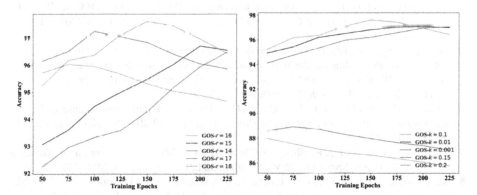

Fig. 3. Performance results of NT-UP under different parameters in GOS.

6 Conclusion

This paper proposes the NT-UP model for the fake news detection, which consists of text representation learning module, text feature extraction module and user propagation module. The text learning module encodes news texts and user tweets into news vector and user nodes, respectively. The channel attention mechanism is employed in the text feature extraction module to strengthen the model to extract news texts features. The graph contrast learning is designed in the user propagation module to improve the model's extraction of graph features in the news propagation network, The effectiveness of both mechanisms is verified in the ablation experiment. This paper uses real-world datasets to verify NT-UP, and experimental results show that NT-UP has better abilities in extracting feature for fake news. In future works, the structure of the model will be further optimized to build a multimodal model with better abilities in extracting feature and discriminating fake news.

Acknowledgement. This work was supported by the following funding organizations: National Natural Science Foundation of China (No. 72261018, No. 62062037), Jiangxi Provincial Natural Science Foundation (No. 20212BAB202014), Jiangxi Provincial Department of Education Science and Technology Project (No. GJJ2200830), Jiangxi Provincial Department of Education Youth Project (No. GJJ2200868).

References

1. Shu, K., Sliva, A., Wang, S., Tang, J.L., Liu, H.: Fake news detection on social media: a data mining perspective. ACM SIGKDD Explor. Newsl. **19**(1), 22–36 (2017)
2. Dou, Y.T., Shu, K., Xia, C.Y., Yu, P.S., Sun, L.C.: User preference-aware fake news detection. In: Proceedings of the SIGIR, Virtual, Online, Canada, pp. 2051–2055 (2021)
3. Chen, T., Li, X., Yin, H., Zhang, J.: Call attention to rumors: deep attention based recurrent neural networks for early rumor detection. In: Ganji, M., Rashidi, L., Fung, B.C.M., Wang, C. (eds.) PAKDD 2018. LNCS (LNAI), vol. 11154, pp. 40–52. Springer, Cham (2018). https://doi.org/10.1007/978-3-030-04503-6_4
4. Shu, K., Wang, S.H., Liu, H.: Beyond news contents: the role of social context for fake news detection. In: Proceedings of the WSDM. Melbourne, VIC, Australia, pp. 312–320 (2019)
5. Ni, S., Li, J.W., Kao, H.Y.: MVAN: multi-view attention networks for fake news detection on social media. IEEE Access **9**(11), 106907–106917 (2021)
6. Qian, S.S., Wang, J.G., Hu, J., Fang, Q., Xu, C.S.: Hierarchical multi-modal contextual attention network for fake news detection. In: Proceedings of the SIGIR, Virtual, Online, Canada, pp.153–162 (2021)
7. Zhang, X.Y., Cao, J., Li, X.R., Sheng, Q., Zhong, L., et al.: Mining dual emotion for fake news detection. In: Proceedings of the WWW, Ljubljana, LJU, Slovenia, pp. 3465–3476 (2021)
8. Mansouri, R., Naderan-Tahan, M., Rashti, M.J.: A semi-supervised learning method for fake news detection in social media. In: Proceedings of the ICEE, Tabriz, Iran, p. 1 (2020)
9. Li, X., Lu, P., Hu, L., Wang, X.G., Lu, L.: A novel self-learning semi-supervised deep learning network to detect fake news on social media. Multimed. Tools Appl. **81**(14), 19341–19349 (2022)
10. Wang, Y., et al.: Weak supervision for fake news detection via reinforcement learning. Proc. AAAI Conf. Artif. Intell. **34**(01), 516–523 (2020)
11. Singhal, S., Kabra, A., Sharma, M., Shah, R.R., Chakraborty, T., Kumaraguru, P.: Spotfake+: A multimodal framework for fake news detection via transfer learning. In: Proceedings of the AAAI, New York, NY, USA, pp. 13915–13916 (2020)
12. Tian, L., Zhang, X.Z., Lau, J.H.: Rumor detection via zero-shot cross-lingual transfer learning. Arxiv, abs/2021.0350243 (2021)
13. Ren, Y.X., Zhang, J.W.: HGAT: hierarchical graph attention network for fake news detection. ArXiv, abs/2002.04397 (2020)
14. Wang, Y.Z., Qian, S.S., Hu, J., Fang, Q., Xu, C.S.: Fake news detection via knowledge-driven multimodal graph convolutional networks. In: Proceedings of the ICMR, Dublin, DUB, Ireland, pp 540–547 (2020)
15. Yuan, C.Y., Ma, Q.W., Zhou, W., Han, J.Z., Hu, S.L.: Jointly embedding the local and global relations of heterogeneous graph for Rumor detection. In: Proceedings of the ICDM, Beijing, BJ, China, pp. 796–805 (2019)
16. Sun, L., Rao, Y., Lan, Y.Q., Li, Y., Xia, B.C., Li, Y.Y.: HG-SL: jointly learning of global and local user spreading behavior for fake news early detection. In: Proceedings of the AAAI, Washington, WA, USA, pp. 5248–5256 (2023)

17. Ti Bian, X., Xiao, T.Y.Xu., Zhao, P.L., Huang, W.B., et al.: Rumor detection on social media with bi-directional graph convolutional networks. Proc. AAAI Conf. Artif. Intell. **34**(01), 549–556 (2020)

18. Konkobo, P.M., Zhang, R., Huang, S.Y., Minoungou, T.T., Ouedraogo, J.A., et al.: A deep learning model for early detection of fake news on social media. In: Proceedings of the BESC, Bournemouth, BOH, UK, pp. 1–6 (2020)

19. Shu, K., Mahudeswaran, D., Wang, S.H., Lee, D.W., Liu, H.: FakeNewsNet: a data repository with news content, social context, and spatiotemporal information for studying fake news on social media. Big data **8**(3), 171–188 (2020)

20. Ruchansky, N., Seo, S.Y., Liu, Y.: Csi: a hybrid deep model for fake news detection. In: Proceedings of the CIKM, Singapore, CBD, SGP, pp. 797–806 (2017)

21. Monti, F., Frasca, F., Eynard, D., Mannion, D., Bronstein, M.M.: Fake news detection on social media using geometric deep learning. ArXiv. abs/1902.06673 (2019)

22. Han, Y., Karunasekera, S., Leckie, C.: Continual learning for fake news detection from social media. In: Farkaš, I., Masulli, P., Otte, S., Wermter, S. (eds.) Artificial Neural Networks and Machine Learning – ICANN 2021: 30th International Conference on Artificial Neural Networks, Bratislava, Slovakia, September 14–17, 2021, Proceedings, Part II, pp. 372–384. Springer International Publishing, Cham (2021). https://doi.org/10.1007/978-3-030-86340-1_30

23. Liu, Y., Wu, Y.B.: Early detection of fake news on social media through propagation path classification with recurrent and convolutional networks. In: Proceedings of the AAAI, New Orleans, LA, USA, pp. 354–361 (2018)

Data Analysis of University Educational Administration Information Based on Prefixspan Algorithm

Yiying Xu[1(✉)], Yi Liu[2], and Haili Yu[3]

[1] Academic Affairs Office of Jiangsu University, Zhenjiang, China
xuyiying19821128@sina.com
[2] School of Computer Science and Communication Engineering, JiangSu University, Zhenjiang, China
ly@ujs.edu.cn
[3] Mengxi Honors College of Jiangsu University, Zhenjiang, China
yuhaili@ujs.edu.cn

Abstract. With the rapid development of data mining technology, a large amount of educational administration information data has been produced. How to make full use of these information resources and dig out valuable information is a hot topic in the research of colleges and universities. Compared with positive sequential pattern mining, negative sequential pattern mining considers not only the events that have already occurred, but also the events that have not occurred, and it can assist decision making when simple positive sequential pattern mining may mislead decision making. And the existing sequential pattern mining algorithm has the same importance in each project when applied, which is impractical. In this paper, a data mining algorithm based on Prefixspan is proposed. During the mining process, different weights are set for the items, and the weighted support degree of each sequence is compared with the minimum support degree to obtain frequent sequence patterns. The algorithm is applied to the modeling and simulation of student data by using K-means clustering method. The experimental results show that this method can effectively improve the efficiency and accuracy of data mining, and has strong practicability.

Keywords: Data mining · Weighted positive and negative sequence model · Weight · Modeling and simulation · Data application

1 Introduction

University data, rich in student academic and lifestyle information like performance, card usage, and library borrowing, is increasingly central to the digital evolution in education. Effectively using data mining to uncover hidden insights within this complex data is crucial for enhancing student development and university management. The holistic growth of college students depends on various factors, not just academic performance. Simply enhancing performance in isolated areas doesn't necessarily equate to overall development, as key information remains concealed within the data, eluding basic analysis methods. Previously, university data analysis relied on basic statistical methods [1], inadequate for uncovering deeper insights. Utilizing data mining [2] is crucial for understanding how different student groups can improve. This technology uncovers hidden patterns, offering valuable predictions for decision-making. Its effectiveness is proven in areas like customer behavior analysis, weather forecasting, DNA information analysis, and financial data analysis [3]. Sequential pattern mining, a key area in data mining, extracts valuable information from large datasets [4]. It includes positive sequence mining, which analyzes existing event relationships [5], and negative sequence mining. The latter, augmenting the positive model, reveals connections between unoccurred events, filling a critical analytical gap and gaining wider application in various fields [6]. After examining the issues with positive and negative sequence mining, this paper explores weighted sequence patterns in data mining. Applied to college student data,this method involves analyzing and modeling the results to guide decisions for diverse student groups' comprehensive development.

From the perspective of different performance levels of students, this paper uses weighted positive and negative sequence model for mining and analysis, which has certain changes and innovations in the method of university data mining analysis compared with the past. This paper studies key technical issues such as weighted positive and negative candidate sequence pattern generation method and pruning strategy, and proposes a data mining algorithm based on Prefixspan, which is proposed in this paper and applied to a variety of data in colleges and universities, so as to obtain valuable information hidden behind college data, such as finding frequent sequences related to grades. Provide decision support for improving the way universities train students. However, the application of mining negative frequent sequence in university data is not very mature, and mining valuable information in university databased on weighted positive and negative sequence pattern has important practical significance in practical application.

2   Research Status Analysis

Data mining includes association rules, sequential patterns, decision trees, neural networks, genetics and other classical algorithms. Sequential patterns have become a very mature technique. Agrawal et al. improved Apriori algorithm [8], proposed GSP sequence pattern mining algorithm [7], introduced the concept of sliding time window, reduced the number of sequences used in scanning, and adopted the hash tree structure to store candidate positive sequence patterns, which improved the efficiency of algorithm application and overcame the limitations of basic sequence patterns. Reduce the size of meaningless sequences; Based on the fast mining algorithm of association rules, the SPADE algorithm [9] proposed by Zaki et al., adopts the width and depth first search technology, and divides a problem into multiple smaller-scale sub-problems by the method of sequence lattice, which greatly improves the mining efficiency. Apriori-All algorithm [10] is a method to obtain the largest large sequence from the excavated large sequence by pruning. However, it still has some disadvantages, such as generating a large number of candidate sequences, requiring multiple scanning of the database, and being difficult to mine sequences with a long length. FreeSpan algorithm [11] uses sharding and database projection technologies. Candidate sequences are segmented in the mining process and restricted to a corresponding smaller projection database when tested. Freespan algorithm runs faster than GSP algorithm and can mine complete and frequent positive sequences. However, this algorithm has some disadvantages, such as a large number of projection databases and too much overhead in generating candidate sequences. In order to make up for the shortcomings of the FreeSpan algorithm, Gong Wei et al later proposed the PrefixSpan algorithm [12], which mining frequent positive sequences through prefix projection, by which the suffix subsequence corresponding to the prefix frequent subsequence is obtained into the projection database, and then the corresponding local frequent sequence pattern is obtained for each projection database. However, when the database size is large, a large number of projection databases will be constructed, and the overhead of this algorithm will become large. The algorithms described above all mine positive sequence patterns on the basis of equal importance, and they cannot distinguish items according to their different importance degrees. Reference [13] proposed a MWSP algorithm, which can mine sequence patterns of different importance.

Many scholars use PrefixSpan algorithm and its improved algorithm to mine university data to find the value behind the data, but scholars use either positive sequence or inverse sequence to think about problems. This paper combines positive sequence and inverse sequence to mine university educational administration information data, and has achieved good results.

3   PrefixSpan Algorithm

3.1   Introduction to Algorithm Concepts

The Prefixspan algorithm proposed in this paper firstly excavates the weighted positive sequence pattern, and then obtains the candidate negative sequence based on the 1length positive sequence, and only when the candidate negative sequence meets the pruning condition and constraint condition is the required negative sequence pattern. The following describes the related concepts involved.

1. Define the minimum support for the sequence

   MIS(i), which represents the minimum term support of item i, where i is a positive or negative term;

   The minimum support of the positive element, that is, the event that has occurred, is the minimum support value of item i in the element;

   The minimum support of negative elements, i.e. no event, is calculated using the information of relevant positive elements: $MIS(\neg i) = 1 - MIS(i)$; The element set contains e1, e2... The minimum support value of negative sequence Y of er is the minimum support value of elements in the sequence, where, the minimum support value of Y $minsup(Y) = min[MIS(e1), MIS(e2), ..., MIS(er)]$;

2. Define positive and negative sequence patterns For a sequence Y and its minimum support minsup(Y) and weighted support Wsup, if Y only packages

With positive elements, Wsup(Y) ≥ minsup(Y), then Y is called positive sequence pattern; If Y contains negative elements, Wsup(Y) ≥ minsup(Y), then Y is called negative sequence pattern;

3. Define multi-minimum support

Minimum support,

$$MIS(i) = \begin{cases} M(i), & M(i) > LS \\ LS & Otherwise \end{cases} M(i) = \alpha f(i) \tag{1}$$

Where, LS is the set minimum support degree, parameter a ranges from 0 to 1, $f(i)$ is the support degree of the project. MIS(i) Is the minimum support degree of the item. It should be noted that minsup(Y) in this paper represents the minimum support degree of sequence Y, while MIS(Y) It represents the multi-minimum support of the sequence Y, which is the same.

4 Define negative sequence constraints

Constraint 1: All sets of terms in a single element must be either all positive or all negative. E.g. <b, (c, ¬d), e >Is a violation of the constraint because in the element (c, ¬d), c is a positive term set and ¬d is a negative term set. Constraint 2: No adjacent negative elements are allowed. For example, the sequence <a, ¬c, ¬d >violates the constraint.

3.2 Algorithm Idea

This paper provides a data mining algorithm based on weighted positive and negative sequence pattern. The steps are as follows: Before data mining, the data must be cleaned and filtered, and then used as the data source after data cleaning. Then according to certain rules, the data is transformed to generate the sequence data that can be used directly by MWN-GSP method. MWN - GSP method is used to analyze the data. The specific steps include:

A. Use the GSP algorithm to mine all the frequent positive sequence patterns that meet the requirements, that is, in a certain period of time, a large number of data appears;
B. Based on the frequent 1-length positive sequence mined in step A, the corresponding negative sequence seed set is generated, and negative candidate sequence NSC is generated through connection. The method is mining based on GSP algorithm. The k-length candidate negative sequence pattern is generated by adding a 1-length frequent positive sequence pattern or 1-length frequent negative sequence pattern to the (k − 1-length candidate negative sequence pattern). The negative candidate sequence NSC is used to determine: which data item sets appear more, which data item sets do not appear, and which cases may necessarily not appear or appear after the data does or does not appear;
C. The support degree of negative candidate sequence NSC will be calculated by the corresponding positive sequence data information;

D. Screen out the required negative sequence pattern from the negative candidate sequence NSC generated in step B, that is, mine the negative sequence pattern that is greater than or equal to the minimum support degree set by the user from the negative candidate sequence NSC generated in step B; These negative sequence patterns are used to analyze the results of the data to provide decision-making guidance.

Specifically, Step A includes:

The weighted support of frequent sequence Y is denoted by Wsup (Y)

$$W_{SUP}(Y) \frac{\sum_{Si \in T \& Y \in Si} W_T(S_i)}{\sum_{Si \in T} W_T(S_i)} \tag{2}$$

The calculation method emphasizes the weighting of the sequence.
Where, the weight of the sequence is WT(Si),

$$W_T(S_i) = \frac{1}{|S_i|} \sum_{i,j \in S_i} W_j \tag{3}$$

Step C includes: If both A and B are frequent item sets, a valid negative sequence pattern must satisfy the following three conditions:

(1) $A \cap B = \varphi$;
(2) $Sup(A) \geq$ minsup and $Sup(B) \geq$ minsup;
(3) $Sup(A \cup B)$ or greater minsup(Sup(suchA $\cup$ B) or greater minsup, Sup(suchA $\cup$ B)minsup) or higher.

The formula for obtaining negative sequence pattern support is:

$$Sup(A) = 1 - Sup(\neg A) \tag{4}$$

$$Sup(A \cup \neg B) = Sup(A) - Sup(A \cup B) \tag{5}$$

$$Sup(\neg A \cup B) = Sup(B) - Sup(A \cup B) \tag{6}$$

$$Sup(\neg A \cup \neg B) = 1 - Sup(B) - Sup(A) + Sup(A \cup B) \tag{7}$$

Weighted support for negative sequence pattern

$$Wsup(Y) = Sup(Y) * WT(Y) \tag{8}$$

That is, the weighted support is obtained by multiplying the support of the sequence and the weight, where the weight of the sequence is obtained by the same method as the positive sequence.

3.3 Algorithm Description

The pseudo-code of the implementation algorithm MWN-GSPin this paper is as follows:
Input: D: sequence database; WT(i) : weight value of each item; LS: The minimum support set by the user;
   β : Calculate the parameters of multiple support;
   Output: NSP: a collection of negative sequence patterns used to analyze data;

(1) for each item in D {
(2) if(β * sup(i)>LS).
(3) MIS(i) = β * sup(i);
(4) else MIS(i) = LS;
(5) } //(1) to (5) Calculate the multi-minimum support degree of each item in the database;
(6) for each item in D {
(7) if(Wsup(i) ≥ MIS(i)).
(8) add(L1,i);
(9) }//(6) to (9) Perform mining to obtain all 1-length frequent positive sequences;
(10) Cn = GSP − join(L (n − 1)); //Generate candidate positive sequences by concatenation;
(11) for each Sn in Cn {
(12) if(Wsup(Sn) ≥ MIS(Sn)).
(13) add(Ln,Sn);
(14) } //(11) to (14) are calculated by formula 1 and 2 for each candidate positive sequence. The weighted support degree of the item set and each candidate sequence, and determine whether the weighted support degree is not less than the minimum set by the user Support degree, only meet the conditions as frequent positive sequence;
(15) N Seed1 = Neg - GSP$_G$eneration(L1); //Generate negative sequence seed sets through 1-length frequent positive sequences;
(16) NSC = GSP - join(N Seed); //The concatenation operation is performed on the seed set to generate the candidate negative sequence NSC;
(17) for (each nsc in NSC){
(18) if (nsc.size == 1) {
(19) sup(nsc) = 1 - sup(¬ (nsc));
(20) }
(21) for any nsc A ∪ B = x and A U B = Φ{
(22) sup(A ∪ ¬B) = sup(A) - sup(A ∪ B);
(23) sup(¬A ∪ B) = sup(B) - sup(A ∪ B);
(24) sup(¬A ∪ ¬B) = 1 - sup(A) - sup(B) + sup(A ∪ B);
(25) } //(17) to (25), the support degree of each candidate negative sequence NSC in NSC was calculated by formula 3-6.
(26) if (nsc.Wsup ≥ MIS(nsc)).
(27) NSP.add (nsc);
(28) } //(26) to (28) Candidate negative sequence NSC was obtained from the support degree of candidate negative sequence NSC by equation 7 Candidate negative sequence NSC whose Wsup is not less than the minimum support set by the user is added frequently In the negative sequence;

(29) select in NSP with restrictions; //In the frequent negative sequences obtained above, the final result is obtained by filtering with constraints.

(30) return NSP; //Return the result and analyze the data using the resulting negative sequence pattern.

4 Data Analysis of University Educational Administration Information Based on Prefixspan Algorithm

The data source used the score data of 552 students from the School of Computer Science and Technology of a university in 2013 during eight semesters, with a total of 32,959 records. Each line is a student's score for a course. The data contains 25 attributes such as student number, name, and usual grades, and we only use the content needed for the research, including the score information of students in each subject (including the six attributes of student number, course name, total score, course attribute, academic year, and semester) and the award information of students collected in each semester. Table 1 shows the raw data of student achievement, where "-" indicates that the data does not exist.

Table 1. Part of the original performance data

student number	Name	Course name	Credits	Grade	Make-up test results	Nature of Course	Refixing the markup	School year	term	...
22015001	**	Linear algebra	3.0	88	-	compulsory course	0	2022–2023	2	...
22015001	**	Data structures	3.0	96	-	compulsory course	0	2022–2023	2	...
22015001	**	University Physics	2.0	90	-	compulsory course	0	2022–2023	2	...
22015001	**	University english	3.0	96	-	compulsory course	0	2022–2023	2	...

The first step is to traverse the data and remove any records that are missing the overall score. It turns out that the records of 610 students fit the bill. Then these data are calculated according to the courses taken in different semesters corresponding to the student number, including the number of failed courses and the course weight coefficient setting different course attributes, and the course grade point of each subject is obtained. In the course weight coefficient, the compulsory course is set at 1.5, and the elective course, general elective course and open experiment course is set at 1.0. According to the formula for calculating average grade point GPA $= \sum^{course} GPA \div \sum^{course}$ credit, the result is taken as sequence item 1; The number of failed subjects is counted as item 2. The corresponding values of students' awards are rounded as item 3, and the coding method according to the award level is shown in Table 2 and the sample of Partial score sequence database is shown in Table 3. If a student has won multiple awards in this semester, the corresponding values can be added together. In this way, each student's grade information is processed as sequence data of length 8 with 3 items per element, and "corresponding term number" is added to each processing value. The way to distinguish. For example, the corresponding sequence of a student is <1.3,1.1,1.0 2.3,2.1,2.0... 7.3,7.1,7.0 8.3,8.1,8.0>, indicating that the student's GPA in the second semester is 3, the number of failed grades is 1, and there is no award. Through the same processing method, a score database containing 610 sequences is obtained.

Table 2. Student award code

Flag value	awards
3.5	Provincial inspirational Scholarship, National Inspirational Scholarship, National first prize in mathematical modeling
3.0	The provincial second prize in mathematical modeling, the first prize in software design competition, and the outstanding student of provincial colleges and universities
2.5	Scholarship, second prize in Software Design competition, second prize in College students English Competition
2	Second class scholarship, third prize in Software Design competition, third prize in mathematical modeling
1.5	Learning Progress Award, Mathematical Modeling Excellence Award, mathematics competition excellence award
0	prizeless

Table 3. Partial score sequence database

Student ID	The processed sequence
22015001	<(1.2,1.0,1.0)(2.4,2.0,2.0)(3.4,3.0,3.3)(4.3,4.0,4.0) (5.4,5.0,5.3)(6.3,6.0,6.0)(7.3,7.1,7.0)(8.3,8.0,8.0)>
22015002	<(1.2,1.0,1.0)(2.4,2.0,2.0)(3.3,3.0,3.0)(4.4,4.0,4.0) (5.2,5.0,5.0)(6.3,6.0,6.3)(7.3,7.0,7.2)(8.3,8.0,8.0)>
22015003	<(1.3,1.0,1.0)(2.4,2.0,2.0)(3.3,3.0,3.0)(4.4,4.0,4.0) (5.3,5.0,5.0)(6.4,6.0,6.0)(7.3,7.0,7.0)(8.4,8.0,8.0)>

In order to obtain the rule of student achievement, the following rules are used to set the weights of different elements in the sequence: The 8 elements contained in each sequence are comprehensively considered. The larger the GPA value of item 1 is, the more important it is; the number of failed courses of item 2 is 0 is the most important; similarly, the award handling value of item 3 is also more important. The importance of different items in this semester is analyzed. Finally, considering the different proportion of different semesters in the undergraduate stage, the weight of each element is set. For example, the corresponding elements of a student's first semester are (1.3,1.0,1.0), indicating that he attached importance to the quality of learning in the beginning of the university without passing the subject, and did not get any awards; In addition, the first semester is the adaptation period, which should strengthen the learning of all subjects. Based on the above analysis, we can set the weight of this element higher than the average value of the weight of all students in the first semester. The weights of different elements are shown in Table 4.

Table 4. The weight of each element in the score sequence

Element	(1.6,1.2,1.0)	(2.3,2.0,2.0)	(3.4,3.0,3.0)	(4.3,4.0,4.0)
weight	0.28	0.30	0.32	0.29

4.1 Mining Weighted Positive Sequence Patterns from Sequences

GSP mining algorithm is used to scan the database to get the support degree of each item. The weighted support degree of each project is obtained according to the given project weight and support degree. For items whose weighted support degree is greater than the set minimum support degree, it is regarded as a positive sequence pattern of 1-length and as the candidate item set C1. Through the join and pruning operation of GSP algorithm, other candidate sequences Cn whose length is continuously increased by 1 are generated, and all the sequences whose weighted support degree is greater than the given minimum support degree are selected as weighted frequent positive sequence patterns.

Step1: given minimum support LS = 0.6, parameter = 0.5, of each element and the minimum support, MIS ((1.3, 1.0, 1.0)) = 0.34, MIS ((2.3, 2.0, 2.0)) = 0.2, MIS ((3.4, 3.0, 3.0)) = 0.28,... ..... ;
Step2: Calculate the weights of each sequence according to the weights of items in Table 5. 2 and formula 4.1, for example: Sequence "(1.3, 1.0, 1.0) (2.3, 2.0, 2.0) (3.4, 3.0, 3.3) (4.3, 4.0, 4.0) (5.4, 5.0, 5.3) (6.3, 6.0, 6.0) (7.3, 7.1, 7.0) (8.3, 8.0, 8.0) > weight of $(0.27 + 0.33 + 0.3 + 0.29 + 0.61 + 0.36 + 0.23 + 0.21)/8 = 0.33$.
Step3: According to the weights of each sequence and formula 4.2, the weighted support degree of all 1-length sequences can be obtained as follows:

C1	(1.3,1.0,1.0)	(2.3,2.0,2.0)	(3.4,3.0,3.0)	(4.3,4.0,4.0)	...
Wsup	0.36	0.23	0.25	0.37	...

Will the weighted support all sequence in the table as compared to its minimum support, that is the sequence mode L (1) and the candidate itemsets C1 are: [(1.3, 1.0, 1.0)>, <(2.3, 2.0, 2.0)>, <(4.3, 4.0, 4.0)>, ... ;

Step4: the candidate itemsets C1 connection according to the connection mode of the GSP algorithm for operation, get the candidate itemsets C2: [<(1.3, 1.0, 1.0) (2.3, 2.0, 2.0)>, <(1.3, 1.0, 1.0) (4.3, 4.0, 4.0)>...] ;

Step5: Pruning operation: Remove one item from each candidate sequence to get the corresponding continuous subsequence to determine whether each subsequence exists in the existing frequent positive sequence pattern;

Step6: Calculate the weighted support degree of each candidate sequence after pruning. All the sequences whose support degree is greater than or equal to the minimum value of the included items are positive sequence pattern L(2).

Step7: Repeat the connection and pruning operations to obtain 16 weighted positive sequence patterns that meet the mining purpose, as shown in Table 5.

Table 5. A partially weighted positive sequence pattern derived from student achievement data mining

Positive sequence module Formula L(2)	<(1.2,1.0,1.0)(2.2,2.0,2.0)><(1.2,1.0,1.0)(4.2,4.0,4.3)> <(1.3,1.0,1.0) module (5.2,5.0,5.3)><(1.3,1.0,1.0)(7.2,7.0,7.1)><(2.3,2.0,2.0)(5.3,5.0,5.3) > L(2)<(4.0,4.0,4.3)(5.3,5.0,5.3)> <(5.3,5.0,5.3)(7.2,7.0,7.1)>
Positive sequence module Formula L(3)	<(1.3,1.0,1.0)(2.3,2.0,2.0)(5.3,5.0,5.0)><(1.3,1.0,1.0)(4.2,4.0,4 3)(5 ?,5.0,5.3)> <(1.3,1.0,1.0)(4.2,4.0,4.3)(7.2,7.0,7.1)><(2.3,2.0,2.0)(5.3,5.0,5.3)(1.3,1.0,1.0)> <(2.3,2.0,2.0)(5.3,5.0,5.3)(4.2,4.0,4.3)> <(4.2,4.0,4.3)(5.3,5.0,5.3)(7.2,7.0,7.1)> <(5.3,5.0,5.3)(7.2,7.0,7.1)(2.3,2.0,2.0)>
Positive sequence module Formula L(4)	<(1.3,1.0,1.0)(4.2,4.0,4.3)(5.3,5.0,5.3)(7.2,7.0,7.0)> <(1.3,1.0,1.0) (2.3,2.0,2.0)(5.3,5.0,5.3)(4.2,4.0,4.3)>

Through the analysis of the weighted positive sequence pattern, the relationship between students and grade processing values is found out. We select the sequence as follows: 1 the results <(1.3, 1.0, 1.0) (2.3, 2.0, 2.0) (5.3, 5.0, 5.3) >. For this sequence of results, we know from our representation query for the items that this sequence represents student performance in Terms 1, 2, and 5. Moreover, the weighted support of this series is 0.53, indicating that there is a 53% relationship between the performance processing values of the three semesters. Therefore, we can obtain a law of student training based on this result sequence: strengthen the quality of academic performance in the first semester, GPA at least 3, and appropriately improve the award situation; Maintain good performance in the second semester; In the fifth semester, while attaching importance to academic performance, we will improve students' award-winning performance in all aspects, that is, improve students' practical ability.

4.2 Mining Weighted Frequent Negative Sequences

Step1: Generate a negative sequence seed set based on the positive sequence L[1] of 1-length: [(1.3, 1.0, 1.0) >, <types (1.3, 1.0, 1.0) >, < = (2.3, 2.0, 2.0) >, <types (2.3, 2.0, 2.0) >, < = (4.3, 4.0, 4.0) >, < types (4.3, 4.0, 4.0) >,.... Seed set of minimum support is: MIS(suchI) = 1 − MIS(I), $suchasMIS(rope(1.3, 1.0, 1.0)) = 1 − MIS((1.3, 1.0, 1.0)) = 0.66$;

Step2: Join operation: Use the mining method of GSP algorithm, by adding a 1-length to the candidate negative sequence pattern of (k − 1)-length;

Step3: Pruning operation: Calculate the support degree of all NSCS, and calculate the weighted support degree according to the support degree. If the weighted support of the candidate sequence is greater than its minimum support, it is added to the frequent negative sequence pattern.

Step4: According to the two constraints for the establishment of negative sequence, remove the ones that do not meet the above conditions, and finally get the remaining frequent negative sequence patterns that meet the mining purpose, a total of 23, as shown in Table 6.

Table 6. The partially weighted negative sequence pattern of student achievement data mining

Positive sequence module Formula L(2)	<(1.3,1.0,1.0) ¬(3.1,3.0,3.0)><(1.3,1.0,1.0) ¬(7.2,7.1,7.0)><(3.3,3.0,3.2) ¬(8.1,8.1,8.0)><¬(3.1,3.0,3.0) (5.3,5.0,5.3)><(3.1,3.0,3.0) (7.2,7.0,7.1)><(4.2,4.0,4.3) ¬(8.1,8.1,8.0)><(5.3,5.0,5.3) ¬(7.2,7.1,7.0)><(5.3,5.0,5.3) ¬(8.1,8.1,8.0)><(7.2,7.0,7.0) ¬(8.1,8.1,8.0)>
Positive sequence module Formula L(3)	<(1.3,1.0,1.0) ¬(3.1,3.0,3.0) (5.3,5.0,5.3)><(1.3,1.0,1.0)(4.2,4.0,4.3) ¬(7.2,7.1,7.0)><(1.3,1.0,1.0) (4.2,4.0,4.3) ¬(8.1,8.1,8.0)><¬(1.2,1.1,1.0)(3.3,3.0,3.2) ¬(7.2,7.1,7.0)><¬(1.2,1.1,1.0) (5.3,5.0,5.3) ¬(8.1,8.1,8.0)><(2.3,2.0,2.0) ¬(3.1,3.0,3.0) (4.2,4.0,4.3)><(2.3,2.0,2.0) (4.2,4.0,4.3) ¬(7.2,7.1,7.0)><(2.3,2.0,2.0)(4.2,4.0,4.3) ¬(8.1,8.1,8.0)><¬(3.1,3.0,3.0) (4.2,4.0,4.3) ¬(7.2,7.1,7.0)><¬(3.1,3.0,3.0) (4.2,4.0,4.3) ¬(8.1,8.1,8.0)><(4.2,4.0,4.3) (5.3,5.0,5.3)
Positive sequence module Formula L(4)	¬(7.2,7.1,7.0)><(4.2,4.0,4.3) (5.3,5.0,5.3) ¬(8.1,8.1,8.0)><(1.3,1.0,1.0) ¬(3.1,3.0,3.0) (4.2,4.0,4.3) ¬(7.2,7.1,7.0)><¬(1.2,1.1,1.0)(3.3,3.0,3.2) (5.3,5.0,5.3) ¬(7.2,7.1,7.0)>(2.3,2.0,2.0)(5.3,5.0,5.3)(4.2,4.0,4.3)>

Through the analysis of the weighted negative sequence pattern, the negative relationship between students and grade processing values is found out. 1 the results of the sequence is: <types (1.2, 1.1, 1.0) (3.3, 3.0, 3.2) (8.1, 8.1, 8.0) >. For this negative sequence in the result, we can know that this sequence represents the student's performance in terms 1, 3, and 8, where terms 1 and 8 are predicted. Moreover, the weighted support of this series is 0.42, indicating that there is a 42% relationship between the performance processing values of the three semesters. Through the analysis of this result sequence, we can obtain the law of student training contained in it: the GPA in the first semester is 2, and the situation of failing courses should not occur; Pay attention to learning quality in the third semester, so that the GPA is at least 3, and has a good award performance; The eighth semester is the last college stage for college students, most of the time is spent in job hunting or internship, but the situation of low GPA and failing courses should not appear. Analyzing positive and negative sequences across eight semesters reveals patterns of excellent students: The first three semesters focus on strengthening core studies, avoiding failing grades, and increasing extracurricular activities. Semesters 4 to 6, a growth phase, involve maintaining good study habits and participating in practical competitions to prepare for internships. In the final two semesters, during internships, sustaining course studies and seeking achievements remains important. This analysis, based on nearly 1,000 students' data, helps extract a student development model for data-driven decision support in university teaching.

5 Summarize

Sequential pattern mining plays an important role in data mining technology and is widely used in various fields of society. There is a lot of valuable information hidden in the mass data of colleges and universities. It is a key problem that we are facing at present to mine the information and provide decision support for school teaching, library management and logistics management. Sequential pattern mining is a method that can mine effective information from the data, and this paper can reflect the importance of different attributes in the data through weighted sequential pattern mining.

References

1. Su, Y., He, X., Wang, Z., et al.: Design of a prefixspan algorithm based on prefix position form. ITM Web Conf. **45**, 01006 (2022). https://doi.org/10.1051/itmconf/20224501006
2. Kaewyotha, J., Songpan, W.: Multi-objective design of profit volumes and closeness ratings using MBHS optimizing based on the PrefixSpan mining approach (PSMA) for product layout in supermarkets. Appl. Sci. **11**(22), 10683 (2021). https://doi.org/10.3390/app112210683
3. Iferroudjene, M., Lonjarret, C., Robardet, C., et al.: Methods for explaining Top-N recommendations through subgroup discovery. Data Min. Knowl. Disc. **37**(2), 833–872 (2023)
4. Wang, L., Gui, L., Xu, P.: Incremental sequential patterns for multivariate temporal association rules mining. Expert Syst. Appl. **207**, 118020 (2022)
5. Tanaka, S., Kato, S., Sakaguchi, T., Takimoto, T.: Method for extracting knowledge of train rescheduling from data of operation records. Q. Rep. RTRI **62**(4), 269–274 (2021)
6. Niyazmand, T., Izadi, I.: Pattern mining in alarm flood sequences using a modified PrefixSpan algorithm. ISA Transactions **90**, 287–293 (2019)
7. Kshatriya, K.A., Patel, B.M., Patel, H.B.: Applied Constraints on Sequential Pattern Mining with Prefixspan Algorithm (2018). https://doi.org/10.14445/22312803/IJCTT-V57P108
8. Yuanzhe, L., Chaowei, W., Yue, Z.: Research on an improved approach for network security detection based on data mining and prefixspan algorithm simulation experiment. Int. J. Eng. Modell. **31**(1), 184–193 (2018)

9. Ma, X., Ye, L.: Career goal-based e-learning recommendation using enhanced collaborative filtering and PrefixSpan. Int. J. Mobile Blended Learn. **10**(3), 23–37 (2018)
10. Bunker, R., Fujii, K., Hanada, H., et al.: Supervised sequential pattern mining of event sequences in sport to identify important patterns of play: an application to rugby union. arXiv e-prints (2020). https://doi.org/10.31236/osf.io/g2bj8
11. Chen, X., Xue, Y., Zhao, H., Lu, X., Hu, X., Ma, Z.: A novel feature extraction methodology for sentiment analysis of product reviews. Neural Comput. Appl. **31**(10), 6625–6642 (2019)
12. Wang, Z., Liu, X., Zhang, W., et al.: The statistical analysis in the era of big data. Int. J. Modell. Identi. Control. **92**, 831–841 (2022)
13. Wang, Z., Wei, X., Pan, J.: Research on IRP of perishable products based on mobile data sharing environment. Int. J. Cognitive Inform. Nat. Intell. **15**(2), 139–157 (2021). https://doi.org/10.4018/IJCINI.20210401.oa10

The Application and Exploration of Big Data in College Student Information Management

Jun Zhang[✉] and Yuanbing Wang

Guangdong University of Science and Technology, Dongguan 523083, China
307830335@qq.com

Abstract. With the rapid development of informatization, higher requirements have been put forward for information management in universities. At present, major business platforms in universities have accumulated a large amount of data, which is large in quantity, diverse in types, and updated quickly. Therefore, it is necessary to use data mining methods in big data technology to integrate, analyze, and process data. This article defines the concept of big data in the research and emphasizes the application value of big data technology in college student information management. At the same time, this article also analyzes the current problems in university student information management based on the actual situation, and proposes corresponding improvement measures from the perspective of the application of big data.

Keyword: big data · College students · Information management · application

1 The Concept of Big Data

Big data originated from the development of information technology. With the continuous advancement of the Internet, people's real lives are filled with massive amounts of data, which have a large capacity and cannot be captured, managed, and analyzed through ordinary software tools. Therefore, higher requirements are also put forward for the operation of processing and analysis tools. The use of big data often refers to the exchange, integration, and analysis of massive amounts of data, and its flexible application can also bring more knowledge and create higher value. In the context of big data, the connections between things are not singular, but have multiple characteristics such as massive, high-speed, diverse, and high value.

At present, big data has been widely introduced into the education industry, providing more services for various educational management work. Especially in recent years, digital learning and interactive teaching models have emerged, generating unstructured data through intelligent tutoring systems and personalized learning systems, laying a solid foundation for the innovative development of digital education. Data in educational activities usually refers to explicit data such as exam scores, number of students, school size, and number of courses. However, in reality, student profile information, such as student attendance rate, homework completion status, enrollment rate, dropout rate, and even the frequency of students answering questions and teacher-student interaction in the classroom, are often overlooked. However, the use of big data can effectively collect, classify, store, and analyze such information. By mining student data, personalized learning environments and courses can be tailored for each student. At the same time, an early warning system can be established to identify potential risks, providing students with a challenging and non gradually tiring learning plan, and finally, feedback information for students can be repeated [1].

Based on the actual situation, the application of data in education, especially in student information management, has gone through the following process (Fig. 1).

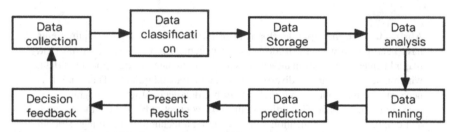

Fig. 1. Big Data Technology Information Processing Process

2   The Application Value of Big Data in College Student Information Management

In the context of big data reality, education management is facing more challenges and opportunities, which have brought many favorable factors to college student information management. Its application value is mainly reflected in:

2.1   Strengthening Psychological Health Education for College Students

At present, China is in a stage of rapid socio-economic development, facing enormous reforms in various aspects such as politics, economy, culture, etc. It has also brought pressure to college students in various aspects such as employment and life. Students not only need to face complex social environments, but also need to face setbacks with a positive attitude in such situations. Therefore, most students find it difficult to adjust their mentality, leading to negative psychological problems such as psychological imbalance. In the process of managing college students, universities can integrate big data technology to grasp the psychological changes of students and actively intervene and assist them; With the support of big data, university administrators can also adopt a one-stop data resource service platform to actively build a sound psychological assistance network. If an emergency occurs, more reasonable measures need to be taken to better leverage the advantages of big data and help students establish correct development goals [2].

2.2 Helps to Conduct Diversified Evaluations

Evaluation is an important means of managing college students. Currently, the evaluation of college students is mainly from the perspective of intelligence and academic performance, but it neglects the cultivation of students' inner character. The application of big data technology can help universities track and investigate the situation of each student, including integrating information data on student attendance and homework completion, so that evaluation teachers can more intuitively see the evaluation results and better carry out evaluation work. This diversified evaluation method has been implemented, which can better guide students, stimulate their inner learning potential, and promote their efforts towards comprehensive development [3].

2.3 Can Help Students Develop Personalized Employment Opportunities

The application of big data can timely and accurately obtain students' information and information. On the one hand, managers can easily grasp students' interests and hobbies, and on the other hand, it can help managers achieve personalized employment guidance. This can also fully target students' actual situation and diverse personality characteristics, guide them to have a better understanding of their career interests and development direction, help students develop career plans that are consistent with their own characteristics, assist students in having a full understanding of the future employment prospects, employment situation, employment methods and skills, and promote college graduates to find satisfactory jobs, And then improve the employment rate of graduates.

3 The Dilemma of Information Management for Current College Students

At present, there are certain problems and difficulties in the information management of college students in universities, mainly reflected in the following aspects:

3.1 Influenced by the Information Characteristics of College Students

The information of college students is directly formed in their daily learning and life, comprehensively recording their moral, intellectual, physical, and aesthetic aspects, which has important reference value for both schools and individuals. As an important information resource, student information can provide reliable information resources for the decision-making of various teaching management departments in schools.

In recent years, with the continuous deepening of the popularization reform of higher education, the number of students has been increasing year by year, resulting in the formation of its own characteristics in the continuous development of university student information management, mainly including: complexity. The number of college students in China is increasing day by day. In the era of rapid development of information technology, the information of college students, including data, text, audio, video, etc., is rapidly showing a comprehensive, multi angle, and wide range trend. Especially, audio and video information has become a new starting point for the explosive increase in the amount of information in universities; Correlation. The correlation of student information data is not equivalent to causality, nor is it simply personalized. For example, the fluctuations in students' academic performance may be related to various factors such as academic pressure, life difficulties, and psychological burden. Therefore, traditional data analysis techniques often cannot comprehensively and effectively collect and analyze information; Dynamicity. The complete and coherent student information is based on the collection and recording of the development status of students' basic qualities. From enrollment to graduation, it is a true record of the development and changes of students' various qualities. Therefore, the gradual evolution of data information is essentially a dynamic process; Finally, the information of college students also has unique characteristics. Student information data includes unstructured data such as students' daily life behaviors, ideological and political expressions, and is also mixed with uncertainty. Student data information presents a colorful and unique nature. The above characteristics have brought certain difficulties to the information management of college students, and advanced technologies need to be introduced to improve the overall level of existing information management [4].

3.2 Lack of a Unified Management Platform

From the above analysis, it can be seen that the increasing amount of data information among college students has brought serious pressure to the implementation of data management work. At present, the data collected by various universities is basically in a state of "data island", with almost zero information and data exchange across departments and schools. Although universities in China have been building information technology since the 1990s, they lack unified planning, centralized management, and a common platform. This fragmented situation continues into the era of modern big data, clearly appearing out of place, and its drawbacks are further highlighted [5].

3.3 The Management Concept is Relatively Backward

Education is a service, and the original intention of information management for college students is to serve teachers and students in teaching and research activities. But from the development history of information management for college students, it is undoubtedly a history of continuously strengthening administrative leadership in student management. In the information age, we cannot recognize the concept that the more open, the more progressive. For example, in traditional data-driven models, students' GPA basically determines the direction of scholarships for this year, which can easily lead to students being only willing to study hard and dampen their enthusiasm for extracurricular exercise activities that are difficult to evaluate. In this way, these collected student data have instead become shackles that constrain students' free development. Therefore, the traditional student information management model is increasingly unable to adapt to the development of society, and is gradually distancing itself from the "people-oriented" educational concept advocated by society [6].

3.4 Lack of Professional and Technical Talents

The role of big data in student information management is closely related to the level of information management workers in universities. At present, the introduction of student information management personnel in Chinese universities mainly includes: the first type is technical talents mainly focused on computer data management; The second type is counselors from different professional backgrounds. For the first type of technical management personnel, data is data, pure management data. In the context of big data, it is clearly not in line with the requirements to use data mining to serve and manage student information. For the second type of counselor, due to the uneven level of computer data management technology, most can only process student information based on basic structural data. Most work still follows traditional processing methods, such as exam scores, credits, attendance rates, etc. Most of them are only reflected in simple text operations and digital report processing. Intelligence has instead promoted mechanization, and data has become a cold existence [5].

3.5 Stubborn Information Management

In traditional college student information management work, teachers, students, and managers each spare no effort to collect data for their goals, in order to complete annual evaluations, maintain high employment rates, and achieve improved academic performance as the purpose of data management. However, this management goal distorts the concept of refined services in the era of big data, hinders the comprehensive development of students, and has a negative impact on China's social construction [7].

4 The Application Countermeasures of Big Data in College Student Information Management

From the above analysis, it can be seen that the application of big data in college student information management plays an important role that cannot be ignored. However, there are serious shortcomings in college student information management at present. To improve this situation, the following aspects can be taken into consideration:

4.1 Building a Unified Information Management Platform

The application of big data largely benefits from the development of cloud computing models in various fields. The essence of cloud computing is to use the Internet as a carrier, utilizing distributed computers with non local or remote servers (clusters) to provide users with large-scale resource sharing services. Cloud computing and cloud storage are like railway communication between various "university data islands" and "professional data islands", while big data is a train shuttling through highways. Therefore, the goal of informationization in student management in universities is to achieve data sharing, smooth channels, establish efficient and fast information feedback channels, and better serve students' learning and life. To this end, it is crucial to connect, integrate, and unify the information management technology standards and norms of various universities,

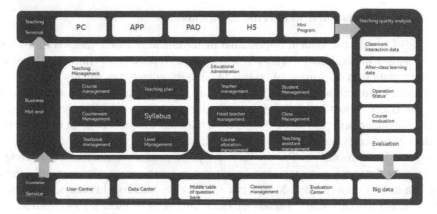

Fig. 2. Example of a teaching management system

achieve interconnectivity, build a unified student management information platform, and achieve standardized management of student data (Fig. 2).

In short, seize the good opportunity of the national development big data strategy, accelerate the reform of student information management, and strengthen the scientific, standardized, and intelligent construction of information management.

4.2 Update Existing Information Management Concepts

Traditional student information is mostly established under an experiential management model, while the prerequisite for the development of modern information technology is that technology is "people-oriented", while traditional student information is mostly established under an experiential management model. In the mindset of big data, the truly important factor comes from data mining rather than taken for granted experience. From this perspective, the introduction of big data into the field of student information management in universities will gradually weaken the management function of teachers and strengthen the service function for students. Victor proposed three major changes in thinking: not precision, but hybridity; Not random samples, but overall data; It is not a causal relationship, but a correlation relationship. Therefore, university information managers in the context of big data should break traditional management thinking, establish awareness of big data service management, timely position and change roles, in order to improve the level of application of big data [8].

4.3 Accelerate the Construction of the Teaching Staff

Against the backdrop of the rapid development of modern information technology, talent is the primary competitive advantage, determining the development and application of material resources. And university data itself has its own particularity, requiring relevant personnel to have a reasonable understanding of education. Firstly, as a special service function, education targets students and requires university administrators to master basic demonstration skills; Secondly,the data collection, classification, storage, and analysis process of big data itself has strong professionalism, requiring talents to master specialized data processing techniques. Therefore, in the management of big data in universities, talent cultivation is the foundation: based on the existing personnel system, carry out organizational unified learning and training, master the basic skills of student information management in the context of big data, transform work models, innovate work thinking, continuously improve one's own quality, stimulate work enthusiasm, and accelerate talent growth. Talent introduction is the key: optimizing the team requires a large, capable, and skilled young backbone teacher team, developing an introduction plan, supplementing fresh blood, and maintaining team vitality [9].

4.4 Conduct Diversified Evaluation of Data Information

With the development of society, especially the deepening of human understanding of oneself, exam oriented education has begun to focus too much on cognitive fields such as knowledge mastery and intellectual development, while neglecting the criticism of students' ideological and moral character, personality, and personality development. It requires a comprehensive and diversified evaluation of students' comprehensive quality development. Evaluating the development of students' comprehensive qualities not only involves evaluating their cognitive development in areas such as knowledge, skills, intelligence, and abilities, but also evaluating the development of non cognitive factors such as emotions, personality, will, and personality. Therefore, with the assistance of certain observation techniques and equipment, data collection, classification, organization, statistics, and analysis of phenomena that occur at every moment in students' daily learning and life can be conducted for diversified evaluation. By analyzing the micro and individual student information situation, timely adjusting educational behavior, achieving formative evaluation of students, avoiding "one size fits all" chaos, and preventing purpose-oriented management in order to achieve data indicators, big data can provide higher quality services for college students [6].

At present, big data technology is widely applied in various fields, such as economy, culture, education, etc. It also effectively promotes the development of various fields. The information management of university students plays a very important role in university management work. So the management of student information in universities has received attention, and the application of big data technology involves various aspects. It not only involves financial management, but also library management and daily management of students, so it also has obvious advantages. Especially since the implementation of China's enrollment expansion policy, more students have gained opportunities in higher education. Therefore, in order to better manage and utilize student teaching information, it is also an important choice for current teaching quality. The application of big data technology can strengthen the management of student teaching information and further reduce the pressure and burden on staff [10] (Fig. 3).

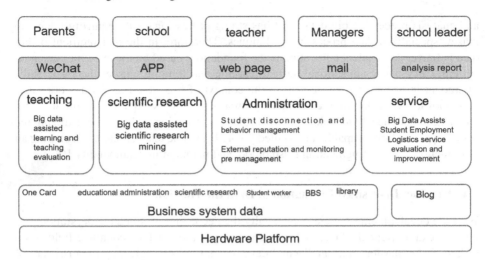

Fig. 3. Big data assists colleges and universities to realize scientific management and intelligent decision-making model

5 Conclusion

The application and exploration of big data in college student information management. With the continuous development of the current information age, big data technology has gradually been widely applied in various fields. This not only further promotes the development of various fields of teaching, but also has very important value and significance for the healthy development of education. Therefore, universities should pay more attention to the application of big data and apply it to student teaching management, reducing the pressure on faculty and improving the overall quality of services.

References

1. Xia, C.: Exploring the innovative path of university student education management model from the perspective of big data. Intelligence **28**, 173–176 (2023)
2. Dong, W.: Optimization path for "student education management" in universities in the era of big data. J. Hubei Open Vocat. College **36**(18), 148–149+153 (2023)
3. Zhao, J.: Transformation and strategies of student education management models in universities in the era of big data. Shanxi Youth **16**, 175–177 (2023)
4. Qiu, W.: College Students Based on Big Data Technology
5. Li, J., Wang, Y.: The application of grid management mode in the management of college students supported by big data. J. Jilin Inst. Educ. **39**(10), 58–62 (2023). https://doi.org/10.16083/j.cnki.1671-1580.2023.10.009
6. Tong, X., Xu, T., Zhao, X.: Research on the application of big data in college students' education management. Knowl. Library (04) (2019), 156.
7. Li, Y.: Challenges and responses to the management of college Students in the era of Big Data. Ind. Technol. Forum **19**(23), 251–252 (2020)
8. Huiying, G.: Current situation and Countermeasures of informatization construction of college Student Management in the era of Big Data. New Industrialization **10**(10), 137–138 (2020). https://doi.org/10.19335/j.cnki.2095-6649.2020.10.060

9. Dejun, S.: Analysis on the new path of college management in the era of big data. J. Chongqing Second Normal Univ. **32**(05), 123–126 (2019)
10. Liu, Y., Lin, O., Zeng et al.: Research on the Information Platform based on cloud computing. Comput. Knowl. Technol. **13**(06), 2–5 (2017). https://doi.org/10.14004/j.cnki.ckt.2017.0451

Research on Precision Marketing Strategy of Guangdong Characteristic Products Enabled by Big Data in Rural

Hua Wang and Yuming Sun[✉]

Guangdong University of Science and Technology, Dongguan 523000, China
76793194@qq.com

Abstract. With the deepening of the rural revitalization strategy, in order to accelerate the implementation of rural revitalization and development, promote the adjustment of industrial structure, and deepen the innovation of management mechanism, it is particularly necessary to accelerate the reform of management mode and promote the implementation of precision marketing strategy under the existing development mode of characteristic product industry. However, due to the restriction of marketing mode and the influence of traditional production mode, the sales channels of Guangdong featured products are limited and the market share is low. In order to promote the development of rural revitalization and enhance the development of rural economy and industry, based on big data technology, this paper makes a detailed exploration of the existing precision marketing methods of Guangdong characteristic products. Both the analysis of marketing methods and the exploration of marketing modes have realized the landing of precision marketing methods. It has not only accelerated rural revitalization and development, but also achieved the goal of industrial empowerment.

Keywords: Big data · Rural revitalization · Precision marketing

1 Introduction

The development of information technology has accelerated the implementation of the rural revitalization strategy, especially the addition of diversified marketing methods, enabling the diversified precision marketing strategy that was difficult to achieve offline, the joint work of online and offline, and the comprehensive transformation and upgrading of the channel environment of featured products that was difficult to promote effectively. Therefore, under the existing development mechanism, it is necessary to deepen the reform and innovation of the internal management mechanism by integrating the development characteristics of The Times and all-round working mechanism [1]. Under the structure of big data, it gives full play to the role of network information, explores the way of precision marketing of featured products by means of information calculation and adjustment, and proposes a special working mechanism. In order to better play the industrial advantages, let the industry to expand the promotion channels, for the development of the industry to create greater market value.

K. Li and Y. Liu (Eds.): ISICA 2023, CCIS 2147, pp. 262–269, 2024.
https://doi.org/10.1007/978-981-97-4396-4_24

2   Big Data Enables the Significance of Precision Marketing of Guangdong Products in Rural Revitalization

In the original industrial development, although Guangdong Province is near the sea, it is far away from the inland, especially many provinces far away from the inland. Although it can realize export, it is relatively difficult to sell in the internal market. Especially in the marketing model, the main advocate is the offline marketing mechanism. Due to the influence of the environment in Guangdong Province, the storage difficulty of many products is relatively large, especially the marketing of featured products. Therefore, only by exploring new marketing strategies and marketing ideas, and carrying out comprehensive marketing management from multiple angles, can the marketing advantages be better played and the innovation and development of the industry be promoted.

The advent of the era of big data has promoted the penetration of information in various industries. No matter industrial products, digital products, chemical products or specialty products, they are constantly exploring new development paths and marketing mechanisms under this mode of information development, realizing the implementation of online and offline integrated marketing tasks, and creating a distinctive marketing mode. Promote the full implementation of precision marketing tasks. First of all, under the existing development mode, a unique development mechanism of rural revitalization precision marketing has been constructed [2]. From the perspective of management, a special organization team of rural revitalization marketing management has been established, which is started by the government management agency to supervise and manage the market, laying a solid foundation for the integrity of the whole market environment. Not only that, many special personnel of management units fully combine the market situation, make overall planning for different kinds of product information, formulate online marketing norms and standards, and put forward suggestions and evaluation standards for the later marketing promotion and production development (Fig. 1).

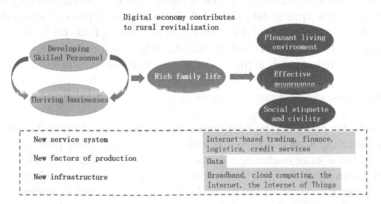

Fig. 1. Digital economy contributes to rural revitalization

In this mode, the difficulty of marketing supervision of Guangdong products has been gradually weakened, and the quality of management has been significantly improved.

Secondly, at the same time of market regulation analysis, each industrial structure according to the actual characteristics and content of their own products, with marketing development as the core, to establish a special marketing gimmick, whether from the product advantages or from the product service, have put forward targeted marketing plans, the implementation of this plan, increase the construction of product marketing mode, break the original industrial marketing deadlock, Make the whole marketing work more specific and comprehensive [3], but also make the marketing work more smooth. Finally, in different market environments, industries rely on big data to build information databases, which are mainly for consumers to have a comprehensive understanding of the product itself. The realization of this open and transparent database presentation is in the public data platform. Different industrial structures will form different open data information platforms, which mainly record the development mode and product content within the industry, and then publicize it in the website platform. Consumer groups that have a demand for products will complete data analysis on the platform, ensuring the accuracy of data and realizing the purpose of precision marketing.

3   There Are Problems in the Precision Marketing of Guangdong Products that Big Data Enables Rural Revitalization

3.1   Low Utilization of Data

The application of big data is the driving force to achieve industrial development, although in the new historical development stage, information big data technology has gradually penetrated into all aspects of industrial development. In the process of Guangdong product marketing management based on rural revitalization, it also starts to realize industrial transformation based on the application of big data, upgrade the product management mode and adjust the marketing mode. However, in many product marketing environments, the significance of data has not been fully understood, resulting in a relatively low application rate of data. First of all, in the process of product quality detection and analysis of the application of data is relatively less, many products of control analysis knowledge by reference to this batch of analysis, judge whether the quality of a batch of product marketing process is in line with the marketing gimmick of featured products, quality is qualified [4]. However, data and information are not integrated and production links in the whole marketing process are analyzed, resulting in relatively low marketing quality.

Secondly, the data analysis of the marketing process is not in place. The implementation of marketing work is achieved in many aspects of marketing promotion, and the development of a lot of marketing work also needs to be evaluated by referring to the marketing status under the same time environment. However, many marketing management personnel of enterprises are particularly poor in data analysis ability, especially for rural revitalization products, and do not pay attention to the marketing data of each link. Although they feel strong marketing ability, the final sales results are not good. Moreover, many managers do not pay attention to the changes in market risk data, and fail to establish an overall analysis mechanism for the contents of rural revitalization products in different regions, resulting in the same goods, the same product structure, and the same production and promotion stunts. Therefore, to do a reasonable job in data analysis and application and accelerate precision marketing data analysis, we can accelerate rural revitalization and development and promote the implementation of product management.

3.2 Failure to Build a Professional Regulatory Team

In order to ensure the effective implementation of the marketing of rural revitalization products, the local government of Guangdong Province has improved its understanding of the existing marketing supervision and management and organized professional personnel to carry out market supervision based on the development status of The Times and the characteristics, advantages and working characteristics of the current industrial development of Guangdong Province. However, these regulatory professionals can only supervise and manage the market environment, and fail to establish a systematic supervision and management mechanism for the marketing strategy, concept and product structure, resulting in insufficient supervision and relatively low marketing quality. First of all, professional marketing management teams do not combine different market environments for content construction. For example, rural revitalization products not only include agriculture, agricultural and sideline industries, but also industrial and chemical industries. If the supervision team lacks the understanding of different industries, the effectiveness of the supervision will be reduced. The market analysis mechanism is not fully understood, and the management function cannot be fully played, resulting in the loss of precision marketing. Secondly, on the basis of the construction of professional supervision and management team, the marketing and management standards and norms for different rural revitalization products have not been formed. In the process of implementation of supervision and management, blindly implement in accordance with the original management ideas, does not conform to the characteristics of the information industry marketing management, resulting in difficult to play the effect of supervision and management, marketing implementation quality is insufficient [5].

Independent data analysis platform has not been formed. Based on the existing marketing model and the establishment of precision marketing ideas, the rural revitalization industry in Guangdong Province has established clear working ideas and working paths, and built a working system and structure by relying on the big data information platform, but it has not built an accurate independent information analysis platform structure, which leads to the weak analysis ability of information data. For the difficult problems in the process of marketing can't be judged in the first time, and put forward targeted marketing management optimization ideas. First of all, most of the self-service data analysis platforms are third-party data analysis and observation platforms, which can only analyze the data information in accordance with the marketing links in different stages provided. If the provided data is biased in the input, it will inevitably not affect the marketing data of the marketing link. Moreover, if these data and information are to be sorted out again for data summary, analysis and judgment, they need to be sorted out again and planned into another data module structure in the form of summary, which makes it difficult to carry out precision marketing analysis effectively. Secondly, because there is no independent data analysis platform, many data have little reference value, and many marketing managers cannot judge the relevant information content in the parameter information, resulting in the final marketing decision mistakes.

4 Big Data Enables the Precision Marketing Strategy of Guangdong Characteristic Products for Rural Revitalization

4.1 Live Streaming Platforms Speed up Precision Marketing

The emergence of live marketing is the main idea of the new marketing at the present stage, and also the main way to carry out the marketing work of the featured products of rural revitalization. According to the data analysis and observation, in 2022, the featured products of rural revitalization in Guangdong Province, with the support of major "Internet celebrities" through the live streaming platform, have been enjoying a boom in their marketing data, with the online marketing data repeatedly hitting a record high. No matter featured products or featured information products, industrial products or other products, they are constantly promoted in this way of network marketing, realizing a systematic marketing channel, which not only speeds up the transformation of the industry, but also improves the quality of marketing work. Through the construction of the live streaming platform, a lot of information and data can be directly obtained from the database of the platform, and the actual data of the current marketing of Guangdong featured products in rural revitalization can be directly observed, so as to better explore new marketing direction and ensure the best marketing effect.

4.2 Establish and Develop Township e-Commerce Service Management Center

For the product operation industry with regional characteristics, as well as some competitive e-commerce service platforms and brand enterprises, special audit, analysis and evaluation are carried out, and special service management center is built, which is mainly for special marketing management audit. Not only that, it provides free working equipment and working place, and provides free station and sharing platform and introduction platform. Provide government services free of charge, enjoy high – quality service mode. The training service mode is provided free of charge. The personnel who enter the enterprise should actively participate in the training service work. Especially in the process of refined service, we should try our best not to copy the experience easily and seek a new management mode unilaterally, which will lead to the inadequate implementation of supervision and management functions. We should optimize service management, enhance service awareness, analyze from the basis of standardized service management, and establish a systematic central management platform. Ensure that the quality of e-commerce operation service management meets the demands of modern development.

4.3 Give Full Play to the Government's Role in Supervising the Economy

The economic function of government is mainly to accelerate the implementation of macro-management, supervision and service management on the basis of comprehensive management of social economic life in order to promote the steady development of national economy. The full play of the government's role in economic management can accelerate the realization of supervision and control and work upgrading in different environmental development fields, and realize the benign promotion of marketing development. For precision marketing services, the government functions of local governments play a relatively high importance. Therefore, it is necessary for relevant management units to strengthen the implementation of infrastructure construction management, optimize the government environment, build a new e-commerce service management mechanism and work management platform, create a comprehensive development of modern rural revitalization service management and supervision network environment, build a management mechanism based on the grassroots masses, further accelerate rural revitalization and development, promote the development and transformation of big data. We will improve the quality of precision marketing.

4.4 Accelerate the Mutual Association with Industry

Under the mutual integration of rural revitalization and big data technology, the operation and development mode of Guangzhou's characteristic industries also show obvious market transformation. However, many enterprises do not have a comprehensive understanding of the transformation of precision marketing in the network environment. They believe that the basic needs of Internet transformation can be met as long as the Internet infrastructure is upgraded and optimized. Blindly optimistic about market development prospects, resulting in the Internet in the environment of industrial structure, it is difficult to achieve accurate marketing integration analysis. In order to meet the deep integration between digital management and industrial structure, it is necessary to analyze various problems faced by industrial structure under the era of development, accelerate the integration and adjustment of market environment, accelerate the construction of agricultural big data engineering system, establish and perfect data information system, focus on promoting new data application service mode, and further improve the development mode of logistics system infrastructure construction. In particular, strengthen the construction and development of logistics management system of characteristic products and improve the quality of logistics work. We will strengthen the training and guidance of agricultural professionals and accelerate the development and application of diversified agricultural production technologies.

5   Precision Marketing Service Platform Construction

Through the collection, analysis and processing of a large amount of historical data, the platform extracts customer attributes and purchasing behavior information as metadata, and builds a relationship model between customer attributes and products to achieve distinguishable and customizable product information accurately. At the same time, a three-level feedback mechanism is established according to the delivery effect to promote the optimization and adjustment of the model and make the product delivery more accurate (Fig. 2).

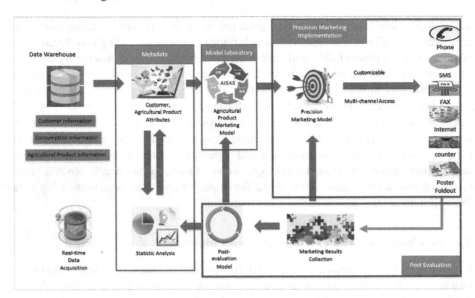

Fig. 2. Precision marketing service platform based on big data analysis

The Platform is Mainly Divided into the Following Four Parts:

The metadata analysis. Platform processes and maps the multi-dimensional historical data information of customers, such as customer information, consumption details, channel access records, etc., into metadata that can reflect the general profile of customers, such as gender, region, salary income, asset status, risk preference, interests, and consumption habits. At the same time, through the analysis of historical sales data of agricultural Bank of China products, the common features of customer groups are generated to reflect the metadata of product sales characteristics.

Marketing model construction. According to the analysis of product sales history information, analyze the special points of target customers, and build customer-product marketing model and product-product cross-marketing model through the mapping of customer metadata, so as to achieve accurate sales of products to customers and improve marketing efficiency.

Precision marketing execution. After the precision marketing execution model is built, customers are identified according to channel access information, and precise product information is delivered to customers through product marketing model matching. At the same time, social network sharing mechanism is provided to amplify marketing effects.

Marketing effect post evaluation. After the marketing effect post-evaluation information is released, take the initiative to collect marketing results, such as whether the customer has clicked on the relevant information, whether the purchase behavior has occurred, whether the customer has recommended to other friends through social sharing, and how long the customer stays on the product page, etc., and form a three-level feedback mechanism according to these customer behavior data: First-level feedback revises marketing activities to make marketing activities more targeted (such as increasing marketing efforts for a certain channel, etc.); The second-level feedback drives the optimization and adjustment of the product marketing model according to the post-evaluation model (such as narrowing the scope of target customers, adjusting the cross-marketing mapping rules, etc.); The three-level feedback increases the metadata dimension and optimizes the statistical analysis method of data.

6 Conclusion

Through the adjustment of industrial structure and the reform of working technology, in order to realize the rapid development of rural revitalization under the existing working mechanism, it is necessary to combine the actual development characteristics, actively accelerate the reform and implementation of diversified management mechanism, and construct the precision marketing strategy of featured products from the perspective of the existing rural revitalization and development. The emergence of big data technology makes cloud technology, intelligent technology, Internet of Things technology and industrial production and promotion more closely connected. Therefore, only by continuously promoting the strategic thought of rural revitalization enabled by big data, fully integrating information development with industrial structure, and accelerating technological R&D and promotion, can high-quality product research and analysis be realized, diversified industrial development be realized, and new ideas be provided for the transformation of industrial structure and innovation of market development mode in the future.

References

1. Chen, G., Gu, H., Wang, L., et al.: Research on the development of brand communication under the background of Internet. Shopping Mall Modernization (19) (2018)
2. Zhang, L.: Research on user portrait assisted precision marketing based on big data analysis. Telecommun. Technol. (1) (2017). https://doi.org/10.3969/j.issn.1000-1247.2017.01.014
3. Cao, C.: Brief discussion on precision marketing under B2-C model. J. Harbin Univ. Commer. (Soc. Sci. Edn.) (3) (2010). https://doi.org/10.3969/j.issn.1671-7112.2010.03.007
4. Liu, M.: Research on improvement of marketing strategy of fresh agricultural products under new retail model – a case study of a company (2020)
5. Chen, N: Research on Strategic Transformation of M Eco-Agriculture Company under the background of New retail (2021)

Research and Application of Offline Log Analysis Method for E-commerce Based on HHS

Haoliang Wang[✉], Kun Hu, Lili Wang, and Jingtong Shang

Dongguan City College, Guangdong 523000, Dongguan, China
haoliang9333@163.com

Abstract. Today, with the rapid development of Internet technology, the storage, analysis and mining of massive data have become an important technical topic in the era of information explosion. The big data technology architecture based on Hadoop, Hive and Sqoop (HHS) has become a popular general architecture for the analysis and processing of massive data. With the support of HHS open source technology, the processing of massive data can effectively promote the speed and quality of information analysis. Based on the HHS open source suite, a complete e-commerce offline log analysis system is designed to conduct targeted analysis of business and user behavior, providing a sample case for general e-commerce offline analysis service.

Keywords: Hadoop · Hive · Massive data · Offline Log Analysis

1 Introduction

Nowadays, with the rapid development and popularization of global Internet technology, the amount of data in all major industries is growing at an explosive rate, and the world has entered the "big data era" [1]. In the global "big data era", big data technology has also been rapidly developed, and is applied in various industries to provide technical support for the development of other industries [2]. In the field of e-commerce, online shopping has been popularized, prompting the user data of major e-commerce enterprises in the background to increase sharply [3], Enterprises can mine and analyze users' behavior habits and interests from such huge and complex data to help enterprises make smarter marketing decisions. In traditional e-commerce data warehouses, data is generally stored in relational databases, but user log data is often unstructured data. Obviously, this makes it impossible for relational databases to analyze and store this data. The use of big data technology can not only solve the storage problem of unstructured data, but also analyze and process the data for different business scenarios and different business needs, dig out user data rich in business value, and provide technical support for enterprises to make smarter business decisions.

2 Related Technology

- Hadoop

Hadoop is the core framework technology of current big data and an open source project of Apache Foundation, which can be provided for free and save costs for system development. In the process of big data processing, Hadoop realizes the storage, analysis and calculation functions of massive data. It is also rich in internal components, the most important of which are HDFS, MapReduce and Yarn, and these three components respectively realize the storage of massive data, analysis, and system resource scheduling and management functions [4].

- Hive

Hive is a data warehouse in the Hadoop ecosystem and a mainstream data warehouse framework in big data development. Its internal framework is shown in Fig. 1.

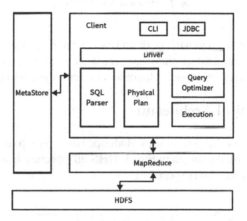

Fig. 1. Hive architecture diagram.

In the original Hive data warehouse, the original MapReduce data computing framework is used at the bottom. The original MapReduce computing framework relies on the disk-based computing framework. However, the original MapReduce computing framework requires a large number of disk I/O operations and the running speed is relatively slow. Therefore, the system replaces the original MapReduce computing framework with the Spark computing framework [5], The Hive On Spark computing mode is adopted, as shown in Fig. 2.

Spark is a memory-based iterative computing framework that avoids a large number of disk I/O operations. By executing Spark jobs on Hive, you can take full advantage of Spark's memory computing and distributed processing capabilities to accelerate data query and analysis.

- Sqoop

Sqoop is an open source data transfer tool for efficient data transfer and integration between Apache Hadoop and relational databases such as MySQL, Oracle, PostgreSQL, etc. Its name comes from the abbreviation "SQL to Hadoop", and its purpose is to simplify the process of importing structured data from relational databases into the Hadoop ecosystem for analysis and processing, as well as the ability to import data

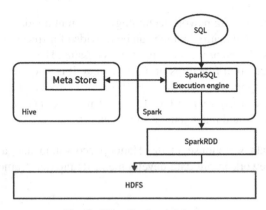

Fig. 2. Hive On Spark Mode.

from relational databases into the Hadoop Distributed File System (HDFS) or Hive. It makes large-scale data analysis more convenient and provides a bridge for enterprises to seamlessly connect traditional relational data with the Hadoop ecosystem.

3   System Technical Architecture

The core architecture of the system uses Hadoop, Hive, Sqoop technology, referred to as HHS architecture [6]. As shown in Fig. 3. HHS architecture is used to complete data storage, data analysis and data export.

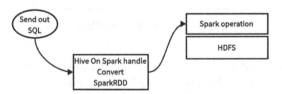

Fig. 3. HHS Architecture.

Based on the HHS architecture, it also combines the relevant technologies of big data ecology to improve the required functions of the system. Figure 4 shows the overall framework diagram of the system.

Therefore, the offline log analysis system of e-commerce based on Hadoop has three main functions: data acquisition, data analysis and data visualization. The overall function module of the system is shown in Fig. 5.

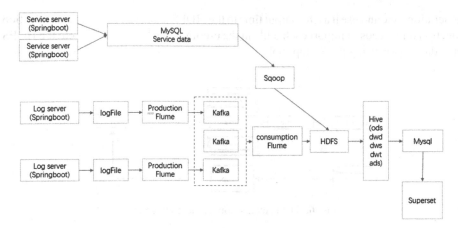

Fig. 4. System Architecture Diagram.

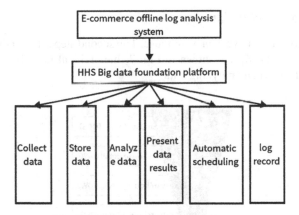

Fig. 5. Function Module of the System.

4   Main Function Modules Design

4.1   Data Acquisition Design

The system uses Flume-Kafka-Flume mode to collect user behavior log data, and uses Sqoop to import service data directly into HDFS [7]. As shown in Fig. 6.

For the collection of user behavior log data, Flume-Kafka-Flume mode is adopted, which is divided into two layers of Flume. In the first layer of Flume, a custom log type interceptor is required. The log interceptor mainly determines the log data format and intercepts and deletes the data that does not conform to the Json data format. The data is then written to the Kafka Channel and then transferred to Kafka. During the data collection at Layer 2, Flume uses the Linux system time as the time for output to the HDFS path by default. If data is generated at 23:59, certain events will be consumed during data transmission, and this part of data will be sent to the HDFS path of the next day. Therefore, you need to customize an interceptor at layer 2 to obtain the data generation time and use it as the output time to the HDFS path. The service data collection function only needs to import each table in the e-commerce business data into the HDFS on a daily basis using the Sqoop tool.

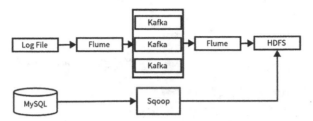

Fig. 6. Data Acquisition Module Diagram.

4.2 Data Analysis and Design

The first step is to build a Hive data warehouse. The second step is to conduct hierarchical modeling of Hive data [8]. As shown in Fig. 7, hierarchical modeling can make data easier to understand when analyzing data.

Fig. 7. Data Warehouse Layering.

In the modeling process, shell scripts need to be written, and HQL statements are injected into shell scripts to realize data analysis at each layer. The third step is to use Azkaban to automatically schedule shell scripts written during data analysis. The fourth step is to import the analyzed data into MySQL for subsequent data visualization. Figure 8 shows the Hive data processing process.

The first step is to import data stored in HDFS to Hive tables [9]. The second step involves developing Hive's permanent functions in Java [10]. Use a permanent function to analyze and process user offline log data. Package the developed permanent function into a jar package and upload it to the HDFS. Create a permanent function in Hive and specify the address where the jar package resides. The third step is to import the original

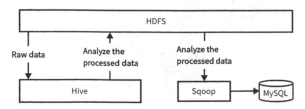

Fig. 8. Data Analysis Flow.

data in the HDFS to the Hive table and analyze the imported data at each layer in Hive. Finally, the data after the analysis is imported into MySQL to provide data source for subsequent data visualization.

4.3 Data Visualization Design

This system uses the Superset visual BI tool. Using Superset to complete the report development, you need to install the Superset installation package on any node, and then configure the MySQL database in the Superset Web page to achieve the data source required for report development, and finally you can directly develop various business requirements reports through the Superset Web page. The Superset visualization tool also provides a dashboard function, which can be used to view multiple indicators on a single page and achieve a large-screen display of data indicators. At the same time, Superset can also refresh the data visual chart page every day through the data refresh function to ensure the authenticity of daily data.

5 Experimental Result

Through the system built with HHS architecture, the existing user behavior log data and e-commerce business data were analyzed, and the total amount of daily order transactions, the distribution of national order transactions, the proportion of order transactions in each channel, the daily user statistics bar chart, the daily number of visitors and the change trend of page visits, and the user access path Sankey chart were counted. The active degree of user logins developed within seven days for three consecutive active users is shown in Fig. 9.

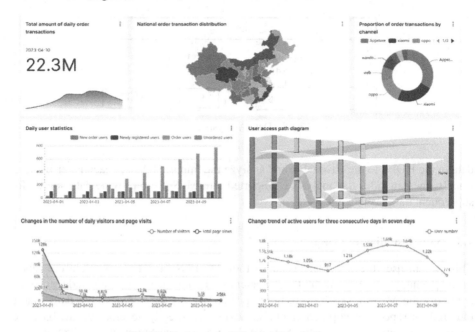

Fig. 9. Dashboard of Various Data Indicators.

6 Conclusion

Building an offline log analysis system using HHS can facilitate the expansion and storage of massive user behavior log data and service data. In addition, Hive data warehouse can be used to analyze unstructured user behavior log data. Finally, by building the Hive on Spark mode, it is found that the data processing efficiency is higher than that of the original MapReduce mode for the same amount of data.

Acknowledgments. This work is supported by Social Development Science and Technology Project of Dongguan City (20211800900282); Young Teacher Development Fund of Dongguan City University (2023QJZ001Z); Quality Engineering Project of Dongguan City University (2021zlgc204); Training Program of Innovation and Entrepreneurship for Undergraduates (202113844022), and supported by Young Teacher Development Fund of Dongguan City University (2020QJZ002Z).

References

1. Cheng, X.: Technological progress and trends of big data. Sci. Technol. Rev. **34**(14), 49–59 (2016)
2. Song, X.: Technological progress and trends of big data. Electron. Technol. Softw. Eng. **143**(21), 145–146 (2018)
3. Li, L.: Research on cost management of the S e-commerce value chain in the context of big data. Advisor: Yao, G. Nanjing University of Posts and Telecommunications (2022)

4. Li, B., Yan, J.: Traffic log analysis system based on Hadoop. J. Guilin Aerosp. Ind. Inst. **26**(04), 412–420 (2021)
5. Wang, J.: Implementation and optimization of log analysis system based on Hive. Nanjing University of Posts and Telecommunications (2017)
6. Li, X., Zhu, H.: Design and implementation of an offline data analysis platform based on HHS. Comput. Knowl. Technol. **19**(10), 75–77 (2023)
7. Chu, J.: Design and implementation of an e-commerce big data analysis system based on Hadoop. Qinghai Normal Univ. (2021). https://doi.org/10.27778/d.cnki.gqhzy.2021.000216
8. Jia, Y.: Design and implementation of an e-commerce multidimensional analysis system based on Hive. Zhejiang Univ. Technol. (2020). https://doi.org/10.27463/d.cnki.gzgyu.2020.000148
9. Zhang, Y., Wu, H.: Data import and application of Hive data warehouse in Hadoop big data environment. Comput. Programm. Skills Maintenance **450**(12), 97–99 (2022). https://doi.org/10.16184/j.cnki.comprg.2022.12.006
10. Zhang, Y., Chen, Z.: Java Programming: Introduction and Practice (Micro Course Edition). People's Posts and Telecommunications Press (2023)

Research on Communication Power of Cross-Cultural Short Video Based on Qualitative Comparative Analysis

Wen Meng[✉] and Chao Yu

Chengdu Neusoft University, Chengdu 611844, China
mengwen@nsu.edu.cn

Abstract. With the continuous cultural communication of Internet bloggers on overseas platforms, exploring the influential factors of short videos on cultural communication became more meaningful. Taking 10 Chinese culture-related bloggers on the YouTube platform as the research object, this paper filters 10 conditional variables from the aspect of the communication attributes of videos and obtains the combination of multiple concurrent factors affecting the communication effect of cross-cultural bloggers on YouTube by using the Qualitative Comparative data configuration analysis. By using this method, the paper explores the multiple concurrent causal mechanisms that lead to the strong communication power of cross-cultural bloggers on YouTube from the perspective of communication, which is a beneficial supplement to the current issues related to the effect of short videos on overseas cultural communication.

Keywords: Communication Power of Cross-Culture · Short Video · Qualitative Comparative Analysis

1 Introduction

As one of the indicators to measure the core competitiveness of the country, the communication of excellent traditional Chinese culture has become an important topic in the development of Chinese culture and the enhancement of national strength [1]. At present, the external communication matrix of China is dominated by the official power which is represented by the mainstream media while the civil discourse power has not been fully utilized. The communication channel has gradually shifted from traditional radio, television, and organizations to overseas social platforms, which constructed a more diversified and interactive way [2].

With the rapid development of mobile Internet technology, social platforms at home and abroad are constantly updated and iterated. Compared with traditional social platforms dominated by graphic forms, short video platforms continue to rise due to their fragmented push methods and more intuitive forms of content presentation, which formed an emerging hot industry [3]. By the end of 2020, the number of online video users in China reached 888 million, accounting for 94.5% of the total Internet users. There are 818 million short video users which account for 87.0% of the total Internet users. As a new way of external communication, short video is frequently discussed by the academic community, providing a broader thinking space for the academic community and the industry.

The qualitative comparative analysis, abbreviated as QCA, known as a case-oriented analysis method, was proposed by American sociologist Charles C Ragin in the 1980s. As a multi-case comparative analysis method, the QCA method goes beyond the boundary of qualitative and quantitative analysis and integrates the advantages of qualitative analysis and quantitative analysis by treating cases as conditional configurations, replacing independent variables with conditional configurations, swapping net effect thoughts for configuration thoughts, and exchanging correlation relations for set relations [4–6]. With the help of the thought of set theory, the QCA method, known as an integrational method of both quantitative and qualitative orientation, investigates the cause of combinational path and influence mode of complex social phenomena, which transforms the linear analysis into "set" analysis. The application of this method is limited in the field of news communication, and the research results on the use of the QCA method in the topic of "short video external communication" are even less.

In this context, by analyzing the Chinese cultural bloggers on overseas short video platforms, this paper attempts to introduce Qualitative Comparative analysis methods to discuss the reasons for the phenomenon that the original Chinese culture video attracted a large number of fans and hotly spread overseas. It is hoped to scientifically and systematically summarize the set of successful factors for overseas Chinese culture communication and provide an efficient path for the communication of Chinese Culture.

2 Analysis of Sample Index

By analyzing the data from the social media website *Social Blade*, 10 representative Chinese cultural bloggers are selected as the research samples for this paper, according to the number of fans, total view counts, etc.

Table 1. Chinese cultural bloggers table.

Case No	Name of Blogger	Case No	Name of Blogger
1	Liziqi	6	The Food Ranger
2	Dianxi Xiaoge	7	Interestingcn
3	Magic Ingredients	8	Magic Travel of Xiao Bai
4	Grandpa Amu	9	Travel Notes of Mr. Liu
5	Chef Wang	10	Zi De Guqin Studio

The purpose of this paper is to analyze the path that affects the development of Chinese cultural bloggers on the YouTube platform since the behaviors affecting the development are sponsored by bloggers. Therefore, when analyzing the selected indicators, it is found that the indicators caused by the behaviors are different, which can be categorized into those caused by the behaviors of bloggers and those caused by the behaviors of audiences. In this paper, blogger behavior refers to the specific and actionable behavior of YouTube bloggers. Since the research object of this paper is the blogger, the research effect is directed to the audience, so the indicator referred to the blogger's behavior is more suitable as the condition variable, while the indicator caused by the audience's behavior is suitable as the result variable (Table 2).

Table 2. Filtered Conditional Variables.

Variable type	Behavior originator	Variables
Result variables	Audience	Fans, Views, Average view duration, Likes rate, Comments, likes, dislikes, Liquidity
Condition variable	Bloggers	Number of videos, multichannel, register length, theme category, video duration, update frequency,language

As a social platform for releasing videos, the communication effect of YouTube bloggers is actually related to the communication effect of videos. Therefore, the indicators for measuring the communication effect of YouTube Chinese cultural bloggers should be closely linked to the communication effect of videos released by the bloggers. Meanwhile, according to Table 1, the indicators generated by audience behavior are suitable for measuring the video transmission effect of Chinese cultural bloggers on YouTube. Based on the characteristics of the YouTube platform, view counts and the number of fans are the most intuitive indicators. In this paper, the number of fans is selected as the result variable.

After repeated exploring of relevant theories and experience, as well as climbing measurement of sample data, 10 condition variables and 1 result variable of communication attribute are finally assigned. Detailed data are shown in Table 3.

Table 3. Variable Scoring.

Variables	Sub variables	Simple names	Scoring criteria	Weight	Scorning
Result variables	Number of fans	FANS	$\geq$200,000	60%	1
			<200,000	40%	0
Condition	Video duration	MINS	Average duration$\geq$5 mins	67%	1
variables			Average duration<5mins	33%	0

(*continued*)

Table 3. (*continued*)

Variables	Sub variables	Simple names	Scoring criteria	Weight	Scorning
	Update frequency	FREQUENCY	Per month≥6times	55%	1
			Per month<6times	45%	0
	Operation Mode	MODE	Team operation	70%	1
			Individual operation	30%	0
	Multichannel distribution	CHANNEL	Multichannel distribution	55%	1
			YouTube distribution	45%	0
	Release time	TIME	times from Fri. to Sun. ≥60%	48%	1
			Times from Fri. to Sun. <60%	52%	0
Narrative attributes	Theme category	CATEGORY	Chinese culture related	87%	1
			Non-Chinese culture	13%	0
	Video language	LANGUAGE	Foreign language	64%	1
			Chinese	36%	0
	Narrative Scene	SCENE	City	42%	1
			farm	58%	0
	Narrative feature	FEATURE	Images dominated	40%	1
			Characters dominated	60%	0
	Culture category	CULTURE	Modern culture	62%	1
			Ancient culture	38%	0

3 Data Analysis of Communication Power Attributes

3.1 The Construction of a Truth Table

According to the binary assignment standard, the original data of 10 samples collected was encoded and imported into fsQCA. Five condition variables including Video duration MINS, update FREQUENCY, operation MODE, multi-channel distribution CHANNEL, release TIMING, and the result variable-number of FANS were selected to set the case frequency threshold as 1 and the consistency threshold as 0.8. The truth value table of the communication attribute. See Table 4 for details.

Table 4. Truth Table of Communication Attributes.

MINS	FREQUENCY	MODE	CHANNEL	TIME	FANS
1	0	1	0	0	1
1	1	1	1	1	1
1	0	1	1	0	1
0	0	1	0	1	0
1	1	0	1	1	0
0	1	1	0	1	1
0	0	0	1	0	1
0	1	1	0	1	1
1	1	0	1	1	0
1	1	0	1	0	0

3.2 The Necessity Analysis of Single-factor

The process of QCA method includes the necessity analysis of a single factor and the sufficiency analysis of conditional configuration. Before analyzing the configuration, the single factor necessity should be measured, which selects necessary conditions in the fsQCA, and pitches on the existence and non-existence of the five conditional variables of the communication attribute, as well as the existence of the result variable "FANS", to conclude the results of the single factor necessity analysis. See Table 5 for details.

It can be shown from Table 5 that the consistency scores of the presence and absence of each condition of the communication attributes are less than 0.9, so the absence of a single condition variable is a necessary condition for configuration. After deeply analyzing Table 5, the result shows that the consistency of "MODE of operation" is the highest at 0.87, which is slightly lower than the standard value of 0.9, with a coverage of 0.84. It can be concluded that 84% of Chinese cultural bloggers on YouTube are affected by cases of strong communication power, which may have a greater impact on the configuration results. For the other conditional variables, there is no variable with a consistency score close to 0.9 appeared, indicating that these variables are not necessary conditions to affect the result configuration, and a further configuration analysis is required.

3.3 Analysis of Configuration Results

After analyzing the single factor necessity, the configuration results analysis is conducted. Running the fsQCA software, there are three different solutions, including complex solution, parsimonious solution, and intermediate solution. In general, the intermediate solution with reasonability and moderate complexity is considered to be the first

Table 5. Single Factor Analysis of Communication Attribute.

Condition variables	Consistency	Coverage
MINS	0.68	0.69
~MINS	0.32	0.66
FREQUENCY	0.45	0.68
~FREQUENCY	0.55	0.73
MODE	0.87	0.84
~MODE	0.13	0.31
CHANNEL	0.86	0.90
~CHANNEL	0.14	0.34
TIMING	0.53	0.82
~TIMING	0.47	0.63

choice for reporting and interpretation in QCA research. In this paper, the intermediate solution is the only solution to be analyzed, which is the main viewpoint since using the current mainstream view in the academic field.

The final configuration results are shown in Table 6.

Table 6. Configuration Analysis of Communication Attributes.

Conditions		Configuration	of	Communication	Attributes
	S1	S2	S3	S4	S5
MINS		⊗	●	⊗	●
FREQUENCY		●		⊗	⊗
MODE	⊗	⊗	●	●	
CHANNEL	●			●	●
TIMING			●	⊗	⊗
CONSISTENCE	1	1	1	1	1
ORIGINAL COVERAGE	0.73	0.06	0.34	0.09	0.18
UNIQUE COVERAGE	0.47	0.05	0.16	0.04	0.05
OVERALL CONSISTENCY			1		
OVERALL COVERAGE			0.94		

Notes: ● refers to core condition contain, ⊗ refers to core condition default, ● refers to boundary condition contain, ⊗ refers to boundary condition default, blank refers to these conditions are not relevant.

According to Table 6, in terms of communication attributes, there are 5 groups of configurations that lead to strong communication power of YouTube Chinese cultural bloggers. The 5 conditional variables in each configuration are statistically analyzed, the result is shown in Table 7.

Table 7. Frequency and Coverage of Communication Attributes in Configuration.

Condition	attributes	Times in configuration	Result coverage
MINS	MINS$\geq$5 mins	1	0.20
	~MINS<5 mins	2	0.07
FREQUENCY	FREQUENCY: per month$\geq$6 times	1	0.05
	~FREQUENCY:per month<6 times	2	0.07
MODE	MODE: team operation	2	0.67
	~MODE: individual operation	1	0.04
CHANNEL	CHANNEL: multichannel distribution ~ CHANNEL: YouTube	2 0	0.55 0
TIMING	TIMING: Times from Fri. to Sun.$\geq$60% ~TIMING: Times from Fri. to Sun.<60%	2 1	0.21 0.03

By discussing the result from Tables 6 and 7, it can be concluded that the conditional variable "MODE" (team operation) appears most frequently in the result configuration with the highest sample coverage, which indicates that this conditional variable is the key factor in affecting the configuration result. This is consistent with both the theoretical basis and empirical judgment. The professional team operation not only guarantees the production of quality content but also provides valuable overseas communication experience, effectively aiming at the target audiences, which saves time and energy than the individual operation.

The conditional variable "CHANNEL", distributed on other major foreign platforms besides YouTube, appeared twice in the result configuration, with a coverage of 51%. The result above shows that, compared with other conditional variables, "CHANNEL" affects the configuration result as well, in addition to the conditional variable "MODE". This condition reflects a way of expanding publicity, which uses different social platforms to release content synchronously in order to expand the communication coverage and audience coverage. Many scholars believe that the release of short videos on multiple platforms is beneficial to attract more audiences [7]." Currently, social media developed vigorously because of its direct docking, high occupancy, and high viscosity, which needs us to "seize the opportunity of media convergence, make full use of advantages of Internet communication so as to transform the single traditional communication to the integration of traditional communication with Internet communication. Therefore, the communication of key social platforms such as YouTube, Facebook, Twitter, Instagram, TikTok, etc. [8]. should be highly valued. The corresponding conditional variable "~CHANNEL" that was not distributed to other foreign mainstream platforms did not appear in the result configuration, which indicates this conditional variable poses no effect on the scope of video communication of the bloggers.

In terms of communication attributes, among the five paths leading to strong communication power of YouTube Chinese cultural bloggers, the one with the highest coverage is S1 (MODE*CHANNEL = team operation * distribution by other foreign mainstream platforms), which explained 70% of the sample cases with a strong explanation. This path shows that regardless of whether other conditions exist or not, as long as a professional team operation (core condition), coupled with the simultaneous distribution of communication on other foreign mainstream platforms (core condition), the communication breadth of YouTube videos for Chinese cultural bloggers will be effectively expanded.

In addition to path S1, both S3 and S5 have a high coverage, explaining 32% and 19% of the sample cases respectively. The Path S3 (MINS*MODE*TIMING = video duration $\geq$ 5 minutes * team operation * Friday to Sunday release) shows that when the blogger is operated by a professional team (core condition), under the three conditions including the duration of most released videos exceeds 5 minutes (core condition), and released on Friday to Sunday, the video communication breadth of bloggers will be also effectively expanded, whether updated frequently or not and whether is synchronized in other platforms. The path S5 (MINS*~FREQUENCY*CHANNEL*TIMING = Video duration $\geq$5 minutes * Monthly update < 6 times * Distribution by other foreign mainstream platforms * Friday to Sunday release) shows that if the blogger meets four requirements, including releasing most videos in the three days from Friday to Sunday (core condition),the video duration exceeds 5 minutes (core condition), the update frequency is less than 8 per month (edge condition), and synchronized distribute on other foreign mainstream platforms except YouTube (core condition), therefore the breadth of the video communication of the blogger can be effectively expanded regardless of operation mode.

As can be shown from Table 6, the coverage of path S2 and S4 is relatively low, as well as the interpretation, which covers only 5% and 8% of the sample cases. In path S2 (~MINS*FREQUENCY*~MODE = Video duration < 5 minutes * monthly updates $\geq$8 times * Individual operation), the conditional variable "MODE" (team operation) does not appear, while being replaced by "~MODE" (individual operation). It shows that under the circumstance of an individual account operating (core condition), keeping the duration of the majority of videos exceeds 5 minutes (core condition) and the update frequency exceeding 6 times per month (core condition) will likely lead to a blogger's communication ineffectively.

Path S4 (~MINS*~FREQUENCY*MODE*~TIMING = Video duration < 5 minutes * Monthly update < 6 times * Team operation * non-Friday to Sunday release) shows that if the account was operated by the team (core condition), most of the blogger's video duration is less than 5 minutes (core condition). Also, the update frequency is less than 6 times per month (core condition), and not released in the three days from Friday to Sunday (core condition), which will likely lead to the blogger's communication ineffectively.

4 Conclusion

At present, the related issues of Chinese bloggers' overseas cultural communication have been widely discussed by domestic academic circles [9, 10]. This is a beneficial supplement to the important topic of "How to tell Chinese stories well". Compared with the external communication mode dominated by official power, "we media" was regarded as a kind of "soft communication", which can better reduce the cultural cognitive gap and build a more diversified and effective communication matrix. The analysis shows that among the communication attributes of video, "MODE team operation" and "CHANNEL multi-channel distribution" are the key factors affecting the communication effect of bloggers. Through the data configuration of QCA, this paper obtains the path model of the hot spread of Chinese cultural bloggers on YouTube, and summarizes and classifies it. The above conclusions play an important role in the overseas communication of Chinese culture.

Acknowledgement. This paper is supported by the Key Research Base of Humanities and Social Sciences of Sichuan Province-Sichuan International Education Development Research Center "Research on the Communication of Traditional Chinese Culture from the Perspective of Chinese International Education(SCGJ2022-22)" and Key Research Base of Humanities and Social Sciences of Sichuan Province- Sichuan Leisure Sports Industry Development Research Center under Grant XXTYCY2023B11.

References

1. Zhang, Z., Li, H.: Subjects, characteristics and influences of public communication in overseas social media. Int. Commun. **12**(05), 7–10 (2020)
2. Huang, X., Su, H.: How can Chinese culture effectively go global? – Qualitative comparative analysis based on 20 cultural going-out cases. National Culture a Tourism, J. Southwest Univ. Nationalities **41**(8), 46–54 (2020)
3. Manovich, L.: The practice of everyday (media) life: from mass consumption to mass cultural production? Crit. Inquiry **35**(2), 319–331 (2009). https://doi.org/10.1086/596645
4. Thiem, A.: Clearly crisp, and not fuzzy: a reassessment of the (Putative) pitfalls of multi-value QCA. Field Methods **25**(2), 197–207 (2013)
5. Thiem, A., Dusa, A.: QCA: a package for qualitative comparative analysis. The R J. **5**(1), 87–97 (2013)
6. Schneider, C.Q., Wagemann, C.: Set-Theoretic Methods for social sciences: A guide to Qualitative Comparative Analysis. Cambridge University Press, Cambridge (2012)
7. Song, Y.: The traditional cultural communication by Li Ziqi's short videos. Young J. **673**(17), 10–11 (2020)
8. Liu, Y.: Short video going to sea: analysis of YouTube operation based on the perspective of overseas audiences – starting from Li Ziqi's overseas popularity. Medium **321**(4), 42–44 (2020)
9. Ziyu, Y.: Analysis of the image of Chinese journalists in international news. Int. J. Front. Soc. **5**, 51–56 (2023)
10. Mansoor, I.: YouTube Revenue and usage statistics (2023). https://www.businessofapps.com/date/youtube-statistics/BusinessofApp,2023-8-2

Multi-recipient Public-Key Authenticated Encryption with Keyword Search

Kejin He, Sha Ma$^{(\boxtimes)}$, and Hao Wang

South China Agricultural University, Guangzhou 510642, China
martin_deng@163.com

Abstract. In cloud computing, private data is usually encrypted and uploaded to a cloud server, which is possibly available to multiple recipients. The traditional schemes fail to account for scenarios with multiple recipients, resulting in low encryption efficiency. In order to better accommodate the multi-recipient scenario, multi-recipient public key authenticated encryption with keyword search (MR-PAEKS) was developed. However, the security of existing MR-PAEKS schemes is not sufficiently secure as they are susceptible to multi-choice ciphertext attacks and multi-choice trapdoor attacks. In order to improve the security of MR-PAEKS in the literature, this paper proposes the first multi-recipient public key authenticated encryption with keyword search scheme satisfying the securities of fully multi-ciphertext indistinguishability (Fully MCI) and fully multi-trapdoor indistinguishability (Fully MTI), and proves its security within the standard model. Compared with related work, our scheme achieves significant improvements in security at the cost of a slight reduction in efficiency.

Keywords: public key authenticated encryption with keyword search · fully multi-ciphertext indistinguishability · fully multi-trapdoor indistinguishability · multi-recipient

1 Introduction

As cloud computing evolves, it brings convenience alongside challenges in security and privacy. Individuals and businesses increasingly store sensitive information in the cloud, necessitating encryption to safeguard against potential malicious tampering. However, encryption complicates data retrieval, making efficient and accurate searches of encrypted data a critical and intriguing issue. Balancing efficient search with privacy preservation in encrypted data remains a key focus.

To solve this problem, Boneh et al. [1] introduced public-key encryption with keyword search (PEKS) in 2004, which later faced keyword guessing attacks (KGA) [2] due to predictable keyword usage. In 2017, Huang and Li [9] developed public-key authenticated encryption with keyword search (PAEKS) to overcome these vulnerabilities by including both the sender's private key and recipient's public key in the encryption, enhancing security against KGA.

© The Author(s), under exclusive license to Springer Nature Singapore Pte Ltd. 2024
K. Li and Y. Liu (Eds.): ISICA 2023, CCIS 2147, pp. 287–296, 2024.
https://doi.org/10.1007/978-981-97-4396-4_27

According to the system model, the traditional PAEKS scheme usually only considers one recipient, and its basic framework is shown in Fig. 1(a). Encryption efficiency is low in the single-recipient PAEKS schemes, where a data sender encrypts and sends data to a cloud server, recipients send search requests, and the server returns encrypted data. The encryption workload grows linearly with the number of recipients.

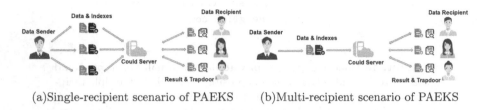

(a)Single-recipient scenario of PAEKS   (b)Multi-recipient scenario of PAEKS

Fig. 1. System model

In scenarios like industrial IoT with multiple data users, single-recipient encryption schemes are inefficient, as the number of ciphertexts increases with the number of recipients, burdening limited-resource IoT devices. Thus, Yang et al. [18] proposed a multi-recipient searchable encryption scheme for better efficiency. Fig. 1(b) illustrates that a company sending sensitive data to multiple leaders via email needs just one encryption operation, thus reducing overall encryption operations and improving efficiency by minimizing repetitive calculations.

According to the security, Huang and Li's PAEKS [9] scheme satisfies both ciphertext indistinguishability and trapdoor privacy, thereby preventing keyword disclosure. However, Qin et al. [15] identified its lack of multi-ciphertext indistinguishability (MCI), where an adversary cannot guess whether multiple ciphertexts contain the same keyword. Consequently, they proposed a new PAEKS scheme to achieve MCI security. Pan et al. [14] proposed a PAEKS scheme addressing both MCI and multi-trapdoor indistinguishability (MTI) security, the latter ensuring indistinguishability among multiple trapdoors. However, Cheng et al. [4] noted that Pan et al.'s scheme [14] fails to meet MCI security. Qin et al. [16] introduced Fully MCI, an advanced form of MCI allowing adversaries more capabilities in post-challenge phase, such as querying a ciphertext oracle. They suggested Fully MTI as future work, namely enhancing MTI by permitting post-challenge trapdoor queries. However current MR-PAEKS schemes don't meet both Fully MCI and Fully MTI. This requirement motivates our research.

1.1   Contributions

Our contributions are summarized as follows.

- We propose the first MR-PAEKS scheme that satisfies both Fully MCI and Fully MTI security, for resisting against multi-choice ciphertext attack and multi-choice trapdoor attack.

- We provide a rigorous security proof in the standard model based on Hash Diffie-Hellman(HDH) assumption.
- Our scheme trades efficiency for enhanced security, offering stronger protection despite being less efficient than existing schemes.

1.2   Related Work

PEKS Scheme. The concept of public key searchable encryption (PEKS) was first defined by Boneh et al. [1] in 2004, who also proposed the initial PEKS scheme in the symmetric group. Since then, numerous schemes and variants of PEKS [3,11] have subsequently been introduced. Unfortunately, all the aforementioned PEKS schemes are vulnerable to KGA; therefore, adopting an KGA-resistant PAEKS scheme would provide a more secure alternative for retrieving ciphertext in cloud storage.

PAEKS Scheme. To resist KGA, Huang and Li [9] introduced PAEKS, an innovative scheme that builds upon PEKS and adds authentication. During ciphertext generation, the sender uses their private key and the recipient's public key to encrypt the keyword. Likewise, recipients use their private key and the sender's public key for keyword encryption during the trapdoor generation process.

Lu et al. [12] introduced a no-pairing PAEKS scheme, which improves efficiency and reduces computational costs, making it ideal for low-powered mobile devices. Huang et al. [8] proposed an efficient PAEKS scheme using inverted indexes to enhance file retrieval. Li et al. [10] proposed public key authenticated encryption with ciphertext update and keyword search (PAUKS) to reduce trapdoor communication overhead.

To enhance encryption efficiency in the multi-recipient scenario, Lu et al. [13] proposed a multi-recipient PAEKS (MR-PAEKS) scheme for IoTs. For multiple users, Lu et al.'s scheme [13] only needs to generate ciphertext once for the same keyword to send to multiple users, which reduces the number of encryptions and hence is suitable for IoTs environment. Yang et al. [18] proposed an MR-PAEKS scheme in traditional public key environments, which can reduce the number of encryptions in multi-recipient environments.

MCI and MTI of PAEKS. To enhance PAEKS security, Qin et al. [15] introduced the concepts of MCI and MTI security. MCI resists multi-choice ciphertext attacks, where adversaries obtain multiple ciphertexts without knowing if they contain the same keywords. MTI resists multi-choice trapdoor attacks, where adversaries obtain multiple trapdoors without knowing if they contain the same keywords. Qin et al. [15] designed a PAEKS scheme satisfying MCI. Pan et al. [14] claimed their PAEKS scheme meets MCI and MTI security, but Cheng et al. [4] disagreed about MCI. Yang et al. [17] proposed a certificateless PAEKS scheme claiming MCI and MTI security, but Cheng et al. [5] found it lacking MTI security. Yang et al. [19] later proposed a PAEKS scheme without a secure channel that satisfies both MCI and MTI security.

Qin et al. [16] introduced Fully MCI, an enhanced security concept based on MCI. It allows the adversary to query challenge keywords using a ciphertext oracle after the challenge phase, thereby enhancing security. Cheng et al. [5] proposed a certificateless PAEKS scheme satisfying both Fully MCI and Fully MTI security, where Fully MTI permits the adversary to query challenge keywords using a trapdoor oracle after the challenge phase, further enhancing security. Cheng et al. [6] also designed a PAEKS scheme with Fully MCI and Fully MTI security in the traditional public key setting.

2   Preliminary

2.1   Asymmetric Bilinear Pairing [7]

A prime p is the order of $\mathbb{G}_1$, $\mathbb{G}_2$, and $\mathbb{G}_T$. Let g_1, g_2 be generators of $\mathbb{G}_1$ and $\mathbb{G}_2$, respectively. a and b are integers in $\mathbb{Z}_p$. An asymmetric bilinear pairing mapping $\hat{e} : \mathbb{G}_1 \times \mathbb{G}_2 \to \mathbb{G}_T$ satisfies the following properties, where $\mathbb{G}_1 \neq \mathbb{G}_2$ and there is no computable homomorphism between $\mathbb{G}_1$ and $\mathbb{G}_2$:

- Bilinearity: $\hat{e}(g_1^a, g_2^b) = \hat{e}(g_1, g_2)^{ab}$;
- Non-degeneracy: $\hat{e}(g_1, g_2) \neq 1$;
- Computability: $\hat{e}(g_1, g_2)$ can be computed.

2.2   HDH Assumption

Let $\mathbb{G}$ be a cyclic group of prime p and g be a generator of $\mathbb{G}$. $H : \{0,1\}^* \to \{0,1\}^l$ is a hash function, where l is a binary number. Given hash function H and tuple $(g, g^a, g^b, Z) \in \mathbb{G}^3 \times \{0,1\}^l$, the HDH problem is to judge whether $Z = H(g^{ab})$, where $a, b \in \mathbb{Z}_p$ and Z is a random element of $\{0,1\}^l$.

3   Definition and Security Model

3.1   Definition of MR-PAEKS

- Setup (λ): Given the security parameter λ, it returns the public parameter pp.
- KeyGen (pp): Given pp as input, it returns a sender's key pair (pk_S, sk_S) and a recipient's set of key pairs $\{(pk_{R_1}, sk_{R_1}), ..., (pk_{R_n}, sk_{R_n})\}$, where n indicates the number of recipients.
- MR-PAEKS ($sk_S, \{pk_{R_1}, ..., pk_{R_n}\}, w$): Given a sender's secret key sk_S, a recipient's public key set $\{pk_{R_1}, ..., pk_{R_n}\}$ where n indicates the number of recipients, and a keyword w as input, it returns the keyword w's ciphertext C.
- Trapdoor (pk_S, sk_{R_i}, w): Given an input consisting of a sender's public key pk_S, the secret key sk_{R_i} of the i-th recipient and a keyword w, it returns a corresponding trapdoor T.
- Test (C, T): Given an input consisting of C and T, it returns 1 indicating C and T have the same keyword, and 0 otherwise.

3.2 Security Models

Game 1. Fully Multi-ciphertext Indistinguishability

1. **Setup.** Given a security parameter λ, the challenger initializes by running the Setup algorithm to generate the public parameter (pp) and subsequently uses KeyGen(pp) to create key pairs for both sender (pk_S, sk_S) and recipient's key pair set $\{(pk_{R_1}, sk_{R_1}), ..., (pk_{R_n}, sk_{R_n})\}$, where n indicates the number of recipients. Finally, it gives $(pp, \{pk_{R_1}, ..., pk_{R_n}\}, pk_S)$ to the adversary.

2. **Phase 1.** The adversary can adaptively query the following two oracles with polynomially many times.
   - **Ciphertext Oracle** $\mathcal{O}_C$. For any keyword $w \in \{0,1\}^*$, the challenger computes $C \leftarrow$ MR-PAEKS$(\{pk_{R_1}, ..., pk_{R_n}\}, sk_S, w)$ and returns C to the adversary.
   - **Trapdoor Oracle** $\mathcal{O}_T$. For any keyword $w \in \{0,1\}^*$, the challenger computes $T_w \leftarrow$ Trapdoor(pk_S, sk_{R_i}, w) where i is i-th recipient and returns T_w to the adversary.

3. **Challenge.** At some point, the adversary sends the challenger two tuple words $\overrightarrow{W}_0(w_{0,1}, ..., w_{0,n}), \overrightarrow{W}_1(w_{1,1}, ..., w_{1,n})$ as the challenge keywords. The only restriction is that for any $i \in [n]$, the adversary never query the trapdoor oracle with $w_{0,i}$ or $w_{1,i}$ previously. The challenger chooses a random bit $b \in \{0,1\}$, computes $C_i^* \leftarrow$ MR-PAEKS$(\{pk_{R_1}, ..., pk_{R_n}\}, sk_S, w_{b,i})$ for $i = 1$ to n, and returns the challenge ciphertexts $\overrightarrow{C}^* = (C_1^*, ..., C_n^*)$ to the adversary.

4. **Phase 2.** $\mathcal{A}$ continues to ask for S adaptively, but with the restrictions that $\mathcal{A}$ can not query $(w_{0,1}, ..., w_{0,n}, w_{1,1}, ..., w_{1,n})$ in the trapdoor oracle.

5. **Guess.** Eventually, the adversary outputs a bit $b' \in \{0,1\}$ as the guess of b. It wins the game if $b' = b$. The advantage that any PPT adversary $\mathcal{A}$ wins the game Fully MCI is defined as below:

$$\mathsf{Adv}_{\text{MR-PAEKS},\mathcal{A}}^{\text{Fully MCI}}(\lambda) = |\Pr[b' = b] - \frac{1}{2}|. \tag{1}$$

Definition 1 (Fully MCI-security). For any PPT adversary $\mathcal{A}$ in the security parameter λ, an MR-PAEKS scheme achieves Fully MCI-security if the advantage $\mathsf{Adv}_{\text{MR-PAEKS},\mathcal{A}}^{\text{Fully MCI}}(\lambda)$ is negligible.

Game 2. Fully Multi-trapdoor Indistinguishability

1. **Setup.** Same as **Game 1**.
2. **Phase 1.** Same as **Game 1**.
3. **Challenge.** At some point, the adversary sends the challenger two tuple words $\overrightarrow{W}_0(w_{0,1}, ..., w_{0,n}), \overrightarrow{W}_1(w_{1,1}, ..., w_{1,n})$ as the challenge keywords. The only restriction is that for any $i \in [n]$, the adversary never query the ciphertext oracle with $w_{0,i}$ or $w_{1,i}$ previously. The challenger chooses a random bit $b \in \{0,1\}$, computes $T_i^* \leftarrow$ Trapdoor$(pk_S, sk_{R_i}, w_{b,i})$ for $i = 1$ to n, and returns the challenge trapdoor $\overrightarrow{T}^* = (T_1^*, ..., T_n^*)$ to the adversary.
4. **Phase 2.** $\mathcal{A}$ continues to ask for S adaptively, but with the restrictions that $\mathcal{A}$ can not query $(w_{0,1}, ..., w_{0,n}, w_{1,1}, ..., w_{1,n})$ in the ciphertext oracle.

5. **Guess.** Eventually, the adversary outputs a bit $b' \in \{0,1\}$ as the guess of b. It wins the game if $b' = b$. The advantage that any PPT adversary $\mathcal{A}$ wins the game Fully MTI is defined as below:

$$\mathsf{Adv}_{\mathsf{MR\text{-}PAEKS},\mathcal{A}}^{\mathsf{Fully\ MTI}}(\lambda) = |\Pr[b' = b] - \frac{1}{2}|. \tag{2}$$

Definition 2 (Fully MTI-security). For any PPT adversary $\mathcal{A}$ in the security parameter λ, a MR-PAEKS scheme achieves Fully MTI-security if the advantage $\mathsf{Adv}_{\mathsf{MR\text{-}PAEKS},\mathcal{A}}^{\mathsf{Fully\ MTI}}(\lambda)$ is negligible.

4   Our Scheme

1. Setup(λ): It takes a security parameter λ as input, and selects a bilinear map $\hat{e} : \mathbb{G}_1 \times \mathbb{G}_2 \to \mathbb{G}_T$ where $\mathbb{G}_1$, $\mathbb{G}_2$ and $\mathbb{G}_T$ are groups with order p. Then it randomly selects two generators g_1, g_2 as generators of $\mathbb{G}_1$, $\mathbb{G}_2$, respectively. H_1, H_2, H_3, H_4 are hash functions: $H_1 : \mathbb{G}_1 \to \mathbb{Z}_p$; $H_2 : \{0,1\}^* \times \mathbb{G}_1 \to \mathbb{Z}_p$; $H_3 : \mathbb{G}_T \to \mathbb{Z}_p$; $H_4 : \mathbb{G}_1 \times \mathbb{G}_1 \times \{0,1\}^* \times \mathbb{Z}_p \to \{0,1\}^l$, where l denotes the binary length of hash values. Finally it outputs public parameter $pp = \{\hat{e}, \mathbb{G}_1, \mathbb{G}_2, \mathbb{G}_T, p, g_1, g_2, H_1, H_2, H_3, H_4\}$.

2. KeyGen(pp): It takes the public parameter pp as input and generates (pk_U, sk_U) as a public/secret key pair of user U. Then it chooses $x, y_i \in_R \mathbb{Z}_p$ where i represents i-th recipient and computes $pk_S = g_1^x$ and $pk_{R_i} = g_1^{y_i}$, and returns $(pk_S = g_1^x, sk_S = x)$ and $(pk_{R_i} = g_1^{y_i}, sk_{R_i} = y_i)$ as the key pairs of sender and recipients, respectively.

3. MR-PAEKS($\{pk_{R_1}, pk_{R_2}, ..., pk_{R_n}\}, sk_S, w$): It takes pp, pk_{R_i}, sk_S, and w as input, where i represents the i-th recipient, and n represents the number of recipients. The sender selects $r \in \mathbb{Z}_p$ randomly and computes $K_i = pk_{R_i}^{sk_S} = g_1^{xy_i}$ for each $i \in [1,n]$, where n is the number of receivers. Select a random integer $\beta \in \mathbb{Z}_p$ and then define a degree n polynomial

$$f(x) = \prod_{i=1}^{n}(x - v_i) + \beta = x^n + \beta_{n-1}x^{n-1} + ... + \beta_1 x + \beta_0, \text{ where } \beta_i \in \mathbb{Z}_p \text{ and } v_i =$$

$H_3(\hat{e}(pk_{R_i}, g_2)^{r \cdot H_2(w, H_1(K_i))})$. Compute $C_1 = g_1^r$, $C_2 = \{C_{2_1}, C_{2_2}, ..., C_{2_n}\} = \{g_1^{r \cdot H_1(K_1)}, g_1^{r \cdot H_1(K_2)}, ..., g_1^{r \cdot H_1(K_n)}\}$, $C_3 = (\beta_0, \beta_1, ..., \beta_{n-1})$, and $C_4 = H_4(C_1, C_2, C_3, \beta)$. Finally, return $C = (C_1, C_2, C_3, C_4)$.

4. Trapdoor(pk_S, sk_{R_i}, w'): The i-th recipient chooses $s \in_R \mathbb{Z}_p$ and computes $T_1 = g_2^{sk_{R_i} \cdot H_2(w', H_1(K_i'))} \cdot g_2^{s \cdot H_1(K_i')}$, $T_2 = g_2^s$, where $K_i' = pk_S^{sk_{R_i}} = g_1^{xy_i}$. Finally, set trapdoor $T = (T_1, T_2)$.

5. Test(C, T): The cloud sever executes as below:
   - Parse ciphertext C_3 as $(\beta_0, \beta_1, ..., \beta_{n-1})$ and reconstruct the polynomial $f(x) = x^n + \beta_{n-1}x^{n-1} + ... + \beta_1 x + \beta_0$;
   - Compute $v_i' = H_3(\hat{e}(C_1, T_1)/\hat{e}(C_{2_i}, T_2))$, where C_{2_i} is the ciphertext of the corresponding recipient. Let $\beta' = f(v_i')$, and check whether $C_4 = H_4(C_1, C_{2_i}, C_3, \beta')$ holds. If it does, output "1" or "0" otherwise.

5 Security Analysis

Theorem 1. *The MR-PAEKS scheme has Fully MCI security under standard model, if $H_1 \sim H_4$ are collision-resistant hash functions and HDH assumption is intractable.*

Proof. Suppose that $\mathcal{A}$ is an adversary against the Fully MCI security of the proposed Fully MCI game in polynomial time. We construct a simulator $\mathcal{S}$ to run $\mathcal{A}$ for breaking the HDH assumption. We prove the theorem 1 via five following game programs Game-$j(j = 0, 1, 2, 3, 4)$, and define Y_j is the event of $\mathcal{A}$ guessing correctly in Game-j, namely $b = b'$.

Game-0: Game-0 is the initial Fully MCI game. So we claim that advantage $\text{Adv}_{\text{MR-PAEKS},\mathcal{A}}^{\text{Fully MCI}}(\lambda) = |\Pr[Y_0] - \frac{1}{2}|$.

Game-1: In this game, $\mathcal{S}$ picks $a, c_i \in \mathbb{Z}_p$ randomly to calculate $pk_S = g^a$ and $pk_{R_i} = g^{c_i}$ for each recipients $R_i(i \in [1, n])$, where g is the generator of group $\mathbb{G}$. Other parameters are the same as Game-0. Obviously, Game-0 and Game-1 are indistinguishable from $\mathcal{A}$. So, we have $\Pr[Y_1] = \Pr[Y_0]$.

Game-2: The only difference between Game-2 and Game-1 lies in the query response and challenge phases executed by $\mathcal{S}$. $\mathcal{S}$ does the following queries:

-$\mathcal{O}^{Ciphertext}$: $\mathcal{A}$ submits a keyword w to $\mathcal{S}$, then $\mathcal{S}$ picks a random integer $r \in \mathbb{Z}_p$ and returns $C = (C_1, C_2, C_3, C_4)$ to $\mathcal{A}$.

-$\mathcal{O}^{Trapdoor}$: $\mathcal{A}$ submits a keyword w' to $\mathcal{S}$, after then $\mathcal{S}$ computes $T_1 = g_2^{SK_{R_i} \cdot H_2(w', H_1(K_i'))} \cdot g_2^{s \cdot H_1(K_i')}$, $T_2 = g_2^s$, where $K_i' = pk_S^{sk_{R_i}} = g_1^{xy_i}$, and returns $T = (T_1, T_2)$ to $\mathcal{A}$.

Challenge: $\mathcal{A}$ submits two different tuples of keywords $\overrightarrow{W}_0^*(w_{0,1}^*, ..., w_{0,n}^*)$, and $\overrightarrow{W}_1^*(w_{1,1}^*, ..., w_{1,n}^*)$, where $\overrightarrow{W}_0^*$ or $\overrightarrow{W}_1^*$ are not challenged in previous phase. $\mathcal{S}$ chooses $r^*, \beta^* \in_R \mathbb{Z}_p$ and $b \in_R \{0, 1\}$, then computes $f(x) = \prod_{i=1}^{n}(x - v_i^*) + \beta^* = x^n + \beta_{n-1}^* x^{n-1} + ... + \beta_1^* x + \beta_0^*$, where $v_i^* = H_3(\hat{e}(pk_{R_i}, g_2)^{r^* \cdot H_2(w_{b,i}^*, H_1(K_i))})$ and $K_i = pk_{R_i}^{sk_S} = g_1^{xy_i}$ for $R_i(i \in [1, n])$. Finally, $\mathcal{S}$ computes $C_1^* = g_1^{r^*}$, $C_2^* = \{C_{21}^*, C_{22}^*, ..., C_{2n}^*\} = \{g_1^{r^* \cdot H_1(K_1)}, g_1^{r^* \cdot H_1(K_2)}, ..., g_1^{r^* \cdot H_1(K_n)}\}$, $C_3^* = (\beta_0^*, \beta_1^*, ..., \beta_{n-1}^*)$, $C_4^* = H_4(C_1^*, C_2^*, C_3^*, \beta^*)$. Finally, $\mathcal{S}$ returns ciphertext $C^* = (C_1^*, C_2^*, C_3^*, C_4^*)$.

From the adversary's view, it is impossible to distinguish between a challenge ciphertext and a real ciphertext. Therefore, the challenge ciphertext $C^* = (C_1^*, C_2^*, C_3^*, C_4^*)$ is the correct ciphertext of the keyword $w_{b,i}^*$. So, Game-1 and Game-2 are indistinguishable and we have $\Pr[Y_1] = \Pr[Y_2]$.

Game-3: In this game, the challenger performs identically to that in Game-2, and $\mathcal{S}$ does not abort Game-3 unless the following events occur.

Event E_1: $\mathcal{A}$ submits w to $\mathcal{S}$ in $\mathcal{O}^{Ciphertext}$, where the keyword's input satisfies $w \neq w_{b,i}$, but $f(x) = f(x)^* = \prod_{i=1}^{n}(x - v_i^*) + \beta^*$ for $C_3^* = (\beta_0^*, \beta_1^*, ..., \beta_{n-1}^*)$, where $v_i = v_{i_{b,i}}$ and $\beta^* \in \mathbb{Z}_p$, and $C_4^* = H_4(C_1^*, C_2^*, C_3^*, \beta^*)$.

Event E_2: $\mathcal{A}$ submits w to $\mathcal{S}$ in $\mathcal{O}^{Trapdoor}$, where the keyword's input satisfies $w \neq w_{b,i}$, and $H_2(w, H_1(K_i)) = H_2(w_{b,i}, H_1(K_i^*))$

Obviously, Game-2 and Game-3 are indistinguishable to $\mathcal{A}$ unless the event $E_1 \vee E_2$ occurs. Due to Difference Lemma, we have $|\Pr[Y_2] - \Pr[Y_3]| \leqslant \Pr[E_1 \vee E_2]$. Furthermore, if the event E_1 occurs, the adversary $\mathcal{A}$ has the advantage $negl^H$ of winning, if $(negl^H)^{n+1} \cdot \frac{1}{p} \geqslant \Pr[E_1]$, where n is the number of recipient and p is random number of $\mathbb{Z}_p$. Similarly, if the event E_2 occurs, the adversary $\mathcal{A}$ has the advantage $negl^H$ of winning, if $negl^H \geqslant \Pr[E_2]$. Therefore, we induce the equation $|\Pr[Y_2] - \Pr[Y_3]| \leqslant negl$.

Game-4: Game-4 is the same as Game-3, except that $\mathcal{S}$ picks a random element $Z \in \{0,1\}^l$ instead of $H_1(g^{ac_i})$ when generating the challenge of ciphertext. Obviously, $\mathcal{S}$ responds queries and challenge via HDH tuples $(H_1, g, g^a, g^{c_i}, Z)$ without revealing the integer of a and c_i. In consequence, Game-3 is equivalent to Game-4. The adversary $\mathcal{A}$ distinguishes the element of $K_i' = H_1(g^{ac_i})$ (for $i = 1, 2, ..., n$) and Z with non-negligible advantage, if the HDH problem is breached. Hence, $\mathcal{A}$ has the advantage $\mathsf{Adv}^{\mathsf{HDH}}(\lambda)$ to win Game-4. We have $|\Pr[Y_3] - \Pr[Y_4]| \leqslant \mathsf{Adv}^{\mathsf{HDH}}(\lambda)$. Z is a random integer of $\mathbb{G}$, so $\mathcal{A}$ has the advantage of winning with $\Pr[Y_4] = \frac{1}{2}$. Next, $\mathcal{A}$ can guess correctly in the above sub-games with the advantage

$$
\begin{aligned}
\mathsf{Adv}^{\mathsf{Fully\ MCI}}_{\mathsf{MR\text{-}PAEKS},\mathcal{A}}(\lambda) =& |\Pr[Y_0] - \frac{1}{2}| \leqslant |\Pr[Y_0] - \Pr[Y_1]| \\
& + |\Pr[Y_1] - \Pr[Y_2]| + |\Pr[Y_2] - \Pr[Y_3]| \\
& + |\Pr[Y_3] - \Pr[Y_4]| + |\Pr[Y_4] - \frac{1}{2}|.
\end{aligned} \tag{3}
$$

Based on the triangle inequality, the above sub-games induce as follow:

$$
\mathsf{Adv}^{\mathsf{Fully\ MCI}}_{\mathsf{MR\text{-}PAEKS},\mathcal{A}}(\lambda) = \mathsf{Adv}^{\mathsf{HDH}}(\lambda). \tag{4}
$$

Due to the collision resistance property of the hash function H, and the HDH problem are complex, the $\mathsf{Adv}^{\mathsf{Fully\ MCI}}_{\mathsf{MR\text{-}PAEKS},\mathcal{A}}(\lambda)$ is negligible in Theorem 1.

Theorem 2. *The MR-PAEKS scheme has Fully MTI security under standard model, if $H_1 \sim H_4$ hash functions are collision-resistant and HDH assumption is intractable.*

The proof of theorem 2 is similar to that of theorem 1, which is omitted here.

6   Performance and Experiments

Table 1 compares computational and communication costs among different schemes. It uses E_1, E_2, and E_T for exponential operations in $\mathbb{G}_1$, $\mathbb{G}_2$, and $\mathbb{G}_T$ groups, respectively, H for the hash function, P for asymmetric bilinear pairing, and n for the number of recipients. Yang et al.'s scheme exhibits the highest efficiency in encryption and test, while Lu et al.'s scheme matches Yang et al.'s in trapdoor generation efficiency. However, our scheme has the lowest efficiency overall. The symbols $|\mathbb{G}_1|$, $|\mathbb{G}_2|$, and $|\mathbb{Z}_p|$ denote the lengths of elements in $\mathbb{G}_1$, $\mathbb{G}_2$, and $\mathbb{Z}_p$, respectively, while $|C|$ and $|T|$ represent the lengths of the ciphertext

Table 1. Computation and Communication

Scheme	Computation			Communication	
	Encrypt	Trapdoor	Test	$|C|$	$|T|$
Lu et al. [13]	$(2+3n)E_1+4H$	$2E_1+2H$	$2E_1+2H$	$|G_1|+n|Z_p|$	$|G_1| + |Z_p|$
Yang et al. [18]	$(2n+1)E_1+(2n+1)H$	$2E_1+2H$	E_1+2H	$(n+1)|Z_p|$	$|G_1|$
Ours	$(2n+1)E_1+nE_T+(3n+1)H+nP$	E_1+2E_2+2H	$2P_2$	$(n+1)|G_1|+|Z_p|$	$2|G_2|$

Table 2. Security comparisons

Scheme	MCI	Fully MCI	MTI	Fully MTI	Model	Assumption
Lu et al. [13]	×	×	×	×	ROM	CDH
Yang et al. [18]	×	×	×	×	SM	HDH&CDH&DL
Ours	✓	✓	✓	✓	SM	HDH

C and the trapdoor T, Yang et al.'s scheme consumes the least communication resources, followed by Lu et al.'s scheme, while our scheme exhibits the highest consumption.

Table 2 compares the security levels of various schemes. Lu et al.'s scheme and Yang et al.'s scheme both fail to meet the MCI and MTI security requirements, let alone the Fully MCI and Fully MTI security standards. Our scheme, however, satisfies all four security metrics, demonstrating significant security advantages.

7   Conclusion and Future Work

We introduced an MR-PAEKS scheme in this work, achieving both Fully MCI and Fully MTI security, significantly outperforming previous multi-recipient PAEKS schemes in security. However, the scheme's efficiency is currently suboptimal. Future efforts will aim to develop a more efficient, no-pairing PAEKS scheme while maintaining Fully MCI and Fully MTI security.

Acknowledgments. This work is supported by the Guangdong Basic and Applied Basic Research Foundation (2024A1515012666) and the National Natural Science Foundation of China (61872409).

References

1. Boneh, D., Di Crescenzo, G., Ostrovsky, R., Persiano, G.: Public key encryption with keyword search. In: Cachin, C., Camenisch, J.L. (eds.) EUROCRYPT 2004. LNCS, vol. 3027, pp. 506–522. Springer, Heidelberg (2004). https://doi.org/10.1007/978-3-540-24676-3_30
2. Byun, J.W., Rhee, H.S., Park, H.-A., Lee, D.H.: Off-line keyword guessing attacks on recent keyword search schemes over encrypted data. In: Jonker, W., Petković, M. (eds.) Secure Data Management, pp. 75–83. Springer Berlin Heidelberg, Berlin, Heidelberg (2006). https://doi.org/10.1007/11844662_6

3. Chen, R., Mu, Y., Yang, G., Guo, F., Wang, X.: Dual-server public-key encryption with keyword search for secure cloud storage. IEEE Trans. Inf. Forensics Secur. **11**(4), 789–798 (2015)
4. Cheng, L., Meng, F.: Security analysis of pan et al.'s public-key authenticated encryption with keyword search achieving both multi-ciphertext and multi-trapdoor indistinguishability. J. Syst. Architect. **119**, 102248 (2021)
5. Cheng, L., Meng, F.: Certificateless public key authenticated searchable encryption with enhanced security model in IIOT applications. IEEE Internet of Things Journal (2022)
6. Cheng, L., Qin, J., Feng, F., Meng, F.: Security-enhanced public-key authenticated searchable encryption. Inf. Sci. **647**, 119454 (2023)
7. Galbraith, S.D., Paterson, K.G., Smart, N.P.: Pairings for cryptographers. Discrete Applied Mathematics **156**(16), 3113–3121 (2008), applications of Algebra to Cryptography
8. Huang, Q., Huang, P., Li, H., Huang, J., Lin, H.: A more practical public-key authenticated encryption with keyword search scheme. Available at SSRN 4226763
9. Huang, Q., Li, H.: An efficient public-key searchable encryption scheme secure against inside keyword guessing attacks. Inf. Sci. **403**, 1–14 (2017)
10. Li, H., Huang, Q., Huang, J., Susilo, W.: Public-key authenticated encryption with keyword search supporting constant trapdoor generation and fast search. IEEE Trans. Inf. Forensics Secur. **18**, 396–410 (2022)
11. Liang, K., Susilo, W.: Searchable attribute-based mechanism with efficient data sharing for secure cloud storage. IEEE Trans. Inf. Forensics Secur. **10**(9), 1981–1992 (2015)
12. Lu, Y., Li, J.: Lightweight public key authenticated encryption with keyword search against adaptively-chosen-targets adversaries for mobile devices. IEEE Trans. Mob. Comput. **21**(12), 4397–4409 (2021)
13. Lu, Y., Li, J., Zhang, Y.: Privacy-preserving and pairing-free multirecipient certificateless encryption with keyword search for cloud-assisted iiot. IEEE Internet Things J. **7**(4), 2553–2562 (2019)
14. Pan, X., Li, F.: Public-key authenticated encryption with keyword search achieving both multi-ciphertext and multi-trapdoor indistinguishability. J. Syst. Architect. **115**, 102075 (2021)
15. Qin, B., Chen, Y., Huang, Q., Liu, X., Zheng, D.: Public-key authenticated encryption with keyword search revisited: security model and constructions. Inf. Sci. **516**, 515–528 (2020)
16. Qin, B., Cui, H., Zheng, X., Zheng, D.: Improved security model for public-key authenticated encryption with keyword search. In: Huang, Q., Yu, Yu. (eds.) ProvSec 2021. LNCS, vol. 13059, pp. 19–38. Springer, Cham (2021). https://doi.org/10.1007/978-3-030-90402-9_2
17. Yang, G., Guo, J., Han, L., Liu, X., Tian, C.: An improved secure certificateless public-key searchable encryption scheme with multi-trapdoor privacy. Peer-to-Peer Networking and Applications, pp. 1–13 (2022)
18. Yang, N., Zhou, Q., Huang, Q., Tang, C.: Multi-recipient encryption with keyword search without pairing for cloud storage. J. Cloud Comput. **11**(1), 1–12 (2022)
19. Yang, P., Li, H., Huang, J., Zhang, H., Au, M.H.A., Huang, Q.: Secure channel free public key authenticated encryption with multi-keyword search on healthcare systems. Futur. Gener. Comput. Syst. **145**, 511–520 (2023)

Analysis of Information Security Processing Technology Based on Computer Big Data

Hua Wang and Fuyu Zhu(✉)

College of Computer Science, Guangdong University of Science and Technology, Dongguan,
China
604114104@qq.com

Abstract. Information security technology is the most important technology in
the world today. It can effectively reduce users' exposure to network viruses and
hacker attacks, maintain network information security, and build a safe and stable
network environment. Because the traditional information security technology is
a passive protection technology, under the rapid development of Internet technology, it has certain defects and lags, which make it unsatisfactory in terms of
efficiency and quality of information processing, and can no longer meet Today's
information explosion network environment. Therefore, new information security
processing technology must be adopted to improve the work efficiency and quality
of information security, and this is exactly the problem to be solved in this paper.

Keywords: computer big data · information security · processing technology

1 Introduction

With the development of big data technology, people pay more and more attention to
the management of information security, and in this process, the way of computer information processing has also changed. Using big data technology, an information security
system can be effectively established to ensure information security, reliability, and
uninterrupted work. Big data analysis [1] technology can combine traditional security
technology with cloud computing technology, and combine it with security technology to
promote the vigorous development of security technology [2]. In the computer network,
due to the threats of viruses and information security, the traditional network security processing technology is facing problems and lags behind. Therefore, this paper focuses on
the combination of computer network big data analysis and traditional network security
processing technology. It constitutes a whole network security management technology.

2 General Overview of Big Data

2.1 The Concept of Big Data

Big data refers to computer-based, large-scale data that cannot be stored, extracted,
retrieved, and analyzed in a timely and effective manner through conventional data
methods. The current data analysis and processing methods are very different from the

previous ones. The current information processing methods require faster and more efficient methods, and they can quickly find what they want in a large amount of data.

2.2 Characteristics of Big Data

Big data analysis has the characteristics of large scale, data diversity and complexity. Big data analysis has architectural complexity and data type complexity. It can not only process general data, but also process information such as video and audio. The characteristic of big data processing is a large amount of information, but the amount of available data is very small, which shows that the value density of big data is very low, so it must be filtered to obtain useful information. Due to the existence of massive data, the requirements for data processing capabilities are getting higher and higher. Therefore, how to efficiently process massive data is a key issue. Accelerating the speed of obtaining and processing information has become a prominent feature.

2.3 The Role of Big Data

The further integration of data processing and network can promote the better performance of the network. The data types of big data are relatively complicated. Using big data analysis, we can clearly grasp the problems that arise in different stages of the company's operations, so that we can judge the company's internal development and understand the internal development trends of different companies, so as to shape the industry. Further understanding. At the same time, based on the analysis results of big data, it can predict the future development trend, effectively improve the strategic decision-making, and contribute to the continuous development and growth of the enterprise.

3 Security Management Technology System of Computer Big Data

In order to realize the safe use of information technology, it is necessary to use more information technology in safety management, and closely combine various big data and computers to form a stable safety technology system. Because only by making full use of big data technology and establishing a variety of effective and complete information security disposal mechanisms can we truly solve and deal with these problems. The system architecture of security management technology based on computer big data analysis is shown in Fig. 1.

4 Research on Security Management of Computer Big Data

4.1 Encryption Technology

The current computer security technology is roughly divided into two categories, namely virtual private network technology and public network technology, but there is also a big difference between them, that is, data on the Internet can be transmitted to the corresponding server with the help of virtual private network technology on, implement

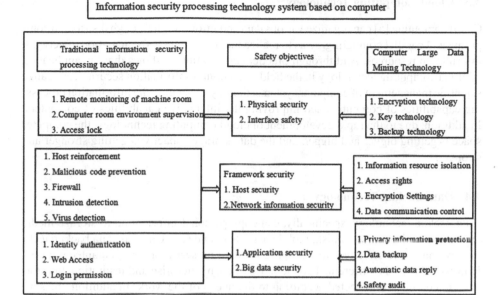

Fig. 1. Framework of information security processing technology system based on computer big data

encryption. Encryption technology [3] can prevent malicious users from maliciously attacking personal or company computers, prevent personal or company personal data from being leaked, and thus effectively protect personal and company network assets. The principle of encryption is very simple. It is to set a login password for the account, and then match the account number with the account number. Network isolation [4] forms a completely closed network, preventing hackers or malicious programs from invading, stealing information, and protecting user data.

4.2 Software Processing Technology

In the processing of mass data using computers, various processing methods must be used to ensure the safety and accuracy of its application. The most common application is the software protection system installed for the computer system, mainly including 360, mobile phone housekeeper and so on. Although some applications are paid, they are basically free, and its use also places more emphasis on the protection of the system. In specific applications, it can prevent attacks such as Trojan horses, ensure the normal operation of the computer, and ensure the security of the system. Information security. In use, when the computer is invaded or invaded, the system will give corresponding prompts according to the actual situation, and make correct decisions under this prompt, avoiding various information security and network security problems. When using software processing technology, it is necessary to analyze the functions of the software in detail to achieve the purpose of anti-virus, achieve in-depth detection of computers, eliminate potential safety hazards in the system, and ensure data security.

4.3 Cloud Computing Technology

Cloud computing [5] can organize various data and information through distributed computing and parallel computing, so as to realize the integration and calculation of network information resources. And through practical application, it shows that the application of cloud computing technology in the field of computer information security can realize scientific management of computer systems, thereby enhancing the information processing capabilities of computers and improving the interactive capabilities of computers. In addition, due to the rapid development of cloud computing technology, the computer space is getting bigger and bigger, and the data storage capacity is getting stronger and stronger.

4.4 Data Backup Technology

Data backup technology, specifically, can copy personal information to storage media such as Baidu hard disk, U disk, mobile hard disk, and CD. Utilizing data backup technology can create effective storage space for data preservation and guarantee security. However, due to the different characteristics of each enterprise and institution, the storage method must be selected according to the needs of the work. In addition, the staff should understand that unexpected situations such as power outages and network outages [6] will cause information loss, and this phenomenon can be effectively prevented by using information backup technology quickly. It can be seen that the reasonable use of information backup technology can provide a safe storage for information and provide a guarantee for safety.

4.5 Data Mining and Acquisition Technology

Data mining is an in-depth study of data and obtains relevant technologies from it, including data preparation, rule search, rules, etc. Through data mining, the quality of decision-making can be optimized. Most enterprises deal with data centrally and discover problems such as insufficient data, which affects the accuracy of data. If a single data record is used, it will be difficult to find valuable information and laws of data. As long as these data can be analyzed in depth, useful data can be found from them. In the information collection system, how to accurately collect data according to the needs of users is a major problem in current data collection technology. Today, with the rapid development of information technology, our country has made corresponding adjustments in 2020, with digital as the main development direction. As an important branch of information processing technology, information collection technology plays an important role in the entire information industry chain. It can improve subsequent data storage and processing capabilities by efficiently collecting and utilizing information. If there is a problem in the collection process, it will have a direct impact on the efficiency of information processing, which in turn will affect the application of big data. Once a problem occurs in the collection process, it will have a direct impact on the processing effect of information security, thereby endangering the use of big data. The analysis of security processing technology for big data analysis needs urgent analysis, as shown in Figure 2. In the era of big data, more abundant information and data have emerged, and useful data can only be obtained through information collection technology. However, with the development of society, it is becoming more and more difficult to obtain information. It is necessary to rely on artificial intelligence to realize information collection.

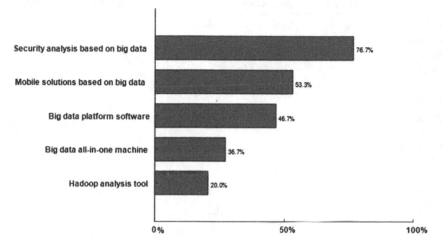

Fig. 2. The demand for big data security analysis solutions is urgent

5   Data Comparison and Control Experiment

5.1   Test Preparation

Through the comparison test with the traditional information security technology system, the accuracy of the information security technology system based on computer big data is verified. More than a thousand virus information attacks were carried out on network environments using two operating systems, more than one thousand times of network information transfers were carried out on two different network environments, and five attacks were carried out on two thousand different forms of network information. Each analysis is calculated once per minute, testing the application capability of the network security technology system based on network big data analysis.

5.2   Experimental Results

The following are the test results of the two systems.

According to Table 1, the computer-based network security system has five processing performances of more than 95%, while the five processing performances of general systems are less than 65%, and the overall performance is insufficient. Sixty-five percent, which shows that the computer-centric system can effectively enhance the management function of network security.

Table 1. Experimental results of two information security processing technology systems

System	First information processing efficiency	Efficiency of the second information processing	Third information processing efficiency	Fourth information processing efficiency	Fifth information processing efficiency
Traditional information security processing technology system	64%	58%	46%	53%	49%
Information security processing technology system based on computer big data	97%	96%	98%	95%	96%

6   Ensure the Information Security Processing Strategy of Computer Big Data

6.1   Using Big Data Encryption Technology to Realize Information Security Processing

In order to realize the security management of computer big data, it is necessary to improve the security operation ability of each operator. Because big data analysis is closely related to our work and daily life in practice, we must pay attention to the integrity of information and encrypt it in practice. Because there are various hidden dangers in actual use, the operator should pay attention to avoid clicking unknown links at will, so as to avoid virus intrusion. In daily applications, it is necessary to prevent random login to illegal websites to ensure the safety of operations and ensure the reliability and security of data. At the same time, in the application, it is necessary to pay attention to the timely storage of data, and scientifically store and encrypt the information processed by big data to ensure data security. Since the massive amount of computer information has its own unique value, special management is still required during actual data processing and storage. If the security of data storage is high, then these data can be stored and processed separately to Further guarantee data security.

6.2   Create a Security Service Background

As a platform for computer data security, the security service platform can apply big data information technology to the security technology system, and then complete the integration and processing of data and data, and then realize the real-time monitoring of data security. In particular, due to the use of big data processing technology, it can deal with the data base problem and the situation of a large amount of heterogeneous data, and then improve and improve the security capability of the system, so that the user experience and experience are better. In addition, by setting a security password, security protection can be performed for data messages, log messages, and the like. Practice has proved that when using a security password, the possibility of data being stolen is higher than that of unset data. Therefore, the relevant departments must distinguish the passwords according to certain methods. Therefore, relevant staff need to classify passwords according to specific patterns. Because it not only guarantees the independence of the account, but also avoids the situation of data theft caused by password leakage to the greatest extent. At the same time, relevant personnel can allow users to use big data backup technology to perform regular data backup. In addition, users can upload important data on the cloud platform, and can extract data from the cloud platform when in use.

6.3 Establish and Improve Network Security Proxy Server

The network security proxy server is the key to ensure the security of network information, and it is directly related to the quality of computer data processing and transmission. Although there are some differences between the two, from the perspective of maintaining network information, it has the same function and role as the host server. In the transmission of data information between the extranet and the intranet, as a transfer station, it can enable all data and information between the extranet and the intranet to be exchanged by the proxy server and the extranet. After receiving the information of the extranet, the proxy server will process the information of the extranet and send it to the intranet, thereby effectively preventing harmful data from flowing into the intranet, thereby ensuring the security of network information. By using the network security Agent server, data can be effectively filtered and processed, effectively preventing viruses and Trojan software from threatening the security of network information, preventing hackers from invading, destroying and stealing data, and ensuring the security of data information. In addition, on the network security proxy server, security protection systems such as firewalls and alarm systems should also be established to add a layer of security barriers to protect network information and reduce network security risks.

6.4 Increase the Ability of Relevant Units to Protect the Network Environment

Due to the open nature of the Internet, a large number of traps and viruses lurk in various network platforms and various types of information, making netizens easily infected when using the Internet. Therefore, relevant departments should strengthen the investigation of various types of illegal software and illegal websites on the Internet, and speed up the research and development and innovation of search technology. It is necessary to quickly check and block all kinds of illegal software and networks lurking in the network, so as to reduce the information security risks caused by network users being attacked by viruses or hackers. For example: search for illegal activities in various websites and software, and stop them immediately if any illegal activities are found. For example, the current signature analysis technology can quickly and accurately detect illegal intrusions, thereby purifying the network environment, allowing users to search in a cleaner and more stable network environment, thereby protecting everyone's personal information.

6.5   Efficient Use of Protection Technology

The security protection of computer network information must not only be carried out from the legal level, but also effectively use technology, that is, a technology of network information security, and its function is to protect the security of the entire computer network. In daily use, computer network engineers must carry out antivirus on the entire computer system, not only to avoid virus intrusion, but also to eliminate viruses more effectively, such as 360 security guards, such as Master Lu, such as Tencent Butler, these are Specially designed for ordinary Internet users, users can protect their own computers through a simple system, thereby protecting their own network information.

6.6   Strengthen the Prevention of Hackers and Viruses

In the big data environment, computer security protection is very necessary, and optimizing it can improve the security of the network. First of all, we must actively deal with a large number of viruses on the Internet, strengthen prevention, strengthen prevention, regularly use anti-virus software to kill viruses, solve computer security problems in a timely manner, and establish a firewall to prevent viruses. Secondly, because the hacker's attack is relatively hidden, if there is an attack, it will cause great damage, so the defense measures should be strengthened to imitate the hacker's behavior and improve the hacker's response ability. In addition to setting up a firewall, the internal and external networks are also isolated, reducing the probability of intrusion. Finally, use digital identity verification technology to effectively control access data, prevent illegal access, and prevent hackers from the root.

7   Conclusion

To sum up, in the era of big data, the security of computer networks has become an urgent problem that needs to be solved. In the face of huge amounts of data and information, computer network systems are easily affected by various aspects, which in turn leads to the failure of Internet information systems. Security issues. The advent of the era of big data processing has greatly promoted the application of computer information management science and technology, and at the same time improved the data processing capabilities. At present, in the field of information processing, there are also problems in information security, talent shortage, and technology optimization. In the new era, the information security of the computer system is threatened by viruses, hackers, etc., and its intrusion methods are becoming more and more complicated. For data monitoring, establish a sound network security proxy server, and reasonably install firewalls and anti-virus software to ensure the security of network information.

Acknowledgments. This work is supported by the Characteristic Innovation Project of Guangdong Universities under no.2022WTSCX133.

References

1. Ma, Q.: Analysis of computer information processing technology in the era of big data. Electroacoust. Technol. **42**(10) (2018)
2. Han, Y.: Analysis of computer information processing technology in the era of big data. Sci. Technol. Commun. **10**(18), 040 (2018)
3. Li, B.: Research on big data and computer information security technology. Inf. Comput. (Theor. Ed.) **18**, 185–186 (2018)
4. Li, S.: Analysis of information security processing technology based on computer big data. Inf. Comput. (Theor. Ed.) **32**(16), 237–239 (2020)
5. Liu, F., Cao, J.: Analysis of information security processing technology based on computer big data. Comput. Knowl. Technol. **16**(01), 21–22 (2020). https://doi.org/10.14004/j.cnki.ckt.2020.0009
6. Chen, B.: Research on computer information security processing technology under the background of big data. South Agric. Mach. **50**(06), 169 (2019)

Intelligent Application of Computer

Research on the Quality Evaluation and Optimization of Ideological and Political Education in Universities Driven by Artificial Intelligence

Yuanbing Wang and Jun Zhang(⊠)

Guangdong University of Science and Technology, Dongguan 523083, China
315724104@qq.com

Abstract. For a long time, how to track the teaching situation of ideological and political course teachers, accurately evaluate the degree of students' mastery and understanding of different knowledge points in the teaching process, and horizontally compare and evaluate the effectiveness of different teachers teaching the same course and knowledge point has been one of the key issues in ideological and political education work in universities. Currently, China has entered a digital society, and artificial intelligence has had a profound impact on the development of education work. Starting from the perspective of artificial intelligence technology, this article explores how ideological and political education in universities can improve and optimize existing quality evaluation methods, thereby improving the overall level of education work.

Keywords: artificial intelligence · Universities · Ideological and political education · Quality assessment

1 The Significance of Applying Artificial Intelligence to the Quality Evaluation of Ideological and Political Education in Universities

Artificial intelligence is a product of the fourth information technology revolution, referring to the intelligence created by humans, also known as "machine intelligence". Artificial intelligence technology has exceptional information retrieval capabilities, text reading and writing abilities, and natural language processing capabilities, all of which are beyond the reach of human intelligence (see Fig. 1).

The evaluation of the quality of university classroom teaching is based on the teaching of teachers and the learning of students, with a focus on improving teachers' educational and teaching abilities, improving teaching quality, and evaluating teaching design, process, and results. Traditional university classroom teaching quality evaluation mainly relies on on-site observation, student evaluation, and peer evaluation to evaluate teacher teaching. With the empowerment of artificial intelligence in university classroom teaching quality evaluation, the evaluation of university classroom teaching quality has begun to show diversification of evaluation subjects, restoration of authenticity in evaluation

K. Li and Y. Liu (Eds.): ISICA 2023, CCIS 2147, pp. 309–318, 2024.
https://doi.org/10.1007/978-981-97-4396-4_29

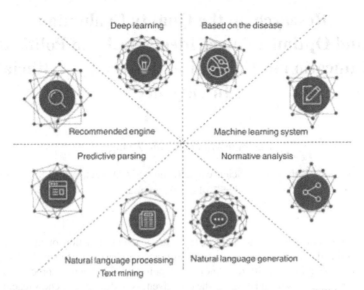

Fig. 1. Analysis diagram of ai key technologies and core capabilities

methods, and more attention to the characteristics of teachers and students in evaluation results. Therefore, the application of artificial intelligence in the evaluation of the quality of ideological and political education in universities has important value, mainly reflected in the following aspects:

1.1 Assist in Data Collection for Teaching Evaluation

Artificial intelligence can use artificial intelligence technology to collect multi-dimensional data such as students' academic and psychological data, enabling the mastery of students' multifaceted data and obtaining the most authentic feedback from multiple sets of data. In particular, the establishment of artificial intelligence algorithms can integrate modules such as intelligent perception, intelligent algorithms, and data decision-making into an educational diagnosis, evaluation, and intervention function system, utilizing artificial intelligence to obtain facial data, voice data, interaction data, and human form data. In summary, the data obtained by artificial intelligence can overcome the subjectivity and narrow focus of human intelligence, making the data more objective and authentic, providing effective reference for teaching evaluation [1].

1.2 Helps to Strengthen In-Depth Analysis of Students' Learning Situation

Artificial intelligence integrates multiple technologies such as natural language understanding and knowledge reasoning, and can participate in classroom tests, homework assignments, etc. By comparing and analyzing students' learning situations, accurate diagnosis can be made, which can reduce the mechanical workload of teachers and give them more creative teaching space. In the context of ideological and political classroom teaching, combined with artificial intelligence technology, a classroom intelligent learning situation analysis system was built based on the intelligent recognition of student behavior. By intelligently identifying student classroom behavior, a systematic quantitative evaluation of student classroom participation status was achieved, achieving the goal of effective learning situation analysis for classroom teaching [2].

1.3 Helps Optimize Learning Paths

The knowledge graph construction technology of artificial intelligence can provide personalized learning path planning for individual students, diagnose learning situations based on process, and provide precise and personalized learning solutions. Through the application of artificial intelligence technology, it is possible to flexibly record classroom process data such as the number of times students raise their hands to speak, the time they raise their heads to listen to classes, their participation in discussions, and their concentration, and provide timely feedback to students. This facilitates students to intuitively understand their learning status, adjust and improve in a timely manner, and conduct formative evaluations. At the same time, relying on large storage space and high-speed computing capabilities, periodic classroom performance data can be generated, To track the growth trajectory of teachers' professional development and students' comprehensive development (see Fig. 2).

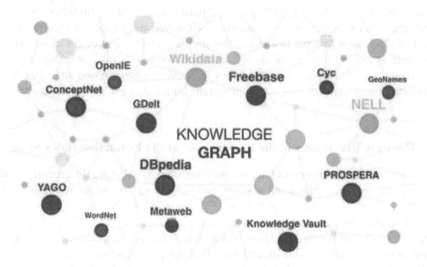

Fig. 2. Artificial intelligence knowledge chart

2 The Problems in the Quality Evaluation of Ideological and Political Education in Universities at Present

Ideological and political education in universities plays an important role in cultivating excellent talents, and the construction of evaluation systems affects the effectiveness of ideological and political education. However, considering the actual situation, there are significant problems in the quality evaluation of ideological and political education in universities, mainly reflected in the following aspects:

2.1 Lack of Awareness of the Importance of Quality Evaluation

There are two main purposes for evaluating the quality of ideological and political education in universities. One is to understand students' mastery of ideological and political theories, and the other is to evaluate the impact of these scientific theories on the formation and development of students' personal moral character and ideological quality. These two goals interact and influence each other, jointly forming the elements for measuring and evaluating the quality of ideological and political education in universities. At present, many universities often prioritize the theoretical knowledge mastered by students in the process of evaluating the quality of ideological and political education, while the evaluation of the impact on students' personal growth and progress is secondary and even completely ignored. The quality evaluation of ideological and political education shows a trend of "pan knowledge-based", which clearly seriously deviates from the essential laws of ideological and political education, and the evaluation results are naturally not scientific Accurately reflect the current situation and overall situation of ideological and political education. The reason for this is that there are problems with the orientation of the evaluation system construction itself, equating the quality assessment and evaluation of ideological and political education with other disciplines, showing a serious "simplification" tendency. However, the impact of theoretical education of ideological and political education on the improvement of students' personal moral quality has not been given enough attention, which itself is a behavior of putting the cart before the horse, It reflects that many universities have a biased understanding of the importance of building an evaluation system for ideological and political education [3].

2.2 There are Deficiencies in the Construction of the Evaluation Index System

Ideological and political courses have unique characteristics that are different from other courses, and are more general and abstract, making it more difficult to establish a scientifically sound evaluation system. Therefore, it is necessary to refine these abstract and conceptual contents based on the essential laws and characteristics of ideological and political courses, and transform them into a series of intuitive, concrete, and operable indicator systems that meet the essential requirements of assessment and evaluation. In addition, these assessment and evaluation systems must also be highly close to the teaching reality and the actual needs of teachers and students. However, based on the specific situation of universities that have already carried out quality assessment and evaluation of ideological and political education at this stage, the vast majority of universities have a homogenization problem in the construction of evaluation index systems. That is, the evaluation index system mainly focuses on teaching attitude, teaching effectiveness, teaching methods, teaching content, and other aspects, without fully highlighting the particularity of ideological and political education [4].

2.3 The Evaluation Method is Relatively Single

At present, many universities adopt methods such as expert course evaluation, student discussions, and filling out survey questionnaires to assess the quality of ideological and political education. This approach has strong subjectivity and contingency, making it difficult to truly reflect the real situation and prominent problems of ideological and political education in universities. Due to the unique understanding and understanding of the evaluation indicator system among the participants in the evaluation, it is difficult to achieve a high degree of uniformity in the evaluation and assessment standards, which leads to significant differences in the evaluation results. Even when facing the same evaluation indicator, students with the same major, background, or even living school conditions often give significantly different evaluation results, which clearly violates the principle of objectivity in evaluation.

2.4 The Application Efficiency of Evaluation Results is Relatively Low

Firstly, the professional abilities of the participants in the evaluation and assessment may not fully meet the standards and requirements, and there is a phenomenon of non-standard assessment and evaluation processes. Secondly, due to the correlation between the assessment and evaluation results and the promotion of ideological and political course teachers' professional titles and performance work benefits, some universities still have the problem of giving a "favor score" in the process of conducting assessment and evaluation, and the assessment and evaluation results are not fair and accurate enough. Finally, the assessment and evaluation results of ideological and political courses in some universities are only provided as feedback to ideological and political course teachers, and there are no strict requirements on how to solve the existing related problems. The binding force of the assessment and evaluation results is insufficient, and they cannot truly become the driving force for promoting the reform of ideological and political course education [5].

3 Optimization Measures for the Quality Evaluation of Ideological and Political Education in Universities Driven by Artificial Intelligence

With the advancement of digital education, artificial intelligence and big data technology have been flexibly applied in ideological and political education in universities, demonstrating strong vitality and creativity, and also emitting new vitality in quality evaluation. From the above analysis, it can be seen that there are still many problems in the quality evaluation of ideological and political education in universities. In order to improve this issue, the driving role of artificial intelligence should be fully utilized to further optimize and improve the quality evaluation of ideological and political education in universities:

3.1 Human-Computer Collaboration for Personalized Evaluation

Personalized evaluation is a teaching evaluation model designed to address the uniqueness and differences of students. In this model, individual students are the center of evaluation, and personalized evaluation can be conducted for individual students, providing targeted evaluation results and measures. In the evaluation of the quality of ideological and political education classroom teaching in universities, the power of teaching evaluation cannot be directly transferred to artificial intelligence technology. Artificial intelligence technology can only be the auxiliary of teaching evaluation, and teachers are the main body and core force of teaching evaluation. In addition to exploring and discovering the uniqueness and creativity of students in the teaching process, teachers also need to conduct human evaluation based on the immeasurable abilities of students, so as to intervene, Provide personalized evaluations for students to promote their comprehensive ideological and political development.

In this process, it is necessary to advocate for teachers not to lose their subjectivity. Teachers and teaching evaluators can explore the effectiveness of teaching evaluation through qualitative evaluation methods such as natural observation, interview surveys, and descriptive analysis. While machines provide data, qualitative methods are used to participate in teaching evaluation, constructing a teaching evaluation system that integrates qualitative and quantitative evaluation, and promoting collaborative evaluation between teachers and artificial intelligence technology, Putting people first and using machines as a supplement, let teaching evaluation fall on every student.

3.2 Multiple Evaluations to Ensure Evaluation Effectiveness

Multi evaluation includes not only the diversity of teaching evaluation subjects, but also the diversity of teaching evaluation tools and teaching evaluation standards.Multiple evaluation can better guarantee the fluency and completeness of teaching evaluation, but also has an important guiding role in promoting the comprehensive development of students and promoting the work of teaching evaluation.

Firstly, the subject of teaching evaluation should be diversified. In the process of teaching evaluation in the era of artificial intelligence, we cannot simply let machines and teachers become the subjects of teaching evaluation, nor can we forget the application of traditional teaching evaluation. We should involve parents, teaching supervisors, managers, and peers in teaching evaluation, and evaluate ideological and political education and related course teaching activities from multiple perspectives and aspects under the joint efforts of multiple subjects, grasping students' values The development of emotional cognition and psychological state.

Secondly, the diversification of teaching evaluation tools. Don't simply use grades as qualitative indicators for teaching evaluation. You can use data samples provided by machine data as reference for teaching evaluation. You can also use methods such as student evaluation, online/offline portfolio evaluation, etc. By comparing and analyzing students' ideological concepts, value cognition, and other aspects before and after, you can more intuitively see the effectiveness of teaching, participate in teaching evaluation from a multidimensional perspective, and comprehensively evaluate teaching effectiveness, Do a good job of evaluation and feedback, so as to further improve the quality of teaching.

Finally, the diversification of teaching evaluation standards. Artificial intelligence technology can easily lead teachers into the misconception of using exam scores in ideological and political courses as evaluation criteria. In the teaching process, more attention should be paid to various evaluation criteria such as student ability cultivation and thinking innovation. Different methods based on intelligent technology, such as test questions, practical operations, interview and defense, can be used for teaching evaluation.

3.3 Breaking Reason and Persisting in Establishing Virtue and Cultivating People

Firstly, in the process of applying artificial intelligence technology, it is necessary to break away from the technical tone of data being the king. Do not let the data results evaluated by machines become the core of teaching evaluation, learn to use data as a reference, rather than absolutely believing in the results given by data. Data should be regarded as an indicator of teaching evaluation, and each set of data should also be viewed with a skeptical attitude, Furthermore, by combining machine data with human data, we can make relevant value judgments and be careful not to be blinded by data to the purpose of teaching evaluation.

Secondly, to cultivate teachers' technical rationality and ethical awareness of technology, break free from tool dependence psychology, and reverse the tool rationality of technology being the king and efficiency being the main focus. In the process of using tools for teaching evaluation, the idea that tools and technology are not omnipotent should be established, and the meta evaluation of university teaching should be integrated into the evaluation of university classroom teaching quality. Only with a reasonable university teaching meta evaluation can a qualified teaching evaluation be obtained. Teaching meta evaluation refers to the evaluation of the teaching evaluation itself, and its main purpose is to test various errors that may appear in the teaching evaluation, and then make corrections and improvements, so that its positive role can be fully played.

Especially in the process of evaluating ideological and political education courses, we should adhere to the fundamental principle of cultivating morality and people, take students as our own responsibility, establish a general tone of development as the core, call for a return to value rationality, understand that "all mechanisms of machine civilization must be subject to human goals and needs", and involve multiple subjects in the process of teaching evaluation. We should integrate quantitative data evaluation and qualitative evaluation, In order to make the teaching evaluation results more authentic and reliable, do a good job in supervising the teaching evaluation, and promote the development of students' ideological and political education and the improvement of teachers' professional abilities in ideological and political education through teaching evaluation.

3.4 Cultivate Talents and Improve Evaluation Techniques

The advancement of artificial intelligence technology directly affects the development of related evaluation activities. In this process, it is necessary to establish the concept of lifelong education. Ideological and political teachers need to take the initiative to be the supervisor of teaching evaluation, maintain a learning attitude of keeping up with the times in the application of technology, and learn new knowledge, teaching methods, and evaluation methods in the new context. For example, in the context of artificial intelligence, teachers should learn to use Chat GPT The new technologies represented by deep learning and adaptive learning based on big data experience the changes brought by artificial intelligence technology to education, thus maintaining a youthful development trend in the education industry, adapting to new environments, and nurturing new vitality.

Secondly, universities should initiate corresponding teacher training. In the face of the lack of AI intelligence literacy among university teachers, courses on artificial intelligence technology tools can be conducted during winter and summer vacations, as well as pre and post job training on artificial intelligence skills. Throughout the process, teachers can master basic AI technology and better apply intelligent technology to the teaching evaluation system.

3.5 Innovative Carrier Platform for Ideological and Political Education in Universities

Innovative ideological and political education in universities cannot be separated from the innovation of carrier platforms. The carrier of ideological and political education in universities is an intermediary connecting educators and educational objects, as well as a platform that includes information on ideological and political education. It is the key to two-way interaction with educational subjects. This section is based on the flow direction of ideological and political education data, and sequentially builds IoT platforms, big data platforms, and virtual simulation ideological and political experience platforms to construct a carrier platform model for ideological and political education in universities,

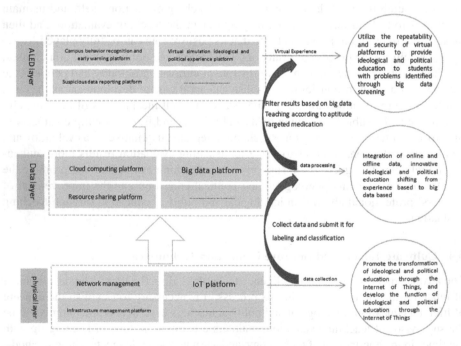

Fig. 3. Model diagram of the carrier platform for ideological and political education in universities

in order to promote innovation in ideological and political education in universities (Fig. 3).

One is to build an IoT platform for ideological and political education in universities. The architecture of the Internet of Things is mainly divided into three layers, among which the perception layer mainly includes the processor of the device, various sensors, embedded operating systems, etc., which plays the role of sensing surrounding environmental information, and then processing it by the processor to transmit the data out. The network layer mainly includes communication protocols used to support the Internet of Things, such as Zig Bee, Wi Fi protocol, etc., which are guarantees of data transmission security and quality. The application layer mainly includes end-to-end products, software development platforms, and other application systems. It is the existence of the "perception connection intelligence" framework in the Internet of Things that connects "people and things" and "things and things", thereby achieving monitoring, control, automation, and intelligence functions.

The second is to build a big data platform for ideological and political education in universities. Before constructing a big data platform for ideological and political education in universities, it is necessary to first clarify the classification of data. From the perspective of ideological and political education in universities, the data generated by educational entities can be divided into teacher data and student data. Teacher data includes courseware, lecture notes, papers, works, reports, etc., while student data includes one card consumption, class situation, homework situation, family information, exercise situation, etc. Most of these data are collected through IoT platforms built on the physical layer. To fully utilize the value of these data and transform them into effective information for decision-making, it is necessary to use big data technology for deep mining.

The third is to build a virtual simulation platform for ideological and political education in universities. The virtual simulation ideological and political experience platform can be seen as the "software" at the application layer of intelligent systems, which is inseparable from the infrastructure support of the physical layer IoT platform and the data support of the data layer big data platform. With a school wide Internet of Things and big data platform, personalized teaching can be developed based on the actual performance of college students to create the necessary virtual environment. Only by mastering the initiative and targeting of ideological and political education in universities can we tailor measures to local conditions and further promote the improvement of its effectiveness.

4 Conclusion

In summary, there are a series of problems in the implementation of ideological and political education and related courses in current universities, especially in the construction of evaluation systems. Due to the lag in the construction of evaluation index systems, there is a lack of comprehensive attention to the development of students' ideological and political abilities and emotional values, which has brought negative and negative impacts on the efficient implementation of ideological and political education work. In this regard, universities should improve their existing technological means, actively utilize the technological advantages of artificial intelligence, update and optimize the existing evaluation system of ideological and political education, in order to systematically follow up on the growth and development of every college student, and cultivate more outstanding talents.

References

1. Fan, C.: The value and path of artificial intelligence empowering ideological and political education. Ref. Middle Sch. Polit. Teach. **33**, 86–87 (2023)
2. Yin, L.: Research on the high quality development of artificial intelligence empowered ideological and political course teaching in universities. Heihe Acad. J. **05**, 68–73 (2022)
3. Zhao, L., Wu, W.: The practical application of big data technology in the quality evaluation of vocational education under artificial intelligence. Economist **09**, 211–212 (2022)
4. Gao, J.: Research and practice on the reform of education philosophy and education quality evaluation in the context of the development of new generation artificial intelligence. New Intell. **12**, 27–28 (2021)
5. Bian, H., Wang, X., Yang, Y.: The application of artificial intelligence technology in the quality evaluation of ideological and political course teaching. Drama House **16**, 114–116 (2020)

Research on the Routing Protocol Algorithm Driven by the Dedicated Frequency Points of the Internet of Things to Build a Network

Lingwei Wang and Hua Wang[✉]

College of Computer Science, Guangdong University of Science and Technology, Dongguan, China
460864372@qq.com

Abstract. In order to build a special networking model system, form a diversified network comprehensive model, improve the work quality of the structural environment, and balance the energy consumption status of various node environments. When the number of nodes increases and link failures are prone to occur, It is particularly necessary to effectively design a routing protocol algorithm model that optimizes the Internet of Things dedicated frequency points to build a network-driven algorithm. Therefore, under the basic mode of the existing structural system, it is necessary to form a model structure body based on the dedicated frequency points of the Internet of Things to build a network-driven model to ensure the effectiveness of the entire environment.

Keywords: Internet of Things · dedicated frequency point · ad hoc network

1 Introduction

The network-driven model structure of the dedicated frequency points of the Internet of Things mainly relies on the relevant content of multimedia printing to synthesize the existing data information and build a diversified node working mechanism. In this working mode, the technological reform and the upgrading of working methods are further accelerated. Under the mode of optimizing management of the existing technology, the comprehensive analysis of the objective function is accelerated to ensure that the actual energy consumption inside the ad hoc network is always maintained in a relatively stable situation. . The optimization and adjustment of the protocol algorithm implemented by the network router is mainly to speed up the repair and adjustment of the network-driven link fault structure problem of the dedicated frequency point of the Internet of Things under the mode structure of the minimum coverage. After repeated exploration and analysis, it can be seen that the network-driven algorithm structure of the optimized Internet of Things dedicated frequency points has a more intuitive and efficient work ability for the adjustment and forwarding of the basic link information of the router, which can improve the transmission effect of network data and can optimize The survival time of the node ultimately better guarantees the calculation and storage capacity of the data and the control and management capabilities of the data.

K. Li and Y. Liu (Eds.): ISICA 2023, CCIS 2147, pp. 319–325, 2024.
https://doi.org/10.1007/978-981-97-4396-4_30

2   The Basic Theory of the Routing Protocol Algorithm Driven by the Dedicated Frequency Point of the Internet of Things

The Internet of Things dedicated frequency point network driver itself is a network system derived from the development of modern industries set in the existing mobile communication and computer networks. This network access environment structure speeds up the analysis of diversified frequency point information. It has also received extensive attention from various groups, because the construction of the Internet of Things dedicated frequency point network is related to the transmission of information networks and the coordination of the network environment, and my country's research and analysis on the driving force of the Internet of Things dedicated frequency point network is also comprehensive. In order to speed up the research and analysis quality of the Internet of Things and ensure the competitiveness of the self-organizing network. The routing protocol mode is mainly to better speed up the allocation of resources and adjust the modern management mechanism [1]. Due to technical problems, its own work efficiency often directly affects the working status of the entire network environment. There are many problems between the traditional routing protocol algorithm structures due to technical limitations. For example, the routing protocol algorithm based on the sixth version of the Internet Protocol can speed up the realization of diversified analysis, judgment and repair of link failures under the structure of the dedicated frequency point ad hoc network of the Internet of Things under the existing working structure environment, but the actual The complexity of the algorithm operation mode is relatively large, and the volume of the operation itself is relatively large. Some routing protocol algorithms based on low-power self-adaptive clustering layered protocols cannot better deal with the influence from the circuit structure to the link structure of the dedicated frequency point of the Internet of Things in the actual environment. , which speeds up the analysis of regulation patterns, and also makes practical problems with relatively low resolution become the main factor restricting the development of industrial structure [2]. For this reason, on the basis of the original work, the professional technicians proposed a design pattern based on the dynamic routing information protocol. This design mode is mainly to better ensure the intelligent fault diagnosis and analysis of the link structure of the Internet of Things dedicated frequency point ad hoc network. The multi-node content can speed up the accurate fault location analysis in the node environment, and at the same time, the accuracy is optimized while the specific work content is realized. However, the calculation process of this method is relatively complicated, and the possibility of implementation is low. It is easy to cause the collapse of the network environment due to improper handling, and it is difficult to advance in the work itself.

Under the comprehensive model of the network structure system, in order to speed up the technical individuality, it is necessary to take the network link structure as the main content and implement the work based on multimedia printed books. In the process of specific business execution, the node operation mechanism should be selected in accordance with the basic form of the work characteristics of the dedicated frequency point ad hoc network of the Internet of Things, and the existing management mode and work system should be integrated to form the transformation of the management and operation system of the Internet of Things and strengthen the Internet of Things. The transmission effectiveness of the ad hoc network structure of dedicated frequency points for networking. In the normal network environment, the network routing is modified and optimized, so as to strengthen the optimization and adjustment of the protocol, and accelerate the data control of network transmission on the basis of realizing the upgrade of the dedicated frequency point ad hoc network of the Internet of Things [3].

3 Optimizing Algorithm for the Driving Protocol of the Internet of Things Dedicated Frequency Ad Hoc Network

3.1 The Structure Design of the Dedicated Frequency Point Ad Hoc Network Model for the Internet of Things

In order to ensure the high-quality operation of the dedicated frequency point ad hoc network for the Internet of Things, on the basis of the design and planning of the dedicated frequency point ad hoc network for the Internet of Things, it is mainly necessary to analyze the driving system for the dedicated operating system of the Internet of Things. The algorithm analysis is carried out below. Especially under the network model structure, the diversified management operation system and driving mode are formed. Especially in the structural state of the sensing network system, the configuration and adjustment of diversified sensing devices should be comprehensively integrated, and the characteristics of the diversified network structure and internal management system monitoring and analysis and autonomous coordination management information processing, etc. should be formed when the network is interrupted. All information and data in the environment are transmitted to the collection location, and the terminal equipment is used to transmit the data information to the remote control operation platform, and finally transmitted to the user through the construction and decomposition of the platform [4]. The basic structure of the dedicated frequency point ad hoc network of the Internet of Things we mentioned currently covers the node link and the communication cycle matrix. In the process of structural design, through the use of basic network coverage devices, the content that appears in the most original topology is integrated, so as to fully adjust the information network transformation model. Speed up the comprehensive analysis of the sensor network system, and in the state of the network path, speed up the automatic detection, analysis, judgment and management, and speed up the repeated detection of the link structure. The link analysis management involved in the Internet of Things dedicated frequency point self-organizing network is mainly constructed under the structure of different algorithms to make a good judgment and observation of differences and form a comprehensive coordination of the communication cycle during the structural form analysis process.

3.2 Algorithm Formula

Frequency Point Allocation Strategy Formula
Suppose that there are N IoT devices, and each device needs M frequency points. Let F be the total resource set of frequency points, and f_i be the set of frequency points assigned to by the i th device. The goal of the frequency point allocation strategy is to achieve the efficient utilization of frequency points and avoid conflicts.

Frequency Point allocation policy formula:

$$f_i = \text{AllocateFrequencies}(F, N, M) \tag{1}$$

AllocateFrequencies is a function that determines how to assign M frequency points to N devices based on the values of F, N, and M.

Route Selection Mechanism Formula
Suppose that there are P possible paths from the source node D to the target node D, and the cost of each path pj is cost (p_j).
Route selection mechanism formula:

$$\text{Pbest} = \text{SelectBestPath}(\{P1, P2, \ldots, Pp\}) \tag{2}$$

SelectBestPath Is a function that selects the optimal path Pbest according to the cost cost (p_j) of each path.

Path Cost Calculation Formula
The path cost may include multiple factors, such as energy consumption, communication delay, and path stability. Path-cost formula:

$$\text{cost}\left(p_j\right) = \alpha \cdot \text{EnergyCost}\left(p_j\right) + \beta \cdot \text{DelayCost}\left(p_j\right) + \gamma \cdot \text{StabilityCost}\left(p_j\right) \tag{3}$$

Among them, α, β and γ are the weighting factors used to adjust the proportion of each factor in the total cost.

4   Experimental Operation Analysis

4.1   Experimental Analysis

In order to more comprehensively analyze the communication transmission signal status and link forwarding operation status, a comparative experiment was conducted. The following is a comparative experimental design.

By comparing the performance of the IoT dedicated frequency ad hoc network with traditional network structures in terms of communication transmission and link forwarding, the advantages and limitations of the IoT dedicated frequency ad hoc network are evaluated. The experimental steps are as follows:

(1) Construct an experimental network based on the dedicated frequency ad hoc network of the Internet of Things to ensure its stable operation. Build a comparison network with a similar structure to a traditional network to ensure comparability in hardware and software configurations.
(2) Conduct data transmission tests in the two networks respectively to record data such as transmission speed and transmission stability. Compare the performance differences between the two network structures in data transmission.

(3) Simulate different link forwarding scenarios in the two networks, such as simultaneous forwarding of large amounts of data, burst data transmission, etc. Record data such as link forwarding delay and success rate, and evaluate the performance of the two network structures in link forwarding.

(4) Combined with the main body of the network model structure built previously, the algorithm analysis and judgment of the two network structures were carried out. Evaluate the performance of the IoT dedicated frequency ad hoc network routing protocol algorithm in complex network environments.

(5) Organize and analyze the collected experimental data to form a comparative report. Demonstrate the performance differences between the two network structures in terms of communication transmission and link forwarding, as well as the advantages and limitations of ad-hoc networks with dedicated frequency points for the Internet of Things.

(6) Through comparative experiments, it can be expected that the dedicated frequency point ad hoc network for the Internet of Things will have certain advantages in communication transmission speed and stability, link forwarding efficiency, etc. At the same time, we can also find the possible limitations of the IoT dedicated frequency ad hoc network in some aspects, providing a basis for further improvement and optimization.

In short, comparative experiments can more comprehensively evaluate the performance differences between the IoT dedicated frequency ad hoc network and traditional network structures in communication transmission and link forwarding, providing useful reference for practical applications and development.

4.2 Node Lifetime Status

By closing the gaps in the existing Internet of Things dedicated frequency point ad hoc network routing protocol algorithm, we can accurately obtain the survival status of network nodes in the Internet of Things dedicated frequency point ad hoc network network nodes, and judge the relationship between the number of dead nodes and the time. Inner relationship. In the case of one dead node, the follow-up response is about 410 rounds. Through the analysis and observation of establishing a clear connection with different algorithms, it can be seen that the main reason is that the routing protocol algorithm of the Internet of Things dedicated frequency point ad hoc network The node itself has a comprehensively selected node operation management mode. Under the system of energy cost analysis, the energy information fed back by different nodes in the network structure can be optimized in a balanced working mode. The problem of reducing energy consumption will also be evenly distributed under different boundary information systems, effectively speeding up the optimization and transformation of survival time nodes. However, the fluctuations in the other two main methods are relatively obvious, and the consumption of the entire working node of the Internet of Things is relatively large, and the survival probability of the node itself will be relatively low.

4.3 Network Life Analysis

By analyzing and comparing the working status of the routing protocol algorithm of the dedicated frequency point ad hoc network of the Internet of Things, it can be known that under the original algorithm structure, because there is no turning node environment analysis and no adjustment and upgrading of intelligent working methods, the network life is actually Relatively low, it is difficult to meet the working basic needs of modern web services. However, the routing protocol algorithm of the dedicated frequency point ad hoc network for the Internet of Things we are currently mentioning can ensure that the working modes of different nodes are optimized in a short time environment, reducing the occurrence of dead nodes, and the network life is gradually reduced from 425 The number of rounds has been reduced to 230 rounds. This reduction is not only a change in the technical characteristics, but also shows that under the proposal of this algorithm, the work efficiency of the network has been upgraded in an all-round way, and the work mode has been systematically changed. Transformation, work level improved significantly.

4.4 Performance Analysis

Using the existing network software system, the simulation system under the topology structure of the network work is deepened to explore the multiple aspects of the simulation network, and the earphone operates the network structure environment, and then conducts an all-round and diversified analysis of the performance operation effect under different algorithm environments. In the process of analyzing different algorithm operations, judge the working status of data arrival rate and routing overhead performance. Synthesize the current network management mode in the ad hoc network topology of the dedicated frequency point of the Internet of Things, analyze the working status of the modern information network structure, set the topology working mechanism and working system mode, and accelerate the formation of diversified working status and the inherent relationship between nodes Link changes have an intuitive impact. Compared with other working algorithms, the data working status and response speed reflected by the routing protocol algorithm of the Internet of Things dedicated frequency point ad hoc network proposed by us are relatively more, and the speed-up status is about 60%, and the final working effectiveness Also relatively high. By analyzing different algorithm structures, it can be known that there is no precise algorithm presentation method that can reach the working state of the Internet of Things dedicated frequency point ad-hoc network routing protocol algorithm, and the difference rate reaches about 31%. As for the internal connection and change state between the control quantity nodes, there are also obvious differences between the actual control quantity of the two parties under different reaction states of the working nodes. However, the control quantity of the Internet of Things dedicated frequency point ad hoc network routing protocol algorithm is relatively efficient, which is far higher than other different algorithm modes. This also shows that under the working mechanism of this algorithm, the overhead of IoT dedicated frequency point ad hoc network routing is relatively efficient. The main reason for this situation is that under the existing working mechanism, due to the influence of algorithm conditions and network environment, the energy consumption inside the dedicated frequency point ad hoc network of the Internet of Things has always been maintained in a relatively balanced development mode. In this way, the occurrence of the loss problem is reduced. At the same time, the operation mode of the existing network routing protocol algorithm has been transformed and upgraded, and the internal response state and working mode can be analyzed and repaired in a timely manner. Circuit faults of various links within the network to ensure the effect of circuit operation. Not only that, the link response state of the Internet of Things dedicated frequency point ad hoc network is relatively good, and the actual operating performance in a specific working environment is far superior to the traditional operating method.

5 Conclusion

Through the refined research, analysis and judgment of the current Internet of Things dedicated frequency point network-driven routing protocol algorithm, a complete set of link green structure is constructed to repair the current internal network structure model of the Internet of Things dedicated frequency point network. Under the coordination of the operating environment of this model and the development management mechanism, the upgrading and transformation of the industrial working mechanism will be continuously accelerated, and the adjustment and optimization of the management method will be accelerated to ensure a better working model and new working system. Under different working mechanisms of nodes, it is possible to analyze the actual consumption management status of the energy driven by the networking at dedicated frequency points for the Internet of Things, to better speed up the adjustment and upgrade of the environmental operation mode, to speed up repair and adjustment, and to strengthen the network construction at dedicated frequency points for the Internet of Things. In the driving channel system, the data transmission operation quality of the main body of the network structure improves the survival time in the network node environment and the actual quantity data of network control management, and guarantees.

References

1. Duan, Y.-Y., Chen, G.-F.: An energy-efficient clustering routing protocol based on LEACH. J. Changchun Univ. Sci. Technol. (Nat. Sci. Ed.) (3) (2018)
2. Luo, D.-J.. Application of routing algorithm in energy consumption analysis of balanced Internet of Things sensor nodes. Sci. Technol. Eng. (25) (2018)
3. Xie, L., Zhou, W.-B., Wang, Y.: Multi-path routing protocol for mobile ad hoc networks based on optimal path strategy. Comput. Eng. Des. (7) (2017)
4. Xiong, Q.-L., Qiu, Q.-M., Wang, Y.: Design of adaptive frequency hopping in mobile ad hoc networks. Aerosp. Electron. Technol. (4) (2017)

Artificial Intelligence in Intelligent Clothing: Design and Implementation

Ping Wang and Xuming Zhang[✉]

Guangdong University of Science and Technology, Dongguan 523083, China
zhangxuming@pukyong.ac.kr

Abstract. In the context of the rapid development of global fashion technology, intelligent clothing has demonstrated significant potential in areas such as fashion design, smart sensing, and health monitoring. However, practical applications of intelligent clothing encounter numerous challenges, including the design and optimization of artificial intelligence algorithms, the selection and integration of smart materials and sensors, and issues related to comfort and wearability. Addressing these challenges, this paper begins with the concept of intelligent clothing and employs a combination of literature review and case analysis to explore the application and implementation of artificial intelligence in intelligent clothing design. By examining the implementation methods of smart technology from various perspectives, the study aims to highlight the advantages of artificial intelligence in intelligent clothing design and to master technical operation methods to foster industry development and progress. The findings reveal that the proposed approach significantly enhances the intelligence and practicality of intelligent clothing, offering innovative design ideas and practical guidance for the field of intelligent garments.

Keywords: Artificial intelligence · Intelligent clothing · Design focus · Implementation pathway

1 Introduction

Amidst the rapid global advancement of digitalization and artificial intelligence technologies, China's fashion industry is significantly transforming from traditional textile techniques to smart clothing. Particularly since the introduction of the "New Generation Artificial Intelligence Development Plan" in 2017, intelligent clothing has emerged as a burgeoning and hot field within fashion technology, as shown in Table 1. It has demonstrated immense potential in areas such as fashion design, the Internet of Things, and smart living, marking a new direction for China's traditionally technology-reliant garment industry.

Scholars widely regard intelligent clothing as an innovative product that integrates artificial intelligence. Existing research has advanced the field of intelligent clothing, particularly in smart materials, sensors, and systems. However, there is a notable gap in the exploration of the integration of intelligent clothing design with artificial intelligence, especially in terms of in-depth discussions on the design and implementation of intelligent garments. While some scholars have recognized the proliferation and development of intelligent clothing technology, a comprehensive exploration of its design in conjunction with relevant technologies and algorithms is lacking. In response, this paper, through literature review and case analysis, commits to an in-depth study of the design and implementation of intelligent clothing. It aims to answer key questions: What are the concept and classification of Intelligent Clothing? What are the critical issues in integrating intelligent clothing design with artificial intelligence? What are the implementation technologies for intelligent clothing? Adopting methods of literature review and empirical research, this study delves into the latest trends and methods in the design and implementation of intelligent clothing. The innovation of this research lies in the deep integration of intelligent clothing design with artificial intelligence technology, offering a more intelligent and personalized approach to garment design and implementation. This has significant value and implications for the advancement of the fashion industry and the promotion of smart living.

Table 1. Annual Publication Trends in the Topic of Intelligent Clothing

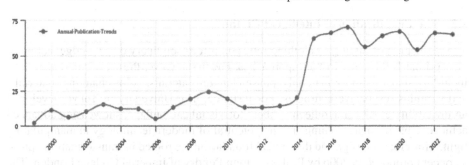

2   Concept and Classification of Intelligent Clothing

2.1   The Concept of Intelligent Clothing

Intelligent clothing represents a groundbreaking fusion of technology and textiles, embodying a life-like system with capabilities for both perception and response. This innovative form of apparel is designed to meet the evolving needs of consumers, harmonizing a variety of perspectives with the latest in technological innovation. Distinct from traditional clothing, intelligent garments are equipped to actively respond to external environmental stimuli. They are adept at adjusting their functionality based on the physiological and psychological states of the wearer, offering a personalized experience [1].

Bridging the gap between concept and reality, the integration of advanced technologies in intelligent clothing allows for a more nuanced and responsive interaction with the user. This adaptability makes it particularly appealing to a wide range of consumers, each with their own unique requirements and preferences. As a testament to its growing relevance, in the modern world, intelligent garments are becoming an increasingly common aspect of our daily lives. They are no longer just concepts but real, tangible products that enhance our daily experiences. For instance, smart shirts are now capable of monitoring vital signs such as heart rate, body temperature, and respiratory rate, providing valuable health insights. Jackets equipped with embedded technology can automatically play music based on the wearer's mood or environment, while T-shirts with integrated displays can show text and images, reflecting the wearer's preferences or mood. This seamless integration of technology and fashion marks the rapid advancement of big data, artificial intelligence, and interactive technology as a catalyst in the transformation of intelligent clothing. These technological strides have not only revolutionized the way intelligent clothing functions but have also had a profound impact on our lifestyle and the way we interact with our environment. Signifying a major paradigm shift, this evolution in the traditional textile and fashion industry ushers in a new era where technology and fashion converge. Looking towards the future, intelligent clothing is set to redefine the boundaries of fashion, functionality, and personal expression, making it an integral part of our future daily lives [2].

2.2 The Classification of Intelligent Clothing

The realization of intelligent clothing relies on interdisciplinary cutting-edge technologies, primarily through two main approaches. The first involves the use of smart materials, including shape memory materials, phase change materials, color-changing materials, and stimulus-responsive hydrogels. For instance, The University of Tokyo developed an invisibility cloak using reflective fibers for visual camouflage, showcasing advanced achievements in smart clothing and the potential of modern technology in manipulating light. This cloak's design and theoretical foundation are rooted in transformation optics, a concept proposed in 2006 by Professor John Pendry of Imperial College London. The principle involves using coordinate transformation to expand a zero-volume point into a region with actual volume, concealing objects within. Since a zero-volume point doesn't affect light propagation, the transformed region allows light to bypass the object, rendering it invisible. The technique's key lies in precisely calculating specific refractive index parameters derived from coordinate transformations (see Fig. 1).

The second major category in intelligent clothing is the integration of information technology and microelectronics. Incorporating technologies like conductive materials, flexible sensors, wireless communication, and power sources into everyday garments has enabled the smart transformation of clothing. For example, robotic exoskeleton technology, which combines sensing, control, information processing, and mobile computing, offers a wearable mechanical structure for the wearer. This technology is primarily aimed at applications in elderly care, disability assistance, firefighting, and police work, among other high-risk professions, (see Fig. 2).

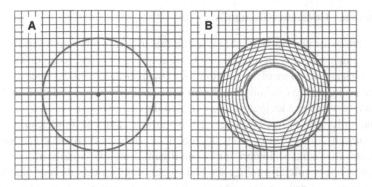

Fig. 1. Spatial Transformation: Converting a Point into a Region

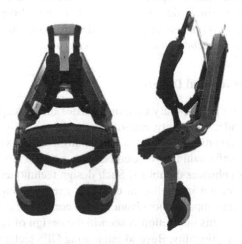

Fig. 2. Rendering of Wearable Booster Product

3 Critical Issues in Integrating Intelligent Clothing Design with Artificial Intelligence

The essence of intelligent clothing design is rooted in addressing the diverse needs and characteristics of individuals, thereby emphasizing the importance of personalized design. This approach integrates scientific technology and materials, merging the realms of clothing with cutting-edge science and technology. Intelligent devices are seamlessly incorporated into clothing designs to cater to this modern demand. In today's rapidly evolving social economy, where market competition across various industries is intensifying, distinctiveness becomes crucial for survival and success. The application of artificial intelligence (AI) in the realm of intelligent clothing design is a testament to this trend. AI's role in this field typically encompasses several key aspects, each contributing uniquely to the sophistication and functionality of intelligent garments [3].

3.1 Following Consumer Demands

The cornerstone of intelligent clothing design lies in its unwavering focus on customer needs, encompassing a deep understanding of consumer preferences and consumption patterns. This approach involves a meticulous consideration of these needs, integrating intelligent features into clothing to enhance user experience. For instance, in Northeast China's harsh winters, a temperature-controlled down jacket exemplifies this principle. It automatically adjusts heating levels in response to ambient temperatures, ensuring optimal warmth. This not only meets the practical needs for thermal protection but also demonstrates the seamless integration of intelligent design with daily life.

Incorporating artificial intelligence (AI) technology in this domain necessitates a keen awareness of consumer demands. This process can leverage advanced technological tools, supported by big data and cloud computing, to conduct comprehensive market analyses. By statistically analyzing current market trends and consumer preferences, designers can pinpoint critical design elements. This data-driven approach not only informs the design process but also ensures that the resulting intelligent clothing aligns closely with consumer expectations and requirements [4].

3.2 Emphasize Personalized Design

Personalized design in contemporary clothing marries the concepts of modern market trends, intelligence, and individuality, ensuring that clothing design not only aligns with the zeitgeist but also maximizes the utility of personalization. This approach is evident in the design of versatile soft clothing, where embedding stainless steel materials connected via micro connectors enhances durability. Such design techniques enable the seamless transmission of information in intelligent clothing, exemplifying the fusion of smart technology with personal apparel and advancing the concept of clothing intelligence.

A prime example of this innovation is seen in the design of rescue clothing, which demands heightened functionality. Beyond integrating GPS technology, these garments employ artificial intelligence to relay environmental data to the wearer through sensors. This capability allows for immediate hazard alerts, leveraging AI and related technologies for enhanced safety. Upon market introduction, such practical and technologically advanced designs are poised to cater to a broad public demand, reflecting a significant stride in intelligent clothing [5].

The personalized design of AI clothing aims to provide customized clothing solutions according to consumers' personal characteristics and needs. By combining artificial intelligence technology and design innovation, more efficient, more creative and more comfortable clothing design and service can be achieved. There are some aspects in the following several aspects specifically. The following Table 2 gives the specific information analysis of the intelligent clothing consumer survey.

3.3 Applying Artificial Intelligence Technology to the Drawing Stage of Design Drawings

In the dynamic market of smart clothing, designers often face the challenge of repeatedly modifying their designs, a process that is both time-consuming and labor-intensive,

Table 2. Consumer demand questionnaire in intelligent clothing design tables.

Personalized design aspects	Depict	example
Design of personal preferences and body data	According to consumers' personal preferences and body data, to generate clothing design schemes that meet personal characteristics and needs	The AI algorithm analyzes consumers' body scan data and preferences, and designs clothes that fit the body shape and style
Intelligent pattern generation	By learning a large number of pattern samples, unique clothing patterns are automatically generated	The AI design tools automatically generate innovative pattern designs according to the parameters and requirements entered by the designer
Intelligent color matching	Provide a personalized color scheme based on consumers' skin tone, preferences and trends	AI analyzes consumers' skin tones and preferences, and combines fashion trends to give appropriate color matching advice
Smart fitting and adjustment	Combined with augmented reality (AR) technology, it provides a virtual fitting experience and intelligently adjusts clothing size and details based on feedback	With AR technology, consumers can try on clothing in a virtual environment, and the AI adjusts the design according to the feedback to ensure the fit and satisfaction
Smart fabric and intelligent functions	Integrated intelligent fabric and intelligent functions, such as temperature regulation, exercise monitoring, health management, etc.	Smart clothing automatically adjusts temperature, monitors movement data, and provides health management functions

yet does not always guarantee successful outcomes. With the rapid advancement of technology, the integration of artificial intelligence (AI) into the design phase of clothing has emerged as a solution, offering significant convenience in the creation of design drafts [6].

The role of AI in fashion design is not to supplant human designers but to augment their creative process. Designing often involves repetitive tasks, and AI's application here aims to alleviate this burden. AI can take over routine tasks and generate preliminary designs based on the designer's initial concepts. These AI-generated drafts are not final products but serve as a source of creative inspiration, sparking new ideas for designers. By reducing the workload and enhancing efficiency, AI allows designers more time to focus on innovation and strategic planning. Consequently, a collaborative approach between designers and AI in drafting designs is becoming a trend. This synergy not only saves time but also fosters creative inspiration, offering advantages over traditional design methods.

3.4 Provide Convenience for Adjusting Specification Parameters

In the fashion industry, the roles of designers and plate makers are distinctly separate yet interconnected. Designers are responsible for creating the initial designs, while plate makers translate these designs into practical patterns, determining the correct proportions and sizes for clothing. This process lays the groundwork for producing sample garments. The integration of artificial intelligence (AI) in intelligent clothing design can significantly enhance the efficiency of developing these sample garment patterns.

Traditionally, plate makers rely on their experience to approximate and iteratively refine the placement and dimensions of various garment components. This method, while effective, can be time-consuming and prone to inaccuracies. AI, with its capability to analyze and compile extensive pattern-making data, offers a solution by providing precise predictive data for the pattern-making process. The self-learning nature of AI means that it continually improves in accuracy and efficiency, reducing the time required for pattern making [7].

Moreover, the application of AI in pattern making is not limited by professional boundaries. Designers can directly utilize AI tools in pattern making, creating a more seamless and cohesive design process. This not only streamlines the workflow but also allows designers to better leverage their creative skills. Consequently, the application of AI in the pattern-making process of intelligent garments holds the potential to significantly shorten the design-to-production timeline.

4 Implementation Technologies for Intelligent Clothing

4.1 Understanding Consumer Needs and Technological Trends in Intelligent Clothing Design

With the progress of science and technology and the improvement of people's living standards, consumers' demand for clothing is also constantly changing. In the intelligent clothing design, the consumer demand mainly includes the following aspects:

Comfort: consumers have high requirements for the comfort of clothing, especially in the intelligent clothing design, comfort is crucial. Consumers hope to get a better wearing experience and quality of life through smart clothing. Functional: Intelligent clothing design not only needs to meet the basic needs of consumers, but also needs to have more functionality. Consumers hope to get more intelligent services through intelligent clothing, such as intelligent temperature control, intelligent monitoring of health status, etc. Fashion: consumers also have high requirements for the fashion of clothing. In intelligent clothing design, designers need to pay attention to the appearance design and material choice of clothing on the premise of satisfying the functionality, so as to meet the fashion needs of consumers. Personalization: Consumers' personalized demand for clothing is also increasing. In intelligent clothing design, designers need to provide customized clothing design according to consumers 'personal preferences and needs to meet consumers' personalized needs [8].

In the intelligent clothing design, the technology trend mainly includes the following aspects: Intelligent technology: With the development of the Internet of Things, artificial intelligence and other technologies, the design of intelligent clothing will become more and more intelligent. For example, the Internet of Things technology can realize the connection between clothing and intelligent devices, and the intelligent wearing experience through artificial intelligence technology. Flexible electronic technology: The development of flexible electronic technology provides more possibilities for intelligent clothing design. By combining flexible electronic technology with clothing, more intelligent functions can be realized, such as intelligent monitoring, intelligent heating, etc. The following Table 3 the demand survey data analysis of smart clothing consumers for smart clothing is conducted.

Table 3. Consumer demand questionnaire in intelligent clothing design tables.

Consumer demand	Depict	Data statistics
Health control	Through intelligent clothing design, monitoring of physical fitness, managing fitness programs, etc.	70% of consumers said they need smart clothing designs to monitor their health
Convenience	Through intelligent clothing design, intelligent control, intelligent home interconnection and other functions can be realized to improve the convenience of life	85% of consumers said they need smart clothing design to improve life convenience
Fashion sex	Intelligent clothing design needs to conform to the fashion trend and meet the consumers' pursuit of fashion	60% consumers say the fashion of smart clothing design is important

Biotechnology: The development of biotechnology has also provided new ideas for intelligent clothing design. By combining biotechnology with clothing, a more personalized intelligent clothing design can be realized, such as automatically adjusting the comfort level of clothing according to the physical conditions of consumers. Environmental protection technology: With the improvement of environmental awareness, environmental protection technology will also become an important trend of intelligent clothing design. Designers will pay more attention to the use of environmentally friendly materials and environmental protection processes to achieve more environmentally friendly intelligent clothing design. Cross-border integration: In the future, intelligent clothing design will pay more attention to cross-border integration, and integrate technologies and creativity in different fields together, so as to achieve more innovative and diversified intelligent clothing design. For example, the design concepts and elements of art, technology, fashion and other fields are integrated into intelligent clothing to create more attractive and practical intelligent clothing.

Data-driven: With the advent of the era of big data, data will become an important driving force of intelligent clothing design. Designers will collect and analyze consumer data, understand consumer needs and behavior habits, so as to achieve more accurate personalized design and optimize the wearing experience. The development of Wearable devices: Wearables are an important part of smart clothing, and the development of wearable devices in the future will also promote the progress of smart clothing design. For example, wearable devices can achieve more accurate health monitoring, exercise management and other functions, as well as more convenient and intelligent services [9]. Development of high-performance chemical fibers and intelligent wearable technologies.

The advent of artificial intelligence (AI) has revolutionized this landscape. AI conducts comprehensive analyses based on vast datasets, processing user data and research information available on the internet through natural language processing (NLP). NLP, a facet of computer technology, enables machines to understand and interpret human language. Post-processing, machine learning algorithms are employed to establish causal relationships between events, forecast consumer demands, and keep abreast of technological progress. This objective approach to predicting consumer demand is invaluable for designers, aiding them in planning styles that resonate more effectively with the market. Moreover, staying updated with cutting-edge technological developments empowers designers to expand their creative horizons, fostering interdisciplinary integration and innovation (see Fig. 3).

Fig. 3. The application of artificial intelligence (AI) in the apparel industry

4.2 The Integration of Artificial Intelligence in Fashion Design: Enhancing Creativity and Efficiency

Fashion designers are pivotal in bridging the past and future of fashion, embodying the essence of creativity in the industry. They craft their design styles by synthesizing market research outcomes, brand image, personal artistic sensibilities, and an acute awareness of contemporary trends. Once a design concept is finalized, designers select appropriate surface materials considering aspects like shape, quality, price, color, and texture. Subsequently, pattern makers bring these designs to life, meticulously crafting samples that align with the envisioned sizes and placements.

Artificial intelligence (AI) plays a transformative role in this creative process. Leveraging its self-learning capabilities, AI integrates and analyzes a wealth of empirical data, enhancing the fluidity and efficiency of the design workflow. Computer vision technologies interpret the colors and shapes in designers' sketches, while machine learning algorithms refine these designs, offering novel perspectives. Additionally, AI, in conjunction with robotic technology, ensures precise assembly of garment sections, minimizing the need for post-error corrections.

4.3   The Multifaceted Nature of Fashion Design: Integrating Diverse Talents in the AI Era

Fashion design transcends mere aesthetic creation, encompassing production, management, marketing, and other critical domains. The flourishing of the fashion design industry hinges on the synergistic contributions of professionals across these varied fields. Particularly in the realm of intelligent clothing, there is a pressing need for not just designers but also for skilled individuals in production, management, and marketing who possess a deep understanding of design principles.

Universities play a crucial role in this landscape, tasked with nurturing a new generation of diverse talents. This involves a multifaceted educational approach, offering specialized training tailored to distinct career paths, thereby cultivating a versatile team of professionals. In the age of artificial intelligence, the formation of comprehensive teams is paramount. These teams, led by fashion designers and bolstered by AI technology, should integrate a wide array of talents from production, marketing, and management. Such interdisciplinary collaboration is pivotal for the dynamic and effective progression of the fashion design industry.

Therefore, in the pursuit of cultivating talents across various disciplines, a balanced emphasis on both theoretical knowledge and practical application is essential. This balance should extend to harmonizing design principles with artificial intelligence insights. By fostering an environment where knowledge is applied iteratively, we can cultivate not only innovative thinking and individuality but also enhance the collective strength and capabilities of the team.

4.4   Advancing Intelligent Manufacturing and Wearable Technology in China's Clothing Industry

The fusion of artificial intelligence and robotics in intelligent manufacturing is poised to redefine future production landscapes. In sectors like chemical manufacturing, intelligent systems have already demonstrated their prowess, enabling unmanned operations that significantly reduce error rates and enhance operational speed. This paradigm shift presents a vital opportunity for China's clothing design industry. By strategically integrating advanced intelligent equipment, the industry can transition towards a more efficient, intelligent manufacturing model specifically tailored for clothing production.

Moreover, the intersection of artificial intelligence and biotechnology is forging new frontiers in intelligent wearable devices, a domain with immense potential for the future of clothing design. These devices transcend traditional clothing functionality, offering integrated health monitoring features. They are capable of real-time tracking of physiological changes, early detection of health irregularities, and facilitating timely lifestyle adjustments to promote well-being. Therefore, it is imperative for China's intelligent clothing design sector to focus on this interdisciplinary synergy. By combining apparel design with electronic information technology, biotechnology, and networked communications, the sector can drive forward the research and development of cutting-edge intelligent wearable technologies.

5 Conclusions

Artificial intelligence (AI), a cornerstone of modern technological advancement, has significantly influenced various sectors of Chinese society, including production and manufacturing. In the realm of the clothing industry, AI's integration into design and production processes has been a recent yet impactful development. This integration has given rise to a new generation of intelligent clothing, characterized by enhanced functionality and innovative design. However, the success of these advancements hinges on their alignment with market trends and consumer preferences. To truly excel, it is crucial for the industry to apply AI and related technologies adaptively across different stages of intelligent clothing design. This approach will not only cater to personalized consumer needs but also elevate the overall standard of intelligent clothing design in China.

Acknowledgement. This paper is part of the phased research results of the exploration and practice of fashion design and engineering integration mode in applied undergraduate universities under the background of "14th Five-year Plan" planning project of Guangdong Higher Education Association (Item No.: 21 GYB 62).

References

1. Xu, Q., Zhou, J., Liu, Y.: Design and implementation of intelligent luminescence in daily clothing based on LED. J. Beijing Inst. Fashion (Nat. Sci. Ed.) **43**(01), 30–34+88 (2023)
2. Xin, Y.: Research on intelligent design technology of three-dimensional clothing three-dimensional cutting pattern. J. Nanchang Normal Univ. **43**(06), 52–56 (2022)
3. Jiang, J.: The integration mode and requirements of artificial intelligence and fashion design. Volkswagen Standard. **19**, 119–121 (2022)
4. Zhu, G., Zhou, X., Guo, J.: The current status of intelligent clothing standardization and the construction of a standard system. Cotton Textile Technol. **50**(07), 74–78 (2022)
5. Artificial Intelligence Technology Column in the Textile and Clothing Industry. Wool Textile Technol. **49**(04), 8 (2021)
6. Su, Y., Zhao, Q.: Modern fashion design from the perspective of intelligent textiles. Woolen Textile Technol. **48**(11), 102–106 (2020)
7. Yi, L.: The integration mode and requirements of artificial intelligence and fashion design. Woolen Textile Technol. **45**(10), 81–85 (2017)
8. Ma, Y.: Intelligent clothing design and application based on interactive technology. Dyeing Finish. Technol. **39**(09), 11–12+22 (2017)
9. Tian, M., Li, J.: Design patterns and development trends of intelligent clothing. J. Textile Sci. **35**(02), 109–115 (2014)

Edible Oil Price Forecasting: A Novel Approach with Group Temporal Convolutional Network and BetaAdaptiveAdam

Lei Yang[✉], Huade Li, Rui Xu, Zexin Xu, and Jiale Cao

School of Mathematics and Informatics, South of China Agricultural University,
Guangzhou 510642, China
yanglei_s@126.com

Abstract. Edible oil, a fundamental food commodity, plays a pivotal role in the economic progression of a nation. The precise forecasting of edible oil prices is of utmost significance to a broad spectrum of stakeholders, including investors, policymakers, and researchers. Over recent years, a myriad of factors, including international influences, have led to substantial price fluctuations and irregular cycles in edible oil markets. As a result, the task of achieving accurate and robust forecasts of edible oil prices has emerged as a daunting challenge. In response to this challenge, this study introduces a novel forecasting framework specifically designed for the prediction of Chinese edible oil prices. This framework incorporates Group Temporal Convolutional Networks (GTCN) and BetaAdaptiveAdam. The proposed methodology employs Singular Spectrum Analysis (SSA) to decompose the original data into multiple subseries. It then utilizes GTCN to extract temporal features from each subseries and integrates BetaAdaptiveAdam to enhance model generalization. The framework subsequently generates predictions for each subseries and ultimately amalgamates the results of each component to produce the final price predictions for the original series. Extensive experiments have been conducted to validate the proposed forecasting framework. The results underscore the exceptional performance of the proposed framework in terms of regression accuracy and directional prediction precision.

Keywords: Forecasting of edible oil prices · Group Temporal Convolutional Networks · BetaAdaptiveAdam

1 Introduction

Edible oils, which remain liquid at ambient temperatures, are integral to food preparation and can be derived from both animal and plant sources. The market for these oils has experienced growth, propelled by enhanced economic conditions, heightened health awareness, and technological advancements. The burgeoning consumer demand has now brought the edible oil market to a pivotal juncture [1].

Edible oil price forecasting is a complex task due to interrelated factors [2]. Recent global events, such as the Russia-Ukraine conflict and rising oil prices, have caused

© The Author(s), under exclusive license to Springer Nature Singapore Pte Ltd. 2024
K. Li and Y. Liu (Eds.): ISICA 2023, CCIS 2147, pp. 337–350, 2024.
https://doi.org/10.1007/978-981-97-4396-4_32

great turmoil in oils and fats markets [3]. Fluctuations in oil prices can trigger an edible oil shortage crisis, impacting the global crude oil market and related industries [4]. Accurate forecasting of edible oil prices is crucial for national economy and investment policy decisions [5, 6]. A robust method can aid purchasers and investors in planning and scheduling edible oil.

Edible oil price forecasting has gained much academic attention in the past decade, with various modeling approaches employed. Due to its non-stationary nature, traditional methods may result in misleading forecasts [7, 8]. To address this, machine learning and deep learning methods are becoming popular in predicting changes and trends in edible oil prices.

Several studies have been conducted on predicting edible oil prices using various methods. Myat et al. proposed training palm oil price data using C4.5 random forest classification algorithm, while hyperparameter tuning technique was used to analyze whether the prediction performance could be improved [9]. Priyanga et al. analyzed the time series data of monthly wholesale prices of coconut oil in Cochin market of Kerala and used Box-Jenkins autoregressive integrated moving average method for modeling and forecasting coconut oil prices [10]. Karia et al. studied five edible oil price datasets with long memory behavior and compared them with two different models, ARIMA and ARFIMA, using Box-Jenkins model [11]. The results showed that the ARIMA and ARFIMA models gave mixed prediction results. Kanchymalay et al. proposed using support vector regression, multilayer perceptron, and Holt-Winter exponential smoothing to predict gross palm oil prices using multivariate time series [12]. The results showed that support vector regression had higher prediction accuracy. Shamsudin et al. proposed a MARMA model for short-term forecasting of Malaysian palm oil prices, which integrates a normal ARIMA model with residuals into a pre-estimated econometric equation [13]. Singh et al. used an artificial neural network to forecast the prices of several edible oils, namely mustard oil, peanut oil, and soybean oil in the Indian market, and compared it with ARIMA, which showed that the artificial neural network has excellent performance [14]. Xu et al. study explored the effectiveness of nonlinear autoregressive neural network (NARNN) and NARNN with exogenous inputs (NARNN-X) in predicting problems in daily price datasets of soybean and soybean oil [15]. However, none of the above studies considered the combination of data feature decomposition and TCN-based prediction models, and TCN with other deep learning techniques to predict edible oil prices and trends. This presents a research gap that needs to be addressed.

Temporal Convolutional Network (TCNs) are a recent deep learning technique [16] that has emerged as a strong competitor to traditional recurrent neural networks (RNNs) such as LSTM and GRU. TCN have demonstrated significant performance improvements in sequence modeling tasks, including action segmentation [17, 18], speech analysis [19], image classification [20], and medical time series prediction [21, 22]. Li et al. proposed a wind speed trend prediction model based on a TCN architecture and a hybrid decomposition method that integrates empirical modal decomposition with adaptive noise [23]. Zhao et al. developed a deep learning framework based on TCN for shortterm citywide traffic prediction and used the Taguchi method to optimize the model's structure for better performance [24]. Wang et al. proposed a short-term load forecasting model for industrial users that combines TCNs and LightGBM and demonstrated accurate load forecasting results [25]. Meka et al. developed a TCN-based short-term prediction model for wind turbine power generation and optimized the model's hyper parameters using the orthogonal array tuning method, showing superior performance over LSTM and other prediction models [26]. However, TCNs are highly sensitive to parameters, as noted by Lara et al. in their study of the applicability of various models for different forecasting tasks [27].

Hybrid models are becoming increasingly popular for predicting the prices of agricultural products like edible oils [28], with the "decomposition and integration" framework being one of the most widely used approaches [29]. The decomposition process involves breaking down complex time series data into several interdependent components to make it easier to analyze and forecast trends and fluctuations in subseries, reducing the apparent complexity of the forecasting process. On the other hand, integration combines predictions from subseries to produce predictions of the original data [30]. Singular spectrum analysis is a powerful method for processing nonlinear time series data that can decompose trends, oscillatory components, and noise from a time series [31], making it a useful tool for analyzing and predicting time series. Hybrid models that combine SSA with other machine learning methods have been proposed for a variety of applications, including day-ahead hourly load forecasting [32], stock closing price prediction [33], and stock price prediction [34]. Afshar et al. proposed an improved SSA method for data analysis and short-term load forecasting in the Iranian electricity market [35]. The authors used SSA to perform data decomposition and denoising to obtain a small number of independent and interpretable components.

This paper proposes a method for edible oil price prediction based on Group Temporal Convolutional Neural Networks (GTCN) and the BA-Adam optimization algorithm, which is compared with other deep learning techniques. The main contributions of this paper are as follows:

(1) Proposing an Improved Model: Introducing the Grouped Temporal Convolutional Neural Network (GTCN) to address the issues present in traditional temporal convolutional neural networks. GTCN improves network complexity, stability, and the ability to extract local features by enhancing the internal structure.
(2) Proposing the BA-Adam Optimization Algorithm: Presenting the Variable Factor Optimization Algorithm BA-Adam, also known as BetaAdaptiveAdam, specifically designed to adapt to the characteristics of the gradient. This is crucial because the data is decomposed into multiple different trend sub-sequences. BA-Adam adjusts the β parameter adaptively, enhancing the model's generalization ability. This adaptation allows the model to fit various data distributions better, reducing the need for hyperparameter tuning and avoiding training instability issues.
(3) Developing Evaluation Metrics: Introducing regression and classification metrics to comprehensively assess the horizontal prediction accuracy and directional prediction accuracy of the data. Numerical experiments demonstrate that the proposed model exhibits higher horizontal and directional prediction accuracy compared to other forecasting methods.

In summary, this paper delineates an innovative methodology that augments the accuracy of edible oil price predictions. The proposed GTCN model, in conjunction with the BA-Adam optimization algorithm, demonstrates superior efficacy in both horizontal and directional prediction accuracy relative to existing methods.

2 Method

2.1 Singular Spectrum Analysis

SSA, a data analysis method for time series problems. It involves two phases, decomposition and reconstruction, and requires four steps.

Step 1: The original time series is mapped into a sequence of vectors of length L, forming K vectors of length: $X_i = (x_1, x_1, \cdots, x_{i+L-1})^T$, where $1 \leq i \leq K$. These vectors form the following trajectory matrix:

$$X = [X_1, X_1, \cdots, X_K] = (x_{ij})_{i,j=1}^{L,K} = \begin{bmatrix} x_1 & x_2 & \cdots & x_K \\ x_2 & x_3 & \cdots & x_{K+1} \\ \vdots & \vdots & \vdots & \vdots \\ x_L & x_{L+1} & \cdots & x_N \end{bmatrix} \tag{1}$$

where L is the window length and $K = N_L + 1$. N_L is the length of the time series.

Step 2: In this step, X is decomposed into the following forms:

$$X = U\Sigma V^T \tag{2}$$

where is the U left matrix, Σ is the diagonal matrix with only singular values on the matrix, and V is the right matrix. In addition, U and V are both unit orthogonal matrices.

Since it is difficult to perform the singular spectrum decomposition (SVD) directly on the trajectory matrix X. Thus the covariance matrix $S = XX^T$ of X is calculated first. $\lambda_1, \lambda_2, \cdots \lambda_L$ is the eigenvalues of S and $\lambda_1 \geq \lambda_2 \geq \cdots \geq \lambda_L \geq 0$, while $U = U_1, U_2, \cdots U_L$ is the standard orthogonal vector of matrices S corresponding to these eigenvalues. X is decomposed into the following forms:

$$X = \sqrt{\lambda_i} U_i V_i^T \tag{3}$$

where $V_i = X^T U_i / \sqrt{\lambda_i}$, $i = 1, 2, \cdots, d$. In this case, the singular spectral decomposition of X can be written as:

$$X = X_1 + \cdots + X_d \tag{4}$$

where $d = rank(X) = max\{i, \lambda_i > 0\}$, In the actual sequence, usually $d = L^*$, where $L^* = min\{L, K\}$.

Step 3: In sequence reconstruction, the projection of the hysteresis sequence U_m on X_i is first calculated as:

$$a_i^m = X_i U_m = \sum_{j=1}^{L} x_{i+j} U_{m,j} 0 \leq i \leq N - L \tag{5}$$

where a_i^m is the weight of the time-evolving type reflected by X_i at time $x_{i+1}, x_{i+2}, \ldots, x_{i+L}$ of the original series, called the temporal principal component (TPC), which is next reconstructed by means of the temporal empirical orthogonal function and the temporal principal component, as:

$$
x_i^k = \begin{cases} \frac{1}{i} \sum_{j=1}^{i} a_{i-j}^k U_{k,j} & 1 \le i < L - 1 \\ \frac{1}{L} \sum_{j=1}^{L} a_{i-j}^k U_{k,j} & L \le i \le N - L + 1 \\ \frac{1}{N-i+1} \sum_{j=i-N+L}^{L} a_{i-j}^k E_{k,j} & N - L + 2 < i \le N \end{cases}
\tag{6}
$$

Finally, the sum of all reconstructed sequences is the original sequence. The formula is as:

$$
x_i = \sum_{k=1}^{L} x_i^k \quad i = 1, 2, \cdots, N
\tag{7}
$$

2.2 Group Temporal Convolutional Network

TCN, proposed by Bai et al. [16], analyzes time series data using causal and inflationary convolutions, as well as residual connectivity. Its structure and principles can be described as follows.

GTCN expands its receptive field by adding hidden layers. Each hidden layer consists of three feature extraction layers, each of which includes a one-dimensional grouped convolution layer, a weight normalization function, a ReLU activation function, and a Dropout layer. When input data is fed into the first feature extraction layer, the convolution layer divides the input channels into n groups and independently performs convolution operations on each group, followed by passing the output data to the second feature extraction layer, repeating the same process. The weights and biases for each group are learned independently, enabling the capture of different features within the input channels. To avoid the issue of gradient vanishing due to excessively deep hidden layers, GTCN employs residual connections within the hidden layers. Figure 1 illustrates the details of GTCN.

The output from the hidden layers enters the subsequent hidden layers. Additionally, due to the structural characteristics of dilated convolution, the data's feature information benefits from inter-group interactions, further enhancing the model's feature extraction capabilities. Furthermore, this paper utilizes ECANet (Efficient Channel Attention Network) [36] to further improve the model's performance.

In order to solve the problem of restricted receptive fields, GTCN uses interval sampling of convolutional inputs for dilation convolution [37]. For one-dimensional timeseries data input $X = (x_1, x_2, \ldots, x_T)$ and filter $f : \{0, 1, 2, \ldots n - 1\} \to R$, the dilation convolution operation $F(\cdot)$ is defined as:

$$
F(T) = \sum_{i=0}^{n-1} f(i) \cdot x_{T-d \cdot i}
\tag{8}
$$

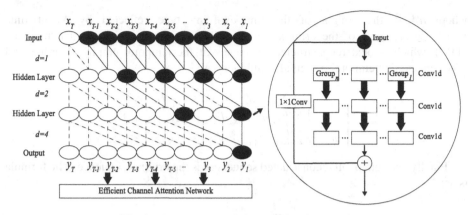

Fig. 1. Group Temporal Convolutional Network

The structure of Bottleneck is shown in Fig. 3. This structure is divided into two types. The first type is an inverted residual structure with a stride of 1, which is different from the traditional residual structure that first reduces dimensionality and then increases dimensionality. This structure first uses pointwise convolutions for dimensionality expansion, applies depthwise convolutions on the expanded feature maps, and finally uses pointwise convolutions for dimensionality reduction. The second type of structure has a stride of 2 and does not use inverted residual structures. These two types of Bottlenecks are alternately used in the 17 bottleneck stacking process to form the main part of the MobilenetV2 network.

2.3 BA-Adam

Adam is a powerful tool for optimizing machine learning algorithms. It combines two key optimization concepts: momentum and adaptive learning rates. Adam maintains the momentum of gradients by exponentially moving averages and simultaneously adapts the learning rates for each parameter. This makes Adam widely popular in fields like deep learning. However, Adam still has certain issues. Through the research [38], Xie et al. discovered that Adam can quickly escape saddle points during training but is not adept at finding flat minima with good generalization performance, implying a problem with generalization performance. The BetaAdaptiveAdam algorithm proposed in this paper possesses an adaptive nature, allowing it to automatically adjust the learning rates based on the gradient variances of each parameter. This means that on parameters with larger gradient variances, the learning rate decreases, leading to more cautious parameter updates, which aids in locating flat minima. This contributes to improving Adam's performance in finding flat minima with good generalization performance. The specific formula is as follows:

$$m_t = \beta_1^t * m_{(t-1)} + (1 - \beta_1) * g_t \tag{9}$$

$$v_t = \beta_2^t * v_{(t-1)} + (1 - \beta_2) * g_t * g_t \tag{10}$$

$$w_t = w_{(t-1)} - lr * m_t / (\sqrt{v_t} + \varepsilon) \tag{11}$$

After one time step at t, β_1 and β_2 undergo adaptive changes based on the current gradient variance, as follows:

$$\beta_1^t = \beta_1^{(t-1)} * \left(1 - \beta_1^{(t-1)}\right) / g_{var} \tag{12}$$

$$\beta_2^t = \beta_2^{(t-1)} * \left(1 - \beta_2^{(t-1)}\right) / g_{var} \tag{13}$$

Here, g_{var} represents the gradient variance at the current time step t. m_t is the firstorder exponential smoothing of historical gradients, used to obtain the gradient value with momentum, serving as the first-order moment. v_t is the first-order exponential smoothing of historical squared gradients, used to obtain the learning rate weight parameters for each weight parameter, serving as the second-order moment. w_t represents the update value of the weight variable. From the formula, it is evident that wt is directly proportional to m_t and inversely proportional to v_t. By adaptively adjusting the range and values of β_1^t and β_2^t based on the gradient variance, the optimization algorithm can adjust itself according to the current gradient's variations.

2.4 Ensemble Paradigm

This paper propose a hybrid prediction model framework combining BA-Adam method with GTCN models. SSA decomposes time series into components for accurate feature learning. The prediction framework, SSA-GTCN-BA, is illustrated in Fig. 2.

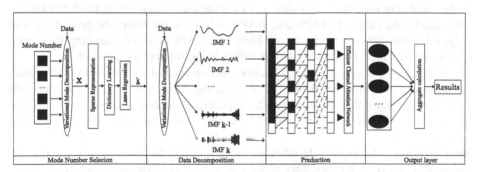

Fig. 2. Framework of SSA-GTCN-BA

3 Special Typefaces

3.1 Data Description

This paper evaluates the proposed model using four datasets obtained from the Wind website (https://wind.com.cn), including average prices of palm oil, soybean oil, peanut oil, and canola oil in selected cities and regions in China. This paper uses these four datasets to comprehensively and systematically evaluate the proposed model's effectiveness and usefulness. The experiments utilize the first 80% of each dataset as the training set and the last 20% as the test set, and evaluate the model's performance using error evaluation index and classification evaluation index. Palm oil prices are observed from February 29, 2008, to September 30, 2022, with a total of 3567 daily observations recorded, excluding public holidays. Soybean oil prices are observed from March 3, 2008, to September 30, 2022, with a total of 3616 daily observations recorded, excluding weekends. Peanut oil prices are observed on a weekly basis, with 953 observations recorded each week from June 4, 2004, to September 23, 2022. Finally, canola oil prices are observed from June 25, 2015, to September 29, 2022, with a total of 1798 daily observations recorded, excluding public holidays.

3.2 Performance Evaluation Criteria

The error evaluation in this paper uses MAPE, WAPE. Among the three statistical methods, MAPE is the mean absolute percentage error, WAPE is the weighted absolute percentage error. The specific calculation formula is as follows.

$$MAPE = \frac{1}{m} \sum_{t=1}^{m} |\frac{R_t - P_t}{R_t}| \tag{14}$$

$$WAPE = \frac{\sum_{t=1}^{m} |R_t - P_t|}{\sum_{t=1}^{m} |R_t|} \tag{15}$$

where R_t is the true value of the timestamp t, P_t is the corresponding predicted value, m is the number of predicted outcomes, and $\overline{P}_t$ is the mean value of the predicted values. The smaller the value of MAPE and WAPE indicators, the better the prediction. The large difference between the values of MAPE and WAPE can reflect the large difference between the predicted and true values at different moments in the forecast.

In this paper, to compare the prediction performance of the model in more detail, we compare the accuracy of the model in predicting the directionality of the price with respect to the true value. In more detail, we predicted the price of the $(F+1)$-th edible oil price by analyzing the F previous edible oil prices, while using the classification evaluation index to determine whether the $(F+1)$-th price is rising or falling relative to the F-th price, and using the classification index to obtain the accuracy of predicting the rising or falling trend. We used three performance indicators: accuracy (ACC), sensitivity (Sen), and specificity (Spe). With the following calculation formula:

$$ACC = \frac{TP - TN}{TP + FP + FN + TN} \tag{16}$$

$$Sen = \frac{TP}{TP + FN} \tag{17}$$

$$Sen = \frac{TN}{FP + TN} \tag{18}$$

the larger the three performance indicators, the higher the accuracy rate is represented. Moreover, ACC reflects the comprehensive directional forecasting ability of the forecasting model, Sen presents the forecasting ability of the uptrend of the forecasting model, and Spe indicates the forecasting ability of the downtrend of the forecasting model.

3.3 Data Decomposition and Parameter Setting

The model proposed in this study decomposes the original price series into multiple trend signal series using Singular Spectrum Analysis (SSA), which can be used for better feature extraction and improved prediction results. However, determining the appropriate number of decomposed sequences (window length L) is crucial. If L is too small, the features may be too concentrated, making extraction and prediction difficult, while over-decomposition leads to decreased accuracy and increased computational costs. Each trend sequence corresponds to a singular value, and their relative size represents the interpretation degree of the trend sequence in the original data [39]. This study observes the variation of singular values for L between 5 to 20 and selects an appropriate window length. Through singular spectrum analysis, Table 1, Table 2, Table 3 and Table 4 show the partial singular values of different window lengths at different price series.

Table 1. Singular value of Peanut oil price

L	Singular value
10	2290.26, 13.64, 3.73, 2.31, 1.89, 1.74, 1.63, 1.59, 1.40, 1.39
11	2400.85, 15.49, 4.23, 2.52, 1.99, 1.84, 1.64, 1.60, 1.56, 1.39, 1.34
12	2506.36, 17.39, 4.78, 2.73, 2.10, 1.96, 1.64, 1.64, 1.57, 1.55, 1.34, 1.34
13	2607.40, 19.33, 5.38, 2.96, 2.24, 1.96, 1.83, 1.64, 1.59, 1.55, 1.48, 1.34, 1.28

Table 2. Singular value of Palm oil price

L	Singular value
10	1326202.36, 25457.99, 10982.35, 8066.26, 5777.70, 4604.21, 3712.93, 3181.76, 2940.79, 2840.69
11	1390489.39, 28396.97, 11852.56, 8769.21, 6389.65, 5093.41, 4151.93, 3449.71, 3090.21, 2909.61, 2822.24
12	1451853.81, 31446.11, 12708.70, 9404.45, 7043.32, 5539.43, 4592.21, 3812.92, 3275.548, 3055.01, 2855.68, 2816.26
13	1510656.30, 34571.32, 13573.72, 10019.14, 7694.03, 5945.70, 5008.5, 4197.56, 3579.22, 3146.81, 3042.63, 2815.09, 2797.45

As described in the table above, Table 1, Table 2, Table 3 and Table 4 show that for each sequence, a window length L of 12 is sufficient for proper trend component

Table 3. Singular value of Soybean oil price

L	Singular value
10	1474853.82,20136.61,8532.58,5844.46,4014.93,3332.45,2871.93, 2620.47,2477.39,2420.86
11	1546251.47,22413.20,9312.635,6483.151,4493.15,3544.16,3129.35, 2725.22,2582.76,2449.78,2419.06
12	1614400.66,24733.16,10077.58,7093.43,5030.56,3770.24,3329.65, 2952.32,2642.07,2560.24,2418.883,2418.59
13	1679697.69,27086.84,10845.97,7663.36,5578.15,4090.89,3479.54, 3179.28,2785.27,2601.12,2545.54,2418.48,2393.01

Table 4. Singular value of Peanut oil price

L	Singular value
10	8671.60, 11.82, 4.89, 3.48, 2.54, 2.12, 1.92, 1.78, 1.64, 1.59
11	9091.77, 13.23, 5.32, 3.81, 2.80, 2.26, 2.01, 1.87, 1.75, 1.62, 1.59
12	9492.82, 14.68, 5.77, 4.11, 3.11, 2.41, 2.12, 1.94, 1.83, 1.71, 1.60, 1.60
13	9877.08, 16.15, 6.24, 4.37, 3.42, 2.60, 2.23, 2.03, 1.89, 1.81, 1.68, 1.60, 1.59

extraction. Further decreases in singular values indicate the subsequent trend series can be treated as noise. Using a window length of 12, this study removes the last two trend subsequences as noise and uses the first 10 subsequences for the experiment.

The network includes an input layer, a GTCN layer, and an output layer. A step size of 12 is set for the input layer. Hyperparameters are determined by trial-and-error as shown in Table 5, and ReLU is the selected activation function. All models use ADAM optimization in addition to the methods proposed in this article. The number of training iterations for all models is 100. Batch sizes for palm oil, peanut oil, soybean oil, and canola oil prices are 20, 10, 20, and 10, respectively.

Table 5. Hyper-parameters

num channels	kernel size	kernel size	groups
36	2	0.2	2

In this paper, the predictive power is evaluated by comparison with benchmark models, and the performance of individual components is analyzed by comparison with compatible models. The cross-validation method was used for other benchmark models. Our experiments show that the neurons with 64 hidden layers produce the best prediction results for LSTM in 32, 64, and 128, and other models are trained by using the above approach. The number of training for all these models was 100 at the same learning rate.

In Table 4, the training dataset undergoes haze removal through a GCAnet algorithm, followed by training using MobilenetV2. Top-1 is an important indicator for determining the model's performance, and combined with Table 3, the accuracy of the models trained with dehazed images shows a slight improvement compared to models trained without dehazing, using the same dataset. However, only when performing dehazing operations on the test set, could the model achieve good performance, indicating that the dehazing algorithm played a certain role. Nonetheless, the results also show that using dehazing algorithms on clear image datasets still resulted in poor performance, indicating that a single model cannot be used universally for detection.

3.4 Results

Experimental results are presented to demonstrate the effectiveness of the proposed SSA-GTCN-BA model. Four edible oil price datasets (palm, peanut, soybean, and canola) were used for testing, and compatible experiments were conducted to compare and analyze module performance. Two types of evaluation metrics were used to analyze forecasting performance, and Table 6 and Table 7 compare the results of different forecasting models. Taking palm oil as an example, SSA-GTCN-BA outperforms other models for error (MAPE, WAPE) and classification (ACC, Sen, Spe) evaluation metrics. Single-model prediction deviations are similar, with TCN having higher directional accuracy, but MAPE and WAPE have different deviations, with LSTM being the most significant. This suggests that errors in different cases differ significantly from true values, while SSA-GTCN-BA performs well.

The SSA-based hybrid model outperforms the original model in directional accuracy due to efficient data decomposition. The GTCN network as the core show superiority, highlighting GTCN's advantages in time series handling. Similar results are obtained

Table 6. Evaluation criteria of Peanut and Palm

	Peanut oil price					Palm oil price				
	MAPE	WAPE	Sen	Spe	ACC	MAPE	WAPE	Sen	Spe	ACC
SSA-GTCN-BA	**0.121**	**0.121**	**0.897**	**0.821**	**0.866**	**1.150**	**1.211**	**0.943**	**0.961**	**0.952**
SSA-TCN-GRU	0.250	0.249	0.887	0.753	0.833	1.854	1.801	0.831	0.845	0.838
TCN-GRU	0.314	0.314	0.775	0.397	0.622	2.534	2.678	0.554	0.534	0.545
TCN-SSA	0.243	0.241	0.897	0.726	0.827	2.269	2.476	0.842	0.803	0.824
TCN	0.330	0.329	0.747	0.369	0.594	3.014	3.006	0.612	0.483	0.552
LSTM-SSA	0.382	0.377	0.887	0.630	0.783	2.727	2.662	0.791	0.711	0.754
LSTM	0.428	0.421	0.579	0.438	0.522	3.488	3.928	0.548	0.537	0.543
GRU-SSA	0.326	0.322	0.869	0.616	0.766	2.262	2.168	0.817	0.747	0.784
GRU	0.343	0.340	0.579	0.438	0.522	3.375	3.852	0.548	0.537	0.543

for peanut oil, soybean oil, and canola oil prices, demonstrating the methods' robustness and applicability.

Table 7. Evaluation criteria of Soybean and Canola

	Soybean oil price					Canola oil price				
	MAPE	WAPE	Sen	Spe	ACC	MAPE	WAPE	Sen	Spe	ACC
SSA-GTCN-BA	**0.710**	**0.724**	**0.955**	**0.951**	**0.953**	**0.169**	**0.168**	**0.961**	**0.935**	**0.955**
SSA-TCN-GRU	2.564	2.571	0.810	0.839	0.823	0.208	0.210	0.845	0.729	0.812
TCN-GRU	3.185	3.198	0.573	0.482	0.532	0.344	0.344	0.867	0.256	0.694
TCN-SSA	3.118	3.248	0.833	0.784	0.811	0.240	0.244	0.904	0.513	0.793
TCN	3.237	3.361	0.607	0.437	0.530	0.330	0.350	0.888	0.216	0.698
LSTM-SSA	4.254	4.405	0.819	0.801	0.811	0.386	0.395	0.898	0.567	0.805
LSTM	4.588	4.835	0.553	0.493	0.526	0.378	0.378	0.670	0.297	0.564
GRU-SSA	3.732	3.891	0.790	0.815	0.801	0.297	0.294	0.946	0.486	0.816
GRU	4.434	4.684	0.548	0.472	0.513	0.379	0.384	0.680	0.337	0.583

4 Conclusion

Edible oil is a vital commodity within the food industry, and price fluctuations can exert a significant influence on business operations and the national economy. In order to enhance the accuracy of edible oil price predictions, this paper proposes a forecasting framework based on Grouped Convolutional Neural Networks (GTCN) and BetaAdaptiveAdam. Specifically, we employ Singular Spectrum Analysis (SSA) signal decomposition to reduce the complexity of the original data and use GTCN for feature extraction from each sub-series. The optimization of the model training process is achieved through the utilization of BetaAdaptiveAdam, and the individual predictions from each sub-series are amalgamated into an ensemble result. Our proposed model outperforms existing methods across all evaluation criteria for palm oil, soybean oil, peanut oil, and canola oil prices.

Our research on edible oil price prediction contributes to the enhancement of businesses' capacity to manage price risks, making a meaningful contribution to the sustained and robust development of China's edible oil industry. In our future work, we plan to improve the model's accuracy by integrating other factors and higher-frequency data.

References

1. Meier, M.A., Metzger, J.O., Schubert, U.S.: Plant oil renewable resources as green alternatives in polymer science. Chem. Soc. Rev. **36**(11), 1788–1802 (2007)

2. Yu, T.H.E., Bessler, D.A., Fuller, S.W.: Cointegration and causality analysis of world vegetable oil and crude oil prices. Technical report (2006)
3. Muller, C.L.: The effect of the Russia-Ukraine conflict on world edible oil prices. Oilseeds Focus **8**(2), 43–47 (2022)
4. Yip, P.S., Brooks, R., Do, H.X., Nguyen, D.K.: Dynamic volatility spillover effects between oil and agricultural products. Int. Rev. Financ. Anal. **69**, 101465 (2020)
5. Russo, D., Dassisti, M., Lawlor, V., Olabi, A.: State of the art of biofuels from pure plant oil. Renew. Sustain. Energy Rev. **16**(6), 4056–4070 (2012)
6. Ng, S.W., Zhai, F., Popkin, B.M.: Impacts of china's edible oil pricing policy on nutrition. Soc. Sci. Med. **66**(2), 414–426 (2008)
7. Brandt, J.A., Bessler, D.A.: Price forecasting and evaluation: an application in agriculture. J. Forecast. **2**(3), 237–248 (1983)
8. Myat, A.K., Tun, M.T.Z.: Predicting palm oil price direction using random forest. In: 2019 17th International Conference on ICT and Knowledge Engineering (ICT&KE), pp. 1–6. IEEE (2019)
9. Priyanga, V., Lazarus, T.P., Mathew, S., Joseph, B.: Forecasting coconut oil price using auto regressive integrated moving average (ARIMA) model. J. Pharmacogn. Phytochem. **8**(3), 2164–2169 (2019)
10. Karia, A.A., Abd Hakim, T., Bujang, I.: World edible oil prices prediction: evidence from mix effect of ever difference on box-jenkins approach. J. Bus. Retail Manag. Res. **10**(3) (2016)
11. Kanchymalay, K., Salim, N., Sukprasert, A., Krishnan, R., Hashim, U.R.: Multivariate time series forecasting of crude palm oil price using machine learning techniques. IOP Conf. Ser. Mater. Sci. Eng. **226**, 012117 (2017). IOP Publishing
12. Shamsudin, M.N., Arshad, F.M.: Short term forecasting of Malaysian crude palm oil prices (2000)
13. Singh, A.: Comparison of artificial neural networks and statistical methods for forecasting prices of different edible oils in Indian markets. Int. Res. J. Modern. Eng. Technol. Sci. **3**, 1044–1050 (2021)
14. Xu, X., Zhang, Y.: Soybean and soybean oil price forecasting through the nonlinear autoregressive neural network (NARNN) and NARNN with exogenous inputs (NARNN–X). Intell. Syst. Appl. **13**, 200061 (2022)
15. Bai, S., Kolter, J.Z., Koltun, V.: An empirical evaluation of generic convolutional and recurrent networks for sequence modeling. arXiv preprint arXiv:1803.01271 (2018)
16. Lea, C., Flynn, M.D., Vidal, R., Reiter, A., Hager, G.D.: Temporal convolutional networks for action segmentation and detection. In: Proceedings of the IEEE Conference on Computer Vision and Pattern Recognition, pp. 156–165 (2017)
17. Kim, T.S., Reiter, A.: Interpretable 3D human action analysis with temporal convolutional networks. In: 2017 IEEE Conference on Computer Vision and Pattern Recognition Workshops (CVPRW), pp. 1623–1631. IEEE (2017)
18. Oord, A.v.d., et al.: Wavenet: a generative model for raw audio. arXiv preprintarXiv:1609.03499 (2016)
19. Pelletier, C., Webb, G.I., Petitjean, F.: Temporal convolutional neural network for the classification of satellite image time series. Remote Sens. **11**(5), 523 (2019)
20. Moor, M., Horn, M., Rieck, B., Roqueiro, D., Borgwardt, K.: Early recognition of sepsis with Gaussian process temporal convolutional networks and dynamic time warping. In: Machine Learning for Healthcare Conference, pp. 2–26. PMLR (2019)
21. Catling, F.J., Wolff, A.H.: Temporal convolutional networks allow early prediction of events in critical care. J. Am. Med. Inform. Assoc. **27**(3), 355–365 (2020)
22. Li, D., Jiang, F., Chen, M., Qian, T.: Multi-step-ahead wind speed forecasting based on a hybrid decomposition method and temporal convolutional networks. Energy **238**, 121981 (2022)

23. Zhao, W., Gao, Y., Ji, T., Wan, X., Ye, F., Bai, G.: Deep temporal convolutional networks for short-term traffic flow forecasting. IEEE Access **7**, 114496–114507 (2019)
24. Wang, Y., et al.: Short-term load forecasting for industrial customers based on TCN-lightGBM. IEEE Trans. Power Syst. **36**(3), 1984–1997 (2020)
25. Meka, R., Alaeddini, A., Bhaganagar, K.: A robust deep learning framework for short-term wind power forecast of a full-scale wind farm using atmospheric variables. Energy **221**, 119759 (2021)
26. Lara-Benítez, P., Carranza-García, M., Riquelme, J.C.: An experimental review on deep learning architectures for time series forecasting. Int. J. Neural Syst. **31**(03), 2130001 (2021)
27. Wang, L., Feng, J., Sui, X., Chu, X., Mu, W.: Agricultural product price forecasting methods: research advances and trend. Br. Food J. (2020)
28. Yu, L., Wang, S., Lai, K.K.: Forecasting crude oil price with an EMD-based neural network ensemble learning paradigm. Energy economics **30**(5), 2623–2635 (2008)
29. Song, G., Dai, Q.: A novel double deep elms ensemble system for time series forecasting. Knowl.-Based Syst. **134**, 31–49 (2017)
30. Vautard, R., Yiou, P., Ghil, M.: Singular-spectrum analysis: a toolkit for short, noisy chaotic signals. Physica D **58**(1–4), 95–126 (1992)
31. Stratigakos, A., Bachoumis, A., Vita, V., Zafiropoulos, E.: Short-term net load forecasting with singular spectrum analysis and LSTM neural networks. Energies **14**(14), 4107 (2021)
32. Fathi, A.Y., El-Khodary, I.A., Saafan, M.: Integrating singular spectrum analysis and non-linear autoregressive neural network for stock price forecasting. IAES Int. J. Artif. Intell. **11**(3), 851 (2022)
33. Fenghua, W., Jihong, X., Zhifang, H., Xu, G.: Stock price prediction based on SSA and SVM. Procedia Comput. Sci. **31**, 625–631 (2014)
34. Afshar, K., Bigdeli, N.: Data analysis and short term load forecasting in Iran electricity market using singular spectral analysis (SSA). Energy **36**(5), 2620–2627 (2011)
35. Wang, Q., Wu, B., Zhu, P., Li, P., Zuo, W., Hu, Q.: ECA-net: efficient channel attention for deep convolutional neural networks. In: Proceedings of the IEEE/CVF Conference on Computer Vision and Pattern Recognition, pp. 11534–11542 (2020)
36. Yu, F., Koltun, V.: Multi-scale context aggregation by dilated convolutions. arXiv preprint arXiv:1511.07122 (2015)
37. Xie, S., Girshick, R., Dollár, P., Tu, Z., He, K.: Aggregated residual transformations for deep neural networks. In: Proceedings of the IEEE Conference on Computer Vision and Pattern Recognition, pp. 1492–1500 (2017)
38. Xie, Z., Wang, X., Zhang, H., Sato, I., Sugiyama, M.: Adaptive inertia: disentangling the effects of adaptive learning rate and momentum. In: International Conference on Machine Learning, pp. 24430–24459. PMLR (2022)
39. Palomo, M., Sanchis, R., Verdu, G., Ginestar, D.: Analysis of pressure signals using a singular system analysis (SSA) methodology. Prog. Nucl. Energy **43**(1–4), 329–336 (2003)

Design and Application of a Teaching Evaluation Model Based on the Theory of Multiple Intelligences

Luyan Lai(✉)

School of Business, Guangdong Polytechnic, Foshan 528041, Guangdong, China
332075619@qq.com

Abstract. This study designs a teaching evaluation model based on the theory of multi intelligence, and applies it to the "New Media Marketing" course in vocational colleges. The objective is to comprehensively evaluate the performance and development of students in different fields of intelligence, and provide basis and guidance for personalized teaching. On the basis of the research background, a systematic and quantitative research method are adopted to conduct empirical research on the model. The research results showed that the teaching evaluation model guided by the theory of multiple intelligences can comprehensively evaluate the development of students in different fields of intelligence, reveal their intellectual strengths and weaknesses, achieve value-added evaluation, and comprehensively assist students in diversified growth and development. This provides new ideas, methods, demonstrations, and references for the reform and innovation of education and teaching evaluation.

Keywords: Multiple intelligences · Teaching evaluation · Evaluation model · Indicator system

1 Introduction

The current evaluation system for vocational education courses in China typically comprises two components: process-oriented and summative assessments. The former generally pertains to students' classroom performance, including attendance and participation, while the latter refers to their final examination results. These components are weighted differently to derive a comprehensive score. However, the weighting can vary across different schools, courses, and instructors, leading to a lack of standardization. This empirically-based evaluation has been somewhat effective in traditional teaching contexts, but it falls short in achieving truly "personalized teaching".

In October 2020, the Central Committee of the Communist Party of China and the State Council issued the "Overall Plan for Deepening the Reform of Education Evaluation in the New Era". This plan advocates for the establishment of an evaluation system that encourages diverse participation from schools and society, and promotes a more varied range of evaluation methods to holistically assess student development.

The curriculum serves as a crucial vehicle for deepening educational evaluation reform and is a starting point for a comprehensive overhaul of educational evaluation. Reforming curriculum evaluation is a key strategy to advance educational equity, genuinely implement a student-centered approach, and foster the comprehensive development of students in moral, intellectual, physical, aesthetic, and labor aspects. Concurrently, it aids in enhancing the quality of teacher education and instruction.

Professor Gardner, an educator and psychologist at Harvard University in the United States, pointed out that traditional educational evaluation does not truly examine the strengths and weaknesses of students. Instead, it tends to focus on students' deficiencies and shortcomings, primarily due to the lack of comprehensive and diverse evaluations. Gardner [2] proposed a novel theory of human intelligence structure, known as the Multiple Intelligences (MI) theory, which posits that human cognition and thinking are diverse. This theory provides a theoretical foundation for constructing multifaceted evaluations and offers fresh insights and inspiration for deepening the reform of educational evaluation in the new era.

2 Research Background and Significance

The theory of multiple intelligences has emerged as a prominent research direction in contemporary educational studies, with the performance and development of students across various intelligence domains forming a crucial aspect of educational evaluation. Traditional evaluation models, which typically focus on a single intelligence, often fail to provide a comprehensive assessment of students' development across multiple intelligences. This limitation hinders the provision of a scientific foundation for personalized and differentiated instruction. Consequently, there is a pressing need to develop a teaching evaluation model that can holistically assess students' performance and development across diverse intelligence domains.

The purpose of this study is to propose a teaching evaluation model based on the theory of multiple intelligences. This model aims to enhance our understanding of students' intellectual characteristics and potential across different intelligence domains, thereby providing a scientific foundation and guidance for personalized instruction. Furthermore, this study seeks to investigate the practical value of the theory of multiple intelligences in education and the efficacy of the proposed teaching evaluation model, with the goal of offering novel insights and methodologies for educational reform and innovation.

The significance of this study lies in its potential to enrich both the theoretical framework and practical application of intelligence evaluation research, and to provide valuable references for enhancing the scientific rigor and accuracy of teaching evaluation. By employing systematic research methods and scientific evaluation models, we can more effectively identify students' strengths and weaknesses across different intelligence domains. This approach enables the provision of targeted instructional interventions and personalized teaching strategies, thereby offering a scientific foundation and guidance for personalized instruction. Moreover, it provides valuable insights and inspiration for teaching evaluation across various disciplines, and offers novel ideas and methodologies for educational reform and innovation.

3   Overview of Multi Intelligence Theory

3.1   The Development of the Theory of Multiple Intelligences

The theory of multiple intelligences, first introduced by American psychologist Howard Gardner in his 1983 book "Multiple Intelligences", posits that human intelligence is multifaceted rather than monolithic [1]. Gardner's theory has evolved over time, initially encompassing seven types of intelligence: linguistic, logical-mathematical, spatial, musical, bodily-kinesthetic, interpersonal, and intrapersonal. Subsequently, he expanded this framework to include naturalistic and existential intelligences, thereby establishing a model of nine distinct intelligences.

This theory offers a novel educational philosophy, advocating for teaching and evaluation practices that align with students' diverse intellectual tendencies. In the context of teaching, the application of the theory of multiple intelligences enables educators to better comprehend students' intellectual predispositions, thereby facilitating the adoption of varied teaching strategies to cater to the learning needs of different students. For instance, students with pronounced linguistic intelligence might benefit from a focus on language expression and text reading, while those with strong musical intelligence might be more engaged through music-based teaching. This approach allows students to leverage their strengths, thereby enhancing their learning outcomes. In terms of evaluation, the theory of multiple intelligences provides a comprehensive framework for assessing student abilities. Unlike traditional evaluation methods, which primarily focus on students' written exam scores, the theory of multiple intelligences promotes the use of diverse evaluation methods, including oral expression, practical operation, and artistic creation. This approach yields more objective and comprehensive evaluation results, offering a more accurate reflection of students' comprehensive abilities.

In summary, the development and significance of the theory of multiple intelligences lie in its potential to better explore and cultivate human potential, thereby fostering more personalized and diversified education. It equips educators with a novel teaching philosophy and evaluation method, and provides students with greater opportunities for development. Future educational practice should continue to research and apply the theory of multiple intelligences to further promote educational reform and development.

3.2   The Main Content of the Theory of Multiple Intelligences

The theory of multiple intelligences encompasses three primary components: the theoretical foundation, the classification of intelligences, and the educational applications of these intelligences.

The theoretical foundation posits that intelligence is multifaceted rather than unidimensional. The classification of multiple intelligences involves categorizing intelligence into seven principal types: linguistic, logical-mathematical, spatial, musical, bodily-kinesthetic, interpersonal, and intrapersonal. Linguistic intelligence pertains to an individual's capacity for communication and self-expression through language. Logical-mathematical intelligence refers to the ability to perform logical reasoning and mathematical operations. Spatial intelligence involves the understanding and manipulation of space. Musical intelligence relates to the perception and creation of music. Bodily-kinesthetic intelligence involves the coordination and execution of bodily movements. Interpersonal intelligence refers to the ability to understand and process others' emotions and relationships. Intrapersonal intelligence involves the reflection and understanding of one's own emotions and thoughts. Each type of intelligence is independent and does not influence the others, with individuals exhibiting varying strengths and potential across these different aspects of intelligence.

The educational application of multiple intelligences involves incorporating the theory into educational practice. Educators are encouraged to adopt diverse teaching methods and strategies that align with students' different types of intelligence, thereby promoting their comprehensive development. For instance, students with strong linguistic intelligence can enhance their language expression abilities through activities such as reading and writing. Similarly, students with pronounced musical intelligence can cultivate their musical talents through activities like music appreciation and instrument performance. This approach underscores the importance of personalized and differentiated instruction in fostering comprehensive student development.

3.3 Current Research Status of the Theory of Multiple Intelligences and Its Application in Teaching Evaluation

The theory of multiple intelligences offers a novel perspective on intelligence, underscoring the unique strengths and potential of individuals across various intelligence domains. This theory provides significant guidance for educational practice, prompting educators to adopt diverse teaching methods and strategies that align with students' multiple intelligences, thereby fostering their comprehensive development. For instance, Gu [3] incorporated the theory of multiple intelligences into the reform of English major teaching evaluation, establishing a corresponding evaluation system to enhance teaching quality. Similarly, He et al. [4] applied this theory to the evaluation of cell biology experimental teaching, creating a novel teaching evaluation system characterized by diversified evaluation content, a dynamic evaluation process, and varied evaluation subjects. This approach facilitates students' coordinated development in knowledge, abilities, and qualities. Suyangna [5] conducted a flipped classroom learning evaluation based on the theory of multiple intelligences and Moodle platform activity records. This approach enabled the tracking of students' online learning activities and the establishment of a diversified, specific, and process-oriented flipped classroom evaluation index system. Li et al. [6] drawing on Gardner's theory of multiple intelligences, adhered to the principles of diversified evaluation subjects, multi-dimensional evaluation content, and diverse evaluation methods. They constructed a vocational college student evaluation system that relies on the linkage mechanism between government, enterprise, and school, and organically combines internal and external factors. This approach led to the creation of a series of characteristic brand activities, including shaping the image of students in the city college and student community social practice. Wu et al. [7] demonstrated the feasibility of constructing performance evaluation standards for counselors based on the evaluation theory of multiple intelligences with the aim of promoting development, and proposed suggestions for standard construction; Li [8] proposed the construction of an evaluation system for college English self-directed learning based on the theory of multiple intelligences, which encourages students to actively participate in learning activities, effectively evaluates their learning outcomes, promotes the improvement of their self-directed learning ability, and also enhances their comprehensive language proficiency; Xia [9] also constructed a college English evaluation system based on the theory of multiple intelligences. Through experiments, it has been proven that formative evaluation systems are more conducive to promoting the cultivation of language learning strategies and the improvement of English grades for students.

4 Design of a Multiple Intelligence Teaching Evaluation Model

4.1 Build Requirement Analysis

Teaching evaluation is an important part of teaching work, which can effectively evaluate the learning situation of students and the teaching effectiveness of teachers. Under the guidance of the theory of multiple intelligences, the construction of teaching evaluation models has become more comprehensive and scientific. The following will elaborate on the construction requirements of the teaching evaluation model for the theory of multiple intelligences.

The construction of such a teaching evaluation model is rooted in the core tenet of the theory of multiple intelligences, which posits that each student possesses multiple types of intelligence, extending beyond traditional intellectual factors. Consequently, the construction of a teaching evaluation model necessitates a comprehensive consideration of students' diverse intelligences to objectively assess their learning progress.

The multi intelligence theory teaching evaluation model is characterized by its comprehensiveness, personalization, and dynamism. Comprehensiveness is manifested in the model's capacity to consider multiple types of student intelligence, extending beyond the evaluation of subject knowledge. Personalization is embodied in the model's ability to evaluate students based on their specific circumstances, emphasizing the identification and cultivation of their individual strengths. Dynamism is reflected in the model's ability to dynamically adjust and provide timely feedback on students' learning processes.

In the multi intelligence theory teaching evaluation model, the evaluation content includes multiple aspects such as subject knowledge, thinking ability, emotional attitude, and practical ability. The evaluation of subject knowledge is mainly carried out through exams, assignments, and projects, which can objectively evaluate the degree of mastery of subject knowledge by students. The evaluation of thinking ability is mainly carried out through problem-solving, analytical thinking, and innovation, which can objectively evaluate the level of students' thinking ability. Emotional attitude evaluation is mainly conducted through questionnaire surveys and observations, which can objectively evaluate the emotional attitude performance of students. The evaluation of practical ability is mainly carried out through experiments, on-site inspections, and internships, which can objectively evaluate the level of students' practical ability.

The construction of a multi intelligence theory teaching evaluation model requires the use of advanced evaluation tools and methods. Common evaluation tools include course platforms, questionnaire surveys, observation records, and work presentations, which can objectively collect data on student learning processes for authentic evaluation. The commonly used evaluation methods include qualitative evaluation and quantitative evaluation, which can objectively analyze and interpret evaluation data. By selecting appropriate evaluation tools and methods, a scientific, comprehensive, and accurate teaching evaluation model can be constructed.

The construction of the multi intelligence theory teaching evaluation model mainly focuses on comprehensively considering the various types of intelligence of students, as well as the diversity and personalization of evaluation content. The evaluation model can objectively evaluate the learning situation of students and the teaching effectiveness of teachers, providing scientific guidance and reference for teaching work.

4.2 Model and System Design

In recognition of students' diverse learning styles, strengths, and intelligences, a teaching evaluation system predicated on the theory of multiple intelligences is established. During the implementation of classroom instruction, teaching platforms such as the Chaoxing Learning Platform are regularly employed to facilitate a blend of online and offline teaching. Information platforms are utilized to gather learning data, enabling the tracking and recording of students' learning outcomes throughout the process. Leveraging the benefits of school-enterprise collaboration, the evaluation process incorporates self-evaluation, peer evaluation, and teacher feedback. Additionally, enterprise mentors participate in practical training guidance and assessment throughout the entire process, conducting diversified evaluations across four dimensions, each accounting for 25% of the total evaluation. This approach combines subjectivity and objectivity, collecting teaching process data for a comprehensive evaluation. The teaching evaluation comprises pre-class diagnostic evaluation (10%), in-class formative evaluation (50%), and post-class summative evaluation (40%).

In the pre-class diagnostic evaluation, the focus is on assessing students' linguistic intelligence through their completion of research tasks, survey reports, and other self-directed learning activities prior to class. During the in-class formative assessment, the focus is on examining the degree of achievement of knowledge, skills, and literacy. In alignment with the theory of multiple intelligences, students' linguistic intelligence is further evaluated through the completion of knowledge tests such as in-class tests and question answering. Students' logical-mathematical abilities are evaluated through the completion of data analysis reports. Students' intrapersonal and musical intelligences are evaluated through self-evaluation, experience sharing, and popular text creation.

In the post-class summative evaluation, the focus is on examining students' interpersonal and spatial intelligences through tasks such as interview research, collaboration plans, poster design, and popular copywriting design. Through offline live streaming sales and other expansion activities, students' bodily-kinesthetic intelligence is examined. Simultaneously, by comparing and analyzing students before and after practical training, we can accurately gauge the improvement of their new media marketing skills and explore value-added evaluation.

Assuming A represents the platform evaluation, B represents the teacher evaluation, and C represents the four-party evaluation, a teaching evaluation model based on the theory of multiple intelligences is established as depicted in Fig. 1.

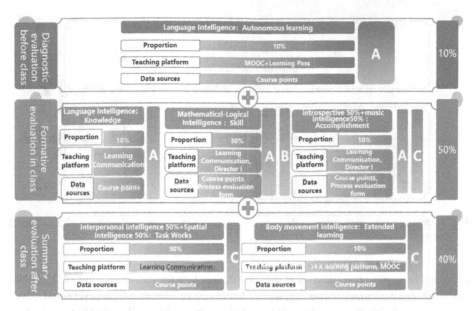

Fig. 1. Multi intelligence teaching evaluation model and indicator system.

5   The Application of Multi Intelligence Teaching Evaluation Model of a Multiple Intelligence Teaching Evaluation Model

An evaluation model and indicator system, grounded in the theory of multiple intelligences, has been developed to provide a framework for course teaching assessment. This model was implemented in the "New Media Marketing" course, with a particular focus on the "XX Brand Clothing" training project, which pertains to a local enterprise's non-legacy cloud yarn series product. On the one hand, in real-life project courses, students need to form small groups according to the working style of the team, and fully tap into their own potential: the ability to learn independently, analyze and solve problems, persuade others to promote themselves, and have hands-on abilities, among other comprehensive abilities; On the other hand, during such practical training, students naturally feel the charm of traditional Chinese intangible cultural heritage culture. At the same time, they can integrate previous knowledge into one furnace and put it into practice. In the application process, they can clearly see their own shortcomings and consciously supplement relevant knowledge, invisibly making the theoretical teaching in the early stage more complete and effective. According to the comparison of the final scores of pre training and post training assessments, students have significantly improved their various new media marketing skills and achieved the three-dimensional teaching objectives efficiently, as shown in Fig. 2.

The implementation of pedagogical models emphasizes the systematic collection of information throughout the teaching process, with assessments and evaluations conducted in alignment with predefined objectives. A dynamic evaluation framework,

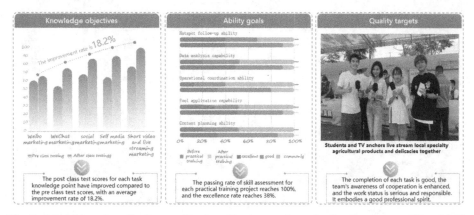

Fig. 2. Model based application - achieving effectiveness in evaluating the course of New Media Marketing.

encompassing pre-class diagnostic evaluation, in-class formative evaluation, and post-class summative evaluation, is employed. This is complemented by a diversified evaluation approach that includes dual mentor evaluation, student self-evaluation, and group peer evaluation. These strategies are integrated organically to foster multifaceted interactions among educational institutions, enterprises, educators, and students, thereby stimulating engagement from all parties. Consequently, students are motivated to participate in various skills competitions and technological innovation activities. This approach significantly bolsters team collaboration awareness and enhances the efficiency of classroom instruction.

6   Conclusion

This article introduces the theory of multiple intelligences and establishes a scientific and reasonable multiple teaching evaluation model and evaluation index system based on course objectives. The evaluation methods are diverse (online and offline evaluation, objective and subjective evaluation), the evaluation content is diverse (evaluating student knowledge, skills, qualities, etc.), and the evaluation subjects are diverse (including student self-evaluation, student mutual evaluation, teacher evaluation, and enterprise mentor evaluation). Through quantitative empirical research, multiple intelligent evaluations have been carried out, while exploring value-added evaluations, respecting personalized differences among students, and helping them grow and become successful.

This study provides a new theoretical and practical framework for intelligent evaluation research, as well as new ideas, methods, references, and insights for educational reform and innovation. However, there are also some shortcomings that need further research. For instance, in the construction of the teaching evaluation model, the study solely considered the theoretical framework of multiple intelligences, neglecting the potential impact of other factors on student development. Future research could address this by incorporating additional factors to further refine and deepen the teaching evaluation model based on the theory of multiple intelligences.

Acknowledgement. This work is supported by the 2022 Guangdong Province Continuing Education Quality Improvement Project with the Grant No. JXJYGC2022GX174, the 2022 Education and Teaching Reform Project of the Guangdong Vocational and Technical College Business Education Commission with the Grant [2023] No. 5, the 2023 Guangdong Vocational and Technical College Course Ideological and Political Demonstration Course Project with the Grant No. XJKC202305, and the 2023 Guangdong Vocational and Technical College Course Ideological and Political Demonstration Plan Project with Grant [2023] No. 45.

References

1. Gardner, H.: Multivariate Intelligence (Trans. by Z. Shen). Shanghai Science and Technology Press, Shanghai (1980)
2. Tao, X.: Interpretation of Multiple Intelligence Theory. Kaiming Publishing House, Beijing (2003)
3. Gu, H.: Research on multiple intelligences in teaching evaluation of English majors in the new era. based on the perspective of the undergraduate teaching guidelines for English majors. J. Foreign Lang. **02**, 126–130 (2021)
4. He, Y., Li, Z., Liu, J., Dong, M.: Evaluation of experimental teaching in cell biology based on the theory of multiple intelligences. Chem. Life **40**(06), 965–968 (2020)
5. Suyangna: Research on flipped classroom learning evaluation based on the theory of multiple intelligences and Moodle platform activity recording: a case study of "Multimedia Courseware Design and Development" course practice. Electron. Educ. Res. **37**(04), 77–83 (2016)
6. Li, Y., Wang, L., Li, N.: Research on the growth evaluation system of vocational college students based on multiple intelligence theory. J. China Inst. Labor Relat. **29**(05), 114–117 (2015)
7. Wu, J., Tu, M.: Performance evaluation of counselors based on the theory of multiple intelligences. Sch. Party Build. Ideol. Educ. **28**, 73–75 (2012)
8. Li, H.: Construction of an evaluation system for autonomous learning of college English based on the theory of multiple intelligences. J. Shanxi Univ. Finance Econ. **33**(S2), 82+84 (2011)
9. Xia, X.: Construction of a college English evaluation system under the theory of multiple intelligences. China Adult Educ. **17**, 168–169 (2010)

Study on TNM Classification Diagnosis of Colorectal Cancer Based on Improved Self-supervised Contrast Learning

Tao Lai and Kangshun Li[✉]

School of Mathematics and Informatics, South China Agricultural University,
Guangzhou 510642, China
likangshun@sina.com

Abstract. TNM classification of colorectal cancer is of great significance for doctors to make clinical decision, evaluate patient prognosis and improve treatment. The diagnosis results of TNM classification of colorectal cancer combined with multi-modal medical data are often more accurate than those based on single modal medical data. However, how to balance the redundancy and complementarity of multimodal medical data in deep learning is a difficult problem. Considering the expensive collection and labeling of medical data, we propose an improved self-supervised contrastive learning method for TNM classification of colorectal cancer. Self-supervised contrast learning guides the feature representation of the learning data according to its own supervised information. In the feature extraction stage, we used a deep convolutional neural network with a spatial-channel attention module to extract the features of MRI images, and incorporated clinicopathological parameters in the fully connected layer to achieve multi-modal data fusion. In the comparison of feature similarity, Mahalanobis distance measurement learning method is used to eliminate the scale interference of features between different models. In the contrast loss stage, we design a contrast loss function suitable for multimodal feature representation for model training. In order to verify the effectiveness of the proposed method, experiments were carried out on the collected data sets. The experimental results show that compared with traditional methods, the TNM staging diagnosis technique proposed in this study based on improved self-supervised comparative learning has achieved significant improvement in accuracy and recall rate.

Keywords: Self-supervised Learning · Contrast learning · Multimodal feature study

1 Introduction

Colorectal cancer refers to malignant tumors of the colon and rectum. It is the third most common cancer in the world and the second most common cause of death from cancer [1]. The incidence and mortality of colorectal cancer increase over the past 20 years in our country. The incidence and mortality of colorectal cancer take the second place and the fourth place in our country. According to the statistical report of the National Cancer Institute of the United States, the 5-year survival rate of colorectal cancer is 64.4%, while the 5-year survival rate of Chinese patients with colorectal cancer is 48% [2]. Therefore, the study of TNM staging of colorectal cancer patients is of great significance for doctors to make clinical decisions, evaluate patients' prognosis and improve treatment plans.

The following problems are faced when using multi-modal medical data for learning research:

1) Due to legal and ethical factors, it is difficult to collect medical data, which leads to a small amount of medical image data, and fewer medical images are sketched by radiologists, and the correct delineation and labeling of medical images depends on the professional level of radiologists;

2) Unbalanced categories are common in medical images. For example, in the TNM staging of cancer, the number of patients in stage T3 is the largest, accounting for 70% in some data sets, while patients in stage T1, T2 and T4 only account for 30%. Overfitting is easy to occur when deep learning is used to train data sets with unbalanced categories. The result may be that when external verification set is used to verify the deep learning model, it can be found that the model's performance is not as good as that in our training data set.

3) Data integrity is poor. In the process of sample collection, there are few valid multi-modal samples with both clinical data and imaging omics information of patients;

4) Due to the redundancy and complementarity of multi-modal medical data, in the process of fusion of multi-modal medical data, the deep learning model needs to combine these two factors in the fusion of medical characteristics;

In view of various problems arising in the process of the above research, this paper intends to diagnose the TNM staging of colorectal cancer by comparing the similarity of multi-modal medical data features among different patients through an improved self-supervised comparative learning method. The model improved the contrast learning method in three aspects. Firstly, the deep neural network added a spatial attention module, which increased the weight of the tumor area of interest by setting arbitrary parameters during feature extraction of MRI images in order to extract MRI images that could better distinguish different TNM stages. Second, the Mahalanobis distance based measurement learning method measures the intrinsic correlation between the features and clinical features extracted from MRI images through covariance to deal with the problem of non-independent co-distribution among the dimensions in the high-dimensional linear distribution data. In the third aspect, an improved comparative learning loss function is used.

Contrast learning method is an effective self-supervised learning method [3]. Different from supervised learning, which requires training data with strong supervision information, self-supervised learning method uses the information of the data itself to form supervisory signals, which can be used to guide the feature expression of the learning data, so that samples can be learned in the absence of a large number of labeled data.

The core idea of contrast learning is to compare positive and negative samples in feature space, so as to learn the feature representation of samples, so as to make it as close as possible to the features of positive samples and as different as possible from the feature representation of negative samples. The way to determine positive and negative samples by contrast learning is defined by the agent task, which defines the similarity between samples. For a given sample, those that are similar to it are positive samples, while those that are not similar are negative samples. Data enhancement is a common means to realize the agent task, such as the classic individual discrimination agent task [4].

Unsupervised presentation learning is often used to learn discriminative features from unlabeled data, which helps with downstream tasks. The benefits of unsupervised presentation learning come from two aspects [5]: (1) The pre-training model of unsupervised presentation learning provides good model parameter initialization. We also think of it as a regularization of the model. (2) Unsupervised representation learning General learning mapping from input to output. Therefore, what is learned from the unsupervised model is also applicable to the supervised model. In recent years, unsupervised presentation learning is very popular in the society, and some unsupervised presentation learning models even exceed the performance of supervised learning models. Wu et al. [6] trained an unsupervised learning model to distinguish each sample in a data set. In this way, the unsupervised model can learn image features. K-nearest neighbor (KNN) algorithm can be used as a classifier of learning features, and its performance is better than some supervised learning models. Chen et al. [7] conducted a large number of experiments to explore the key issues of contrast learning. They propose a comparative learning framework called SimCLR. He et al. [8] proposed an unsupervised comparative learning model based on momentum parameter updating. The algorithm maintains a large feature dictionary in the training process, which is advantageous to the unsupervised algorithm requiring a large feature dictionary for comparison learning of a large number of negative samples. Chen et al. [9] proposed an unsupervised representation model based on Siamese network. They constructed an asymmetrical Siamese network to avoid the collapse of the solution. Unsupervised representation learning is also a hot topic in the field of medical imaging. Haghighi et al. [10] and Zhou et al. [11] designed a series of image enhancement methods based on the characteristics of medical images, and then the autoencoder learned the features from the processed medical images. However, the above model is only trained on the single mode and does not induce the information of the multi-mode data.

To overcome the limitations of a single mode, some researchers use relevant data from different sources (i.e., multimodal data) to improve the performance of disease prediction models. Multimodal model is widely used in RGB-D semantic segmentation [12–14], visual question answering [15–17], action recognition [18–20] and other fields. It can be used to integrate different modal data (such as natural image and depth image, text data and image or video data and voice data). In the medical field, according to whether the multi-modal data types are the same, the multi-modal models are divided into homogeneous fusion (such as MRI and CT images) [21–23] and non-homogeneous fusion (such as CT images and clinical data) [24, 25], [26]. In this paper, we are interested in models that fuse medical images and clinical data. Xu et al. [24] combined cervigram and clinical data in a multimodal model, which used convolutional neural networks (CNN) to extract image features and fused image features with clinical data. Spasov et al. [25] used MRI and clinical data as inputs to a predictive model of Alzheimer's disease. The difference between [24] and [25] is that [25] uses a fully connected layer to extract features from clinical data. Guan et al. [26] transforms clinical features into shapes that are identical to image features and can therefore be merged using the self-attention module. In addition to building an end-to-end model by explicitly combining multi-modal features, Reda et al. [27] trained each sub-model of different modes independently, and then aggregated their output through an additional classifier model. Parisot et al. [28] established the interaction model between medical images and clinical data through graph neural network (GNN). In the figure, medical images and clinical data are represented by nodes and edges, respectively. Where, nodes are image features obtained by feature selection strategy, and edges are similarity between nodes measured by clinical data information.

2 Methods

2.1 An Improved Deep Convolutional Neural Network for Feature Extraction in MRI Images

Traditional convolutional neural networks often lack effective integration of local spatial information and channel information when processing image information, resulting in limited model performance. In order to further improve the performance of convolutional neural networks, the researchers proposed an attention mechanism, which allows the model to focus more on key information by assigning different weights to different parts of the feature map. Among them, the spatial-channel attention module combines spatial attention and channel attention to capture key features in images more comprehensively, thus improving the accuracy of the model.

Spatial-channel attention module is a kind of module which combines spatial attention and channel attention to improve the processing ability of convolutional neural networks. By introducing an attention mechanism, the module enables the model to adaptively adjust the weights on different Spaces and channels, thereby highlighting key features and suppressing irrelevant information. Spatial attention is mainly concerned with the spatial position information in the image. By assigning different weights to different positions of the feature map, the model can pay more attention to the key areas. This attention mechanism helps the model accurately identify the target object in the complex background. Channel attention is mainly concerned with channel information in feature graphs. By assigning different weights to different channels, the model can pay more attention to the channels with large amount of information. This attention mechanism helps the model to extract key features more efficiently when processing multi-channel images.

The application of spatial-channel attention module in deep convolutional neural networks can effectively improve the performance of the model. Specifically, the module can be embedded in different layers of the network, and by adjusting the weights adaptively, the model focuses more on key information when extracting features. The introduction of this attention mechanism can not only improve the accuracy of the model, but also reduce the amount of calculation and improve the efficiency of the model to a certain extent. In the image classification task, by introducing the spatial-channel attention module, the model can more accurately identify the target object in the image, so as to improve the accuracy of classification.

2.2 A Metric Learning Method Based on Mahalanobis Distance

In the field of artificial intelligence and machine Learning, Metric Learning is an important research direction aimed at learning an appropriate distance metric in which similar data points are closer together and dissimilar data points are farther apart. The measures obtained from this learning can be used for a variety of tasks, such as classification, clustering, information retrieval, etc.

The metric learning method based on Mahalanobis Distance is one of them. Mahalanobis distance is a distance measure that takes into account the covariance structure of data, not only the absolute distance between data points, but also the correlation between different dimensions. Compared with Euclidean distance, Maanobis distance is more flexible and can better adapt to the distribution characteristics of data. Euclidian distance are shown in Fig. 1. Mahalanobis distance are shown in Fig. 2.

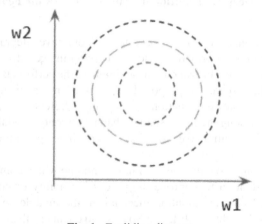

Fig. 1. Euclidian distance.

Mahalanobis distance-based metric learning methods usually include the following steps. Feature extraction: First, the features used to measure learning are extracted from raw data. These features can be designed by hand or learned automatically through methods such as deep learning. Mahalanobis distance calculation: For two given data points, the Mahalanobis distance between them is calculated. Metric learning: Learning a suitable Mahalanobis distance measure by optimizing the algorithm. This usually involves minimizing a loss function that measures the consistency between the learned distance measure and the data label. Common Loss functions include Contrastive Loss and Triplet Loss. Application and optimization: Apply the learned Mahalanobis distance measure to a specific task.

The metric learning method based on Mahalanobis distance shows good performance in many tasks, especially when dealing with data with complex distribution characteristics. By learning a suitable Mahalanobis distance measure, the intrinsic structure of the data can be better captured, thus improving the accuracy and efficiency of the task.

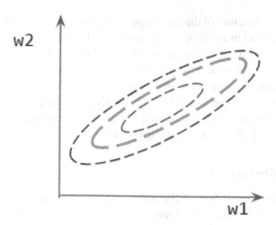

Fig. 2. Mahalanobis distance.

2.3　Dimensional Adaptive Contrast Loss Function

In deep learning, the choice of loss function has a decisive effect on the performance of the model. In view of the complexity and multi-dimensionality of the data, we design an adaptive dimensional contrast loss function, which is designed to dynamically adjust the loss weights in different dimensions to optimize the performance of the model. For a given positive sample pair (x_i, x_j) and negative sample pair (y_i, y_j), the adaptive dimensional contrast loss function is defined as:

$$L = \sum_{1}^{n} w_d * [max(0, margin - D_d(f(x_i - f(x_j)))) + max(0, D_d(f(y_i) - f(y_j)))]$$

(1)

where $\sum_{1}^{n}$ means summing over all dimensions d. w_d is the weight of the d-th dimension, calculated from the variance of the data on that dimension. $D_d f(x_i) - f(x_j)$ means to calculate the Mahalanobis distance between $f(x_i)$ and $f(x_j)$ in the d-th dimension. Margin is a preset minimum distance threshold.

Weight adaptive adjustment: The core of the loss function is the adaptive adjustment of the weight w_d. To achieve this, we introduce the variance of the data on each dimension as the basis for the weights. Specifically, we calculate the variance of the model's output on the (d) dimension of all samples and take its reciprocal as the weight in that dimension.

$$w_d = \frac{1}{v_d(x_i)}$$

(2)

Here, $v_d(x_i)$ represents the output variance of the sample x_i in dimension (d). Because greater variance means more drastic changes in the data on this dimension, a smaller weight is assigned; A smaller variance means that the data in this dimension is more stable, so a larger weight is assigned.

By introducing the adaptive dimension contrast loss function, the model can deal with different dimensions differently in the learning process. This helps the model better capture the intrinsic structure of the data, especially when dealing with high-dimensional and complex data. In addition, by dynamically adjusting the weights of each dimension, the model can be more flexible to adapt to different tasks and data distributions, thereby improving its generalization ability and performance. In short, the adaptive dimensional contrast loss function is a loss function designed for multidimensional data, which optimizes the performance of the model by dynamically adjusting the weights on different dimensions. This design allows the model to better adapt to the characteristics of the data and show better performance in a variety of tasks.

2.4 Model Architecture

The model architecture consists of the following key components to achieve accurate prediction of the TNM stage of colorectal cancer. Improved self-supervised contrast learning is the foundation of the entire model architecture. Through self-supervised learning, the model can be pre-trained using unlabeled pathological image data to learn the feature representation of the lesion area. Contrast learning mechanism enhances the model's ability to recognize lesion features by comparing the similarity between different images. In the feature extraction network of MRI images, we introduce a spatial-channel attention module. This module can automatically learn the spatial and channel dependencies of the input data, thereby enhancing the model's ability to extract key features. By focusing on important spatial locations and channels, the model can better understand the internal structure of the data and improve the quality of representation learning. In self-supervised contrast learning, we use Mahalanobis distance to measure the distance between positive and negative sample pairs. By calculating the distance of sample pairs in Markov space, we can more accurately judge the degree of similarity between them, so as to optimize the representation learning of the model. The traditional contrast loss function usually fixes the distance threshold between the positive and negative sample pairs. However, in practical applications, the dimensions and distribution of the data may change. To solve this problem, we design a dimensionally adaptive contrast loss function. The function can dynamically adjust the distance threshold according to the dimension and distribution of data, so that the model can maintain good performance under different conditions. Model architecture are shown in Fig. 3.

3 Experiment and Result

3.1 Controlled Experiment

The computing environment used in this research experiment is configured as follows: The experiment is conducted based on Ubuntu 20.04.3 LTS, which provides a stable and safe environment and provides good support for deep learning experiments. The CPU model used in the experiment is an Intel(R) Core(TM) i9-10900K central processing unit with a main frequency of up to 3.70 GHZ. This high-performance processor provides powerful computing power for a large number of computing tasks in the experiment, ensuring the smooth progress of the experiment. The experiment is equipped with an NVIDIA Geforce RTX 3090 Founders Edition graphics card, which has powerful graphics processing capabilities and is especially suitable for large-scale matrix computing

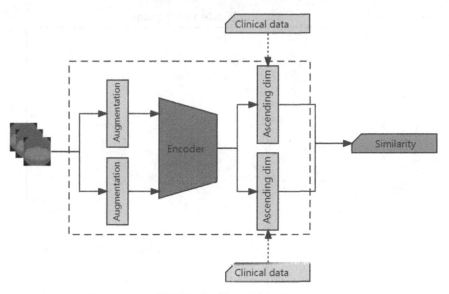

Fig. 3. Model architecture.

and parallel computing in deep learning models, significantly improving the efficiency of the experiment. In this study, PyTorch is adopted as the main deep learning framework. PyTorch has been widely used in academia and industry for its simple and easy to use interface and efficient computing performance. The experiment adopts Python 3.8.0 as the programming language, and Python has become the preferred language in the field of deep learning due to its legibility, simplicity and rich library resources. In order to improve the development efficiency, Visual Studio was adopted as the integrated development environment. Visual Studio provides powerful functions such as code editing, debugging and version control, and provides convenient support for the development and testing of experiments. The advanced configuration of the experimental environment not only meets the high requirements of computing resources for deep learning experiments, but also ensures the development efficiency and convenience. Conducting experiments in such an environment is conducive to obtaining stable and reliable experimental results.

3.2 Experimental Result

The experimental results showed that the accuracy of T stage of colorectal cancer finally converged to 0.8360, the accuracy of M stage to 0.9136, and the accuracy of N stage to 0.7540. The test results are shown in Fig. 4.

3.3 Discussion

This study significantly improved the overall effectiveness of the colorectal cancer TNM stage prediction model based on improved self-supervised contrast learning by integrating deep neural networks of spatial-channel attention modules, multi-modal data fusion

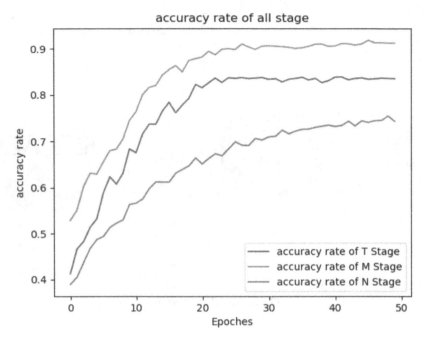

Fig. 4. Accuracy rate of all stage.

techniques, and dimensional adaptive contrast loss functions. Firstly, by introducing the space-channel attention module, the model can better focus on the key lesion areas in the pathological images, improving the accuracy and pertinence of feature extraction. This attentional mechanism enhances the model's ability to process complex image content, which enables it to show higher accuracy in TNM staging prediction. Secondly, the use of multi-modal data fusion technology provides more abundant tumor characteristics information for the model. Combined with pathological image data and clinicopathological parameters, the model can comprehensively analyze tumors from multiple dimensions, enhancing the comprehensiveness and reliability of prediction. This data fusion approach not only expands the input range of the model, but also helps capture the complex biology of the tumor. Finally, the design of dimensionally adaptive contrast loss function further improves the training efficiency and prediction performance of the model. By dynamically adjusting the dimensions of the comparison samples, the loss function enables the model to adapt more flexibly to data of different dimensions, thus optimizing the feature learning process. This loss function not only enhances the generalization ability of the model, but also helps to improve the accuracy and stability of TNM stage prediction. By integrating spatial-channel attention modules, multimodal data fusion techniques, and dimensional adaptive contrast loss functions, our TNM stage prediction model has achieved significant improvements in overall effectiveness. These improvements not only enhance the feature extraction capability and data analysis dimension of the model, but also improve the prediction accuracy and reliability of the model. This provides a new and powerful tool for accurate diagnosis and treatment of colorectal cancer and is expected to bring substantial improvements to clinical practice.

4 Conclusion and Future Work

In this study, we used an improved self-supervised contrastive learning method to predict the TNM stage of colorectal cancer. TNM stage is an important prognostic indicator of colorectal cancer, and accurate prediction is helpful to make personalized treatment plan. Traditional supervised learning methods rely on a large number of labeled data for training, but in practical applications, such data is often difficult to obtain. Self-supervised learning can learn useful feature representations without labeled data, thus solving the problem of insufficient labeled data. On the basis of self-supervised contrast learning, this experiment further improves the effect of feature learning by improving the feature extraction network, integrating multi-modal data and improving the contrast loss function. The experimental results show that the improved self-supervised contrast learning algorithm can effectively use unlabeled data to learn the feature representation of the lesion region, and achieve good performance in the TNM stage prediction task. This method not only alleviates the problem of insufficient labeled data, but also improves the generalization ability and prediction accuracy of the model. Although we have achieved remarkable research results, there are still many directions worthy of further exploration and improvement. First, we plan to collect more types and sources of pathological image data and clinicopathological parameters to expand the training set of the model, thereby further improving the generalization ability of the model. Secondly, we will further study the mechanism of contrast learning and explore more effective feature learning methods to improve the accuracy of the model's recognition of complex lesion areas. In addition, we will consider combining other advanced deep learning techniques, such as domain adaptive, knowledge distillation, etc., with improved self-supervised contrast learning to further enhance the performance of the model. We believe that with the continuous progress of technology and the accumulation of data, the TNM stage prediction method based on improved self-supervised comparative learning will play a more important role in future clinical practice, and provide strong support for the accurate diagnosis and treatment of colorectal cancer.

Acknowledgments. This project relies on Professor Li Kangshun's team, which has 1 national key research and development plan, 1 key research and development project of Guangdong Province and 1 international cooperation project of Huangpu District. The disposable scientific research expense is more than 2 million yuan, which can provide sufficient funding for this project. At the same time, the laboratory conducts academic activities related to this topic every week, forming a good academic atmosphere.

References

1. Wu, Z., Xiong, Y., Yu, S.X., et al.: Unsupervised feature learning via non-parametric instance discrimination. In: Proceedings of the IEEE Conference on Computer Vision and Pattern Recognition, pp. 3733–3742 (2018)
2. Goodfellow, I., Bengio, Y., Courville, A.: Deep Learning. MIT Press, Cambridge (2016)
3. Wu, Z., Xiong, Y., Yu, S.X., Lin, D.: Unsupervised feature learning via non-parametric instance discrimination. In: Proceedings of the IEEE Conference on Computer Vision and Pattern Recognition, pp. 3733–3742 (2018)

4. Chen, T., Kornblith, S., Norouzi, M., Hinton, G.: A simple framework for contrastive learning of visual representations. In: International Conference on Machine Learning, pp. 1597–1607. PMLR (2020)

5. He, K., Fan, H., Wu, Y., Xie, S., Girshick, R.: Momentum contrast for unsupervised visual representation learning. In: Proceedings of the IEEE/CVF Conference on Computer Vision and Pattern Recognition, pp. 9729–9738 (2020)

6. Chen, X., He, K.: Exploring simple SIAMESE representation learning. In: Proceedings of the IEEE/CVF Conference on Computer Vision and Pattern Recognition, pp. 15750–15758 (2021)

7. Haghighi, F., Hosseinzadeh Taher, M.R., Zhou, Z., Gotway, M.B., Liang, J.: Learning semantics-enriched representation via self-discovery, self-classification, and self-restoration. In: Martel, A.L., et al. (eds.) MICCAI 2020. LNCS, vol. 12261, pp. 137–147. Springer, Cham (2020). https://doi.org/10.1007/978-3-030-59710-8_14

8. Zhou, Z., et al.: Models genesis: Generic autodidactic models for 3d medical image analysis. In: Shen, D., et al. (eds.) MICCAI 2019. LNCS, vol. 11767, pp. 384–393. Springer, Cham (2019). https://doi.org/10.1007/978-3-030-32251-9_42

9. Park, S.-J., Hong, K.-S., Lee, S., RDFnet: RGB-D multi-level residual feature fusion for indoor semantic segmentation. In: Proceedings of the IEEE International Conference on Computer Vision, pp. 4980–4989 (2017)

10. Valada, A., Mohan, R., Burgard, W.: Self-supervised model adaptation for multimodal semantic segmentation. Int. J. Comput. Vision **128**, 1239–1285 (2020)

11. Ji, W., et al.: Calibrated RGB-D salient object detection. In: Proceedings of the IEEE/CVF Conference on Computer Vision and Pattern Recognition, pp. 9471–9481 (2021)

12. Antol, S., et al.: VQA: visual question answering. In: Proceedings of the IEEE International Conference on Computer Vision, pp. 2425–2433 (2015)

13. Ilievski, I., Feng, J.: Multimodal learning and reasoning for visual question answering. In: Advances in Neural Information Processing Systems, vol. 30 (2017)

14. Zheng, W., Yin, L., Chen, X., Ma, Z., Liu, S., Yang, B.: Knowledge base graph embedding module design for visual question answering model. Pattern Recogn. **120**, 108153 (2021)

15. Garcia, N.C., Morerio, P., Murino, V.: Modality distillation with multiple stream networks for action recognition. In: Ferrari, V., Hebert, M., Sminchisescu, C., Weiss, Y. (eds.) ECCV 2018. LNCS, vol. 11212, pp. 106–121. Springer, Cham (2018). https://doi.org/10.1007/978-3-030-01237-3_7

16. Ren, Z., Zhang, Q., Gao, X., Hao, P., Cheng, J.: Multi-modality learning for human action recognition. Multimedia Tools Appl. **80**, 16185–16203 (2021)

17. Song, S., Liu, J., Li, Y., Guo, Z.: Modality compensation network: cross-modal adaptation for action recognition. IEEE Trans. Image Process. **29**, 3957–3969 (2020)

18. Guo, Z., Li, X., Huang, H., Guo, N., Li, Q.: Deep learning-based image segmentation on multimodal medical imaging. IEEE Trans. Radiat. Plasma Med. Sci. **3**, 162–169 (2019)

19. Cheng, X., Zhang, L., Zheng, Y.: Deep similarity learning for multimodal medical images. Comput. Methods Biomech. Biomed. Eng. Imaging Visual. **6**, 248–252 (2018)

20. Dolz, J., Ben Ayed, I., Desrosiers, C.: Dense multi-path u-net for ischemic stroke lesion segmentation in multiple image modalities. In: Crimi, A., Bakas, S., Kuijf, H., Keyvan, F., Reyes, M., van Walsum, T. (eds.) BrainLes 2018. LNCS, vol. 11383, pp. 271–282. Springer, Cham (2019). https://doi.org/10.1007/978-3-030-11723-8_27

21. Xu, T., Zhang, H., Huang, X., Zhang, S., Metaxas, D.N.: Multimodal deep learning for cervical dysplasia diagnosis. In: Ourselin, S., Joskowicz, L., Sabuncu, M.R., Unal, G., Wells, W. (eds.) MICCAI 2016. LNCS, vol. 9901, pp. 115–123. Springer, Cham (2016). https://doi.org/10.1007/978-3-319-46723-8_14

22. Spasov, S.E., Passamonti, L., Duggento, A., Lio, P., Toschi, N.: A multi-modal convolutional neural network framework for the prediction of Alzheimer's disease. In: 2018 40th Annual International Conference of the IEEE Engineering in Medicine and Biology Society (EMBC), pp. 1271–1274. IEEE (2018)
23. Guan, Y., et al.: Predicting esophageal fistula risks using a multimodal self-attention network. In: de Bruijne, M., et al. (eds.) MICCAI 2021. LNCS, vol. 12905, pp. 721–730. Springer, Cham (2021). https://doi.org/10.1007/978-3-030-87240-3_69
24. Reda, I., et al.: Deep learning role in early diagnosis of prostate cancer. Technol. Cancer Res. Treatment 17, 1533034618775530 (2018)
25. Parisot, S., et al.: Disease prediction using graph convolutional networks: application to autism spectrum disorder and Alzheimer's disease. Med. Image Anal. 48, 117–130 (2018)

Construction and Quality Evaluation of Learning Motivation Model from the Perspective of Course Ideology and Politics

Luyan Lai[(⊠)] and Yongdie Che

School of Business, Guangdong Polytechnic, Foshan 528041, Guangdong, China
332075619@qq.com

Abstract. This study is based on the perspective of curriculum ideological and political education, and establishes a learning motivation model consisting of an internal driving force subsystem and an external driving force system. A combination of quantitative and qualitative research methods is used to collect data on the implementation effect of curriculum ideological and political education based on the learning motivation model through questionnaire surveys and field observations. The quality of the model is evaluated. The research results found that learning motivation is of great significance for the effective implementation of ideological and political education in courses. It can improve students' learning enthusiasm and initiative, thereby promoting the comprehensive improvement of their ideological and political literacy, and providing theoretical and practical reference for the implementation of ideological and political education in universities.

Keywords: Course ideological and political education · Learning motivation model · Quality evaluation · Indicator system

1 Introduction

On May 28, 2020, the Ministry of Education issued the "Guidelines for the Construction of Ideological and Political Education in Higher Education Curriculum", which systematically explains the goals, requirements, and content of ideological and political education in the new era, focusing on the fundamental issue of "what kind of people to cultivate, how to cultivate them, and for whom to cultivate them". Building a sound curriculum ideological and political system is a necessary prerequisite for implementing the fundamental task of cultivating morality and nurturing talents, a powerful guarantee for comprehensively enhancing the ideological awareness of college teachers and students, and an important measure to strengthen the construction of college cultural security. The report of the 20th National Congress of the Communist Party of China also pointed out that "education is the great plan of the country and the Party". The fundamental issue of education is what kind of people to cultivate, how to cultivate them, and for whom to cultivate them. The fundamental aspect of education lies in moral character. It can be seen that ideological and political education in the curriculum is an important way to achieve comprehensive education, and it is a major deployment for the Party and the country to fully implement the fundamental task of moral education in the new era and new journey.

In October 2020, the Central Committee of the Communist Party of China and the State Council issued the "Overall Plan for Deepening the Reform of Education Evaluation in the New Era", which pointed out that "fully implement the Party's education policy, adhere to the socialist direction of education, implement the fundamental task of cultivating morality and talents, follow the laws of education, and systematically promote the reform of education evaluation". The evaluation of the quality of ideological and political education in courses has a significant impact on the moral education of universities, and it is particularly important to establish a scientific and reasonable evaluation of the quality of ideological and political education in courses.

The purpose of this study is to establish a learning motivation model based on the perspective of curriculum ideological and political education, and to evaluate its application quality. The aim is to reveal the inherent relationship between learning motivation and curriculum ideological and political effects, improve students' initiative and enthusiasm in learning, and promote the comprehensive improvement of students' political literacy. At the same time, this study not only provides a theoretical basis for the excavation of ideological and political elements in courses, but also establishes an indicator system for quality evaluation, which has important reference and reference value for the research of ideological and political construction in university courses.

2 Research Background

2.1 Current Status of Research on Learning Motivation Models

In recent years, research on learning motivation models has attracted widespread attention from scholars. Based on the theories of group dynamics and social cohesion, Xu et al. [1] constructed an online collaborative learning group dynamics model from four dimensions: realistic purpose, dynamic mechanism, spatial form, and practical dimension. Taking "Innovative Thinking and Its Cultivation Methods" as an example, they designed a case from the perspectives of individual learner task driven and collaborative learner group driven. They proposed strategies to enhance the group dynamics of online collaborative learning, including emphasizing the cultivation of higher-order thinking abilities, creating an online collaborative atmosphere, enhancing teacher emotional and social support services, and conducting interdisciplinary research based on cognitive neuroscience, achieving good teaching results. Ma et al. [2] combined the current grassroots medical service system in China, centered around general practitioners, and established a general practitioner learning motivation model using multidisciplinary comprehensive methods and learning perspectives. This provides a researchable direction for improving the learning motivation of general practitioners and increasing public recognition of community health. Xu et al. [3] established a dynamic model for the evolution of ubiquitous learning resources from the external forces of user selection and environmental selection, as well as the internal forces of self-organizing behavior. Based on the definition of the essence, characteristics, growth, and social construction of knowledge from a postmodern perspective, they argued that ubiquitous learning resources are evolutionary resources. From an ecological theory perspective, they demonstrated that ubiquitous learning resources are a living organism that, together with users and learning environments, constitute the ecological elements of the ubiquitous learning ecosystem. From the perspectives of users and environments, they proposed strengthening natural selection to promote resource evolution, promoting the elimination of inferior resources, and promoting the change and improvement of learning environments. Wen et al. [4] proposed a multi-point and multi-dimensional learning motivation evaluation model that runs through the four-year undergraduate education of students. They summarized and summarized 13 variable factors of learning motivation, and conducted a preliminary analysis of the sensitivity of learning motivation to variable factors among students of different grades. The research results were applied to guide the undergraduate training work of the author's unit, with significant benefits. Ruan [5] conducted an in-depth analysis and synthesis of various motivational factors that constrain learning, proposed a three-layer structural system of learning motivation for college students, established an artificial intelligence probability reasoning mathematical model, and relied on the MYCIN credibility measurement method to analyze the contribution and negative impact of learning motivation and its mutual constraints on learning. She deeply explored the learning motivation, provided scientific evaluation criteria for learning motivation, and provided support for the development of modern educational theory.

2.2 Current Status of Research on Curriculum Ideological and Political Evaluation

In recent years, there has been a lot of research on the evaluation of ideological and political education in courses. The author searched and retrieved nearly 700 records on China National Knowledge Infrastructure using the keywords of ideological and political education and evaluation in courses. This indicates that universities across the country also attach great importance to the evaluation of their effectiveness in implementing ideological and political education in courses. Guided by the theory of curriculum evaluation, Xie et al. [6] embody the basic idea of human-machine collaboration, leverage the role of data elements, construct a human-machine collaborative diagnosis and evaluation model for ideological and political education in university courses, and form a "four-dimensional and multi-modal" new model for data collection and evaluation practice of ideological and political education in university courses. Yang [7] constructed a "two-stage and six dimensional" dynamic evaluation model for professional spirit based on the psychological development laws of college students. In the middle stage of the academic stage, he evaluated professional sentiment, including three dimensions: professional identity, professional learning engagement, and professional psychological contract. In the end stage, he evaluated professional beliefs, including three dimensions: professional knowledge acquisition, professional aspirations, and scientific moral responsibility. This proposed a theoretical attempt for the evaluation of ideological and political education in university courses. Based on the theoretical framework of the CIPP evaluation model, Xu et al. [8] used policy texts such as the Ministry of Education's "Guidelines for the Construction of Ideological and Political Education in Higher Education Curriculum" and qualitative interview results as the basis. On the basis of selecting evaluation indicators for ideological and political education activities in primary courses, they collected research samples through measurement tools (scales). After conducting project analysis, factor analysis, and reliability testing on the sample data, a stable four-dimensional structure of "background evaluation, input evaluation, process evaluation, and result evaluation" was formed. Eleven secondary indicators (public factors) were separated, including political environment, course resources, teaching plans, and teaching effects. At the same time, the factor score coefficients generated during the factor analysis process were used to assign weights to each level of indicators, thus completing the task. Construction of indicator system. Zhang et al. [9] proposed to shift towards interdisciplinary practice, achieve logical transformation of curriculum ideological and political evaluation, form a diversified tendency of evaluation subjects, and build a community of curriculum ideological and political evaluation. Sun et al. [10] used literature research, survey questionnaires, and expert consultation methods. Based on thorough research, three subsystems can be designed according to the roles and identity characteristics of students, peer experts, and teachers in the process of ideological and political education in science and engineering courses. Each subsystem includes evaluation items and evaluation standards, and weights are assigned to each evaluation item using the Network Analytic Hierarchy Process. Together, they form an evaluation index system for ideological and political education in science and engineering courses. The aim is to provide feedback to teachers from different perspectives through multiple evaluation subjects, promote the continuous improvement of ideological and political education abilities of science and engineering professional teachers, and provide reference for the construction of evaluation systems for ideological and political education in other subject courses. Duan [11] constructed an evaluation system for curriculum ideological and political education from the aspects of evaluation subjects, evaluation methods, and evaluation index systems. However, She et al. [12] proposed that the reform and development of ideological and political theory course evaluation should adhere to the principles of "scientific and effective, improving result evaluation, strengthening process evaluation, exploring value-added evaluation, and improving comprehensive evaluation", update the concept of course evaluation, strengthen process evaluation, highlight the role of evaluation subjects, advocate for diversified evaluation, and promote the reform and development of course evaluation. Starting from the practical significance of exploring the effectiveness evaluation system of curriculum ideological and political education, Su et al. [13] used the Analytic Hierarchy Process as a support to construct a curriculum education quality evaluation model through five dimensions: curriculum design, teacher team, student cognition, development evaluation, and institutional design. They also clarified the hierarchical relationship between the various execution elements of curriculum ideological and political education from a micro level, clarified basic responsibilities, and provided reference for curriculum teaching reform and education quality evaluation.

3 Construction of a Learning Motivation Model from the Perspective of Course Ideology and Politics

3.1 Analysis of Learning Motivation from the Perspective of Course Ideology and Politics

(1) Analysis of the Nature and Objectives of Course Ideological and Political Education

In recent years, with the continuous development and progress of the education industry, the importance of integrating ideological and political education into the curriculum has been increasingly recognized. In the context of deepening educational reform and actively promoting the construction of quality education, ideological and political education in the curriculum has become an important component of cultivating students' comprehensive qualities and promoting their comprehensive development. As an important component of the curriculum, ideological and political education in universities aims to guide students to establish correct values, enhance political awareness, and improve political literacy.

However, it is worth noting that the effectiveness of ideological and political education in the curriculum largely depends on the learning motivation of students. Learning motivation reflects the intrinsic motivation of students to participate in learning activities and determines the success or failure of learning. Therefore, exploring the relationship between learning motivation and ideological and political education in the curriculum, constructing a learning motivation model, is of great significance for optimizing the implementation of ideological and political education, improving students' initiative and enthusiasm in learning, and cultivating their political literacy.

Course ideological and political education, as an important component of higher education, has unique characteristics and goals. Below, we will explore the nature and objectives of ideological and political education in the curriculum.

1) The Nature of Course Ideological and Political Education.

Course ideological and political education is not a specific course, it is different from subject courses, focusing on cultivating students' ideological, moral, and social responsibility. Firstly, the ideological and political education curriculum aims to guide students to think about life and cultivate their correct worldview, outlook on life, and values. Secondly, the course of ideological and political education has theoretical significance. By explaining ideological and theoretical knowledge, students can have a deeper understanding of society, history, and human development. Again, the ideological and political education curriculum has practicality, not only in imparting knowledge, but also in cultivating students' practical abilities and innovative spirit. Finally, the course of ideological and political education is comprehensive and covers a wide range of fields, including politics, economy, culture, society, etc., aiming to help students fully understand and grasp the direction and laws of social development.

2) The goal of ideological and political education in the curriculum.

The goal of ideological and political education in the curriculum is to cultivate students' ideological and moral literacy and social responsibility, making them become socialist builders and successors with comprehensive moral, intellectual, and physical development. Firstly, the goal of course ideological and political education is to cultivate students' political awareness. Through studying course ideological and political education, students can establish correct political concepts, enhance their sense of identification and belonging to the Party and the country. Secondly, the goal of ideological and political education in courses is to cultivate students' moral character. Through the study of ideological and political education in courses, students can establish correct moral concepts, cultivate good moral qualities and moral behavior habits. Thirdly, the goal of ideological and political education in the curriculum is to cultivate students' sense of social responsibility. Through the study of ideological and political education in the curriculum, students can recognize their responsibilities and obligations as members of society, actively participate in social practice activities, and make contributions to social development. Finally, the goal of curriculum ideological and political education is to cultivate students' innovative abilities. Through the study of curriculum ideological and political education, students can cultivate innovative thinking and abilities, which can provide intellectual support for the development of the country and society.

Given the ideological, theoretical, practical, and comprehensive characteristics of curriculum ideological and political education, constructing a model suitable for students' learning motivation should focus on leveraging the nature and goals of curriculum ideological and political education to promote the comprehensive improvement of students' ideological and political literacy.

(2) Analysis of the Relationship between Course Ideology and Learning Motivation

In today's higher education, curriculum ideological and political education has been widely recognized as an important way to cultivate students' ideological and moral qualities and shape good personalities. However, many students lack positive learning motivation during the learning process, which poses certain challenges to the effective implementation of ideological and political education in the curriculum. Therefore, studying the correlation between curriculum ideology and student motivation is of great significance.

Course ideological and political education can stimulate students' learning motivation. The ideological and political content of the course covers knowledge of ethics and morality, the spirit of the rule of law, and socialist core values, and has strong ideological and social significance. Through the implementation of ideological and political education courses, students can deeply understand and feel the importance of this knowledge for personal development and social progress, thereby stimulating their learning motivation. Course ideological and political education can enhance students' learning enthusiasm and initiative. The course of ideological and political education emphasizes the cultivation of students' critical thinking ability and innovative consciousness. By inspiring students to think and discuss, it stimulates their interest and initiative in learning. Students can freely express their opinions and ideas in the ideological and political education curriculum, actively participate in classroom discussions and learning activities, thereby improving their learning enthusiasm and initiative.

Course ideological and political education plays an important role in comprehensively improving students' ideological and political literacy. Course ideological and political education is not only about imparting knowledge, but more importantly, it cultivates students' ideological and moral character, moral emotions, and social responsibility. Through the implementation of ideological and political courses, students can understand and master the core socialist values, form correct values and moral concepts, and thereby enhance their own ideological and political literacy.

3.2   Construction of a Learning Motivation Model from the Perspective of Course Ideology and Politics

The principle of Marxist dialectical materialism tells us that the development of things is the result of the joint action of internal and external factors. Internal factors are the basis for the development of things, and they determine the basic trend of their development. External factors are the external conditions for the development of things, which delay or accelerate the process of development. The learning motivation system can only fully exert its effectiveness when internal and external factors work together.

Based on the above analysis, we believe that the learning motivation model from the perspective of ideological and political education in the curriculum is a comprehensive system, consisting of an internal driving force subsystem and an external driving force subsystem, as shown in Fig. 1. The aim is to enhance learning interest, achieve learning goals, cultivate positive learning attitudes of learners, improve learning efficiency, and promote the growth of learning benefits. The internal driving force subsystem is composed of four internal motivations: cognitive motivation, psychological motivation, emotional motivation, and willpower. The external driving force subsystem is composed of external motivations such as survival motivation, learning environment motivation, social environment motivation, and behavioral motivation. Among them, internal influencing factors are a driving force that learners themselves generate to drive learning, known as internal motivation. External influencing factors are another driving force for learning generated by the environment, external stimuli, or pressure, known as external motivation.

(1) Internal motivation
    1) Cognitive motivation: One of the factors driving learning is cognitive motivation, which comes from the desire for knowledge and intellectual growth. The creation of cognition can stimulate learners to pursue new learning opportunities and challenges, new fields and develop new skills. Learners acquire new knowledge and achieve new achievements to stimulate their inner motivation, and gain satisfaction and self-realization through continuous growth.
    2) Psychological motivation: This theory is based on the personality motivation theory of psychologist Freud. The theory of psychological motivation holds that human character defects are not inherent and stable, but can be corrected, which requires the creation of many situational education and training to reduce the incidence of accidents.
    3) Emotional motivation: Learning motivation is closely related to students' emotional experiences. During the learning process, students are influenced by both positive and negative emotions, resulting in varying degrees of learning motivation. Positive emotions such as interest, fun, and pride can enhance students'

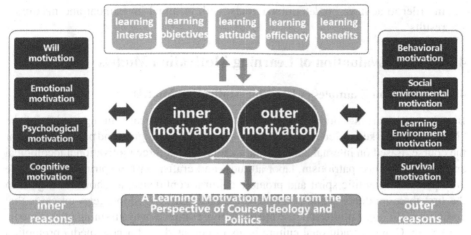

Fig. 1. A Learning Motivation Model from the Perspective of Course Ideology and Politics.

learning motivation. Negative emotions such as anxiety, boredom, and frustration can reduce students' motivation to learn. Therefore, teachers should focus on cultivating students' positive emotions, improving their emotional participation, and promoting the formation and development of learning motivation in teaching.

4) Willpower: Willpower is the conscious determination of purpose, regulation and domination of action, and the achievement of predetermined goals through overcoming difficulties and setbacks. It mainly reflects the purposefulness, hierarchy, intensity, external stability, internal stability, effectiveness, and meticulousness of the learner's behavioral value.

(2) External motivation

1) Survival motivation: usually refers to the purpose for which learners live, the goals and directions for living in the world, etc. The survival motivation may vary at each stage, and it is also in the process of development. Generally speaking, it includes learners' work and employment, economic foundation, survival needs, elderly care and childcare, and so on.

2) Learning environment motivation: Learning environment motivation is one of the important factors in the learning process of learners, including factors such as school, family, and society. As for schools, it includes at least school management, professional settings, teacher level, campus environment, practical training environment, teaching mode, reward and punishment system, various activities, and so on.

3) Social environmental driving force: Social environmental driving force provides various possibilities for individual development, making the development provided by inheritance a reality, and thus driving the physical and mental development of individuals. For example, the power of industry role models, internet influence, social needs, parental expectations, and friend advice.

4) Behavioral motivation: Behavioral motivation refers to various reactions made by individuals to maintain their survival and racial continuity, adapt to constantly changing complex environments, including physical and psychological activities. In the design process, efforts are made to create behaviors that learners can choose and plan

in order to achieve predetermined goals, with minimal investment and maximum results.

4   Quality Evaluation of Learning Motivation Models

4.1   Application Examples of Learning Motivation Models

Based on the learning motivation model, we take the practical training module of the "New Media Marketing" course as an example to create ideological and political goals for the course. Based on internal and external factors, we set three motivational ideological and political points: patriotism, labor attitude, and craftsmanship spirit. The main line is to promote scientific spirit and promote cultural confidence, as shown in Fig. 2. In the first three modules of the course, each introduced case adopts relevant cases that combine traditional culture and new media. In this comprehensive training of new media marketing, Chinese traditional culture is also integrated, and a new media promotion and dissemination project for Foshan's local intangible cultural heritage, the Foshan enterprise women's clothing brand "Tingxi Clothing", is introduced, allowing learners to master new media marketing knowledge and skills while also learning. Being able to deeply feel the vastness and profoundness of traditional Chinese culture, establishing cultural confidence, and having a subtle and influential effect.

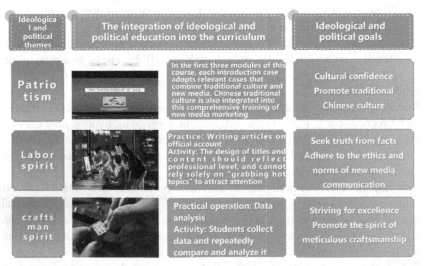

Fig. 2. Integrating Ideological and Political Education into Curriculum Based on Dynamic Learning Model.

4.2   Determination of Evaluation Indicators

In order to accurately evaluate the quality of learning motivation, we need to establish a scientifically effective evaluation index system. Based on relevant literature review and the research results and methods of learning motivation evaluation at home and abroad, this learning motivation evaluation index [14–16] has been determined, which mainly includes two aspects: student learning enthusiasm and student learning initiative. Learning motivation refers to the proactive and proactive attitude and behavior exhibited by students during the learning process, including indicators of learning motivation, learning goals, learning interests, and other aspects. Learning initiative refers to the ability and willingness of students to actively explore and learn independently in their studies, including indicators of learning strategies, self-learning ability, and learning self-awareness.

When determining specific evaluation indicators, we adopted a combination of expert consultation and questionnaire survey methods. Firstly, we invited experts from fields such as education and psychology for evaluation, and based on their professional knowledge and experience, we reviewed and supplemented the preliminarily determined indicators. Secondly, we designed a questionnaire survey targeting students, which further improved and refined the evaluation indicators through statistical analysis of a large number of student opinions and feedback

Finally, we have established an evaluation index system that includes six aspects: learning motivation (10%), learning objectives (10%), learning interest (20%), learning strategies (20%), self-learning ability (20%), and learning self-awareness (20%). Each indicator has undergone strict screening and argumentation, which can accurately reflect the learning motivation of students and provide scientific basis for the quality evaluation of course ideological and political education.

4.3 Evaluation Methods and Techniques

In this study, we adopted a combination of quantitative and qualitative research methods, collecting a large amount of data through questionnaire surveys and field observations, in order to accurately evaluate learning motivation.

We used a questionnaire survey to collect data on students' learning motivation. In the questionnaire design, we fully considered the personal characteristics of students and the requirements of ideological and political education in the curriculum to ensure the effectiveness and reliability of the questionnaire. We also used statistical analysis methods to process and analyze the questionnaire data in order to draw accurate conclusions.

In order to comprehensively evaluate learning motivation, we conducted field observations. By observing students' performance and participation in the classroom, we can have a more intuitive understanding of their level of learning motivation. We also evaluated the ideological and political literacy of students in combination with the teaching objectives of the course, to ensure the comprehensiveness and accuracy of the evaluation.

In terms of evaluation methods, we have adopted various evaluation techniques. For example, we use quantitative evaluation methods to quantitatively evaluate the learning motivation of students, and obtain their learning motivation level through statistical analysis. At the same time, we also adopted qualitative evaluation methods to observe and analyze students' learning attitudes, learning methods, and learning outcomes, in order to draw more in-depth conclusions.

5 Conclusion

This article is based on the perspective of ideological and political education in the curriculum, constructing a learning motivation model and conducting quality evaluation. Through in-depth analysis, the inherent connection between learning motivation and curriculum ideological and political education has been revealed. Research has found that learning motivation is of great significance for the effective implementation of ideological and political education in courses, as it can enhance students' learning enthusiasm and initiative, thereby promoting the comprehensive improvement of their ideological and political literacy.

This study provides important references for the construction of learning motivation. The construction of a learning motivation model requires a clear understanding of the main elements of the motivation model, and based on this, combined with the nature and goals of ideological and political education in the curriculum, determine its role in the motivation model. By constructing a motivation model, we can better understand and guide students' learning motivation, and promote the implementation effect of ideological and political education in the curriculum.

There is still room for further deepening in the quality evaluation of dynamic models in this article. When determining evaluation indicators, factors such as the degree of achievement of course ideological and political goals, the performance of student learning motivation, and learning outcomes can be comprehensively considered to form a more comprehensive evaluation indicator system. In terms of evaluation methods and techniques, evaluation tools can be further refined and improved, and a combination of quantitative and qualitative research methods can be fully utilized to evaluate learning motivation models from different dimensions.

There are still some shortcomings in research methods and data collection in quality evaluation. Research methods can be more diverse and comprehensive, including the application of mixed methods, the introduction of experimental methods, etc., to further improve the credibility and reliability of research. Data collection can add various forms of data sources, not limited to questionnaire surveys and field observations, but can also be combined with learning logs, interviews, and other methods to obtain more comprehensive and in-depth data.

Future research can be conducted in the following directions. Firstly, it is possible to conduct in-depth research on the differences and characteristics of learning motivation among different student groups, and provide differentiated educational measures for different groups. Secondly, the impact of other factors on learning motivation can be analyzed, such as teacher factors, classroom environment factors, etc., in order to gain a more comprehensive understanding of the formation and development mechanisms of learning motivation. Finally, we can draw on the research results of other disciplines and combine them with the actual needs of curriculum ideological and political education to further deepen the construction and quality evaluation of learning motivation models, in order to improve the practical effectiveness and application value of learning motivation models.

Acknowledgement. This work is supported by the 2022 Guangdong Province Continuing Education Quality Improvement Project with the Grant No. JXJYGC2022GX174, the 2022 Education and Teaching Reform Project of the Guangdong Vocational and Technical College Business Education Commission with the Grant [2023] No. 5, the 2023 Guangdong Vocational and Technical College Course Ideological and Political Demonstration Course Project with the Grant No. XJKC202305, and the 2023 Guangdong Vocational and Technical College Course Ideological and Political Demonstration Plan Project with Grant [2023] No. 45.

References

1. Xu, J., Hu, W., Lu, X.: Theoretical model, case design, and implementation strategy of group dynamics in online collaborative learning. China Electron. Educ. (03), 81–89 (2022)
2. Ma, Z., Zhou, H.: Research on the theory and model construction of learning motivation for general practitioners. Econ. Res. Guid. **08**, 94–97 (2020)
3. Xu, L., Yu, S., Guo, R.: Construction of a dynamic model for the evolution of ubiquitous learning resources. Res. Electron. Educ. **39**(04), 52–58 (2018)
4. Wen, J., Shen, J., He, Z., et al.: A study on the learning motivation evaluation model and mutant factors of undergraduate students in military colleges and universities. J. Sci. Educ. (First Ten Days) **31**, 178–181 (2017)
5. Ruan, S.: Research on the learning motivation model of college students and mining of learning motivation. J. Hubei Normal Univ. (Nat. Sci. Edn.) **36**(04), 87–90 (2016)
6. Xie, Y., Qiu, Y., Zhang, R., et al.: Implementation approach and evaluation innovation of digital transformation empowering college curriculum ideological and political education. China Electron. Educ. **09**, 7–15 (2022)
7. Yang, Y.: Preliminary construction of a curriculum ideological and political evaluation model based on student development. Heilongjiang High. Educ. Res. **40**(01), 115–119 (2022)
8. Xu, X., Wang, J.: Construction of a comprehensive evaluation index system for ideological and political education in university courses: a theoretical framework based on the CIPP evaluation model. High. Educ. Manag. **16**(01), 47–60 (2022)
9. Zhang, R., Qin, Q.: Evaluation of course ideological and political education: connotation, resistance, and resolution. Educ. Theory Pract. **41**(36), 49–52 (2021)
10. Sun, Y., Cao, H., Yuan, X.: Research on the construction of evaluation index system for ideological and political education in science and engineering courses. J. Jiangsu Univ. (Soc. Sci. Edn.) **23**(06), 77–88+112 (2021)
11. Duan, Y.: Construction of an effective evaluation system for "course ideological and political education" in universities. J. Hubei Univ. Econ. (Hum. Soc. Sci. Edn.) **18**(11), 105–107 (2021)
12. She, S., Zhang, Q.: Characteristics and reform path of evaluation of ideological and political theory courses in universities. Ideological Theoretical Educ. **03**, 18–24 (2021)
13. Su, X., Hong, Y.: Exploration of evaluating the effectiveness of course ideological and political education based on the analytic hierarchy process evaluation model. Educ. Teach. Forum **22**, 150–152 (2020)
14. Xia, J.: Construction and implementation of a higher education teaching quality evaluation system from the perspective of curriculum ideology and politics. Sci. Consult. (2023)
15. Huang, Y.: Exploration and practice of integrating "ideological and political elements" into classroom teaching quality evaluation from the perspective of "course ideology and politics". Mod. Vocat. Educ. (2019)
16. Yang, F.: Research on quantitative evaluation of teaching quality in universities from the perspective of curriculum ideology and politics. Univ. Res. Manag. (2021)

The 3D Display System of Art Works Based on VR Technology

Huyuan Lu[1], Qiner Xu[2], Zhiqiang Chen[2] (iD), and Beixin Zhong[2]([envelope])

[1] School of Art, Nantong University, Nantong, China
[2] School of Information Science and Technology, Nantong University, Nantong 226019, China
zpx@ntu.edu.cn

Abstract. The conventional 3D display system of art works mainly uses Vega Prime to render and load 3D display instructions, which is susceptible to the stacking effect of the storage scene data of the display list, resulting in some abnormal operation functions. Therefore, it is necessary to design a brand-new 3D display system of art works based on VR technology. In the hardware part, STM low-power processor, E18-D80N 3D infrared sensor and GM25Q64APIG memory chip have been already designed. In the software part, the 3D display model of art works based on VR technology has also been constructed, and the 3D display function module of art works is designed, thus realizing the 3D display of art works. The system test results vividly show that all functions of the designed 3D display system of art works based on VR technology are running in order, which proves that it has good performance, reliability and certain application value, and has made certain contributions to improving the display effect of art works.

Keywords: VR technology · Art · Works · 3D · Display · System

1 Introduction

Art works condense students' learning achievements, which is undoubtedly significant to the teaching of art majors. Every year, there are as many as dozens of art exhibition activities in only one art college. It can be easily seen that the effectiveness of artwork presentation has a direct impact on student learning outcomes. Due to artwork exhibition space constraints, students have higher limitations in the exhibition, which need to be arranged in accordance with the time of the site planning and is prone to the problem of information islands. In addition, many art works in exhibitions are prone to damage and other problems, resulting in the lag of the exhibition. In order to break the limitation of art exhibition and promote the development of art education, it is necessary to design an effective 3D display system of art works [9].

In fact, the 3D display system of art works is more closely related to the development of the Times [10], which can better meet the learning needs of colleges and universities, and create a good learning and interaction platform for their students [3]. It will be more convenient for some relevant teachers and students to participate in the process of art-work exhibition activities, communicate with students, and optimize the current teaching

K. Li and Y. Liu (Eds.): ISICA 2023, CCIS 2147, pp. 384–393, 2024.
https://doi.org/10.1007/978-981-97-4396-4_36

process. Not only that, the cost of using the 3D display system of art works is relatively lower, and there is no need to be confronted with the problem of damage of exhibits, and often belong to a one-time investment, and the follow-up maintenance is relatively simple [6, 7]. In the information age, digital technology has gradually integrated with people's daily life, and students' learning methods have also changed dramatically [8]. In order to adapt to the requirements of art teaching in the era of digitalization and improve the pertinence of learning, relevant researchers proposed to use the 3D display system to replace the traditional art exhibition method to realize the upgrading of education [4]. Although many colleges and universities have constructed different 3D display systems for art works, most of the 3D display systems need to use Vega Prime rendering to load the 3D display instructions, which is vulnerable to the stacking effect of the stored scene data of the display list, leading to some functions running abnormally, and does not comply with the requirements of the 3D display system's operational smoothness [5]. Therefore, this paper designs a new 3D display system for art works based on VR technology [1, 2].

2 Hardware Design

2.1 The STM Low-Power Processor

In the process of 3D display of art works, it is necessary to continuously obtain the characteristic information of exhibits and transform them into 3D digital images. Therefore, it is necessary to select an effective low-power processor to complete the online interoperability and control the scaling and detail display operation of 3D exhibits. STM low-power processor is a programmable processor, and its main composition structure is shown in Fig. 1 below.

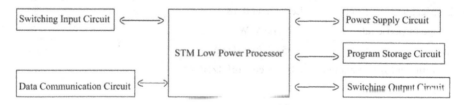

Fig. 1. Composition structure of the STM low-power processor

As can be seen from Fig. 1, the STM low-power processor is mainly composed of input circuit, communication circuit, power supply circuit, etc., which can collect the 3D input state of art works in real time and generate reasonable control programs.

The STM low-power processor has an operating frequency of up to 72 MHz with excellent computing performance, and can handle complex commands through the internal high-speed Cortex core processing center. In addition, STM low power processor has built-in multiple timers, which can realize high-speed counting with lower overall power consumption and stabler performance, to meet the operational reliability requirements of the 3D display system of art works. It requires a high working voltage, so it is necessary to use a series voltage divider circuit to reduce signal noise interference during its operation. The 3D display system of art works designed in this paper uses the ULN2803 output pin to reduce the transient high voltage to avoid the damage of the processor's processing chip. The STM low-power processor uses the RS-485 serial interface for communication, and the internal setup of the low-power transceiver can reduce the terminal matching reflection and improve the communication efficiency of processor's operation transmission.

2.2   E18-D80N 3D Infrared Sensor

There are different imaging characteristics of art works in different imaging environments. Responding to the above issues, the 3D display system of art works designed in this paper selects E18-D80N 3D infrared sensor for imaging interaction to obtain the interactive gesture signal of the exhibition. E18-D80N 3D infrared sensor is a diffuse reflective sensor, which has a special NPN type photoelectric switch. Once the interaction target is detected, the sensing state can be immediately adjusted to change the distance between exhibits. The technical parameters of E18-D80N 3D infrared sensor are shown in Table 1 below.

Table 1.   Technical parameters of E18-D80N 3D infrared sensor

Technical indicators	Index parameters
Index parameters	Φ 3.5 mm
Sensitivity element area	$1.9 * 1.9 \text{ mm}^2$
Compensation	With compensation
With compensation	With compensation
Electrical time constant	Less than 60 ms
Voltage response rate	>90000 V/W
Noise	<65 μV/(sqr[Hz])
Proportion measurement rate	6.0 + 8 cm(sqr[Hz])/W
6.0 + 8 cm(sqr[Hz])/W	2.7 V~8 V
Package	3TO8
Working temperature	−40 °C~+85 °C
Window material	Silicon-based narrow-band filter: 5.0 μm, 4.48 μm, 3.9 μm, 2.7 μm, and window materials can be customizable
Effective perspective	120 °C

As can be seen from Table 1, when the 3D display system is in operation, the optical properties of the emitted signals will change to some extent. The above sensors can effectively carry out signal enhancement, reduce the interference of visible light, and improve the comprehensive operation performance of the system.

2.3 TGM25Q64APIG Memory Chip

The 3D display system of art works generates more parameter data in the process of operation. If some parameter data are stacked, it will seriously affect the smoothness of the system's operation, causing part of the system functions to run abnormally. In order to solve the above problems and improve the operation reliability of the system, the GM25Q64APIG chip is selected as the core memory chip of the system in this paper, and the package schematic of the chip is shown in Fig. 2 below.

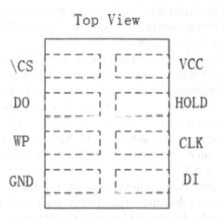

Fig. 2. Schematic diagram of the GM25Q64APIG memory chip package

As can be seen from Fig. 2, the GM25Q64APIG memory chip mainly adopts the serial SPI interface for docking, which supports Daul-I with the highest frequency of 108 MHz. The memory chip has excellent data and strong temperature resistance, which can execute the SPI code in real time and randomly access different storage locations. Therefore, the memory chip is highly reliable and the speed of the data exchange is fast, which satisfies the storage requirements of the designed system.

3 Software Design

3.1 Constructing a 3D Display Model of Art Works Based on VR Technology

VR technology is a kind of virtual reality technology, which is mainly based on the computer, combined with 3D image technology to create a simulation of 3D virtual space and improve a sense of experience in the display of art works. Therefore, the 3D display system of art works designed in this paper constructs a 3D display model of art works based on VR technology. First of all, it is necessary to generate a 3D display quadratic error matrix, which contains the quadratic error of the planar set of display vertices Δ (v) as shown in (1) below.

$$\Delta(v) = \sum_{peplanes} \left(p^T V \right)^2 \tag{1}$$

In formula (1), the notation P^T represents the distance from the plane containing the vertex, and the notation V represents the error measure. At this time, the position of the 3D model is updated, and the intersection of the 3D planes is solved, resulting in the error function f (v) as shown in (2) below.

$$f(v) = V^T Q v \qquad (2)$$

In formula (2), the notation P^T represents the optimal approximate parameter, the notation Q represents the new vertex measurement value, and the notation v represents the 3D folding similarity. According to the above calculated quadratic error function, the 3D model construction parameters can be solved, so that the preset values are more similar to the actual 3D display parameters.

The 3D display system designed in this paper uses 3D Max modeling tools to construct a 3D display model of art works. In the actual display process, some models will have light perspective problems. So this paper mixes the model with QBJ expressions, plans the application language through the Windows Presentation Foundation program, changes the model operation, arranges the visual effect control with a richer WPF environment, and dilutes the secondary elements. Multiple 3D scenes are preset in the 3D display system, which can be connected to 2D/3D controls by providing visualization windows through Viewport3D.

Gamera is an important element in the process of 3D display. It is necessary to set a reasonable perspective to determine the display angle attributes, deal with the key objects in 3D scenes, and solve the light source problem in the process of 3D display. Lights is a key factor in the 3D display, which can control the display effect of art works and make it closer to the actual impressions. Therefore, this paper utilizes Geometry grid objects for diffuse reflection processing, and constructs the coordinate system of 3D display model, as shown in Fig. 3 below.

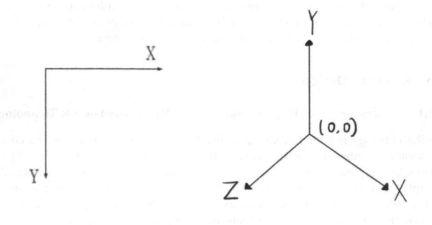

2D coordinate system   3D coordinate system

Fig. 3. 3D display model coordinate system

It can be seen from Fig. 3 that the above coordinate system of the 3D display model can be adjusted at the specified position to make it more consistent with the 3D display relationship of art works.

The 3D display model of the system directly affects the final display effect. Hence, the change factors need to be focused on when constructing the model. In this paper, the perspective and light of the 3D display model are adjusted by Expression Blend 3 software to fit the 3D display adaptability of art works, reducing the problem of system operation lagging caused by unsuitable parameters and improving the operation fluency of the system.

3.2   Designing the 3D Display Function Module of Art Works

The 3D display system of art works designed in this paper mainly includes multiple functional modules such as data management, user login, etc. In order to increase the interactivity of the system, this paper designs the 3D display functional module by the layer-by-layer planning method of art works:

The first is the 3D display data management module of art works, which belongs to the information management module. It can name, update, modify the uploaded art works, and set effective user roles, etc. The important tables involved in the data management module include Object, ObjectForms, ObjectMaterial, ObjectSize, Source, CreateDate, User, Author, and Country. Among them, Object table contains the number, ID and other information of art works. User is mainly used to save users' information, assign relevant using number, and obtain valid using password. ObjectForms, CreateDate, Author, ObjectMaterial, ObjectSize, Country, and Source represent the collection value of art works, realize classification and preservation, complete information change and other operations.

The second is the establishment module of art works, including the establishment of 3D studio, the selection of suitable templates, the generation of art exhibition hall, the optimization of the gallery, etc., and also has the filling function. It can obtain the corresponding catalogue of art works from the database and convert it into the 2D plane. The third is the interactive access module, which is mainly connected with the 3D display model to effectively obtain the relevant information of the art works and improve the interactive performance of the system. The above modules collaborate with each other to improve the comprehensive performance of the system.

4   System Testing

In order to verify the operational performance of the designed 3D display system of art works based on VR technology, this paper configures the basic test platform and conducts the system test as follows.

4.1   Test Preparation

According to the requirements of system testing, in this paper, the Unity3D engine is selected as the test engine, and the 3D Max model is built according to the requirements of 3D display of art works. It also uses Photoshop software for processing. For the fluency requirements of the test, the test environment configured in this paper is shown in Table 2 below.

Table 2. Test environment

Configuration	Parameter
GPU	NVIDIA Ge Fonce GTX 1060\AMD Radeon RX
CPU	Intel 15-4590\AMD FX 8350
RAM	4 GB
Video	HDMI 1.4 or Display Port 1.2
USB Port	1x USB 2.0
Operating System	Windows 7 SPK Windows 8.1

As can be seen from Table 2, based on the basic test environment configured above, this paper sets up the test flow as shown in Fig. 4 below.

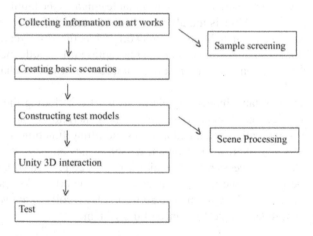

Fig. 4. Test process

As can be seen from Fig. 4, according to the above test process, virtual test scenes can be created to run the 3D display system of VR art works designed in this paper, so as to get the operational effect of different test functions.

4.2   Test Results and Discussion

On the basis of the above test preparation, you can carry out the artwork 3D display system test, that is to say, debugging the test environment is in the default state, and the actual art work 3D display requirements fit, at this point, running the design based on the VR technology of artwork 3D display system, the test results of the various functional modules are shown in Table 3 below.

Table 3. Test results

Test items	Step	Expected results	Test results
Login system	Enter the correct user name and password and click Login	Enter the login interface	pass-test
Use exclusive buttons	Click on the upload model button	Enter the upload model interface	pass-test
Name art works	Click Upload model and enter the model name in the model naming box	The data displayed on the page is consistent with the corresponding data in the database	pass-test
Upload the texture of art works	Click the upload texture button to select the texture file to be uploaded	The texture data displayed on the page is consistent with the corresponding texture data in the database	pass-test
Update the 3D model list of art works	Complete the upload of art works in 3D details	Update to show final upload results	pass-test
Analyze the rendering model	Click on the running button of the model	The model selected for running is shown in the main screen	pass-test
Mouse rotation operation	Press the left mouse button, do not release, and drag the mouse	As the mouse moves, the art work rotates	pass-test
Mouse zoom operation	Roll the mouse roller	Roll up to zoom in, scroll down to shrink	pass-test
Keyboard amplification operation	Press keyboard F, then release, and also press Z, + key for testing	Press the F key of the keyboard, the model starts to enlarge, and release the F key, press the Z key, the test is normal	pass-test
Recovery operation	Press the keyboard a, then release, and press the keyboard r, then release	The 3D display model returns to its original state	pass-test

As can be seen from Table 3, the 3D display system of art works based on VR technology designed in this paper has passed different test items, with no abnormal operation problems, and the overall smoothness of operation is high. The above test results prove that the 3D display system of art works designed in this paper has good operational performance, reliability and certain application value.

5 Conclusion

To sum up, in the era of rapid development of network technology, the national comprehensive quality of China is gradually improving, and the facilities related to education, culture and aesthetics are also becoming increasingly perfect. The research shows that the number of buildings such as multimedia centers, art exhibition venues and other buildings in various cities has surged dramatically, giving important support to the development of the new era. Conventional display of art works is easily restricted by the display venues, display time and other limitations, which affects the display effect, and is prone to information islands. In order to solve this problem, it is necessary to design a 3D display system of art works. The operation fluency of the conventional art 3D display system is low, and some functions are difficult to run. Thus, this paper designs a new 3D display system for art works based on VR technology. After the system test, the results show that the 3D display system of the designed art works has good performance, smooth operation, reliability, and certain application value, which has made certain contributions to solving limitations of the exhibition of art works.

References

1. Anthes, C., García-Hernández, R.J., Wiedemann, M., Kranzlmüller, D.: State of the art of virtual reality technology. In: 2016 IEEE Aerospace Conference, pp. 1–19, March 2016. https://doi.org/10.1109/AERO.2016.7500674
2. Ardiny, H., Khanmirza, E.: The role of AR and VR technologies in education developments: opportunities and challenges. In: 2018 6th RSI International Conference on Robotics and Mechatronics (IcRoM), pp. 482–487, October 2018. https://doi.org/10.1109/ICRoM.2018.8657615
3. Cheng, S., et al.: Solve the IRP problem with an improved discrete differential evolution algorithm. Int. J. Intell. Inf. Database Syst. 12(1–2), 20–31 (2019)
4. Kuroda, T.: Wireless proximity communications for 3D system integration. In: 2007 IEEE International Workshop on Radio-Frequency Integration Technology, pp. 21–25, December 2007. https://doi.org/10.1109/RFIT.2007.4443910
5. Favalora, G.E.: Volumetric 3D displays and application infrastructure. Computer 38(8), 37–44 (2005). https://doi.org/10.1109/MC.2005.276
6. Bruno, F., Bruno, S., De Sensi, G., Luchi, M.-L., Mancuso, S., Muzzupappa, M.: From 3D reconstruction to virtual reality: a complete methodology for digital archaeological exhibition. J. Cult. Herit. 11(1), 42–49 (2010). https://doi.org/10.1016/j.culher.2009.02.006
7. Qiu, Y., Xiao, Y., Jiang, T.: An online college student art exhibition app based on virtual reality technology. IOP Conf. Ser. Mater. Sci. Eng. 750(1), 012132 (2020). https://doi.org/10.1088/1757-899X/750/1/012132
8. Skublewska-Paszkowska, M., Milosz, M., Powroznik, P., Lukasik, E.: 3D technologies for intangible cultural heritage preservation—literature review for selected databases. Heritage Sci. 10(1), 3 (2022). https://doi.org/10.1186/s40494-021-00633-x

9. Loaiza Carvajal, D.A., Morita, M.M., Bilmes, G.M.: Virtual museums. Captured reality and 3D modeling. J. Cult. Heritage **45**, 234–239 (2020). https://doi.org/10.1016/j.culher.2020.04.013
10. Kiourt, C., Koutsoudis, A., Pavlidis, G.: DynaMus: a fully dynamic 3D virtual museum framework. J. Cult. Herit. **22**, 984–991 (2016). https://doi.org/10.1016/j.culher.2016.06.007

Petrochemical Commodity Price Prediction Model Based on Wavelet Decomposition and Bayesian Optimization

Lei Yang[✉], Rui Xu, Huade Li, and Zexin Xu

College of Mathematics and Informatics, South of China Agricultural University,
Guangzhou 510642, China
yanglei_s@scau.edu.cn

Abstract. The current predictions of petrochemical commodity prices are generally inadequate in terms of effective time and frequency domain modeling, and are plagued by issues such as lagging predicted values and memory dependence. To address these problems, this paper proposes an attentional neural network model that utilizes wavelet decomposition and Bayesian optimization. Initially, the intricate petrochemical commodity price data is decomposed into sub-series of varying frequencies, utilizing wavelet decomposition in a divide-and-conquer strategy to extract longitudinal frequency domain features. Subsequently, a neural network that has been improved by an attention mechanism is utilized to extract the transversal time domain features. Throughout this process, a Bayesian computing approach is proposed to optimize the hyperparameters. In order to investigate the forecasting performance of the proposed model, extensive experiments were conducted utilizing the prices of propylene, butadiene, phenol, and toluene, with six other state-of-the-art methods included for comparison purposes. The experimental results demonstrate that the proposed model has superior performance when it comes to petrochemical commodity price forecasting.

Keywords: Petrochemical commodity prices · Wavelet decomposition · Bayesian optimization · Time series prediction

1 Introduction

The petrochemical industry is one of the important components of the national economy [1]. The upstream industry of this sector relies on four basic raw materials including acetylene, olefins, aromatics, and syngas, such as propylene, butadiene, and toluene [2]. The prices of these petrochemical commodities are not only important indicators for predicting trends in the chemical industry, but also reflect changes in market supply and demand and the state of economic growth. However, the prices of these commodities are affected by complex factors such as global economic changes and market supply and demand, resulting in significant price differences at different points in time. These price fluctuations exhibit strong non-stationarity, non-linearity, and stochasticity, which severely affect the market decision-making of the petrochemical industry. Therefore, accurately predicting the prices of petrochemical commodities is of great significance.

However, there are still some problems with time-domain prediction methods in commodity price prediction in the petrochemical industry. Long Short-Term Memory (LSTM) networks proposed by Hochreiter and Schmidhuber are typical deep learning networks [3]. However, LSTM's time-domain prediction has some shortcomings, which affect the prediction effect. For example, Lee et al. predicted the price of butadiene and optimized purchasing decisions, but the predicted values showed obvious lagging phenomena [4]. Ozdemir et al. overlooked the impact of the importance of information at each time point on nickel price data, and there is further room for improvement in prediction accuracy [5]. Moreover, the hyperparameter setting in machine learning methods is of crucial importance, but researchers often use subjective methods such as cross-validation to set hyperparameters [6], which limits the upper limit of the model's predictive ability.

In terms of frequency domain analysis, the Fourier analysis method is not accurate enough for decomposition since it is not sensitive to temporal changes [7]. Empirical mode decomposition-based methods commonly suffer from boundary effects, mode overlapping, and lack of precision [8]. Variational mode decomposition-based methods are not only sensitive to noise but also require a large number of preset parameters [9]. Therefore, choosing a suitable frequency domain exploration method is particularly important.

Therefore, this paper proposes a Wavelet decomposition and Tree-structured Parzen Estimator (TPE) optimized Attention LSTM model (WTA-LSTM) for short-term petrochemical commodity price forecasting. First, based on the divide-and-conquer strategy, the complex commodity price data is transformed by wavelet decomposition and reconstruction to extract longitudinal frequency features, which not only reveal the frequency domain patterns of the prices but also significantly mitigate the issue of lagging forecast values. Second, the attention mechanism is adopted to enhance the native LSTM model to capture critical time-point features, which makes the LSTM network more intelligent in its temporal feature extraction capabilities. Finally, the TPE algorithm based on Bayesian calculation is proposed to optimize the hyperparameters of the attention LSTM, which improves the upper limit of the prediction capability of the petrochemical commodity price forecasting model.

The rest of the paper is organized as follows. Section 2 demonstrates the proposed model and its computational mechanism. Section 3 discusses the experiments and results, followed by conclusion in Sect. 4.

2 Methodology

In this section, a comprehensive overview of the proposed hybrid model is presented, detailing the flow and the operational mechanics of each constituent component.

2.1 Model Overview

This paper proposes a hybrid deep learning model called WTA-LSTM for predicting petrochemical commodity prices. The structure of the entire model is shown in Fig. 1. The model consists of five layers: the input layer, decomposition layer, optimization layer, prediction layer, and output layer. Firstly, price data is inputted into the decomposition layer through the input layer to explore the frequency domain patterns of petrochemical commodity prices. By using a divide-and-conquer approach, the non-stationarity and non-linearity of the original price data are reduced while capturing trend and detail information in the price sequence. Secondly, the decomposed sub-sequences are optimized through the optimization layer to prepare for improving the upper limit of the model's predictive ability. Next, an attention mechanism-improved LSTM network is used to form the prediction layer, which trains the model with the optimized hyperparameters from the optimization layer and assigns weights to time points to predict the complexity and stochastic volatility of petrochemical commodity prices. Finally, the predicted results of each sub-sequence are reconstructed to form the final prediction results and output.

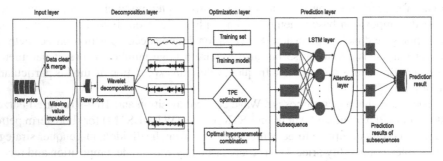

Fig. 1. Structure of the proposed WTA-LSTM model.

2.2 Decomposition Layer

The decomposition layer of the model is assumed by discrete wavelet decomposition. The wavelet basis function exhibits two characteristics, namely volatility and attenuation. The former denotes the presence of both positive and negative oscillations in the amplitude of the function, while the latter refers to the rapid decay of the function's value outside of a predefined region.

The mother wavelet function $\psi(t)$ typically meets the following criteria:

$$\int_{-\infty}^{+\infty} \psi(t)dt = 0 \tag{1}$$

The mother wavelet function is transformed into a wavelet basis function by scaling and translation:

$$\psi_{a,b}(t) = \frac{1}{\sqrt{a}}\psi\left(\frac{t-b}{a}\right) \tag{2}$$

The a represents the scaling factor that determines the degree of expansion and contraction of the wavelet basis function, reflecting the frequency characteristics of the time series. The b represents the translation factor, which regulates the amplitude and direction of the wavelet basis function translation to reflect the time domain properties of the time series. Wavelet analysis consists of two components: wavelet decomposition and wavelet reconstruction [10]. In actual research, wavelet analysis often divides the time series into low-frequency and high-frequency signals and performs discrete wavelet decomposition and reconstruction based on the Mallat algorithm to investigate the traits and laws of sequence changes [11]. The trend of the series becomes smoother and easier to predict than the original series. The frequency of the detail sequence is higher, which characterizes the fluctuation of the original sequence at different scales and in different time periods, thus capturing the longitudinal features of the data [12]. In general, simple sequences are easier for the model to comprehend than complex sequences, resulting in more accurate prediction outcomes [13]. The deconstructed signal is processed and then reconstructed to become the original signal..

The low-frequency approximation sequence is represented by A_n, and the high-frequency detail sequence is represented by D_n. Each wavelet decomposition is based on the most recent approximation of the decomposition value A_i, the depth of decomposition k, and the number of decompositions k. Thus, the wavelet analysis in the model may examine the data not only from a transversal perspective in the time domain but also from a longitudinal perspective in the frequency domain.

2.3 Optimization Layer

Hyperparameters refer to the predetermined parameters set by the user prior to training a model. Choosing appropriate values for these parameters is crucial for deep learning-based models, as it can significantly enhance the model's predictive capability. Sequential Model-Based Global Optimization (SMBO) [14] employs previously evaluated parameters and results to hypothesize unobserved parameters, and the acquisition function utilizes this information to suggest the next set of parameters. One of the SMBO Bayesian rule-based optimization algorithms is TPE [15], which models the loss function using a probabilistic model and creates a plausible estimate for a specified number of iterations in order to discover the optimal set of hyperparameters for the model. TPE has high efficiency, flexibility, and adaptability for deep learning models that have a larger number of hyperparameters [16]. Therefore, we adopt the TPE algorithm as the core algorithm of the optimization layer for the model. The specific steps are as follows:

Step 1: Initialize random sample points, where each sample point represents a combination of hyperparameters.

Step 2: Divide the sample points into a training set and a validation set. Conduct machine learning training on the training set and estimate the probability density function using kernel density estimation on the validation set. The Parzen window method is employed for kernel density estimation in the TPE algorithm.

Step 3: Divide the probability density function into regions of high probability density and low probability density. The high probability density region contains more optimal hyperparameter combinations.

Step 4: Within the high probability density region, sample the next candidate point according to the probability density function and record the latest results.

Step 5: Repeat steps 2–4 until the specified number of optimization iterations is reached or the target performance is achieved.

2.4 Prediction Layer

The prediction layer of the model consists of the LSTMs network improved with attention mechanism, which can capture the influence of historical feature states on the commodity price sequence by analyzing the data sequence.

The proposed model employs a modified LSTM with an attention mechanism as the neural network component. LSTMs are seldom used alone, but rather in a stacked multi-level structure as depicted in Fig. 2. The input data is formatted into a three dimensional matrix. The entire architecture comprises an input layer, multi-layer LSTMs, a Dense layer, and an output layer [17]. To alleviate overfitting, a dropout layer is attached to each LSTM layer. A two layer LSTM model with an attention mechanism is employed as the neural network section. First, the wavelet-decomposed time series are fed into the first LSTM layer to extract features, and then the output of the LSTM layer serves as the input to the attention layer. After the attention mechanism operates, the key feature sequences with high importance are then fed into the second LSTM layer as input.

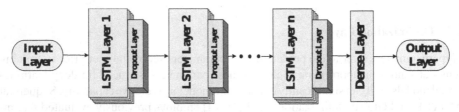

Fig. 2. The hierarchical structure of LSTM.

Then an attention mechanism is proposed to enhance the original model by leveraging valuable information from diverse time periods to construct a robust price forecasting model. The overall architecture of the attention mechanism is depicted in Fig. 3.

The mechanism begin by feeding the sequence $x^k = (x_1^k, x_2^k, ..., x_m^k)^T \in R^m$ into the first layer LSTM for encoding. The output is the hidden state $h_t \in R^s$ of the first layer LSTM, where s represents the dimensionality of the hidden state.

$$h_t = f(h_{t-1}, x_t) \tag{3}$$

In the context of our study, f serves as a non-linear activation function, embodying the encoding of individual elements. Subsequently, our focus shifts to delineating the weight evaluation process. In this regard, we introduce a straightforward neural network to compute the relevance weight for each informational element. For a given element within the sequence x_k, the neural network leverages the hidden state h_{t-1} from the LSTM unit and the long-term memory C_{t-1} to ascertain the weight assigned to the specific element, expressed as follows:

$$e_t^k = \alpha^T \tanh(W_1 \cdot [h_{t-1}, C_{t-1}] + W_2 \cdot x^k) \tag{4}$$

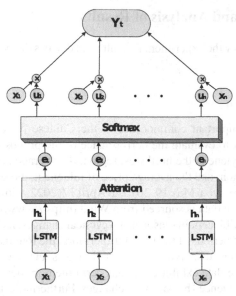

Fig. 3. The hierarchical structure of attention mechanism.

$$u_t^k = \frac{\exp(e_t^k)}{\sum_{i=1}^{n} \exp(e_t^k)} \tag{5}$$

We propose a weight evaluation mechanism using a simple neural network to calculate the weight of each information. Given a specific element in the sequence xk, the neural network utilizes the hidden state h_{t-1} and the long-term memory C_{t-1} in the LSTM unit to calculate the weight of the corresponding element. The vector α and matrices W_1, W_2 are learnable parameters of the model. The weight vector u^k contains scores indicating how much attention should be given to the k-th sequence. These scores are normalized by the softmax function, resulting in a vector of length m that measures the importance of each input sequence. The output of the attention model at time t is the weighted input sequence yt for the next LSTM layer, given by:

$$y_t = (u_t^1 x_t^1, u_t^2 x_t^2, ..., u_t^n x_t^n)^T \tag{6}$$

Then, y_t is passed as the ultimate input sequence to the subsequent layer of LSTM, thereby completing the task of transforming the original time series into an attention-based time series. Conventional RNN-based frameworks for financial time series forecasting typically use the raw time series as input and perform uniform processing on all input sequences. However, the newly obtained y_t can give more emphasis to particular input sequences, extract crucial features effectively, and mitigate the impact of attention weights on redundant sequences. In theory, using yt as the input of the LSTM deep learning model can result in better prediction accuracy.

3 Experiments and Analysis of Results

In this section, we show the experimental results and discuss the possible causes of these results.

3.1 Data Description

The prices of four important commodities in the Chinese petrochemical industry is applied as our dataset to evaluate the performance of the proposed model, namely the average price of propylene in the northwest region, the average prices of butadiene and phenol in the east region, and the average price of toluene in the south region. The time span of the prices ranges from May 19, 2013, to April 17, 2022, with weekly observations. The time series of price data is sourced from Wind (https://www.wind.com.cn/). These four price datasets exhibit certain seasonal and cyclical changes, indicating similar trends during certain specific periods of the year. Furthermore, the four datasets exhibit distinct nonlinear characteristics, displaying strong randomness and uncertainty. Additionally, these four sets of price data exhibit clear long-term memory dependence, with current price changes often influenced by past price changes. Furthermore, these four sets of price data exhibit certain noise and outliers, which may affect the accuracy of the prediction model. For each price dataset, we selected the first 85% of the data as the training set and the remaining 15% as the test set.

3.2 Experimental Results

To confirm the efficacy and superiority of our proposed model, we compare it with several other models, including the conventional prediction model ARIMA, the individual deep learning models RNN and LSTM, as well as three related hybrid deep learning models: attention mechanism-enhanced LSTM (AT-LSTM), TPE algorithm-enhanced LSTM (TPE-LSTM), and wavelet analysis-enhanced LSTM (Wave-LSTM). These models are employed to forecast the four sets of price data, and the corresponding evaluation metrics are calculated to compare their performance with our proposed model WTA-LSTM. The fitted graphs generated by these models for each set of price data are displayed in Fig. 4 and Fig. 5. Evidently, WTA-LSTM provides better fitting of the true values than the other models.

For smooth price fluctuations, such as the average prices of phenol and toluene from March 2021 to October 2021, it is challenging for single models such as native LSTM and ARIMA to capture the small, sawtooth fluctuations in the smooth trend of the series. Wave-LSTM, which incorporates wavelet decomposition, can better fit these sawtooth fluctuations by capturing the longitudinal features of the data series, but its accuracy still requires improvement. Thus, WTA-LSTM is capable of fitting these fine fluctuations more accurately.

For the more severe fluctuations, such as the average propylene price from October 2020 to February 2021 and from October 2021 to April 2022, the average butadiene price from June 2021 to April 2022, and the average toluene price from January 2022 to April 2022, relying on a single model to make predictions is hardly satisfactory. While AT-LSTM, TPE-LSTM, and Wave-LSTM have improved prediction results, they still

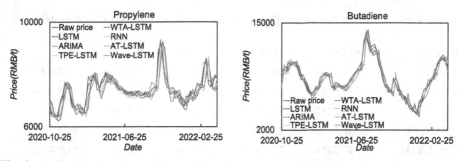

Fig. 4. Predicted prices of propylene and butadiene compared with the corresponding real values under different models.

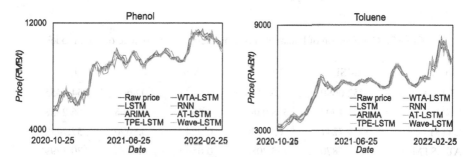

Fig. 5. Predicted prices of phenol and toluene compared with the corresponding real values under different models.

struggle to describe the drastic fluctuations. In contrast, the WTA-LSTM's prediction results can more accurately capture these strongly fluctuating price trends.

The predicted curve of the proposed WTA-LSTM aligns most accurately with the true values, and the hybrid model exhibits superior performance compared to the single models for both intervals characterized by smooth and sharp price fluctuations.

3.3 Performance Comparison

Tables 1, 2, 3 and 4 present the assessment metrics of the various models discussed above for different price data, providing further differentiation and quantitative analysis of each model.

Upon analyzing the errors, the hybrid models outperform the single models, with the proposed WTA-LSTM prediction model exhibiting the lowest error of the three. Wave-LSTM comes second in terms of effectiveness, while TPE-LSTM lags slightly behind AT-LSTM. In the single models, LSTM yields the smallest error, whereas ARIMA yields the largest. The disparity may stem from the fact that ARIMA is more adept at predicting smoother time series; nevertheless, due to the drastic changes in petrochemical commodity prices, ARIMA falls short compared to deep learning models.

To assess the trend prediction performance of the models, WTA-LSTM is found to be the best, followed by Wave-LSTM. However, Wave-LSTM falls short of AT-LSTM and TPE-LSTM in predicting the butadiene price. In general, hybrid models outperform

Table 1. Comparison of propylene price prediction performance of each model.

Models	RMSE	MAE	MAPE	R^2	D_{stat}
ARIMA	414.9744	293.7922	0.0398	0.4414	53.2468%
RNN	374.5754	257.1468	0.0353	0.5051	54.5455%
LSTM	352.2232	241.9117	0.0322	0.5494	40.2597%
AT-LSTM	332.7138	222.8990	0.0298	0.5841	49.3506%
TPE-LSTM	348.7454	226.8090	0.0304	0.5555	51.9481%
Wave-LSTM	184.7558	139.9556	0.0187	0.8778	79.2208%
WTA-LSTM	**124.0412**	**97.8479**	**0.0129**	**0.9467**	**87.0130%**

Table 2. Comparison of butadiene price prediction performance of each model.

Models	RMSE	MAE	MAPE	R^2	D_{stat}
ARIMA	752.0690	516.1114	0.0693	0.8662	54.5455%
RNN	733.4909	508.0381	0.0662	0.8765	66.2338%
LSTM	719.7241	505.9100	0.0647	0.8696	57.1429%
AT-LSTM	681.7870	465.2080	0.0599	0.8875	70.1299%
TPE-LSTM	688.5698	471.9971	0.0615	0.8793	71.7846%
Wave-LSTM	419.8546	328.0588	0.0435	0.9568	67.5325%
WTA-LSTM	**280.8371**	**228.9860**	**0.0283**	**0.9809**	**80.5195%**

single models, but in terms of Dstat for propylene price prediction, AT-LSTM and TPE-LSTM perform slightly worse than ARIMA and RNN.

The assessment of each model's performance can be visually observed in Fig. 6 and Fig. 7, where the difference between the true and predicted values at each time point is plotted. Models with line fluctuations closer to the zero axis have better prediction accuracy. In particular, volatile points in time, such as the butadiene price in July 2021 and the propylene price in October 2021, are examined. Many models show large forecasting errors at these points, highlighting the difficulty of forecasting. Although the hybrid models generally outperform the single models, the improvement in forecasting accuracy at these specific time points is not significant. However, models with a decomposition step exhibit more pronounced improvements and are more robust against risks than other models. In other words, for some challenging time periods, the wavelet decomposition-based models show less fluctuation in prediction error. Among them, the WTA-LSTM model, which incorporates attention mechanism and TPE, performs better than Wave-LSTM, with line fluctuations closest to the zero axis for the majority of time points, despite having a larger prediction deviation than Wave-LSTM for some time points.

Table 3. Comparison of phenol price prediction performance of each model.

Models	RMSE	MAE	MAPE	R^2	D_{stat}
ARIMA	385.5477	295.0557	0.0349	0.9416	51.9481%
RNN	338.6861	272.2242	0.0317	0.9547	61.3377%
LSTM	302.7808	235.4245	0.0276	0.9636	59.7403%
AT-LSTM	297.2512	231.9760	0.0273	0.9638	64.9651%
TPE-LSTM	296.3649	231.1885	0.0268	0.9646	66.2338%
Wave-LSTM	213.4507	159.3709	0.0195	0.9805	79.2208%
WTA-LSTM	**138.3353**	**105.7481**	**0.0127**	**0.9922**	**83.1169%**

Table 4. Comparison of toluene price prediction performance of each model.

Models	RMSE	MAE	MAPE	R^2	D_{stat}
ARIMA	276.5255	199.6003	0.0377	0.9414	41.5584%
RNN	275.1099	209.1756	0.0406	0.9457	35.0649%
LSTM	267.4630	189.5388	0.0348	0.9434	36.3636%
AT-LSTM	254.4255	173.7339	0.0299	0.9467	57.1429%
TPE-LSTM	259.1687	175.3446	0.0304	0.9411	57.5789%
Wave-LSTM	135.4969	105.3975	0.0188	0.9845	75.3247%
WTA-LSTM	**110.9556**	**86.3852**	**0.0152**	**0.9903**	**80.5195%**

3.4 Result Analysis and Discussion

Previous findings have demonstrated that the proposed WTA-LSTM outperforms commonly used single and hybrid models in terms of RMSE, MAE, MAPE, R2 and Dstat predictions, and it is crucial to discuss the factors that account for this superiority.

The experiment start by discussing the three single models, namely ARIMA, RNN, and LSTM, and identify the key factors that contribute to the superior performance of LSTM over the other two. For the conventional prediction model ARIMA, it is more suitable for predicting smoother time series. The price curve of petrochemical commodities exhibits strong nonlinearity and non-stationarity, making it challenging for ARIMA to predict accurately. In contrast, deep learning models are better suited for such complex and dynamic time series data. While RNN has a simpler structure than the other two single deep learning models, it is often plagued by the issue of gradient disappearance or explosion, rendering it less effective. Thus, LSTM emerges as the top-performing single model due to its superior ability to capture long-term dependencies in the input data.

Next we focus on three related hybrid deep learning models, TPE-LSTM, AT-LSTM and Wave-LSTM. The TPE algorithm optimizes the selection of hyperparameters for the native LSTM, resulting in TPE-LSTM achieving higher accuracy than the native LSTM. However, the long-term memory dependence issue of the LSTM still persists.

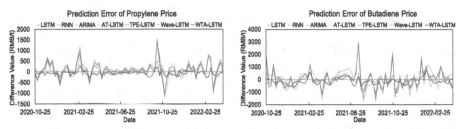

Fig. 6. The differences between the predicted value and the real value of propylene and butadiene price data under different prediction models. The closer the error line is to the 0 axis, the better the prediction effect of the model is.

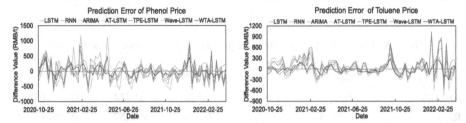

Fig. 7. The differences between the predicted value and the real value of phenol and toluene price data under different prediction models. The closer the error line is to the 0 axis, the better the prediction effect of the model is.

To overcome this, the AT-LSTM incorporates an attention mechanism that focuses more on crucial time series, leading to improved accuracy compared to both the native and TPE-LSTM models. Analysis of Fig. 4 and Fig. 5 reveals that the predicted value curves of the single LSTM exhibit slight sawtooth fluctuations during certain time periods, such as butadiene prices from October 2021 to December 2021 and phenol prices from January to April 2022. Figure 6 and 7 show that the attention mechanism can mitigate these fluctuations, closing the gap between predicted and true values while preserving extreme value points, which are easier to approximate.

Wave-LSTM, on the other hand, incorporates wavelet decomposition and reconstruction, dividing the data into multiple subseries with distinct frequencies. The decomposed subsequences exhibit smoother and more obvious periodicity, allowing the model to analyze the longitudinal features of the original sequences and significantly improve prediction accuracy. Wave-LSTM outperforms other models in predicting intervals with more drastic price fluctuations, such as propylene prices from November 2021 to April 2022 and toluene prices from December 2021 to April 2022. In contrast, other models exhibit divergent and large fluctuations in the predicted value curves, making it challenging to fit the test set curve. Wave-LSTM partly resolves this issue, and its predicted curves are less divergent, fitting the test set curves more accurately. Nonetheless, the overall prediction curve of Wave-LSTM is overly smooth, making it difficult to highlight extreme value points. To address this, optimization using the attention mechanism can enhance the highlighting of extreme value points.

WTA-LSTM stands out as the top performing model, owing to its incorporation of the advantageous elements of the previously mentioned models, particularly in the following aspects:

(1) WTA-LSTM boasts the selective memory of LSTM, which surpasses that of ARIMA and RNN.
(2) Through the utilization of wavelet decomposition and reconstruction, the issue of lagging predicted values can be significantly reduced, while simultaneously decreasing divergence fluctuations and capturing the longitudinal frequency features of time series.
(3) The internal TPE algorithm's computation of the optimal combination of hyperparameters provides stronger prediction performance than the empirical setting of hyperparameters for LSTM.
(4) The attention mechanism's preference for crucial time series effectively alleviates the native LSTM's long-term memory dependency issue, thereby enhancing the prediction of extreme value points. In totality, the proposed WTA-LSTM model outperforms all others in terms of prediction accuracy.

4 Conclusion

This paper proposes a hybrid model, WTA-LSTM, for predicting the prices of petrochemical commodities, which is optimized in both the time and frequency domains. The model captures both the overall trend and local fluctuations of the prices, thus capturing the longitudinal frequency-domain characteristics of the price sequence. Moreover, the model has the ability to focus on key historical information, thereby capturing the horizontal time-domain characteristics of the price sequence. The optimization of hyperparameters reduces the interference of human factors in the modeling process. To verify the predictive performance of the model, we tested it on four different types of petrochemical commodities with distinct features in China, demonstrating the model's universality, and conducted model comparison experiments. The experimental results show that the model has higher prediction accuracy than other models, and the possibility of this result is discussed.

Acknowledgment. This work was partially supported by Natural Science Foundation of Guangdong Province of China (Grant No. 2020A1515010691), Agricultural Science and Technology Commissioner Project of Guangzhou City (Grant No. 20212100036), and National Natural Science Foundation of China (Grant Nos. 61573157 and 61703170). The authors also gratefully acknowledge the reviewers for their helpful comments and suggestions that helped to improve the presentation.

References

1. Kwon, H., Do, T.N., Kim, J.: Optimization-based integrated decision model for smart resource management in the petrochemical industry. J. Ind. Eng. Chem. (2022)

2. Tawancy, H., Ul-Hamid, A., Mohammed, A., Abbas, N.: Effect of materials selection and design on the performance of an engineering product–an example from petrochemical industry. Mater. Des. **28**(2), 686–703 (2007)

3. Hochreiter, S., Schmidhuber, J.: Long short-term memory. Neural Comput. **9**(8), 1735–1780 (1997)

4. Lee, C.Y., Chou, B.J., Huang, C.F.: Data science and reinforcement learning for price forecasting and raw material procurement in petrochemical industry. Adv. Eng. Inform. **51**, 101443 (2022)

5. Ozdemir, A.C., Buluş, K., Zor, K.: Medium-to long-term nickel price forecasting using LSTM and GRU networks. Resources Policy **78** (2022) 102906

6. Wang, Z., Su, X., Ding, Z.: Long-term traffic prediction based on LSTM encoder-decoder architecture. IEEE Trans. Intell. Transp. Syst. **22**(10), 6561–6571 (2020)

7. Gorus, M.S., Ozgur, O., Develi, A.: The relationship between oil prices, oil imports and income level in Turkey: evidence from Fourier approximation. OPEC Energy Rev. **43**(3), 327–341 (2019)

8. Mo, H., Xiong, L., Lu, R.Y.: Material demand combination forecasting model based on EMD-PSO-LSSVR. In: 2018 International Conference on Education Reform and Management Science (ERMS 2018), pp. 347–356. Atlantis Press (2018)

9. Liu, Y., Yang, C., Huang, K., Gui, W.: Non-ferrous metals price forecasting based on variational mode decomposition and LSTM network. Knowl.-Based Syst. **188**, 105006 (2020)

10. Qiao, W., Wang, Y., Zhang, J., Tian, W., Tian, Y., Yang, Q.: An innovative coupled model in view of wavelet transform for predicting short-term PM10 concentration. J. Environ. Manag. **289**, 112438 (2021)

11. Mallat, S.G.: A theory for multiresolution signal decomposition: the wavelet representation. IEEE Trans. Pattern Anal. Mach. Intell. **11**(7), 674–693 (1989)

12. Goodell, J.W., Goutte, S.: Co-movement of covid-19 and bitcoin: evidence from wavelet coherence analysis. Financ. Res. Lett. **38**, 101625 (2021)

13. Rhif, M., Ben Abbes, A., Farah, I.R., Martínez, B., Sang, Y.: Wavelet transform application for/in non-stationary time-series analysis: a review. Appl. Sci. **9**(7), 1345 (2019)

14. Hutter, F., Hoos, H.H., Leyton-Brown, K.: Sequential model-based optimization for general algorithm configuration. In: Coello, C.A.C. (ed.) Learning and Intelligent Optimization. LNCS, vol. 6683, pp. 507–523. Springer, Heidelberg (2011). https://doi.org/10.1007/978-3-642-25566-3_40

15. Bergstra, J., Bardenet, R., Bengio, Y., Kégl, B.: Algorithms for hyper-parameter optimization. In: Advances in Neural Information Processing Systems, vol. 24 (2011)

16. Zhang, J., Meng, Y., Wei, J., Chen, J., Qin, J.: A novel hybrid deep learning model for sugar price forecasting based on time series decomposition. Math. Prob. Eng. **2021** (2021)

17. Fischer, T., Krauss, C.: Deep learning with long short-term memory networks for financial market predictions. Eur. J. Oper. Res. **270**(2), 654–669 (2018)

Mutate Suspicious Statements to Locate Faults

Guangsheng Zhan, Shi Cheng, and Jinbao Zhang[✉]

School of Information Science and Technology, Nantong University, Nantong 226019, China
kingbao@ntu.edu.cn

Abstract. In the process of software development, fault localization is an important part of software quality assurance, which can help developers identify code locations that may cause program failures. For faults in software programs, researchers have proposed many different fault localization techniques, such as spectrum-based fault localization (SBFL), mutation-based fault localization (MBFL), and more. These technologies have shown good results in fault localization, but they also have their limitations. SBFL technology only considers the execution of individual statements, without analyzing the dependencies between contextual statements, potentially leading to the localization of non-faulty statements. Moreover, SBFL technology has relatively high requirements for the quality of test cases. While MBFL technology improves accuracy compared to SBFL, it requires the execution of a large number of mutated versions and test cases for the target program, resulting in higher time costs. Therefore, this paper proposes a new method: Mutate suspicious statements to locate faults. First, SBFL technology is used to identify suspicious statements within the target program. These suspicious statements are then mutated to create mutants that are subsequently tested. Finally, the suspiciousness of these statements is calculated to pinpoint faults. Experiments show that this method enhances localization accuracy, reduces time costs, and proves effective.

Keywords: Fault Localization · Software Testing · Coincidental Correctness · Mutation Test

1 Introduction

The fundamental cause of a software program's inability to function correctly or experiencing errors is referred to as software failure [1]. Since software is a product developed by humans, it often contains defects. During the process of software debugging, a critical task is to identify the cause of the program's failure. However, the fault may not always be immediately apparent within the program's code. For example, when the program crashes or produces error outputs, the specific cause of the fault is often not easily identifiable, making fault localization a time-consuming and labor-intensive task [2]. In the context of software development and maintenance, debugging software is a task that is prone to errors, with fault localization and defect correction being the most critical aspects [3–5]. In comparison to fault correction, fault localization takes precedence, as only by finding the fault can it be rectified. As software programs continue to grow

© The Author(s), under exclusive license to Springer Nature Singapore Pte Ltd. 2024
K. Li and Y. Liu (Eds.): ISICA 2023, CCIS 2147, pp. 407–418, 2024.
https://doi.org/10.1007/978-981-97-4396-4_38

in size and complexity, traditional methods such as setting breakpoints or step tracing are neither effective nor efficient [6]. Therefore, there is an urgent need to enhance the efficiency of fault localization, reduce the workload of developers, and minimize the overall costs of the development process.

Spectrum-based fault localization (SBFL) technology is a typical dynamic analysis technique that primarily relies on two types of information collected during software testing [7]: test execution results and program spectrum analysis. It is used to identify faulty program components such as statements, branches, basic blocks, methods, or predicates. SBFL technology calculates the suspiciousness value of program statements using coverage information, and it is characterized by low computational complexity, simplicity of operation, and broad applicability [8]. Consequently, SBFL technology has been proposed as an automated mechanism for localizing program faults [8–12], and it has consistently been a prominent research focus. Wong et al. [13] made the following conjecture regarding the relationship between statements and test cases: the probability that a statement contains faults is related to the number of successful and failed test cases. This probability is inversely proportional to the number of successful test cases and directly proportional to the number of failed test cases. Building on the same concept, Jones [14] proposed using coverage information and test data to locate faulty statements and developed a visualization tool called Tarantula. Jones et al. [15] also suggested using different colors to represent the suspiciousness of statements and provided a color calculation formula, where red indicates deep suspicion, green represents no doubt, and they further offered the formula for calculating statements suspiciousness scores (as shown in Table 1). Abreu et al. [9], inspired by concepts from the field of molecular biology, such as similarity coefficients and cluster analysis, proposed using the Ochiai [16] and Jaccard [17] formulas to calculate the suspiciousness scores of statements. Statements are ranked in descending order based on their suspiciousness scores, with higher-ranked statements considered more likely to be potential faulty statements [18, 19]. However, coincidental correctness [20] can impact the accuracy of SBFL technology, resulting in the successful execution of test cases that originally failed due to specific reasons; this phenomenon is referred to as coincidental correctness. Furthermore, the above-mentioned approach significantly simplifies the program's execution process and trajectory, overlooking contextual information during program execution, such as the inherent semantic relationships and dependencies between variables [7], which affects the accuracy of fault localization and makes it challenging for developers to rely solely on the suspiciousness of individual program statements to identify the location of faulty code in the program.

By introducing the mutation test, the mutation operator is used to simulate the artificial fault in the program, compared with the traditional test method (based on the program spectrum), it can effectively detect the unknown and real fault statement, this method is called the mutation-based fault location technology. (MBFL) [21]. Mutation analysis was first used in mutation testing, it was proposed by DeMilo [22], and it is a software testing method that simulates program failures and verifies the validity of test cases by modifying source code. The idea of using mutants to help software fault location was first proposed by Papadakis and Le-Traon [23–25], which is called the fault location method based on mutation analysis, which mainly relies on the similarity between mutants to locate fault statements. Empirical studies have shown that the MBFL technique has higher accuracy than the SBFL technique in locating faults, but it incurs a huge time overhead due to the need to execute a large number of mutants of the target program [24, 25]. Subsequent paragraphs, however, are indented.

To overcome the drawbacks of the two aforementioned techniques, in this paper, we propose a new approach that leverages MBFL to enhance the results of SBFL. Initially, we employ SBFL technology to generate a list of suspicious statements in the target program. Then, we perform mutations on the top-ranked statements from this list, generating mutated statements and creating mutants of the target program. Subsequently, we use the same set of test cases to test these mutants, obtaining the execution results for test cases. We calculate the suspiciousness scores of the suspicious statements based on the measurements between these execution results. Finally, we rearrange the statements in a new descending order based on the recalculated suspiciousness scores, forming a fresh statement ranking list where higher-ranked statements are more likely to contain faults. Our approach significantly reduces time costs compared to MBFL technology while maintaining higher accuracy than SBFL technology. To validate the effectiveness of our approach, we conducted experiments using real programs from the SIR benchmark suite [10], and the results confirm the efficacy of our method.

2 Spectrum-Based Fault Localization (SBFL)

Reps and his colleagues introduced the concept of program spectra in 1997 while addressing the Y2K issue [27]. They collected information about the runtime distribution of program paths by inserting probes into the program. Program spectra [25] refer to the information gathered during program execution, describing the program's execution process and characteristics. They are also known as program traces, program behavioral features, and more. The utilization of program spectra for fault localization [14, 18, 29] is due to the fact that they encompass whether program code is executed during runtime, the number of executions, the values of predicates in the program, code branch coverage, and other related information.

The probability that a statement contains a fault is related to the frequency of successful and failed test cases, and is inversely proportional to the frequency of successful test cases and inversely proportional to the frequency of failed test cases [13]. However, when the test results of many test cases are not displayed correctly, the suspiciousness of program statements cannot be accurately calculated. Coincidental correctness occurs when a fault is triggered but the program still runs without showing the fault. Previous studies have shown that it can negatively affect the accuracy of software testing techniques [14, 22]. In this paper, we use the mutation testing technology to solve the negative effects of coincidental correctness, and effectively improve the accuracy of fault localization.

The SBFL method generally executes test cases, obtains the coverage information and execution results of the source program under a given test case set through instrumentation technology [7, 29, 30], and builds the test coverage matrix and execute the result vector, and finally generate a test report. The framework of a spectrum-based error location technique is shown in Fig. 1, where *SIC* represents spectrum information collection.

Table 1. SBFL Suspicion Calculation Formula.

Formula name	Formula expression
Tarantula	$\dfrac{N_{CF}/N_F}{N_{CF}/N_F+N_{CS}/N_S}$
Ochiai	$\dfrac{N_{CF}}{\sqrt{N_F\times(N_{CF}+N_{CS})}}$
Naish1	$\begin{cases} -1 N_{CS} < N_F \\ N_S - N_{CS}N_{CS} = N_F \end{cases}$
Dstar	$\dfrac{N_{CF}}{N_{CS}+N_{UF}}$
Jaccard	$\dfrac{N_{CF}}{N_{CS}+N_{CF}+N_{UF}}$
Wong2	$N_{CF} - N_{CS}$

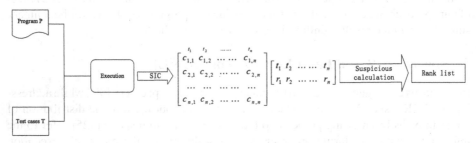

Fig. 1. The framework of SBFL.

Given a target program P, expressed as: $P = \{S_1, S_2, S_3, ..., S_n\}$, it contains n statements, S_i represents the i - th sentence, $1 \leq i \leq n$.

Its set of test cases is expressed as: $T_S = \{T_1, T_2, T_3, ..., T_n\}$. Where m represents the number of test cases in the test case set of the target program P. After running P on T_S, the execution result of statement S_i is obtained. The execution result of a test case can be expressed as a result vector R, the program spectrum can be expressed as a coverage matrix M_{ij}, and M is expressed as a matrix of T_S and P coverage relations, which is an $m * n$ matrix. The i - th row in M represents whether the statement in P is executed when the i - th test case is run. Column j in M represents whether the executable statement in line j is executed when all test cases are running. The execution result of the program statement can be represented by N_{CF}, N_{UF}, N_{CS} and N_{US}, and the suspiciousness of the statement can be calculated by these four elements. Among them, N_{CF} indicates the number of test cases that fail to execute and cover the program statement S_i, N_{UF} indicates the number of test cases that fail to execute and does not cover the program statement S_i, N_{CS} indicates the number of test cases that execute successfully and cover the program statement S_i, and N_{US} indicates the number of test cases that execute successfully and do not cover the program statement S_i. The number of test cases covering the program statement S_i. Usually, N_F and N_S are used to represent the number of all failed test cases and the number of all successfully executed test cases. The suspiciousness of program statements is calculated by calculation formulas to form a suspiciousness ranking list, and developers start checking from the program statement with the highest suspiciousness until the faulty statement is found.

3 Mutation-Based Fault Localization (MBFL)

Mutation analysis, initially proposed by Hamlet [16] and DeMillo [22], is a fault-based testing technique used to assess the effectiveness of a test case. Mutation operators are rules for generating mutants of the original program. These operators encompass various categories, including statement mutation, operator mutation, variable mutation, and constant mutation. The mutation operators in these categories are designed to simulate the faults programmers typically make when constructing expressions, expression functions, and function compositions using iterations and conditional statements [32]. For a target program P, if it fails when executed with certain test cases, it indicates the presence of faults in the program. To mutate P, let m_f be a mutant of P, which will make the fault sentence mutate; the mutant that makes the correct sentence mutate is m_s. According to the research of Seokhyeon et al. [24], we can get that the test cases that fail on P are more likely to succeed on the mutant m_f than on m_s, because m_f is the mutant of the mutation failure statement. In other words, mutating the fault statement is to modify the fault statement so that the fault m_f can be repaired. Failed test cases will fail with a high probability on m_s, because it is impossible for more test cases to succeed by mutating the correct statement. Conversely, a test case that succeeds on P is more likely to fail on m_s than m_f. Because m_s is a mutant that mutates the correct statement, that is modifying the correct statement may introduce faults that cause test cases to fail on m_s. Therefore, mutating the faulty statement may cause the faulty program to be repaired, making more test cases succeed; mutating the correct statement may cause more faults, making more test cases fail.

However, MBFL technology usually requires a large number of mutants to execute the target program [24], so the additional time cost is high. In addition, different programs require different mutation operators, which limits the application of mutation analysis in fault location. Therefore, we choose to use mutation analysis technique to improve the results of SBFL technique, so as to improve the accuracy of fault localization and reduce the time overhead.

4 SIR

SIR (Software-artifact Infrastructure Repository) is a program component library [10] that provides the necessary programs and related materials for research work in program analysis and software testing techniques. SIR has a significant impact in the field of software testing, and many researchers use its programs and datasets for their studies. SIR includes programs in various languages such as Java, C, C++, and C#. Some of these programs have original faults, while most of them are artificially injected with errors or mutated to create different error versions, with each version containing only one fault. Researchers have also designed corresponding test cases for these programs.

All of these, including the programs, error versions, and test cases, can be accessed on the SIR website.

5　Method in this Paper

In this section, we introduce the framework of our method and the process of fault localization.

5.1　Framework

The method in this paper takes the target program P and the test case set T_S as input, and finally obtains the fault location report. The overall framework is divided into three parts: obtaining suspicious statements, mutating suspicious statements, and fault location results, as shown in Fig. 2.

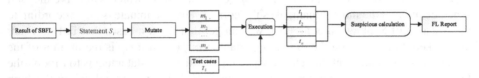

Fig. 2. The framework of our method.

Step 1: We use a test case set T_S to test a target program P, and obtain the coverage information and test case results of each executed statement. Then use SBFL technology to calculate the suspiciousness of the execution statement, and generate a list of statements in descending order of suspiciousness.

Step 2: By selecting several statements from the list obtained in step 1 as suspect statements, we will select statements that have been executed by at least one failed test case and mutate them. Because a statement may produce many different mutants, we adopt the mutant sampling technique [32] to randomly select mutants for execution. Subsequently, we use the same test case set T_S as in step 1 to run these mutants and collect their results.

Step 3: The suspicious degree calculation formula is used to calculate the suspicious degree of each suspicious statement, adjust the statement sorting list in step 1, and obtain the fault localization report.

5.2　Method

We experimented with real programs such as grep, gzip, flex, and sed from the SIR benchmark suite. The SIR program set is widely used in fault location research in software testing, and it is also equipped with a corresponding set of test cases.

We select a fault version from grep and test it with the corresponding set of test cases. Then Gcov is used to measure the statement coverage of a given test case implementation, and the coverage matrix and execution result vector are constructed according to the

Table 2. Test results of target program P and mutant m.

Statements	Result	m	Result Changes				
			T_{S1}	T_{S2}	T_{S3}	T_{S4}	T_{S5}
$S_1 : \max = -x$	F	$m_1 : \max = x - 1$			S → F		
		$m_2 : \max = x$	F → S	F → S			
$S_2 : if\,(\max < y)$	F	$m_3 : if\,(\max > y)$			S → F	S → F	S → F
		$m_4 : if\,(\max == y)$	F → S			S → F	
$S_3 : \max = y$	S	$m_5 : \max = -y + 1$				S → F	S → F
		$m_6 : \max = y + 1$				S → F	S → F
$S_4 : if\,(x * y < 0)$	S	$m_7 : if\,(x * y > 0)$				S → F	S → F
		$m_8 : if\,(x * y == 0)$					S → F
$S_5 : print(\text{"fault"})$	S	$m_9 : return$				S → F	S → F
		$m_{10} ::$			S → F	S → F	S → F

coverage information and execution result. The coverage information and the frequency of failure or success of test cases are analyzed, and the suspicious degree of statements is calculated with the help of SBFL technology. For comparison, this paper adopts various SBFL techniques to calculate the suspiciousness of sentences. Then select the corresponding suspicious sentences, that is, the top-ranked sentences in the sorted list, as shown in Table 2. Because the syntax elements in different sentences are different, the mutation operators applicable to the same statement are also different. Different mutation operators will also generate different amounts of mutants, so we use mutant sampling technology to generate all mutants. Then, a certain number of mutants are randomly selected for execution, and each suspicious statement generates the same number of mutants, and this technique was proposed by Acree and Budd [33]. We employ the mutation tool Proteum [34] to mutate suspicious sentences, generating two mutants for each mutation point of the target sentence of each mutation operator. Finally, run these mutant versions and collect their execution results, compare the SBFL test results with the mutant test results, use the suspiciousness calculation formula to calculate the suspiciousness of suspicious statements, and get the faulty statement location report, as shown in Table 2. We employ the mutation tool Proteum to mutate suspicious sentences, generating two mutants for each mutation point of the target sentence of each mutation operator. Finally, run these mutant versions and collect their execution results, compare the SBFL test results with the mutant test results, use the suspiciousness calculation formula to calculate the suspiciousness of suspicious statements, and get the faulty statement location report, as shown in Table 2.

5.3 A Subsection Sample

Similar to the SBFL technique, the suspicion degree calculation formula of our method also needs the execution results of the test cases, and the suspicion degree measure W

is:

$$W(S_i) = \frac{1}{|m(s)|} \left(\sum_{m \in m(s)} \frac{|N_{CF} \cap S_m|}{|N_F|} - \alpha \frac{|N_{CS} \cap F_m|}{|N_S|} \right) \tag{1}$$

For suspicious statement S_i, N_{CF} represents the number of test cases of execution failure and overwriting program statement S_i; N_{CS} represents the number of test cases of successful execution and overwriting program statement S_i; N_F and N_S represent the number of all test cases of execution failure and all test cases of execution success. For a fixed set of mutation operators, $m(S)$ is expressed as the set of all mutants of suspicious statements, where S_m is the number of successful test cases on mutant m, and F_m is the number of failed test cases on mutant m. In the formula, the first term $\frac{|N_{CF} \cap S_m|}{|N_F|}$ represents the proportion between the number of test cases passed on mutant m after failing on P but mutating the suspect statement S_i and the number of test cases failed on P; the second term $\frac{|N_{CS} \cap F_m|}{|N_S|}$ represents the proportion between the number of test cases failed on mutant m and the number of test cases successful on P but mutating the suspect statement S_i. Since the value of the first item is more likely to be greater than the value of the second item, in order to balance the average value of the two items, we introduce a weight α, which is calculated as:

$$\alpha = \left(\frac{FtS}{|m(s)||N_F|} \cdot \frac{|m(s)||N_S|}{StF} \right) \tag{2}$$

Among them, FtS represents the number of test cases whose test results changed from failure to success after testing mutant m, and StF represents the number of test cases whose test results changed from success to failure after testing mutant m. We use the corresponding test case set to test the grep2.2 real program in the SIR benchmark suite, get the coverage information of each executed statement and the result of the test case, and then use the SBFL technology to generate a suspicious ranking list of the statement. We select the top five suspicious statements $S_i (1 \le i \le 5)$ from the statement ordering list obtained from the real program, and the first statement S_1 is a real fault statement. Five test cases $(T_{S1}, T_{S2}, T_{S3}, T_{S4}, T_{S5})$ are used to test these five statements, in which the test cases T_{S1} and T_{S2} fail, $N_F = 2$ and $N_S = 3$. Mutate each sentence to generate a total of 10 mutants m, $m(S) = 10$. We use the same five test cases to test 10 mutants. According to the execution results of the test cases, we can get $FtS = 2$, $StF = 19$. The test results are shown in Table 2. Thus, from formulas (1) and (2), we can calculate $\alpha = 0.18$ and $W(S_1) = 0.42$, and the same method can calculate other suspicious statements with the suspicious degree of 0.39, 0.08, 0.27, and 0.31. In contrast, when we use Jaccard [10], Ochiai 20, and Op2 19 for fault localization of the same target program P, these techniques regard the third and fourth statements as the most suspicious statements, and do not find the real fault statement S_1. The results of suspicious degree calculation of the four technologies are shown in Table 3. According to the experimental results, it can be found that our method has higher accuracy than SBFL technology, and because it only needs to mutate a small number of suspicious sentences obtained by SBFL technology, the time cost is lower than MBFL technology.

5.4 Result

To assess the effectiveness of our method, we conducted tests and comparisons with three other techniques: Jaccard [10], Ochiai [19], and Op2 [18], all on the same target programs. Table 4 presents the precision assessment of Jaccard, Ochiai, Op2, and our method, along with the proportion of statements checked before locating the erroneous statement. In the best-case scenario, all techniques locate the error without checking any code. Our method is able to locate the error after checking an average of 3.03% of the executed statements. According to Table 4 and 5, the average precision of our method is 5.46 times higher than Jaccard, 5.26 times higher than Ochiai, and 4.84 times higher than Op2. Additionally, in terms of locating erroneous statements, our method outperforms the three mentioned techniques in all cases except for the "sed" program. This indicates that our method is significantly superior to these three techniques.

Table 3. The results of suspicious degree calculation by different techniques.

Jaccard	Ochiai	OP2	Our Method
S/Rank	S/Rank	S/Rank	S/Rank
0.40/4	0.58/4	1.21/4	0.42/1
0.37/5	0.65/3	1.25/3	0.39/2
0.56/1	0.69/2	1.53/1	0.08/5
0.50/2	0.71/1	1.50/2	0.27/4
0.43/3	0.50/5	0.75/5	0.31/3

Table 4. % of executed statements examined.

P	Jaccard	Ochiai	OP2	Our Method
flex	23.46	21.08	17.63	3.54
grep	4.61	4.61	4.61	1.26
gzip	0.99	0.96	0.92	0.13
sed	37.12	37.12	35.47	7.2
Average	16.55	15.94	14.66	3.03

Table 5. Rank of faulty statements.

P	Jaccard	Ochiai	OP2	Our Method
flex	256	197	163	1
grep	64	64	64	13
gzip	19	17	14	1
sed	12	12	12	26
Average	84.75	72.5	63.25	10.25

6 Conclusions and Future Work

Fault location is the key of software debugging and an important part of software quality assurance. When the existing spectrum-based fault location technology (SBFL) is used for fault location, its accuracy will be affected by the quality of test cases and accidental correctness. Mutation-based fault location technology (MBFL) is more accurate than SBFL in fault location, but it has a higher time cost. Therefore, our method combines the advantages of the two technologies, and obtains suspicious statements in SBFL technology through mutation of MBFL technology to locate faults and find fault statements in the program. This method not only improves the accuracy of fault location, but also reduces the time cost. Although we demonstrated the validity of our approach on SIR Benchmark assemblies, future work needs to be carried out on a larger scale of real programs, as well as more programming languages, to further adapt to failure scenarios in real software development.

References

1. Collofello, J.S.: Evaluating the effectiveness of reliability-assurance techniques. J. Syst. Softw. **9**(3), 191–195 (1989)
2. Bill, T.: Software visualization in the large. Computer **29**(4), 33–43 (1996)
3. Vessy, I.: Expertise in debugging computer programs: a process analysis. Int. J. Man-Mach. Stud. **23**(5), 459–494 (1985)
4. Pearson, S.: Evaluating and improving fault localization. In: Proceedings of the 39th IEEE/ACM International Conference on Software Engineering, pp. 609–620 (2017)
5. Wang, K.C.: Key scientific issues and state-art of automatic software fault localization. Chin. J. Comput. **38**(11), 2262–2278 (2015)
6. Wong, W.E.: A survey on software fault localization. IEEE Trans. Softw. Eng. **42**(8), 707–740 (2016)
7. Jiang, S.J.: Fault localization approach based on path analysis and information entropy. Ruan Jian Xue Bao/J. Softw. **32**(7), 2166–2182 (2021)
8. Abreu, R.: An evaluation of similarity coefficients for software fault localization. In: Proceedings of the 12th Pacific Rim International Symposium on Dependable Computing (PRDC 2006), pp. 39–46. IEEE Computer Society, California (2006)
9. Abreu, R.: On the accuracy of spectrum-based fault localization. In: Proceedings of the Testing: Academic and Industrial Conference Practice and Research Techniques - MUTATION (TAICPART-MUTATION 2007), pp. 89–98. IEEE Computer Society, Windsor (2007)

10. Do, H.: Supporting controlled experimentation with testing techniques: an infrastructure and its potential impact. Empir. Softw. Eng. **10**(4), 405–435 (2005)
11. Jia, Y.: An analysis and survey of the development of mutation testing. IEEE Trans. Softw. Eng. **37**(5), 649–678 (2011)
12. Zhang, L.: A theoretical analysis on cloning the failed test cases to improve spectrum-based fault localization. J. Syst. Softw. (JSS), 35–57 (2017)
13. Debroy, W.: The DStar method for effective software fault localization. IEEE Trans. Reliab. **63**(1), 21–32 (2014)
14. Jones, J.A.: Visualization of test information to assist fault localization. In: Proceedings of the 24th International Conference on Software Engineering (ICSE 2002), pp. 19–25. Association for Computing Machinery, Florida (2002)
15. Jones, J.A.: Empirical evaluation of the tarantula automatic fault localization technique. In: Proceedings of the 20th IEEE/ACM International Conference on Automated Software Engineering, pp. 273–282. Association for Computing Machinery, California (2005)
16. Jaccard, P.: Étude comparative de la distribution florale dans uneportion des Alpes et des Jura. Bull. Soc. Vaud. Sci. Nat. **37**, 547–579 (1901)
17. Ochiai, A.: Zoogeographic studies on the soleoid fishes found in Japan and its neighbouring regions. Bull. Jpn. Soc. Sci. Fish. **22**(9), 526–530 (1957)
18. Jones, J.A.: Fault localization using visualization of test information. In: Proceedings of the 26th International Conference on Software Engineering (ICSE 2004), pp. 54–56. IEEE Computer Society, Edinburgh (2004)
19. Santelices, R.: Lightweight fault-localization using multiple coverage types. In: Proceedings of the 31st International Conference on Software Engineering (ICSE 2009), pp. 56–66. IEEE Computer Society, Vancouver (2009)
20. Offutt, A.J.: Mutation 2000: uniting the orthogonal. In: Proceedings of the 1st Workshop on Mutation Analysis (MUTATION 2001), pp. 34–44 (2001)
21. Andrews, J.H.: Using mutation analysis for assessing and comparing testing coverage criteria. IEEE Trans. Software Eng. **32**(8), 608–624 (2006)
22. Demillo, R.A.: Hints on test data selection: help for the practicing programmer. Computer **11**(4), 34–41 (1978)
23. Papadakis, M.: Metallaxis-FL: mutation-based fault localization. Softw. Test. Verification Reliab. **25**(5/7), 605–628 (2015)
24. Yoo, S.: Ask the mutants: mutating faulty programs for fault localization. In: Proceedings of the 2014 IEEE Seventh International Conference on Software Testing, Verification and Validation, pp. 153–162. IEEE Computer Society, Cleveland, OH (2014)
25. Papadakis, M., Traon, Y.L.: Using mutants to locate "unknown" faults. In: Proceedings of the 2012 IEEE Fifth International Conference on Software Testing, Verification and Validation (ICST 2012), pp. 691–700. IEEE Computer Society, Canada (2012)
26. Reps, T.: The use of program profiling for software maintenance with applications to the year 2000 problem. ACM SIGSOFT Softw. Eng. Not. (1997)
27. Chao, L.: SOBER: statistical model-based bug localization. ACM SIGSOFT Softw. Eng. Not. **30**(5), 286–295 (2005)
28. Liblit, B.: Scalable statistical bug isolation. ACM SIGPLAN Not. **40**, 15–26 (2005)
29. Wang, X.: Taming coincidental correctness: coverage refinement with context patterns to improve fault localization. In: Proceedings of the 31st International Conference on Software Engineering (ICSE 2009), pp. 45–55. IEEE Computer Society, Vancouver (2009)
30. Chen, X.: Review of dynamic fault localization approaches based on program spectrum. Ruan Jian Xue Bao/J. Softw. **26**(2), 390–412 (2015)
31. Goues, C.L.: A generic method for automatic software repair. IEEE Trans. Softw. Eng. **38** (2011)

32. Weimer, W.: Automatically finding patches using genetic programming. In: Proceedings of the 31st International Conference on Software Engineering (ICSE 2009), pp. 364–374. IEEE Computer Society, Vancouver (2009)
33. Acree, Jr. A.T.: On mutation, Ph.D. thesis. Georgia Institute of Technology, Atlanta, Georgia, p. 184 (1980)
34. Budd, T.A.: Mutation analysis of program test data. Ph.D. dissertation. Yale University, New Haven, CT, USA, p.155 (1980)
35. Maldonado, J.C.: Proteum: a family of tools to support specification and program testing based on mutation. In: Mutation Testing for the New Century (2001)

Iterative Learning Control with Variable Trajectory Length in the Presence of Noise

Yuangao Yan[1], Xixian Tan[2], and Yunshan Wei[1,2(✉)]

[1] School of Electronic and Communication Engineering, Guangzhou University,
Guangzhou 510006, Guangdong, China
`weiys@gzhu.edu.cn`
[2] Key Laboratory of On-Chip Communication and Sensor Chip of Guangdong Higher
Education Institutes, Guangzhou 510006, Guangdong, China

Abstract. This scientific paper addresses the problem of iterative learning control for the design of linear discrete-time multiple-input-multiple-output (MIMO) systems, where the system contains nonrepetitive (iteration-dependent) load noise and measurement disturbances noise. Under the condition that the trajectory length varies with iteration, the effects of disturbances noise and loss of system output error information are dealt with by mathematical expectation, and the tracking error convergence is achieved by using open-closed-loop law. Among them, the feedforward part ensures the convergence of the ILC tracking error in the sense of mathematical expectation. The feedback control part is used to compensate for the loss of tracking information in previous iterations using the tracking information of the current iteration. Through rigorous mathematical induction analysis, the convergence conditions are derived. Finally, the effectiveness of the control law is proved by the actual simulation experiments for this kind of system.

Keywords: Iterative learning control · nonrepetitive load noise · measurement disturbances noise · open-closed-loop law

1 Introduction

The Iterative Learning Control (ILC) is an intelligent control method that mimics human learning behavior. It utilizes the tracking information from the previous iteration and adjusts the learning control law based on the input information of the current iteration, in order to ensure efficient, effective, and robust for system control. Its paramount benefit lies in the fact that it not necessitates precise mathematical model, it merely requires the formulation of an appropriate learning control law, tailored to the real-world system, to accomplish trajectory tracking (for a detailed discussion, please refer to [1–4]). There are numerous applications for a class of systems designed to repeatedly achieve specific outcomes within a fixed trajectory length [5–9]. The study encompasses nonlinear systems [7, 8] and discrete-time systems [6, 8], etc.

Under fixed trajectory length and without load noise and measurement disturbances noise, in order to achieve perfect tracking performance, the ILC controlled system must

© The Author(s), under exclusive license to Springer Nature Singapore Pte Ltd. 2024
K. Li and Y. Liu (Eds.): ISICA 2023, CCIS 2147, pp. 419–429, 2024.
https://doi.org/10.1007/978-981-97-4396-4_39

be strictly required to keep the repeatability of the dynamic system, that is, the initial and final positions of the system are kept consistent in the iterative evolution. However, in the case of real system control, the requirement of a fixed trajectory length that remains unchanged and the initial state is completely the same is difficult to be achieved, or even impossible. For example, ILC is applied in the field of robot position control [10], the robot's parallel force/position control is used in practice to learn the trajectory of the slowing task and progressively correct this trajectory to the original task velocity. When the original task velocity is reached, an iterative learning control law combined with force/position control is used to further reduce the force error. In special cases, when emergency situations are considered, the robot must stop the current behavior, so that trajectory length is shortened, and the original rule of keeping fixed trajectory length unchanged is broken, which leads to the failure of the ILC control. Therefore, it makes sense that the variable trajectory length is added to the ILC control.

In addition, in the practical application process, the influence of noise on the system output should be considered [11–15]. Convergence analysis of iterative learning schemes for discrete systems with system noise and unknown dynamics was performed in [12], and the process of system noise affecting the learning network system was proved using the λ-norm technique. It is shown that the input steady-state error in the expected sense is proportional to the standard deviation of the system noise. However, the noise for the output considered in [12] is only the noise at the previous time, and the noise at the current time is not considered, that mean only load noise is added to the system. In [14], the paper only considers the noise interference in the output error, and studies the convergence effect of suppressing noise interference on the ILC system control. In [15], both the influence of load noise and measurement disturbances noise are considered, and the tracking convergence of the system is proved by using high-order internal model (HOIM) ILC algorithm, and the convergence condition is obtained. The drawback is that the case where the trajectory length varies randomly with iteration is not considered.

Therefore, this paper studies how to use the ILC control to achieve complete tracking of the desired trajectory under the condition of the system with noise interference under the condition of fixed trajectory length. Bernoulli variable is used to solve the loss of error information caused by different trajectory length, and Mathematical expectation is used to solve the influence of noise on error convergence. An ILC control method is designed which contain feedforward and feedback parts. The feedforward ILC part is used to ensure the iterative convergence of the tracking error in the sense of mathematical expectation, and the feedback control part is used to compensate for the missing tracking information in the first few iterations. Moreover, due to the role of mathematical expectation, the initial condition of the system is no longer strictly required, that is, the initial state of the system does not need to be strictly consistent.

2 Problem Formulation

2.1 System and Assumption

Consider the following linear discrete-time multiple-input-multiple-output (MIMO) system with iteratively variable trajectory lengths

$$\begin{cases} x_k(t+1) = Ax_k(t) + Bu_k(t) + w_k(t), \\ y_k(t) = Cx_k(t) + v_k(t), \end{cases} \tag{1}$$

where $k \in \{0, 1, 2, 3, \ldots\ldots\}$ denotes iteration index and $t \in \{0, 1, 2, \ldots, N_k\}$ is the discrete time index. For all t and k, $x_k(t) \in R^r$, $u_k(t) \in R^p$ and $y_k(t) \in R^l$ are the state, control input and output of the system (1), addition, $w_k(t) \in R^n$ and $v_k(t) \in R^l$ is the nonrepetitive (iteration-dependent) load noise and measurement disturbances noise. $A \in R^{n \times r}$, $B \in R^{n \times p}$ and $C \in R^{l \times n}$ denote the system matrices. The unknown and randomly varying N_k between iterations is the trajectory length of the system at the $k - th$ iteration, with a range of $(\underline{N} \le N_k \le \overline{N})$. In the ILC problem based on iterative variable trajectory length, it is reasonable to assume that there exists a minimum trajectory length $\underline{N} > 0$ and a maximum experimental length $\overline{N} > \underline{N} > 0$. It is worth mentioning that the values of N and $\overline{N}$ are determined independently of the ILC scheme, and the ILC scheme design itself will not be affected by them. Let the desired output of the system (1) be $y_d(t) = Cx_d(t) \in R^l$, $t \in \{0, 1, 2, \ldots, N\}$, where $x_d(t)$ is the desired state. For any realizable output trajectory $y_d(t)$, there exists a unique control input $u_d(t) \in R^p$ such that

$$\begin{cases} x_d(t+1) = Ax_d(t) + Bu_d(t), \\ y_d(t) = Cx_d(t), \end{cases} \tag{2}$$

where $t \in \{0, 1, 2, \ldots, N\}$.

Assumption 1: Consider the iterative randomness of $x_k(0)$, mathematical expectation is applied to define system input $x_k(0)$ and the expected input $x_d(0)$, they satisfy

$$E\{x_k(0)\} = x_d(0). \tag{3}$$

And the nonrepetitive (iteration-dependent) loan $w_k(t)$ and measurement disturbances $v_k(t)$ clearly satisfies

$$E\{w_k(t)\} = 0, \tag{4}$$

$$E\{v_k(t)\} = 0, \tag{5}$$

where $k \in \{0, 1, 2, 3, \ldots\ldots\}$ and $t \in \{0, 1, 2, \ldots, N_k\}$.

The ILC tracking error of the linear discrete-time MIMO system is thus defined as

$$e_k(t) = y_d(t) - y_k(t), \tag{6}$$

where $t \in \{0, 1, 2, \cdots, min\{N_k + 1, N + 1\}\}$. Also, the definitions of input error and state error are

$$\Delta u_k(t) = u_d(t) - u_k(t). \tag{7}$$

$$\Delta x_k(t) = x_d(t) - x_k(t). \tag{8}$$

In this paper, for system (1), under iteratively varying trajectory lengths and Assumption 1, find suitable input $u_k(t)$ to ensure that the system fully tracks expected trajectory (2).

2.2 Open-Close-Loop ILC Laws and Convergence

We define $\varphi_k(t), t \in \{0, 1, \cdots, N\}$ as a random variable that satisfies the Bernoulli distribution, where the value is a binary value of 0 or 1, to solve the ILC problem of a linear discrete-time MIMO system (1), where the trajectory length varies with iteration. $\varphi_k(t) = 1$ means that the control input of system (1) can continue until time point t of $k-th$ iterations, which occurs as a probability function of $p(t), (0 < p(t) \le 1)$. $\varphi_k(t) = 0$ means that the control input of system (1) cannot continue until time point t of $k-th$ iterations, which occurs as a probability function $1 - p(t)$. Obviously, the expectation of $\varphi_k(t)$ is

$$E\{\varphi_k(t)\} = 1 \cdot p(t) + 0 \cdot (1 - p(t)) = p(t). \tag{9}$$

Since the trajectory length of system (1) is iteratively variable, the modified tracking error is defined as

$$e_k^*(t) = \varphi_k(t) \cdot e_k(t), t \in \{0, 1, \cdots, N\}. \tag{10}$$

From the definition of Bernoulli stochastic variable $\varphi_k(t)$, When $N_k < N$ is established, we rewrite (10)

$$e_k^*(t) = \begin{cases} e_k(t), t \in \{0, 1, \cdots, N_k\}, \\ 0, t \in \{N_k + 1, \cdots, N\}. \end{cases} \tag{11}$$

And for $N_k \ge N$,

$$e_k^*(t) = e_k(t), t \in \{0, 1, \cdots, N\}. \tag{12}$$

According to (9) and (10), it can be concluded that

$$E\{e_k^*(t)\} = p(t) \cdot E\{e_k(t)\}. \tag{13}$$

To reach the comprehensive tracking form system (1) to the desired system (2), the following open-close-loop law is applied to the system (1).

$$u_{k+1}(t) = u_{f,k+1}(t) + u_{b,k+1}(t), \tag{14}$$

$$u_{f,k+1}(t) = u_k(t) + Le_k^*(t + 1), \tag{15}$$

$$u_{b,k+1}(t) = Me_{k+1}^*(t). \tag{16}$$

Theorem 1: Under Assumption 1, for the linear discrete-time MIMO system (1) with iteratively variable trajectory lengths and the open-close-loop ILC law (14)–(16), choose the learning gain matrix L and M such that, for any constant $0 \le \varepsilon \le 1$, if the control gain L is chosen to make.

$$\|I - p(t+1)LCB\| \le \varepsilon < 1. \tag{17}$$

Then

$$\lim_{k \to +\infty} E\{e_k(t)\} = 0, t \in \{0, 1, 2, \cdots, N+1\}. \tag{18}$$

2.3 Proof of Assumption 1

Form Assumption 1 of (7) and denoting $\Delta u_{f,k}(t) = u_d(t) - u_{f,k}(t)$, subtracting both sides of (15) from $u_d(t)$, we have

$$\Delta u_{f,k+1}(t) = \Delta u_k(t) - Le_k^*(t+1). \tag{19}$$

Taking $E\{\cdot\}$ on the sides of (19) and considering (13)

$$E\{\Delta u_{f,k+1}(t)\} = E\{\Delta u_k(t)\} - Lp(t+1) \cdot E\{e_k(t+1)\}. \tag{20}$$

According to (1), (2) and (20) become

$$\begin{aligned}
E\{\Delta u_{f,k+1}(t)\} &= E\{\Delta u_k(t)\} - p(t+1)LC \cdot E\{\Delta x_k(t+1)\} - p(t+1)L \cdot E\{v_k(t+1)\} \\
&= E\{\Delta u_k(t)\} - p(t+1)LC \cdot [A \cdot E\{\Delta x_k(t)\} + B \cdot E\{\Delta u_k(t)\}] \\
&\quad - p(t+1)LC \cdot E\{w_k(t)\} - p(t+1)L \cdot E\{v_k(t+1)\} \\
&= (I - p(t+1)LCB) \cdot E\{\Delta u_k(t)\} - p(t+1)LCA \cdot E\{\Delta x_k(t)\} \\
&\quad - p(t+1)LC \cdot E\{w_k(t)\} - p(t+1)L \cdot E\{v_k(t+1)\}.
\end{aligned} \tag{21}$$

Considering (4) and (5) of Assumption 1

$$\begin{aligned}
E\{\Delta u_{f,k+1}(t)\} &= (I - p(t+1)LCB) \cdot E\{\Delta u_k(t)\} \\
&\quad - p(t+1)LCA \cdot E\{\Delta x_k(t)\}.
\end{aligned} \tag{22}$$

Form (14) to (16)

$$\Delta u_k(t) = \Delta u_{f,k}(t) - u_{b,k}(t) = \Delta u_{f,k}(t) - Me_k^*(t). \tag{23}$$

Taking $E\{\cdot\}$ on the sides of (23) and considering (13)

$$E\{\Delta u_k(t)\} = E\{\Delta u_{f,k}(t)\} - p(t)M \cdot E\{e_k(t)\}. \tag{24}$$

Substituting (24) into (22) derives that

$$\begin{aligned}
E\{\Delta u_{f,k+1}(t)\} &= [I - p(t+1)LCB] \cdot E\{\Delta u_{f,k}(t)\} \\
&\quad - p(t)M[I - p(t+1)LCB] \cdot E\{e_k(t)\} \\
&\quad - p(t+1)LCA \cdot E\{\Delta x_k(t)\}.
\end{aligned} \tag{25}$$

Taking norm $\|\cdot\|$ on both sides of (25), we have

$$
\begin{aligned}
\left\|E\{\Delta u_{f,k+1}(t)\}\right\| &\leq \|I - p(t+1)LCB\| \cdot \left\|E\{\Delta u_{f,k}(t)\}\right\| \\
&+ p(t)\|I - p(t+1)LCB\| \cdot \|M\| \cdot \|E\{e_k(t)\}\| \\
&+ p(t+1)\|LCA\| \cdot \|E\{\Delta x_k(t)\}\| \\
&\leq \varepsilon\left\|E\{\Delta u_{f,k}(t)\}\right\| + \beta_1(\|E\{e_k(t)\}\| + \|E\{\Delta x_k(t)\}\|),
\end{aligned}
\tag{26}
$$

where $\beta_1 = max\{\bar{p} \cdot \|M\| \cdot \underset{t}{max}\|I - p(t+1)LCB\|, \bar{p} \cdot \|LCA\|\}$ with $\bar{p} = \underset{i\in\{0,1,\cdots,N+1\}}{max} p(i)$, and $\|I - p(t+1)LCB\| \leq \varepsilon$ by (17) of Theorem 1 Addition, according to (1), (2) and (24), with (4) of Assumption 1, there are

$$
\begin{aligned}
\|E\{\Delta x_k(t)\}\| \\
\leq \|A\| \cdot \|E\{\Delta x_k(t-1)\}\| + \|B\| \cdot \|E\{\Delta u_k(t-1)\}\| - \|E\{w_k(t-1)\}\| \\
\leq \|A\| \cdot \|E\{\Delta x_k(t-1)\}\| + \|B\| \cdot \left\|E\{\Delta u_{f,k}(t-1)\}\right\| \\
+ p(t-1)\|B\| \cdot \|M\| \cdot \|E\{e_k(t-1)\}\|.
\end{aligned}
\tag{27}
$$

And

$$
\begin{aligned}
\|E\{\Delta e_k(t)\}\| \\
\leq \|C\| \cdot \|E\{\Delta x_k(t)\}\| \\
\leq \|C\| \cdot \|A\| \cdot \|E\{\Delta x_k(t-1)\}\| + \|C\| \cdot \|B\| \cdot \left\|E\{\Delta u_{f,k}(t-1)\}\right\| \\
+ p(t-1)\|C\| \cdot \|B\| \cdot \|M\| \cdot \|E\{e_k(t-1)\}\|.
\end{aligned}
\tag{28}
$$

Since $E\{x_k(0)\} = x_d(0)$ from (3) in Assumption 1 and $E\{e_k(0)\} = 0$, it can be derived form (27) and (28) that

$$
\begin{aligned}
\|E\{e_k(t)\}\| + \|E\{\Delta x_k(t)\}\| \\
\leq (\|C\| \cdot \|A\| + \|A\|) \cdot \|E\{\Delta x_k(t-1)\}\| \\
+ (\|C\| \cdot \|B\| + \|B\|) \cdot \left\|E\{\Delta u_{f,k}(t-1)\}\right\| \\
+ (\|C\| \cdot \|B\| \cdot \|M\| + \|B\| \cdot \|M\|) \cdot \|E\{e_k(t-1)\}\| \\
\leq \beta_2(\|E\{e_k(t-1)\}\| + \|E\{\Delta x_k(t-1)\}\|) + \bar{b}\left\|E\{\Delta u_{f,k}(t-1)\}\right\| \leq \cdots\cdots \\
\leq \beta_2^t(\|E\{e_k(0)\}\| + \|E\{\Delta x_k(0)\}\|) + \sum_{s=0}^{t-1} \beta_2^{t-s-1} \bar{b}\cdot\left\|E\{\Delta u_{f,k}(s)\}\right\| \\
= \sum_{s=0}^{t-1} \beta_2^{t-s-1} \bar{b}\cdot\left\|E\{\Delta u_{f,k}(s)\}\right\|,
\end{aligned}
\tag{29}
$$

where $\beta_2 = max\{\|C\| \cdot \|A\| + \|A\|, \|B\| \cdot \|M\| \cdot \bar{p}(1 + \|C\|)\}$ and $\bar{b} = \|C\| \cdot \|B\| + \|B\|$. Substituting (29) into (26), we have

$$
\begin{aligned}
\left\|E\{\Delta u_{f,k+1}(t)\}\right\| \\
\leq \varepsilon\left\|E\{\Delta u_{f,k}(t)\}\right\| + \beta_1 \sum_{s=0}^{t-1} \beta_2^{t-s-1} \bar{b}\cdot\left\|E\{\Delta u_{f,k}(s)\}\right\|.
\end{aligned}
\tag{30}
$$

Letting $t = 0$ in (26) and $t = 1, 2, \cdots, N$ in (30), respectively, and considering (3)–(5) in Assumption 1, there are

$$
\begin{cases}
\left\| E\{\Delta u_{f,k+1}(0)\} \right\| \leq \varepsilon \left\| E\{\Delta u_{f,k}(0)\} \right\| \\
\left\| E\{\Delta u_{f,k+1}(1)\} \right\| \leq \varepsilon \left\| E\{\Delta u_{f,k}(1)\} \right\| + \beta_1 \bar{b} \cdot \left\| E\{\Delta u_{f,k}(0)\} \right\| \\
\cdots\cdots \\
\left\| E\{\Delta u_{f,k+1}(N)\} \right\| \leq \varepsilon \left\| E\{\Delta u_{f,k}(N)\} \right\| + \beta_1 \beta_2^{N-1} \bar{b} \cdot \left\| E\{\Delta u_{f,k}(0)\} \right\| + \cdots \\
+ \beta_1 \bar{b} \cdot \left\| E\{\Delta u_{f,k}(N-1)\} \right\|.
\end{cases}
\tag{31}
$$

Denote $\theta_{f,k} = [\| E\{\Delta u_{f,k}(0)\} \|, \| E\{\Delta u_{f,k}(1)\} \|, \cdots \| E\{\Delta u_{f,k}(N)\} \|]^T$, rewrite (31):

$$
\theta_{f,k+1} \leq \sigma_1 \theta_{f,k},
\tag{32}
$$

where $\sigma_1 = \begin{bmatrix} \varepsilon & 0 & \cdots & 0 \\ \beta_1 \bar{b} & \varepsilon & \cdots & 0 \\ \vdots & \vdots & \ddots & \vdots \\ \beta_1 \beta_2^{N-1} \bar{b} & \beta_1 \beta_2^{N-2} \bar{b} & \cdots & \varepsilon \end{bmatrix}$. And satisfy spectral radius $\rho(\sigma_1) = \varepsilon < 1$.

Consequently, from (32), when $t \in \{0, 1, \cdots, N\}$, Let

$$
\lim_{k \to +\infty} \theta_{f,k} = 0,
\tag{33}
$$

Which clearly infer for $t \in \{0, 1, \cdots, N\}$ $t \in \{0, 1, \cdots, N\}$

$$
\lim_{k \to +\infty} \left\| E\{\Delta u_{f,k}(t)\} \right\| = 0.
\tag{34}
$$

According (26) and (3) in Assumption 1, for $t \in \{0, 1, \cdots, N\}$ $t \in \{0, 1, \cdots, N\}$

$$
\lim_{k \to +\infty} \left(\left\| E\{e_k(t)\} \right\| + \left\| E\{\Delta x_k(t)\} \right\| \right) = 0.
\tag{35}
$$

It is proved that

$$
\lim_{k \to +\infty} E\{e_k(t)\} = 0, t \in \{0, 1, 2, \cdots, N+1\}.
\tag{36}
$$

The proof is completed.

2.4 Illustrative Examples

In this section, in the cause of verifying the feasibility and advancement of the open-closed-loop ILC laws, consider the following MIMO linear discrete-time system with iteratively variable trial lengths and disturbances.

$$
\begin{cases}
x_k(t+1) = \begin{bmatrix} 0.35 & 0 \\ 0.62 & 0.3 \end{bmatrix} x_k(t) + \begin{bmatrix} 0.06 & 0.7 \\ 5.7 & 0.5 \end{bmatrix} u_k(t) + w_k(t), \\
y_k(t) = \begin{bmatrix} 1 & 0 \\ 0 & 1 \end{bmatrix} x_k(t) + v_k(t),
\end{cases}
\tag{37}
$$

The iterative initial state $x_k(0) = \left[x_k^{(1)}(0)\ x_k^{(2)}(0) \right]^T \in R^2$ with mathematical expectation $E\{x_k(0)\} = x_d(0) = \left[0\ 0 \right]^T$ is depicted in Fig. 1. Define the reference output trajectories $y_d^{(1)}(t)$ and $y_d^{(2)}(t)$ of the system (37) as

$$\begin{cases} y_d^{(1)}(t) = 0.016t[1 + \cos(6\pi t/N - \pi)], \\ y_d^{(2)}(t) = 0.5[1 + \sin(4\pi t/N - \pi/2)], \end{cases} \tag{38}$$

where $t \in \{0, 1, \cdots, N\}$ and $N = 100$. The nonrepetitive (iteration-dependent) load $w_k(t) = \left[w_k^{(1)}(t)\ w_k^{(2)}(t) \right]^T$ and measurement disturbances $v_k(t) = \left[v_k^{(1)}(t)\ v_k^{(2)}(t) \right]^T$ vary randomly in each iteration with mathematical expectation $E\{w_k(t)\} = \left[0\ 0 \right]^T$ and $E\{v_k(t)\} = \left[0\ 0 \right]^T$, respectively. Besides, the trial lengths N_k with lower bound $N_{low} = 95$ and upper bound $N_{up} = 116$ varies randomly, as shown in Fig. 2. In order to measure the tracking effect of output trajectories of system (37) to the reference output trajectories $y_d^{(1)}(t)$ and $y_d^{(2)}(t)$, the tracking error indexes is designed as

$$CE_k^{(i)} = \sum_{i=0}^{N+1} \left| E\{y_d^{(i)}(t) - y_k^{(i)}(t)\} \right|, \qquad i = 1, 2. \tag{39}$$

Without losing generality, the initial control input $u_0(t) = \left[0\ 0 \right]^T$, $t \in \{0, 1, \cdots, N\}$. Apply the developed P-type open-closed-loop ILC law (14)–(16) into system (37).

To show the tracking performance with open-closed-loop control, the control gains in (14)–(16) are chosen as

$$L = \begin{bmatrix} 0.03 & 0.06 \\ 0.07 & 0.02 \end{bmatrix}, \qquad M = \begin{bmatrix} 0.03 & 0.02 \\ 0.09 & 0.05 \end{bmatrix}.$$

Meanwhile, in the case of illustrating the effectiveness of P-type open-closed-loop ILC law under interference, the iteratively variable trial lengths-based P-type ILC scheme without feedback control part with control gain $L = \begin{bmatrix} 0.03 & 0.06 \\ 0.07 & 0.02 \end{bmatrix}$ is applied to system (37). In consequence, the profiles of tracking error indexes $CE_k^{(i)}$ ($i = 1, 2$) at difference iterations are provided in Fig. 3. Figure 4 displays the tracking performance of the system outputs $y_k^{(i)}(t)$ ($i = 1, 2$) to the reference trajectories (38) at iterations $k = 20$ and $k = 30$ with the open-closed-loop ILC law (14)–(16).

When the initial state $x_k(0)$ fluctuates around the desired initial state $x_d(0)$ with $E\{x_k(0)\} = x_d(0)$ as displayed in Fig. 1. Figure 3 shows that the tracking error indexes $CE_k^{(i)}$ ($i = 1, 2$) can be driven to zero as the iteration number k increases to infinity. As can be seen from Fig. 3, in the presence of interference, the developed open-closed-loop ILC law (14)–(16) can make the tracking error indexes convergent in less iteration than the P-type ILC scheme without feedback control. Figure 4 implies that although the trial length N_k of system (37) is randomly variable, the nonrepetitive (iteration-dependent) load $w_k(t)$ and measurement disturbances $v_k(t)$ vary randomly in each iteration with mathematical expectation $E\{w_k(t)\} = \begin{bmatrix} 0 & 0 \end{bmatrix}^T$ and $E\{v_k(t)\} = \begin{bmatrix} 0 & 0 \end{bmatrix}^T$, the system outputs $y_k^{(i)}(t)$ ($i = 1, 2$) are progressively closed to the reference trajectories $y_d^{(i)}(t)$ ($i = 1, 2$).

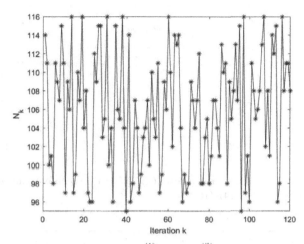

Fig. 1. The values of initial state $x_k^{(1)}(0)$ and $x_k^{(2)}(0)$ at different iterations.

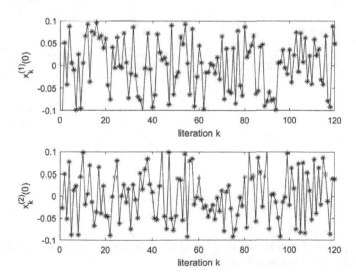

Fig. 2. The iteratively variable trial length N_k in system (37) at different iterations.

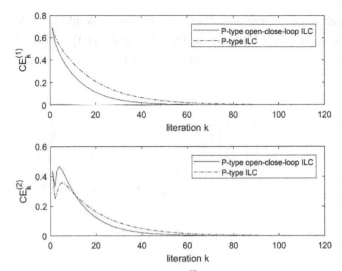

Fig. 3. The profile of ILC tracking error indexes $CE_k^{(i)}$ ($i = 1, 2$) at different iterations by adopting the open-closed-loop ILC law (14)–(16).

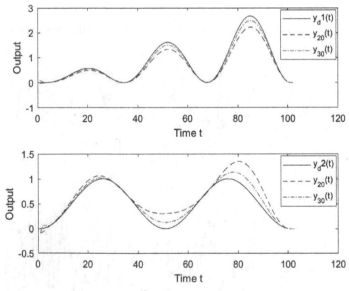

Fig. 4. The profile of system outputs $y_k^{(i)}(t)$ ($i = 1, 2$) at iterations $k = 20$ and $k = 30$ by adopting the open-closed-loop ILC law (14)–(16).

3 Conclusion

In this paper, we design open-closed-loop ILC algorithm for linear discrete-time systems with trajectory length variations, where the system contains the nonrepetitive (iteration-dependent) load noise and measurement disturbances noise. On the one hand, due to the trajectory length variations, the tracking error of the previous iteration will most likely be lost. Aiming to compensation the lost tracking error information, the feedback control is added to P-type ILC. On the other hand, to mitigate the impact of noise on system convergence, the noise is defined using mathematical expectation, and its influence on system tracking performance is eliminated during mathematical analysis and proof.

Acknowledgement. This work is supported in part by Innovation and Entrepreneurship Training Program for College Students (S202311078001) and Key Laboratory of On-Chip Communication and Sensor Chip of Guangdong Higher Education Institutes Guangzhou University, KLOCCSCGHEI (2023KSYS002).

References

1. Riaz, S., Hui, L.: A future concern of iterative learning control: a survey. J. Stat. Manag. Syst. **24**(6), 1301–1322 (2021)
2. Xu, J., Zhou, Lin, N,, Chi, R.H.: High-order model-free adaptive iterative learning control. Trans. Inst. Measur. A Control **45**(10), 1886–1895 (2023)
3. Bristow, D.A., Tharayil, M., Alleyne, A.G.: A survey of iterative learning control: a learning-based method for high-performance tracking control. IEEE Control. Syst. Mag. **26**(3), 96–114 (2006)
4. Shen, D., Wang, Y.: Survey on stochastic iterative learning control. J. Process. Control. **24**(12), 64–77 (2014)
5. Chi, R., Hou, Z.S., Jin, S., Huang, B.: Computationally-light non-lifted data-driven norm-optimal iterative learning control. Asian J. Control **20**(1), 115–124 (2018)
6. Shen, D., Xu, Y.: Iterative learning control for discrete-time stochastic systems with quantized information. IEEE/CAA J. Autom Sinica **3**(1), 59–67 (2016)
7. Wang, J.A., Si, Q.Y., Bao, J., Wang, Q.: Iterative learning algorithms for boundary tracing problems of nonlinear fractional diffusion equations. Netw. Heterogen. Media **18**(3), 1355–1377 (2023)
8. Chien, C.-J.: A discrete iterative learning control for a class of nonlinear time-varying systems. IEEE Trans. Autom. Control **43**(5), 748–752 (1998)
9. Park, K.-H.: An average operator-based PD-type iterative learning control for variable initial state error. IEEE Trans. Autom. Control **50**(6), 865–869 (2005)
10. Parzer, H., Gattringer, H. Müller, A.: Robot force/position control combined with ILC for repetitive high speed applications. In: 25th International Conference on Robotics in Alpe-Adria-Danube Region, vol. 540, pp. 10–17. Belgrade, SERBIA (2017)
11. Zhang, Y.M., Liu, J., Ruan, X.E.. Impact of measurement noise on networked iterative learning control systems. In: 38th Chinese Control Conference (CCC), pp. 2398–2403. Guangzhou, People's Republic of China (2019)
12. Liu, J., Ruan, X.E., Zhang, Y.M.: Convergence analysis of ILC process for networked system with system noise. In: 9th IEEE Data Driven Control and Learning Systems Conference (DDCLS), Liuzhou, People's Republic of China, pp. 486–491 (2021)
13. Yang, X.X., Riaz, S.: Accelerated iterative learning control for linear discrete systems with parametric perturbation and measurement noise. CMES-Comput. Model. Eng. Sci. **132**(2), 605–626 (2022)
14. Huang, LX., Ding, H.G.: Suppress the accumulated effect of channel noise on ILC systems over wireless channels. In: 28th Chinese Control and Decision Conference, Yinchuan, People's Republic of China, pp. 7013–7018 (2016)
15. Meng, D.Y., Zhang, J.Y.: Robust tracking of nonrepetitive learning control systems with iteration-dependent references. IEEE Trans. Syst. Man Cybern.-Syst. **51**(2), 842–852 (2021)

Quality Control Model of Value Extraction of Residual Silk Reuse Based on Improved Genetic Algorithm

Qi Ji[✉], Mingxing Li, and Chao Shen

China Tobacco Zhejiang Industrial CO., LTD., Hangzhou, China
jiqi_zjtb@hotmail.com

Abstract. The recovery and reuse of residual silk in the tobacco industry is a normal work, but the efficiency of this work has always been a weak link in quality control. In this paper, the author proposes to use genetic algorithm to improve the efficiency of residual wire recovery and improve the value of residual wire recovery. For the whole recycling process of residual silk, multiple options in the process (whether to put it into storage on the same day, whether to put it into reproduction on the same day, the selection of input regeneration output, whether the products after reproduction are put into storage on the same day, and the sequence selection of utilization of regenerated cut tobacco) are regarded as a vector of five dimensions. The search space of this vector is the solution space of the problem. Genetic algorithm is used to explore the optimal solution in this search space. The practice shows that the algorithm proposed in this paper can greatly improve the value of residual wire reuse.

Keywords: Genetic Algorithm · Quality Control Model · Residual Silk Reuse

1 Introduce

Tobacco industry recycling and reuse of residual silk is a normal work, in the process of cigarette production due to equipment failure or manual error and other reasons will produce a certain number of quality can not reach the factory standard of defective products, because the quality problem is not the tobacco itself, so it can be recycled to reduce costs. Defective cigarettes are collected and transferred into the next processing process by manual after consideration. Due to simple equipment, the whole process is completed by manual, resulting in low efficiency and consumption of a lot of human and material resources. Therefore, the efficiency of this work has always been a weak link of quality control. In the existing literature, the focus of the research is almost to improve the recovery rate of residual silk by improving the residual silk processing equipment, improve the recovery value, such as Gao Yongliang proposed that by increasing the speed of the dispersion wheel, the tobacco defilement rate has been greatly improved, the silk rate in the paper decreased significantly. According to statistics, about 20 kg more tobacco can be produced every day, and the percentage of silk in paper can be

© The Author(s), under exclusive license to Springer Nature Singapore Pte Ltd. 2024
K. Li and Y. Liu (Eds.): ISICA 2023, CCIS 2147, pp. 430–441, 2024.
https://doi.org/10.1007/978-981-97-4396-4_40

reduced to about 4.5%, saving the value of tobacco up to about 900,000 yuan every year, with obvious economic benefits. Lou Qi et al. improved the FY1115 residual smoke processor, straightened and quantified the residual smoke after screening and removing impurities. After rationalizing the residual smoke, the use of opening knife roller rapid rotation and tobacco roller low-speed rotation to form a rotary cut, the tobacco longitudinal open a seam, through the vertical vibration elevator, so that the tobacco paper, silk primary separation. Then enter the primary screen, screen out the mouthstick and cigarette paper into the playing mechanism, playing after the cigarette paper, silk fall into three screening, respectively, the mouthstick and cigarette paper, filament, broken out. In this way, the recovery rate and whole filament rate are higher. This paper focuses on improving the recycling process of residual silk and improving the recycling value of residual silk through process control.

Genetic Algorithm (GA) is a computational model of biological evolution that simulates the natural selection and Genetic mechanism of Darwin's biological evolution [1, 2, 9, 11]. It is a method to search for the optimal solution by simulating the natural evolution process [3, 4, 10]. It mainly operates directly on the structure object without the limitation of derivation and continuity of function. It has inherent implicit parallelism and better global optimization ability. Using probabilistic optimization method, the optimized search space can be automatically obtained and guided without definite rules, and the search direction can be adjusted adaptively [5, 6, 12, 13]. Genetic algorithm is an effective optimization algorithm, and it is a global optimal algorithm. Genetic algorithm has been applied in many fields [7, 8, 14, 15]. For example, Wang Boquan improved the problem that genetic algorithm is prone to fall into local optimization and precocity in his doctoral thesis, and applied the improved genetic algorithm based on self-organization mapping to the optimal operation of cascade reservoir group and achieved good results [3]. In her master's thesis, Ji Yunxia also improved the genetic algorithm, and then applied the improved algorithm to the optimization of train energy-saving operation, which also achieved good economic efficiency. In addition, many scholars have applied genetic algorithm in many practical fields and achieved objective economic efficiency. It can be seen that the application of genetic algorithm in solving combinatorial optimization problems has its feasibility and high efficiency.

In this paper, an improved genetic algorithm is proposed to improve the efficiency and value of silk recovery. In view of the whole recycling process of silk residue, the multiple selection items in the process (whether to put into storage on the same day, whether to put into reproduction on the same day, the selection of input recycled output, whether to put into storage after reproduction on the same day, the sequence selection of recycled tobacco output) are regarded as a vector, a five-dimensional vector. The search space of this vector is the solution space of the problem, which is a typical optimization problem, and the genetic algorithm can be used to explore the optimal solution in this search space. Through practice, the algorithm proposed in this paper can greatly improve the value of silk recycling.

2   Improved Genetic Algorithm

2.1   Basic Theory of Genetic Algorithm

Darwin's theory of evolution has been recognized as the biological evolutionary theory, the theory of "survival of the fittest, superior bad discard" is an important thought, organisms in the process of survival, involves intraspecific and interspecific struggle and struggle with nature and the environment, in the struggle, biological evolution is to update themselves, namely variation, variation will be for the benefit of the large probability to survive and pass on to offspring; The adverse variants will eventually be eliminated. In this way, in order to thrive, organisms need to constantly renew themselves by evolving in their own favor and by interbreeding within the group in order to survive. Genetic algorithm is a random search algorithm based on the above ideas. The algorithm regards a biological individual as a solution in the optimization algorithm, and the population composed of many individuals is the solution set of the algorithm. The one who can better adapt to the fitness function is the superior individual. Then, through to the population to choose genetic operation such as crossover and mutation, keep out the individual performance is poor, and the favorable individual genes by probability of selection into the next generation, through constant iterative update, when reach the termination conditions, the output of the individual is filtered and the best individual survive, namely the optimal solution of optimization problem. Genetic algorithm is easy to operate and can be applied to both continuous optimization problems and discrete problems. Its multi-direction global optimization performance has a good theoretical value for solving complex optimization problems today.

Genetic algorithms employ three core operations: selection, crossover, and mutation. These crucial operations ensure the algorithm's success in identifying the best solutions and highlight its proficiency in the search process. A brief overview of each operation follows. Selection involves choosing the best-performing individuals and discarding those who underperform. While some of these top performers continue to the next generation as they are, others transmit their superior genes via crossover. There are several common methods for selecting operations.

(1) the roulette selection method is based on the individual fitness proportion of a selection method, through four individual fitness to get their accounts for the proportion of the overall, then according to the proportion of the wheel is divided into four regions, each region represents the proportion of each size, spin, stopped at which region, on behalf of the individual which is selected from the area. It can be seen that the greater the fitness of the individual, the higher the probability of being selected.

(2) The random traversal selection method selects from the population by equidistant selection, calculates the fitness of each individual, and obtains the fitness proportion corresponding to each individual, which is divided on the line segment by proportion accumulation. If the number of individuals to be selected is set as n, a random number is generated on $(0, 1/n)$, which is the starting position of selection. The random number is selected with an equal distance of $1/n$ until the selection is complete.

(3) The best individual preservation method

In this method, individual's fitness is initially assessed. They are then ranked based on their fitness scores, descending from the highest. Those exhibiting the highest fitness levels are directly carried over to the subsequent generation. This selection step is often referred to as "replication."

(4) Tournament selection method

Firstly, the league size is determined, and then the individuals of this size are randomly selected from the population, and their fitness is calculated and sorted. The individuals with the highest fitness levels are chosen, and this procedure continues until the desired count is reached.

Crossover operation, and it is well known that genetic recombination in the process of evolution and development plays an indelible role, it is because of genetic recombination, species to breed and each are not identical, in genetic recombination is the crossover operation of genetic algorithm, the algorithm plays an important role in the process of optimization, the algorithm to obtain the optimal solution is one of essential conditions. In general, crossover is categorized into real recombination and binary crossover.

Mutation entails alterations in certain genes as they transfer from parent to offspring, leading to distinct behaviors in the descendant. This operation is vital for upholding the algorithm's diversity and is crucial for achieving the global optimum within the algorithm. Mutations can be classified into real number and binary types.

Genetic algorithm implementation steps

Genetic algorithm is mainly composed of population individual initialization, fitness calculation, selection, crossover, variation and other parts. The mathematical representation of using a genetic algorithm for optimization tasks can be outlined as followed.

(1) Begin by initializing the population members $(x_1, x_2, \ldots x_i, \ldots x_N)$, where N denotes the population size. Each x_i is the ith member of the population, symbolizing an optimization problem's solution. Each x_i can be expressed as $(x_{i1}, x_{i2}, \ldots x_{ij}, \ldots x_{in})$ with x_{ij} being the j th gene value of the ith individual, signifying the j th variable of the optimization issue, and n indicating the problem's variable dimension.

(2) Set the algorithm's parameters, chiefly the population size N0, iteration count N1, crossover probability Pc, and mutation probability PM.

(3) Compute the fitness for the aforementioned members and employ the selection technique to choose the individuals exhibiting top Nch performance.

(4) For crossover operations, pick two members from the population. Depending on their encoding types, the crossover technique updates these members, yielding Ncr fresh crossover members.

(5) For mutation, choose Nm members from the populace and conduct the mutation process during the mutation calculation phase, resulting in Nm new members.

(6) Determine if the termination criteria have been met. If they have, reveal the best member. If not, revert to step (3) to persist with the algorithm's iterative process.

As can be seen from the above steps, after the algorithm initializes the population, each individual has certain advantages and disadvantages as it is randomly generated. Therefore, through selection operation, the individuals with good performance are saved and the ones with poor performance are eliminated, and then the population diversity is expanded through crossover and mutation: Crossover allows gene recombination between parental individuals to achieve locally searched H. The mutation operation makes the gene change on the individual, thus produces the new individual, achieves the global search effect. When the termination condition is reached, the individual with the best fitness is the optimal solution required.

2.2 Improved Genetic Algorithm

In the late stage of genetic algorithm, the algorithm is close to convergence, and the obtained solution is basically close to the characteristics of optimal solution. However, due to crossover and mutation has a certain randomness, so it often loses the attributes of the original individual due to crossover and mutation update, and the performance is worse than the original, leading to the failure to update the better individual. Therefore, genetic algorithm based on self-organizing mapping to update of the individual, in the late genetic algorithm and the self-organizing mapping attribute characteristics of genetic algorithm can effectively learn the individual, the output is consistent with the data structure and input structure, then enhance the searching capability of the algorithm in the later, to ensure the precision of the algorithm. The individual update strategy for the self-organizing mapping genetic algorithm is designed as follows:

(1) Based on the optimization task, establish multiple neurons j and their associated weight vectors $w_j(j = 1,2,...M)$. Here, M signifies the count of neurons initialized. Together, they constitute the structure network of the self-organizing mapping genetic algorithm. Concurrently, set the iteration count to N0 and initialize t to 0.

(2) Taking individuals produced by the genetic algorithm $(x_1, x_2...x_i, ...x_N)$, an individual x_i is chosen as the input for the self-organizing mapping network. Subsequently, compute the distance between the chosen individual and the neuron using:

$$dis(w_j, x_i) = \sqrt{\sum_{k=0}^{n} \left(w_j^k - x_i^k\right)^2} \tag{1}$$

Here, w_i^k indicates the gene value at position K on neuron J, while w_i^k denotes the gene value at position K on individual i.

(3) Traverse all neurons within the self-organizing mapping network to pinpoint the neuron J* nearest to the current individual, termed as the best matching unit.

(4) Derive the neighborhood radius of the neuron using:

$$\sigma_j = \sigma_0 exp(-t/\varphi) \tag{2}$$

where σ_0 is the initial neighborhood radius and $\Psi = N_0/lg(\sigma_0)$ acts as a time constant.

(5) Identify neurons within this radius and refresh the genes they possess. The gene modification formula is:

$$w_j^k(t + 1) = w_j^k(t) + L(t)\theta_{j*j}(t)\left(x_j^k(t) - x_j^k(t)\right) \tag{3}$$

With $w_j^k(t)$ as the present gene at position K on the neuron, and $w_j^k(t+1)$ being its post-update counterpart. L(t) is the learning rate, defined as $L_0 exp(-t/N_0)$, with L0 as the initial rate.

$$\theta_{j*j}(t) = exp\left(-\left\|w^{j*} - w^j\right\|^2/2\sigma_t^2\right) \tag{4}$$

Here, "$w^{j*} - w^j$" represents the distance between neuron J and the optimal neuron J* in this vicinity.

(6) Assess if the termination criteria have been fulfilled. If not, increment t by one and revert to Step (2) for further computation; otherwise, proceed to Step (7).

(7) Refresh the individual using:

$$x_t = w^{j*} + \sum_{k=0}^{m-1} \alpha_{j*}^k \mu_{j*}^k + N(0, \sigma I) \qquad (5)$$

where w_{j*} is the best matching unit for the individual, a_{j*}^k is the distance between two neighboring meteors, and μ_{j*}^k is the vector of adjacent neurons. $N(0, \sigma I)$ denotes a noise vector with a normal distribution.

3 The Quality Control Model of Value Extraction of Residual Silk Reuse Based on Improved Genetic Algorithm

3.1 Analysis of Residual Silk Recovery Process

See Fig. 1.

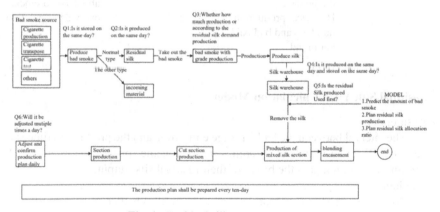

Fig. 1. Residual silk recovery process

3.2 The Quality Control Model of Value Extraction of Residual Silk Reuse Based on Improved Genetic Algorithm

Prediction Model of Cigarette Bad Packet Quantity

(1) Assumptions

This model predicts bad packages according to production schedule. The default bad packet generation is related to the production plan, that is, the more the production, the more bad packets. Cigarette damage during transportation, which is an accident, may not be predicted in the model. The proportion of cigarettes damaged in the process of transportation to cigarettes in storage, if the proportion is large, the damage in the process of transportation should be considered. There is a fixed interval between the production time of cigarettes and the storage time after the bad packets are found. For example, they are all stored on the same day, or they are all stored on the next day.

(2) Input parameter out parameter model (Table 1)

Table 1. Input and output parameters of cigarette bad packet quantity prediction model

model	input parameter	Output parameter	Algorithm cycle
Number of bad packets forecast-per ten days	The production quantity of each brand in the next ten days; Historical production numbers and bad packs per brand	Number of bad bags per brand in the next ten days	Run three times a month on a ten-day schedule
Bad packet number forecast-daily	The production quantity of each brand on the day; Historical production numbers and bad packs per brand	Number of bad bags per brand that day	After the daily schedule adjustment, about 7–9 o'clock, every day operation

Residual Silk Yield Calculation Model

(1) Assumptions

The interval between the bad package entry time and the production time should be fixed. Different brands of cigarettes mixed together production, there will be a certain rule, can not pass the brand of their residual silk output.

(2) problem

What is the production strategy of residual silk? Whether to produce all the bad bales on the same day, or to push back the production of residual silk according to demand The plan? The logic of the algorithms is different.

(3) input and output parameter (Table 2)

Calculation Model of Residual Silk Proportion

(1) Assumptions

The residual silk produced first is put into use first. According to the production of 10 days of residual silk, all mixed with the standard. If not, save it for the next cycle.

(2) problem

Is the residual silk produced on the same day to be used the next day or the same day? If it is produced on the same day and used on the same day, residual silk may

Table 2. Policy model parameters for daily production of all bad packets

model	input parameter	Output parameter	Algorithm cycle
Residual silk yield calculation - per ten days Residual silk yield calculation - daily	In the next 10 days, the number of bad bales of Residual silk output per each brand, and the corresponding relationship grade per day for the next between the brand and the grade of residual silk 10 days		Three times a month when the ten-day plan is issued After the daily schedule adjustment, about 7-9 o'clock, every day operation
	The number of bad bales of each brand on the day, the remaining time in each ten-day, the number of bad bales of each brand on the day, the corresponding relationship between the brand and the grade of residual silk	The daily output of residual silk of each grade	

not reach When produced, provided to the mixing section. Is the fluctuation ratio of mixing ratio calculated between two adjacent tasks? Or all of them. For example, the first batch is 1%, the second batch is 1.1%, and the third batch is 1.21%. The fluctuation of the two adjacent batches is not more than 10%, but the fluctuation of the first and third batches is more than 10%. I don't know how to calculate the mixing ratio, so I feel very complicated. In fact, we should consider both uniformity and full mixing, because if we only do full mixing, the uniformity may be very poor. It may be necessary to develop a set of computational logic that considers both homogeneity and full mixing. The average blending ratio can be calculated according to the predicted residual silk yield in 10 days and the planned yield in the next 10 days. The daily residual silk usage is calculated by multiplying the daily production plan for the next ten days by the average mixing ratio. Then, according to the actual daily data and whether there is residual silk before, the daily mixing ratio of the remaining days of each ten-day is calculated according to the 10-day full mixing rolling. However, the precondition that the stock must be greater than the dosage should be considered.

Algorithm Design

(1) Set the foundational parameters for the algorithm. This encompasses population size N0, SOM-GA iteration count M, crossover probability pc, mutation likelihood pm, SOM learning iteration N2, starting neighborhood radius σ0, beginning learning rate L0, and SOM individual update probability pg (initially set at 0.2 and later adjusted to 0.8). The real encoding mode is chosen for computation (Fig. 2).

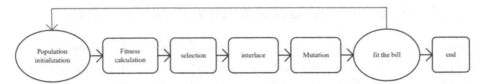

Fig. 2. Residual silk recovery process

(2) Based on constraints, define the feasible domain. Generate N0 individuals randomly within this domain, and evaluate each individual's fitness (here, the fitness function aligns with the objective function).

(3) Selection. Rank individuals based on the objective function type and employ the roulette wheel method for selection. An elitism strategy ensures top-performing individuals transition to the subsequent generation.

(4) Crossover. Execute the crossover operation based on probability pc. To acquire novel individuals and expand the generational space, a multi-parent crossover approach is applied. If K individuals are randomly chosen from the population and represented as $X1, X2,\ldots X_K$, the crossover is done using:

$$\alpha' = \alpha_1 X_1 + \alpha_2 X_2 + \cdots \alpha_K X_K \tag{5}$$

where a_K is a stochastic coefficient with: $\sum a_K = 1$ and $-0.5 \leq a_K \leq 1.5$.

(5) Mutation. Operate mutations with a pm likelihood. For diversity, a multi-gene point mutation is preferred. For an individual xi, S gene locations are reinitialized within the feasible range, followed by individual updates.

(6) Algorithmic individual updates take place using the previously detailed steps with a probability pg.

(7) Rank individuals produced in steps (3) to (6) and identify the best candidates.

(8) Evaluate if the termination criteria are met. If not met, revert to Step (3); otherwise, conclude the computation and reveal the optimal solution.

4 Numerical Experimentation

Some parameters of the genetic algorithm are specified as follows: population size is 150, crossover rate is 0.75, variation rate is 0.25, early penalty rate is 0.25, and delay penalty rate is 1.50. The control parameters are $\alpha = 0.05$, $\beta = 80$, and $\delta = 65$. The solution space consists of a five-dimensional space consisting of whether to put into storage on the same day, whether to put into reproduction on the same day, the choice of input recycled output, whether to put into storage after reproduction on the same day, and the choice of the sequence of application of recycled tobacco.

Figure 3 variation of optimization results of traditional genetic algorithm. It can be seen from Fig. 3 that population variation and iteration times are good, fitness value continues to decline, and solution change is small.

Figure 4 is the change diagram of the solution based on the improved genetic algorithm. It can be seen that the genetic algorithm population using process priority scheduling first drops rapidly at first and then fluctuates slowly up and down, and its change is similar to that of the population mean.

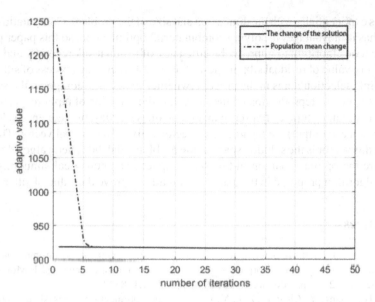

Fig. 3. Algorithm search process

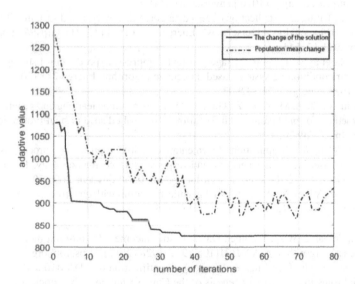

Fig. 4. Improved genetic algorithm

5 Conclusion

The recycling of residual silk is the normal work of tobacco industry, but how to improve the efficiency of residual silk recycling has not been a good solution. The dimension of the solution space causes the difficulty of solving the problem, and the difficulty often increases exponentially with the dimension and size of the problem. Computational intelligence has its advantages in solving combinatorial optimization, so this paper proposes to use genetic algorithm to improve the efficiency of residual silk recovery and improve the recovery value of residual silk. In view of the whole recycling process of silk residue, the multiple selection items in the process (whether to put into storage on the same day, whether to put into reproduction on the same day, the selection of input recycled output, whether to put into storage after reproduction on the same day, the sequence selection of recycled tobacco output) are regarded as a vector, a five-dimensional vector. The search space of this vector is the solution space of the problem, and the genetic algorithm is used to explore the optimal solution in this search space. The numerical simulation shows that the algorithm proposed in this paper can greatly improve the value of silk recycling.

References

1. Weidong, L.I., et al.: Implementation of AdaBoost and genetic algorithm machine learning models in prediction of adsorption capacity of nanocomposite materials. J. Mol. Liq. **350**, 118527 (2022). https://doi.org/10.1016/j.molliq.2022.118527
2. Wu, H., Huang, Y., Chen, L., Zhu, Y., Li, H.: Shape optimization of egg-shaped sewer pipes based on the nondominated sorting genetic algorithm (NSGA-II). Environ. Res. **204**, 111999 (2022). https://doi.org/10.1016/j.envres.2021.111999
3. Aygun, H., Turan, O.: Application of genetic algorithm in exergy and sustainability: a case of aero-gas turbine engine at cruise phase. Energy (2), 121644 (2021). https://doi.org/10.1016/j.energy.2021.121644
4. Huang, X., et al.: A novel multistage constant compressor speed control strategy of electric vehicle air conditioning system based on genetic algorithm. Energy **241** (2022). https://doi.org/10.1016/j.energy.2021.122903
5. Ji, Y., Liu, S., Zhou, M., Zhao, Z., Guo, X., Qi, L.: A machine learning and genetic algorithm-based method for predicting width deviation of hot-rolled strip in steel production systems. Inf. Sci. Int. J. **589** 2022
6. Wang, J., Wan, W.: Optimization of fermentative hydrogen production process using genetic algorithm based on neural network and response surface methodology. Int. J. Hydrogen Energy **34**(1), 255–261 (2009). https://doi.org/10.1016/j.ijhydene.2008.10.010
7. Leung, Y.W., Wang, Y.: An orthogonal genetic algorithm with quantization for global numerical optimization. IEEE Trans. Evol. Comput. **5**(1), 41–53 (2001). https://doi.org/10.1109/4235.910464
8. Yao, L., Sethares, W.A., Kammer, D.C.: Sensor placement for on-orbit modal identification via a genetic algorithm. AIAA J. **31**(10), 1922–1928 (2012). https://doi.org/10.2514/3.11868
9. Li, Q., Xie, J., He, L., Wang, Y., Wang, Q.: Identification of ADAM10 and ADAM17 with potential roles in the spermatogenesis of the Chinese mitten crab, Eriocheir sinensis. Gene **562**(1), 117–127 (2015). https://doi.org/10.1016/j.gene.2015.02.060
10. Cheng, S., Wang, Z.: Solve the IRP problem with an improved discrete differential evolution algorithm. Int. J. Intell. Inf. Database Syst. **12**(1/2), 20–31 (2019). https://doi.org/10.1504/IIDS.2019.102324
11. Chahar, V., Katoch, S., Chauhan, S.S.: A review on genetic algorithm: past, present, and future. Multimedia Tools Appl. **4** (2020). https://doi.org/10.1007/s11042-020-10139-6
12. Deb, K., Pratap, A., Agarwal, S., Meyarivan, T.: A fast and elitist multiobjective genetic algorithm: NSGA-II. IEEE Trans. Evol. Comput. **6**(2), 182–197 (2002). https://doi.org/10.1109/4235.996017

13. Lehmann, A.R.: DNA repair-deficient diseases, xeroderma pigmentosum, Cockayne syndrome and trichothiodystrophy. Biochimie (2003). https://doi.org/10.1016/j.biochi.2003.09.010

14. Fabiani, C., Pizzichini, M., Spadoni, M., Zeddita, G.: Treatment of waste water from silk degumming processes for protein recovery and water reuse. Desalination **105**(1/2), 1–9 (1996). https://doi.org/10.1016/0011-9164(96)00050-1

15. Shamim, A.M., Sultana, S., Mia, R., Selim, M., Banna, B.U.: Reuse of standing dye bath of reactive dyeing with nylon silk. Int. J. Eng. Sci. **8**(8), 45–50 (2019). https://doi.org/10.5281/zenodo.3365385

Key Technology and Application Research Based on Computer Internet of Things

Lingwei Wang and Hua Wang[✉]

College of Computer Science, Guangdong University of Science and Technology, Dongguan, China
460864372@qq.com

Abstract. As a representative of emerging technologies, Internet of Things technology has become an important force to promote social progress, and has played a positive role in the sustainable development of related industries in my country. With the rapid development of computer technology in China, the Internet of Things technology is becoming more and more perfect, and it has been applied in many domestic researches. On this basis, sensors using the Internet of Things technology can realize the connection between people, things, and the Internet, so that related instruments can be better operated, which greatly improves the efficiency of scientific research and greatly facilitates scientific research. People's daily life. Therefore, this paper focuses on the IoT technology and real-life IoT applications.

Keywords: Internet of things technology · computer Internet of things technology

1 Introduction

Since the beginning of the 21st century, due to the rapid development of computers, the rapid development of Internet of Things technology [1] has produced the Internet of Things, which has gradually spread and widely entered people's sight. The Internet of Things solves the problem that the Internet cannot combine physical needs and connections, and truly builds an information channel between things, things and people, and then realizes the "Internet of Everything", which is of great importance to the development of human society and the national economy. Promote meaning. IoT technology has been widely used in all aspects of life and technology, such as: third-party logistics, smart home, positioning and navigation, semiconductor manufacturing, digital medicine, etc. Obviously, computer technology has played an important role in promoting the development of the Internet of Things, and its scope and depth depend on the development of computer technology.

K. Li and Y. Liu (Eds.): ISICA 2023, CCIS 2147, pp. 442–449, 2024.
https://doi.org/10.1007/978-981-97-4396-4_41

2 Overview of Internet of Things Technology

In the 1990s, the design concept of the Internet of Things was also introduced in large quantities. With the vigorous development of the Internet of Things technology, the Internet of Things technology has also been more deeply used in various industries, and has been widely used in various industries. It greatly improves the productivity of the industry and promotes the sharing of information. The use of Internet of Things technology can realize the intelligent interaction between people, and it can make the connection between people and things, things, things and things, which is an important breakthrough in the development of the world today [2]. In the practical application of the information technology of the Internet of Things, it needs to reflect the functions of data processing, signal transmission and perception, which involves sensor technology, smart chip transmission technology, etc. It has been widely used in various countries, and at the same time, many researchers in various industries have begun to expand and use it to enhance the actual application benefits of Internet of Things information technology. In a broad sense, with the development of informatization, the wide application and development of network technology will greatly promote the efficiency of information exchange. Secondly, the Internet of Things technology has the advantages of capture, transmission, and processing. It can sense and measure the target at any time and any place, and transmit the obtained information to the network terminal for sharing. Finally, the intelligent computer can sense and measure the mass of objects. The data is analyzed and processed to lay the foundation for future decision-making and intelligent control.

3 Key Technologies of the Internet of Things

The IoT product system includes four aspects: perception and management, Internet, service technology, and application services. The Internet of Things is not a simple sensor, but a combination of the four layers of the Internet of Things to form a smart Internet of Things management system, which can truly support the industry. The technical system of the Internet of Things is shown in Table 1.

Table 1. IoT technology system

system level	technical analysis
application layer	Communication platform, information sharing,data storage
processing layer	Intelligent technology, GIS/GRS technology, 3C technology, cloud computing
transport layer	Satellite communication, network Internet, 3G/M2M/mobile communication network
perception layer	IC card and barcode, RIFD/sensor, smart machine

3.1 Network Communication Technology

In the Internet age, the function of network communication technology is more obvious. In the operation of the Internet of Things, all kinds of information generated can be applied not only to the internal environment of the enterprise, but also to the external environment. In terms of the Internet of Things, communication equipment is also a very critical aspect, mainly divided into gateways, wired and wireless equipment, etc. M2M network technology is currently the most frequently used technology in the field of communication LAN [3]. It has great universality and superiority. Require. The transfer of information is the exchange between the internal and external systems. Therefore, the network communication technology must be realized based on the high-speed broadband communication network.

3.2 Radio Frequency Identification Technology

RFID technology is the core of the modern Internet of Things, and it can also be called an electronic tag. RF technology mainly uses RF signals to perform a series of application calculations to realize the effective transmission and identification of RF related information. Specifically, RFID technology uses radio frequency to read and write electronic tags, and sends them to the cloud database to achieve corresponding identity authentication [4]. In reality, RFID can be divided into load modulation and backscatter modulation according to its application range. RFID technology includes electronic tags, readers, and data management systems. In an RF identification system, a reader is a read/write device. When the RF identification system is working, the reader will form an electromagnetic field around it, and the size of the electric field determines the power of the radiation. In this case, if the tag is activated within the magnetic field of the reader-writer, the reader-writer will transmit the corresponding data, or by the reader-writer's internal command to rewrite the relevant data. Ultimately, the reader transmits or communicates in real time and/or via a computer system connected thereto.

3.3 Sensing Technology

The role of the Internet of Things should be based on information collection. However, the existing data acquisition methods are mainly sensors, sensor nodes and so on. Sensors are currently the most important detection equipment and the most important collection equipment. The Internet of Things needs the development and breakthrough of the sensing ball in the field of perceptual information to make it intelligent and networked.

3.4 Cloud Computing Technology

Cloud computing is the core technology in the Internet of Things technology. It can perform calculations on massive amounts of data, decompose the complex calculation process into several small software, and analyze and process them through a system composed of multiple servers. It is fed back to the user's mind, and through cloud computing technology, millions of data can be processed in a very short period of time, thereby improving the quality and efficiency of network services.

4 Internet of Things Technology and Computer Internet of Things Technology

4.1 Application of Internet of Things Technology in the Security Industry

With the application of Internet of Things technology in the field of security, people pay more and more attention. The connotation of IoT security is: terminal products with intelligent features. It can connect sensors and other things together, so that it has functions such as monitoring, service, and warning. With the continuous improvement and development of Internet of Things technology, it will provide a huge development space for the entire security industry. According to the prediction of relevant experts, the Internet of Things in the future will be a high-tech high-tech application market, in which security is the main development direction in the future. The Internet of Things security application at Shanghai Pudong International Airport has given us a deeper understanding of the future Internet of Things. The application of the Internet of Things is not only reflected in major events, but also in people's daily life. For example, if someone is outside the house, you can send a text message or turn on the air-conditioning in the room by voice; when you get home, your door lock is smart, and some of your anti-theft devices will automatically turn on your alarm. Bell; when resting at home, a remote control can control all the electricity in the home. When going to sleep at night, if the rest mode is set, all electronic devices in the room will be automatically turned off and the alarm system will be activated.

4.2 Application of Internet of Things Technology in the Field of Medicine

The application of Internet of Things technology in security, health and other fields is more and more extensive. The composition of medical network includes perception layer, transmission layer and application layer. The perception layer is to use various technologies to record the patient's condition in the form of data, so that the patient can deliver the correct information before the operation and prepare for the patient. The transport layer refers to a relatively common transmission method, that is; wireless transmission and Bluetooth transmission. The Internet of Things not only plays a huge role in the medical field, but also plays a key role in drug management, medication, etc. According to relevant surveys, there are many accidents caused by drug abuse in foreign countries. With this experience, in China, the Internet of Things technology can be used to save information such as the name and place of origin of the drug on the REID label, thereby avoiding problems during drug use.

4.3 Application of Internet of Things Technology in the Logistics Industry

With the rapid development of China's market economy, people's demand for logistics distribution is increasing, which also promotes the rapid growth of China's logistics industry. Due to the large number of items that come in and out of the logistics center every day, and the data is very complex, it is difficult to carry out statistics and management, making it difficult for the existing logistics management system to adapt to the development needs of current logistics enterprises. IoT technology is especially suitable for the monitoring and management of transportation and transshipment from suppliers to customers. At present, the informatization level of my country's logistics industry is still very low, and the enterprises themselves lack corresponding information systems, which leads to the inability of logistics enterprises to meet customers' logistics needs in a timely, accurate and effective manner. Therefore, IoT technology can be applied in future work. RFID technology uses Internet of Things technology to realize real-time tracking of each commodity in logistics, and realizes the visualization and intelligent management of logistics process data. With the development of the Internet of Things technology and the continuous improvement of the logistics management system, today, when the level of logistics distribution is further improved, it can also effectively reduce the labor cost of logistics, thereby greatly reducing the cost of logistics, thereby improving the efficiency of logistics. The integration of IoT technology and the logistics industry has also become a major trend in the future development of the logistics industry.

4.4 Application of Internet of Things Technology in Various Intelligent Fire Protection Facilities

In todays rapid economic development, a kind of intelligent fire fighting equipment has appeared. Once a potential safety hazard is found, the intelligent fire extinguishing system will automatically call the police and notify the relevant departments for effective management and maintenance. The application of Internet technology in various fire protection systems can effectively solve the transmission and sharing of various fire information. Starting from the working principle of the system, due to the basic information generated during the operation of the system, the staff not only have to store it in the internal network of the unit, but also actively submit it to the fire department. In addition, in the management of fire-fighting water sources, the commander can view the required information (such as: fire-fighting vehicle performance, water source location, water source storage, water supply methods, etc.) configuration.

4.5 Application of Computer Internet of Things in Family Life

With the popularization of Internet of Things technology and the rise of intelligent services, more and more applications have been applied in people's life and work, which has greatly improved people's living standards. For example, when carrying out home decoration in a high-end residential area, the Internet of Things technology will be used to create a more comfortable and efficient lifestyle through the wiring system, lighting control system, and temperature control system, ensuring the safety of the entire residential system. Secondly, applying computer network technology to residences can effectively integrate various furniture and equipment to create a comfortable living space. The smart home company has carried out intelligent transformation of everything in the home, such as remote control of electrical appliances, remote control of light switches, etc., which greatly improves the comfort and convenience of life. The detection and control method of smart home appliance nodes based on the Internet of Things is shown in Fig. 1.

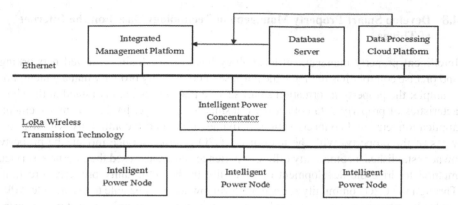

Fig. 1. Scheme of monitoring and control system for intelligent power node based on Internet of Things

4.6 Development and Application of Computer Internet of Things Technology in Rural Applications

Extending computer network technology to agricultural production is not only effective, but also has a lot of room for development. The use of Internet information technology can realize intelligent supervision of rural ecology, agricultural product resources, food safety, rural facilities, etc., give full play to the functions of Internet information technology, improve the overall promotion of rural informatization, drive rural informatization, and promote rural informatization and The integration of agricultural informatization. At present, wireless sensor networks are most used in agricultural production. Through random placement, a large number of sensor nodes can be set within a certain distance, and the data can be effectively identified, collected and processed. The real-time data transmission of the detected target is fed back to the farmer. When the farmer encounters a problem, the positioning system is used to determine the location of the problem, thus achieving an efficient processing effect. Secondly, through the Internet of Things technology, agricultural production can be transformed, and the trace elements, temperature, and humidity in the soil can be controlled through sensors, so as to ensure that crops grow at an appropriate temperature and temperature. At the same time, through the mobile phone app, you can see the growth status of the crops in the farmland. If there is any problem, you can control the surrounding environment through the function of the app, making the farm more intelligent.

4.7 Application of Computer Internet of Things in Intelligent Transportation

With the rapid development of my country's transportation industry, it has been used more and more in intelligent transportation systems, which can effectively reduce the overload problems caused by cities on the road of development. Secondly, the use of computer network technology can effectively reduce urban traffic environmental pollution and improve traffic efficiency and safety.

4.8 Develop Smart Property Management Technology Based on the Internet of Things

Intelligent property management technology has also gradually achieved networking and precise control, thus realizing the purpose of community property management. For example: the property information smart service platform is mainly aimed at the characteristics of property data collection, transmission, storage, hardcover, management, application, etc., and connects various activities of the entire community with life services of the property. With the help of the O2O Internet service model, the property owner established a community offline experience center, and used the Internet+ service method to drive the development of smart life in the community property direction. Through the O2O community service platform, owners will be able to use mobile APPs to check business progress, online shopping, smart parking spaces, pay water, gas, property management fees, scan codes to open doors, face recognition services, smart street lights, and use sensors Technology, testing services for equipment, water and electricity, fire protection, etc., enable owners to fully experience the convenience of intelligent services, and use this to create a new type of smart community.

4.9 Strengthen the Application of Firewall and Intrusion Detection Technology

The information transmission on the network needs to be continuously strengthened, and the characteristics and performance of the network need to be analyzed in detail, so as to establish a practical firewall. Improve the construction of the access mechanism by combining with specific system applications. For different networks, adopting a reasonable state can continuously improve the security of the network. On this basis, each independent network must meet the security requirements of information transmission. At the same time, it is also necessary to ensure that the data transmission efficiency is continuously improved during the entire transmission period, so as to avoid malicious theft. Relevant staff should use monitoring technology to continuously discover and deal with intrusions, improve processing efficiency, and take corresponding countermeasures according to different attacks, thereby enhancing the system's vulnerability repair capabilities. In the field of traffic management, the current situation is that due to the rapid growth of society and national economy, the means of transportation have also been enriched. For this reason, relevant departments pay more and more attention to the construction of urban roads, which makes the number of urban roads more and more, and the traffic conditions are also more and more complicated. Therefore, introducing an effective traffic management system in urban road traffic safety will have a huge positive impact. Computer network technology is an organic combination of computer technology, numerical control technology and communication technology, which can develop the traffic system intelligently, scientifically, systematically and standardizedly according to the actual traffic conditions. At the same time, due to its timely, fast and accurate technical advantages, it can more effectively improve the road network structure of urban traffic, thereby alleviating the city's road traffic problems and improving the city's traffic safety and transportation capacity. Applying computer network technology to urban traffic management can not only promote the orderly development of urban traffic, but also promote the economic development of the city.

5 Conclusion

The application of network technology has brought great changes to people's daily life. In short, it is to build an information interaction system based on "things and things" as a link. With the rapid development of information technology and the rapid development of Internet of Things technology in China, while promoting the production of intelligent equipment, it also promotes the development of various industries to a certain extent, and has made a significant contribution to my country's economic and social development. contribution. The Internet has penetrated into all walks of life, and the Internet of Things technology has gradually become an important system to promote the stability and development of human society. Only by continuously expanding the scope and field of application can the application value of the network be improved and the development level of the network be improved. We believe that with the passage of time, network technology will be more and more effectively used in people's production and life, bringing more convenience to people's life.

Acknowledgments. This work is supported by General Project of Dongguan Social Development Science and Technology under grant no. 20221800902742.

References

1. Zhiyi, Z.: Application of computer internet of things and key technologies of Internet of Things [U/OL]. Electron. Technol. Softw. Eng. **02**, 123 (2019)
2. Wenwen, J.: The key technology of the Internet of Things and the application of the computer Internet of Things U. China's Strategic Emerging Ind. **24**, 105 (2018)
3. Xubing, D., Minhan, Y.: Application and innovation analysis of computer internet of things technology in logistics field U. Logist. Eng. Manage. **40**(10), 15–16 (2018)
4. Sun, Q., Liu, J., Li, R., Fan, C., Sun, J.: Internet of Things: concept, architecture and key technology research review. J. Beijing Univ. Posts Telecommun. (03) (2010)

Research on the Innovative Application of Computer Aided Design in Environmental Design

Lei Wang[✉]

Guangdong University of Science and Technology, Dongguan, China
105772803@qq.com

Abstract. Since the 1980s, The computer-aided design technology has been widely introduced into China, applied in the field of environmental design. The application of this technology has also been rapidly developed and used by many enthusiasts and designers. With the development of China's environmental design industry, professional talents are constantly emerging, traditional hand drawing methods have undergone changes, and the application of computer-aided design technologies (AutoCAD, 3dsmax, Photoshop) is deepening. This technology has advantages such as speed, accuracy, and ease of modification, which has also been recognized by staff and product design enthusiasts. Many design companies and related units regard this technology as a necessary skill for internships. In this situation, schools should strengthen their attention to the application of computer-aided design technology in teaching, enabling students to learn and apply this technology more efficiently and proficiently, and providing support for future work. This article analyzes the problems in the teaching of computer-aided design courses in universities, and explores how to improve existing teaching methods by introducing modern case teaching methods and project-based inquiry teaching models based on relevant course practices. In addition, this article also explores how to promote the teaching of related technologies in the field of environmental design and promote the comprehensive development of students.

Keywords: Environmental design · Teaching innovation · Computer-aided design

1 Introduction

With the continuous advancement of the information age, the application of computer-aided design technology in the design industry has deepened, and the scope of application has been expanded. The use of this technology greatly improves the efficiency of designers and adds more innovative elements to the development of design work. Relatively speaking, environmental design is more complex in course arrangement and teaching content. It is based on design and requires students to be able to flexibly apply and handle environmental design elements and spatial interfaces while mastering certain design methods, creating a good environment that meets people's functional and

K. Li and Y. Liu (Eds.): ISICA 2023, CCIS 2147, pp. 450–458, 2024.
https://doi.org/10.1007/978-981-97-4396-4_42

aesthetic needs. At present, the environmental design major in Chinese universities also recognizes the application value of computer-aided design technology, and hopes to cultivate more outstanding talents through the establishment of technical courses. However, some universities have certain deficiencies in the design of related courses, which has become the focus of this study [1].

2 The Significance of Computer-Aided Design Teaching

Computer assisted technology is an emerging technological means that will replace traditional hand drawing methods in application, bringing more convenience and creativity to design and creation. In the teaching of environmental design and other related design majors in universities, it is necessary to optimize the design of courses for the development of computer-aided design, in order to provide services for students' professional growth.

The scientific implementation of this teaching activity also helps students consolidate their professional knowledge, improve the overall efficiency of environmental design education, and provide more support for their future growth and development. In the field of environmental design, computer-aided design is a highly practical course that not only requires the use of technology, but also requires the application of aesthetic elements and the presentation of art [2].

At present, computer-aided technology is also in a state of development and updating. In addition to several interior design software such as AutoCAD, 3dsmax, and Photoshop, the software and related technologies available to students will be periodically updated, which also requires relevant courses to introduce the latest technical knowledge and improve and optimize teaching content.

3 Problems in Computer Aided Design Teaching

In recent years, universities have gradually strengthened the importance of computer-aided design teaching, but there are still many problems in the implementation of specific work, mainly reflected in the following aspects.

3.1 The Teaching Mode is Too Rigid

At present, the environmental design major in Chinese universities mainly adopts the following two methods in computer-aided design teaching: instruction based on command statements. This teaching mainly discusses the use of instructions in various applications, and the model only stays at the level of technical learning without utilizing technology to implement auxiliary design ideas. Although students can proficiently master various command sentences, they cannot flexibly use these sentences to design excellent works; Teaching based on virtual environment. This teaching creates a certain virtual design scenario for students, requiring them to carry out operational design according to the steps, but it is difficult to grasp the key factors and lacks sufficient experience in expanding design thinking [3].

3.2 Emphasizing Skills While Neglecting Innovative Thinking

Based on the actual situation, many university teachers have a deviation in direction in computer-aided design teaching. They will focus on the operation of application software. Although students can master the operation methods of menu bars, editors, image filters, and other interfaces, they cannot use these software for creative design. In addition, although some teachers have conducted case teaching, they have not closely combined specific cases with skill teaching and have not strengthened the training of students' innovative thinking. Therefore, the implementation of this teaching activity has also transformed into a training course for skill operation. In addition, many teachers only focus on students' mastery of software and the production of renderings, causing them to focus on the use of computer software plugins and effects [4].

3.3 Teaching Methods and Content Need to Be Improved

At present, the vast majority of universities in China offer courses in computer-aided design for environmental art and design majors, which include the application of software technologies such as PS, Auto CAD, and 3dMax. However, considering the actual situation, there are technical difficulties in the application of these software, which also poses certain challenges for teachers' teaching. Usually, teachers need to provide detailed explanations of software instructions to help students engage in autonomous communication activities. But students cannot only master software instructions, otherwise they cannot guarantee the flexibility of software applications or the ability to apply the learned technology to the field of environmental design, which poses a certain obstacle to their future employment. Therefore, teachers need to deeply reflect on past teaching methods and content, identify their shortcomings, and innovate existing teaching methods to further improve students' software application level [5].

4 Innovative Strategies for Computer Assisted Instruction

4.1 Reforming Teaching Concepts and Content

In order to promote the development of computer-aided design teaching activities, it is necessary to update and optimize the existing educational concepts. Teachers need to refine the key and difficult points of relevant courses and use them as the main basis for course design. In addition, in the design of teaching plans and syllabuses, timely adjustments need to be made based on software updates, while paying attention to the student's main position, to help students construct their own knowledge architecture; Provide opportunities for cooperation, allowing students to collaborate with each other and operate software together, and enhance students' aesthetic literacy and software operation skills through full cooperation [6].

In addition, in terms of optimizing teaching content, teachers can use teaching methods such as "startup module", "step-by-step", and "joint penetration" to guide students to gradually learn the operation of software such as AutoCAD, PhotoShop, 3dsMAX, ILLUSTRATOR, SketchUp, V-Ray, etc. In teaching, teachers need to follow the principles of easy to difficult, shallow to deep, and construct teaching modules for A + P,

D + V, A + S + I, D + P + V, P + I + S, A + D + P, A + D + P + I + V, so that students can grasp the differences in the application of various software in their learning and deepen their mastery of various knowledge [7].

The innovative strategy of computer-aided teaching mainly includes the following aspects:

1. Personalized teaching strategy

Use big data and artificial intelligence technology to analyze the personalized learning needs, interests, and styles of individual students.Provide customized learning resources and teaching plans for each student based on the personalized analysis results to meet their individual needs.

2. Interactive teaching strategy

Utilize online discussions, real-time Q&A, online tests, and other methods to enhance interactions between students and teachers and improve students' interest and participation.Interactive teaching allows teachers to understand students' learning situations in a timely manner and adjust teaching strategies to improve teaching effectiveness.

3. Gamification teaching strategy

Combine educational content with gaming elements to create engaging and enjoyable learning experiences for students.Gamification teaching can stimulate students' interest in learning, enhance their motivation, and deepen their understanding and memory of knowledge.

4. Blended teaching strategy

Combine online and offline teaching methods, integrating traditional classroom teaching with online learning to create a blended teaching model.Blended teaching can fully leverage the advantages of both online and offline teaching methods to improve teaching effectiveness and the learning experience.

5. Collaborative teaching strategy

Utilze online collaboration tools to encourage students to collaborate and learn together, working together to solve problems and complete tasks.Collaborative teaching can cultivate students' teamwork and communication skills while deepening their understanding and application of knowledge.

6. Situation simulation teaching strategy

Use virtual reality technology to simulate real scenarios and contexts, allowing students to learn and practice in a simulated environment.Situation simulation teaching can help students better understand and apply knowledge, improving their practical abilities and problem-solving skills.

These innovative strategies can be customized and combined based on specific teaching needs and goals to achieve the best teaching effectiveness. With the continuous development of technology, new teaching concepts and technologies can be continuously introduced to promote innovative development in computer-assisted teaching (see Fig. 1).

Fig. 1. Digital Management of garment design

4.2 Expand Teaching Modes

The discipline of environmental design has a strong intersection and complexity. In practical teaching, teachers can further expand classroom teaching, break through obstacles in classroom teaching, and provide students with a broader knowledge perspective.

For example, teachers can arrange for students to participate in professional conferences, invite enterprise experts to the school, and jointly design courses based on industry talent needs; Teachers can use the project-based teaching model to organize students to visit specific projects in the field of environmental design, and guide students to communicate with environmental designers and construction personnel. In these activities, schools need to play a guiding role and actively cooperate with relevant enterprises to provide a platform for students' internship and employment, in order to promote the improvement of students' career understanding and cognitive abilities [8].

Not only that, teachers can also organize on-site research, market research and other activities, encourage students to boldly use computer-aided design software, carry out environmental design and other related assignments, and enable students to combine environmental design with the application of computer-aided software from a professional perspective, further improving their practical abilities.

In formal classroom teaching, teachers can use tasks as a driving force to guide students to discover their own learning problems based on completing tasks. Teachers need to apply professional knowledge of software operation to specific task projects, and conduct more practical activities to provide detailed explanations and explanations of relevant software operation skills, helping students grasp the cognitive laws of software [9].

In recent years, online teaching and educational software have been widely used in higher education. Teachers should closely combine traditional online and offline teaching, open up a new hybrid teaching mode, and flexibly use the internet to meet teaching needs, in order to promote students' personalized and professional development.

4.3 Emphasize Differentiated Cultivation of Students

In the field of environmental design, teaching the application of computer-aided software technology needs to start from two aspects: operational skills and aesthetic abilities, with a focus on building a dual knowledge structure. Before officially carrying out teaching activities, teachers should strengthen communication and contact with students to understand their actual learning situation and understanding level of professional knowledge. Subsequently, the teacher adjusts the difficulty of the course and the progress of teaching based on the characteristics of the students. After class, teachers should pay attention to students' learning and growth. If students have strong logical reasoning and are proficient in using software, then they can guide students to innovate in design. If students have good innovative thinking, teachers can assign more practical tasks and guide students to efficiently use various technical means.

4.4 Introducing a Project-Based Teaching Mechanism

In order to adapt to the development of the market, teachers should integrate theory and practice into the field of environmental design. In daily teaching, teachers can organize teaching lectures and encourage students to participate in various forms of academic reports, exchange and discussion activities, in order to help students understand the overall development trend of the design market and gradually expand their personal knowledge; In addition, teachers can develop a project-based teaching model, organize students to participate in relevant studios and innovation centers created by the environmental design profession, create more experimental platforms for students, observe construction sites, communicate with construction workers, and explore the possibility of specific implementation plans for engineering projects [10].

In order to promote mutual learning among student groups, teachers can set up different teaching echelons based on age differences, leverage the mutual cooperation between higher and lower grades in teaching, carry out survey activities related to industry design, and comprehensively grasp various design materials and differences between actual design and construction by sorting and archiving the first-hand data collected.

5 The Future of Computer Aided Design and Its Enlightenment for Teaching

As an important tool in the field of modern design, the development trend of computer-aided design (CAD) technology has had a profound impact on the industry and teaching. In recent years, the development trend data of computer-aided design mainly includes the wide application of artificial intelligence technology, the integration of virtual reality technology, the popularization of cloud technology, intelligent design optimization, automatic design process, diversified design tools, efficient design efficiency, accurate design quality and so on.

From the above analysis, it can be seen that the current computer-aided design technology has been efficiently applied in environmental design and other related industries. From the application of existing technology and the implementation of related teaching activities, this article draws the following conclusions.

5.1 The Development of Teaching Activities Must Pay Attention to the Application of Visualization Technology

Visualization essentially means that the design content is visible. This visualization differs significantly from the 2D drawing presentation in traditional interior design courses.

Especially with the use of BIM technology, traditional CAD two-dimensional drawings are transformed into three-dimensional three-dimensional graphics, allowing the public to clearly see furniture, furnishings, materials, lighting, and other contents in interior design, making the design expression more complete, accurate, and clear. Therefore, in the actual teaching process, it is necessary to guide students to flexibly use visualization technology to meet the diverse needs of design and display.

5.2   Grasp the Latest Trends in Technology Teaching

Currently, environmental design is becoming more intelligent, integrated, and diverse. For example, 3D printing technology and VR virtual reality technology are rapidly becoming popular in some regions, which also means that to some extent, the dependence of environmental design on computer-aided design will become more apparent. As a professional teacher, one should closely monitor the development trends in the field of environmental design computer-aided design, continuously explore advanced design concepts and teaching methods, and make the teaching content more adaptable to the development of the times. In addition, the latest developments and knowledge points of other professional courses can be appropriately integrated into teaching to enhance students' interest in learning, deepen their understanding of professional knowledge, and make future course arrangements more scientific and reasonable.

5.3   The Integration of Virtual Reality Technology and the Popularization of Cloud Technology

Virtual reality technology provides a more immersive and interactive design experience for computer-aided design. By applying virtual reality technology to computer-aided design, users can view and modify designs more visually and operatively. At the same time, virtual reality technology can be used to simulate and test design plans, reducing the need for experiments and prototype manufacturing, and lowering costs. Cloud technology provides more efficient and flexible computing and storage resources for computer-aided design. By storing design software and data in the cloud, users can access and use design tools anytime, anywhere, improving the flexibility and efficiency of the design. Additionally, cloud technology enables multi-user collaborative design and data sharing, promoting teamwork and knowledge exchange.

5.4   Intelligent Design Optimization and Automatic Design Process

Intelligent design optimization is another important development trend of computer-aided design. By introducing artificial intelligence technology and optimization algorithm, computers can automatically conduct design optimization and improve design performance and reliability. For example, the genetic algorithm is used to optimize the design scheme to achieve the optimal design scheme. Automated design process is another important development trend of computer-aided design. By introducing automation technologies and process management tools, computers are able to automatically perform repetitive tasks and tedious processes, improving the efficiency and accuracy of design. For example, using automated tools for tasks such as part design and assembly simulation to reduce manual intervention and error rates.

In recent years, the development trend of computer aided design shows that artificial intelligence technology, virtual reality technology and cloud technology have been widely used and developed in computer aided design. The development of these technologies brings more innovation and application opportunities for computer-aided design, and improves the efficiency and accuracy of design. In the future, with the continuous progress of technology and the development of innovative applications, computer-aided design will continue to play an important role and promote the development of the industry.

6 Conclusion

With the continuous progress and development of society, the demand for applied and skilled talents is becoming increasingly prominent. As a professional course that cultivates students' skills, computer-aided design has its unique advantages and huge development potential. In the 21st century, with the arrival of the information and network era, it has provided a unique development soil for computer-aided design. Various advantages and advantages urge us to keep up with the development of the times, innovate concepts, meet the practical needs of students' professional learning and talent cultivation, continuously analyze technology from the people-oriented design principle, extract and select content suitable for students to learn and master software technology. In the actual teaching process, teachers should apply the scientific development concept and keep up with the times, continuously innovate teaching methods and update teaching content to achieve the goal of computer-aided design courses, in order to cultivate applied and skilled talents.

This article is the university-level quality engineering project of Guangdong University of Science and Technology: engineering drawing first-class course (Item No.: GKZLGC2021232); Scientific Research project of Guangdong University of Science and Technology: Urban river landscape design under the background of regional culture (Item No.: GKY-2022KYYBW-83).

References

1. Zhang, Y., Ma, Y., Yang, Y., et al.: Reform and exploration of computer aided design curriculum under the "new engineering". Educ. Teach. Forum (7), 115–116 (2020)
2. Tang, H.: Analysis of the teaching characteristics of art and design in foreign universities. Art Design (Theory) 2(08), 171–173 (2021)
3. Li, H.: The development status and trends of computer aided design. Yihai (12), 98–99 (2019)
4. Liang, F.: Exploration and reform of computer aided design in the teaching of environmental design majors. Comput. Prod. Circulat. (10), 241 (2019)
5. Xu, J., Zhan, H.: From design to construction - teaching notes and reflections on parametric design at Nanjing Academy of Arts. Urban Architect. 10, 74–76 (2021)
6. Chen, N.: Interior design based on computer-aided design software. Inf. Record. Mater. 22(1) (2021)

7. Chen, X.: Application of computer-aided design software in interior design. Inf. Technol. Informatiz. (12) (2019)
8. Xie, L.: Computer aided design software in interior design in teaching, the application of. Inf. Syst. Eng. (09) (2018)
9. Yu, C.: Computer-aided design for environmental art. Xi'an Traffic the University Press (2009)
10. Ma, A.: Analysis of computer-aided design software in interior design apply. In: Coastal Enterprises and Technology (2012)

Research on Intelligent Clustering Scoring of English Text Based on XGBOOST Algorithm

Zhaolian Zeng[1]([✉]), Wanyi Yao[2], Jia Zeng[1], Jiawei Lei[1], Feiyun Chen[1], and Peihua Wen[1]

[1] School of Foreign Studies, South China Agricultural University, Guangzhou 510642 , China
376711579@qq.com
[2] College of Mathematics and Informatics, South China Agricultural University, Guangzhou 510642, China

Abstract. In today's explosive growth of online text data, how to feature mine text data from a wide range of sources, cluster text data with similar features, and classify them according to their features has become a hot issue. In this paper, according to the characteristics of English text data, considering the existence of a large number of redundant features in English text, the existence of features with similar meanings in English text and other problems, the following method is proposed: in the first step, the English text is segmented, data preprocessing is carried out, and then the text data is processed by using the TF-IDF algorithm, to get the roughly selected subset of features; in the second step, the improved Binary Particle Swarm Optimization algorithm is used to In the second step, the improved binary particle swarm optimization algorithm GSBPSO (Global-to-local Searching-based BinaryParticle Swarm Optimization) is used to re-select the features of the roughly selected feature subset obtained in the first step to obtain the optimal feature subset; in the third step, the binary SKM clustering algorithm is used to cluster English text according to the optimal feature subset obtained in the second step. English text is clustered according to the optimal feature subset obtained in the second step, and the classification criteria are formulated for the existing text corpus to categorize the text data. Finally, the XGBOOST algorithm is applied to evaluate the prediction and establish the intelligent clustering scoring model of English text based on XGBOOST.

Keywords: feature selection · text clustering · particle swarm algorithm · dichotomous SKM clustering algorithm · XGBoost algorithm

1 Introduction

In today's rapid development of the Internet, a large amount of text data explodes, and processing, analyzing, extracting and classifying text data has become a hot issue. Text data come from a wide range of sources, with diverse topics and mixed information. Therefore, how to analyze the content of text data, how to extract the characteristic words of text data, and how to categorize text data according to the analysis results have become the basic problems of text data processing. Based on the complexity of

© The Author(s), under exclusive license to Springer Nature Singapore Pte Ltd. 2024
K. Li and Y. Liu (Eds.): ISICA 2023, CCIS 2147, pp. 459–475, 2024.
https://doi.org/10.1007/978-981-97-4396-4_43

text data information, removing useless and redundant information and retaining key and representative information is the most important part of the analysis process. The solution to this link is feature selection. After obtaining the feature words, the obtained feature words will be clustered, how to ensure the accuracy of the clustering results, how to improve the efficiency of the clustering, is the problem we should solve in this link. Dichotomous SKM clustering algorithm is a clustering method that effectively solves the above problems. The model obtained after clustering needs to be trained and tested to obtain a more perfect and accurate clustering scoring model.

The study of feature selection methods began in the 20th century on the problem of signal processing, due to the small number of features designed, often using the exhaustive enumeration method to enumerate all the subsets of features for which there may be an optimal solution can be solved. However, with the exponential growth of text data, the high-dimensionality of features in text data leads to the fact that exhaustive enumeration can no longer meet the requirements of feature selection. Gaoxin et al. [2] proposed a feature selection and text clustering algorithm that combines the dung beetle optimization algorithm to improve the binary sparrow search algorithm, which solves the problem of redundant features in text affecting the clustering accuracy. Experimental results show that the method can effectively reduce the dimension of text features and improve the clustering effect. Dou Xiaofei [3] proposed a population-based clustering method, which solves the problems of feature selection methods in the pre-search process, such as easy premature convergence. Experimental results show that the method can effectively avoid the problem of premature convergence in the pre-search and improve the performance of the classification model. Wang Chen et al. [4] proposed a feature selection and text clustering algorithm based on binary gray wolf optimization, which solves the problems such as the reduction of feature dimensions. The experimental results show that the algorithm can effectively reduce the feature dimension and the clustering index performs better on most datasets. Zeng Hui [5] proposed a classification model for sentiment classification logistic regression and support vector machine for Chinese text. The experimental results show that it is found that feature selection and weighting by Term Frequency-Inverse Document Frequency (TFIDF) algorithm can reduce the difference in the predictive performance of the classifier for different categories and improve the generalization ability of the classifier. In order to deal with unbalanced text dataset to alleviate its skewed distribution, Jiang Wanrong [6] proposed a text unbalanced classification algorithm based on data augmentation and an unbalanced classification algorithm for police text based on feature selection. The experimental results show that this method effectively improves the classification performance of the model compared with other algorithms of feature selection. Tian Xiaoli et al. [7] proposed a text feature selection algorithm based on particle swarm optimization in order to solve the problem of noise features in text information affecting text clustering effect. The experimental results show that the algorithm performs better in a number of evaluation indexes and reduces the initial document feature size in the feature selection scale.

Text clustering is an unsupervised learning text mining technique that aims to cluster texts with similar distances into the same target cluster [8]. Therefore, text clustering has been widely applied to web mining, spam filtering, news text clustering and other fields [9]. At present, clustering technology has become a research hotspot in many fields such as text mining and information retrieval, text clustering can discover potential semantic knowledge and laws from a large amount of text data, not only to obtain the relevant knowledge in the text, but also to process the text. The flexibility of text clustering makes it used in linguistics, natural language processing and other fields, and it has been maturely applied in the comparison between text documents, sorting for text importance and relevance, extracting the salient features of text, automatically generating document query conditions, and fast retrieval of similar texts in databases. Gaoxin et al. [10] proposed a feature selection and text clustering algorithm based on binary mayfly optimization to address the problem of low clustering accuracy caused by redundant text features. Experimental results show that the algorithm can effectively shorten the feature dimension and improve the efficiency of text clustering. In order to improve the speed and visualization effect of clustering while ensuring the quality of clustering, Hao Xiuhui et al. [11] proposed a method combining the TF-IDF algorithm and Latent Semantic Analysis (LSA) algorithm. Experimental results show that the method not only ensures the quality of text clustering, but also greatly improves the speed and visualization of text clustering. Aiming at the problems that clustering as an unsupervised technique makes feature selection more difficult, Feng Ying [12] proposed a text feature selection method based on heuristic search algorithm, which is an unsupervised method and suitable for text clustering. Experimental results show that the feature selection algorithm can effectively reduce feature redundancy and improve clustering accuracy, and the proposed clustering algorithm can improve the stability of text clustering. Wang Mingfeng [13] proposed a method based on PSO algorithm, when the population update of PSO algorithm is stagnant and the search space is limited, the crossover and mutation operations of DE algorithm can be used to perturb the population, increase the diversity of the population, and improve the algorithm's global optimization ability, which solves the problem of poor clustering effect that may occur when solving the text clustering problem of high dimensions with traditional clustering algorithms such as the K-means algorithm and the K-means++ algorithm. When the traditional clustering algorithms such as K-means algorithm and K-means++ algorithm solve the high dimensional text clustering problem, the clustering effect may be poor, and the algorithm is not stable. The experimental results show that the effectiveness and feasibility of the algorithm of this method is very high. Because in text clustering, for large-capacity, high-dimensional, unstructured text data, simple K-Means clustering is ineffective and easy to fall into the local optimal solution, Niu Yongli et al. [14] proposed a method based on the text clustering algorithm with improved particle swarm and K-Means (MPK-Clusters). The experimental results show that the new algorithm outperforms the other two algorithms in terms of accuracy, recall and F-value, and achieves better text clustering results.

This paper proposes an intelligent clustering scoring method for English text based on the XGBOOST algorithm, which is an unguided machine learning process. The method is applicable to English text clustering, which will utilize the existing English text corpus to build the model, and then validate the accuracy of the model by applying the validated English text dataset. Firstly, in this paper, when the feature selection is carried out, considering that the binary particle swarm optimization algorithm has the problems of decreasing the diversity of the population, easy to fall into the local optimum [15], premature convergence, and poor convergence performance [16], which may lead to the inaccurate classification results, the improved binary particle swarm optimization algorithm is used for feature re-selection; secondly, considering that the K-means, spectral clustering, SKM algorithm Secondly, considering the problems of low accuracy and low efficiency of K-means, spectral clustering, SKM algorithm, this paper adopts binary SKM clustering algorithm for English text clustering. Based on this, an intelligent clustering scoring model for English text based on XGBOOST is established. Experimental analysis shows that the model proposed in this paper has better results.

2 Text Pre-processing

2.1 Introduction to Text Preprocessing

Text data comes from a wide range of sources, and the original text has different standards, so the original text data is basically semi-structured or unstructured data with loose structure and poor machine readability when it is not processed. Text preprocessing can extract the feature items in the text data, remove the unimportant or meaningless redundant items in the text data, so that the feature items of the text data are preserved as structured data, which is convenient for subsequent feature selection and clustering. The text preprocessing steps in the article are mainly to segment the English text, remove the deactivated words, remove the feature words whose text frequency DF (Document Frequency) is too high and too low, and use the TF-IDF algorithm to transform the remaining feature subsets into feature matrices.

2.2 TF-IDF Algorithm

The TF-IDF (Term Frequency-Inverse Document Frequency) algorithm, is a common weighting technique used to calculate the weight of feature values in a text. As a statistical method for information retrieval and text mining, TF-IDF uses word frequency TF (Term Frequency) to represent the frequency of words (i.e., features) appearing in a certain text, and inverse document frequency IDF (Inverse Document Frequency) to represent the frequency of features appearing in other texts. Term Frequency and Inverse Document Frequency together determine the weight of a feature in a certain text, the specific formula is as follows:

$$tf_{ij} = \frac{n_{i,j}}{\sum_k n_{k,j}} \tag{1}$$

where tf_{ij} denotes the word frequency of feature i in text j, and $n_{i,j}$ denotes the number of occurrences of feature i in text j, and the denominator $\sum_k n_{k,j}$ denotes the total number of occurrences of all features in text j.

$$idf_i = \log \frac{N}{N_i + 0.01} \tag{2}$$

where idf_i denotes the inverse document frequency of feature i, N denotes the total number of texts in the corpus, and N_i denotes the number of texts in the corpus that contain feature i. And the purpose of adding 0.01 to the denominator here is to prevent the denominator from being zero.

$$w_{ij} = tf_{ij} \times idf_i \tag{3}$$

where w_{ij} denotes the weight of feature i in text j.

3 Feature Selection Based on Particle Swarm Algorithm

In 1995, Particle Swarm Optimization Algorithm (PSO), a bionic stochastic optimization algorithm, was proposed by Eberhart and Kennedy, inspired by the collaborative search for the solution of the problem based on the interactive behaviors such as information sharing among individuals within the group of animals such as a flock of birds or a school of fish. Particle swarm algorithm is relatively simple to operate, no crossover and mutation operations, particles are only updated by speed, and the particle swarm algorithm has memory, so it can retain the excellent particles well. In addition, the easily adjustable parameters allow the particle swarm algorithm to adjust the global and local search ability, so that it can be easily applied to a variety of practical problems. The particle swarm optimization algorithm can be well applied to problems such as feature selection and solution optimization.

3.1 Encoding and Decoding of Particles

For the problem of selecting feature words in the text, the features in the text only have two options, "selected" and "discarded", so the feature selection can be transformed into a simple discrete optimization problem, using "0 "0" and "1" represent the selection of features, if the feature is selected, it is labeled as "1", otherwise, it is labeled as "0". The following is a simple example of particle encoding and decoding:

Assuming that after the pre-selection of text features, there are 10 feature words left, and the particle swarm population size is set to be 5, then the initial population is a matrix with 5 rows and 10 columns, and each position of the matrix is randomly generated as "0" or "1":

$$\begin{pmatrix} 1\ 0\ 0\ 1\ 1\ 1\ 0\ 1\ 0\ 1 \\ 1\ 0\ 1\ 0\ 1\ 0\ 0\ 0\ 0\ 1 \\ 0\ 1\ 0\ 1\ 0\ 0\ 1\ 0\ 1\ 0 \\ 0\ 0\ 1\ 1\ 0\ 0\ 1\ 0\ 0\ 0 \\ 0\ 0\ 0\ 1\ 1\ 0\ 0\ 0\ 1\ 1 \end{pmatrix}$$

Taking the first line as an example,, it means that in this solution, the 1st, 4th, 5th, 6th, 8th, and 10th feature words are selected, while the remaining feature words are discarded. After initialization, the particles are iteratively updated according to the algorithmic process, and finally the globally optimal individual with the best adaptation value is selected as the final solution. Suppose after 100 iterations, the algorithm terminates, at this point, then the algorithm selects the 1st, 3rd, 4th, 6th as well as the 10th feature words as the final feature subset and the rest of the features are discarded.

The SIM value of the feature value is then calculated according to the following formula and the SIM value will be used later to calculate the adaptation value for the text data:

$$D_{euclid}(d_i, d_j) = (\sum_{k=1}^{n} |x_{ik} - y_{jk}|^2)^{\frac{1}{2}} \tag{4}$$

where $D_{euclid}(d_i, d_j)$ denotes the computation of the Euclidean distance between two vectors, the x_{ik}, and y_{jk} denotes two different vectors.

$$\text{SIM}(d_i(\text{lnd}), d_i(j)) \tag{5}$$

3.2 Improved Binary Particle Swarm Optimization Algorithm (GSBPSO)

Since the binary particle swarm optimization algorithm has the problem of decreasing the diversity of the population, which may lead to inaccurate classification results, the improved binary particle swarm optimization algorithm is used to re-select the features to obtain the final feature subset X. The implementation steps are as follows:

The first step is to define the parameters:

(1) Population size N;
(2) Speed range;
(3) Inertia weights, learning factors c1, c2;
(4) Maximum number of iterations MaxIter;
(5) Threshold;

Step 2: Initialization.

Let the current iteration number be $t = 0$. According to the solution range, randomly generate N individuals as initial positions. According to the velocity range, randomly generate the velocity of each particle and initialize the historical optimal position of each particle.

Step 3: Calculate the fitness function. Calculate the respective fitness values of the individuals in the population and initialize the population global optimal solution to the position of the individual with the optimal fitness value.

Step 4: The following formulas are needed:

$$v_{id}^{t+1} = w \times v_{id}^{t} + c1 \times r2 \times (pbest_{id}^{t} - x_{id}^{t}) + c2 \times r2 \times \left(gbest_{d}^{t} - x_{id}^{t}\right) \tag{6}$$

In the formula, the k is the first iteration of the k iteration, the i and d takes the number from 1 to the N (the total number in (population size), and ω are the inertia weights, the c1 and c2 are acceleration coefficients, and r1 and r2 are random numbers between 0 and 1.

The S-mapping (Sigmoid) function is formulated as:

$$s(v_{id}^{t+1}) = \frac{1}{1 + e^{-v_{id}^{t+1}}} \tag{7}$$

The optimization process of the particle swarm algorithm is a process in which the particle velocity decreases gradually and finally converges to 0. When the probability of the particle's position changing is large, it is favorable to the increase of population diversity, but it is unfavorable to the convergence of the algorithm at the later stage, i.e., it is unfavorable to the local search, so scholars proposed a binary particle swarm algorithm with local search ability (New Particle Swarm Optimization, NBPSO), the improvement formula is as follows:

$$s(v_{id}^{t+1}) \begin{cases} 1 - \frac{2}{1+\exp(-v_{id}^{t+1})}, & v_{id}^{t+1} \leq 0 \\ \frac{2}{1+\exp(-v_{id}^{t+1})} - 1, & v_{id}^{t+1} > 0 \end{cases} \tag{8}$$

$$\begin{cases} x_{id}^{t+1} = \begin{cases} 0, rand() \leq s(v_{id}^{t+1}) \\ x_{id}, otherwise \end{cases}, v_{id}^{t+1} < 0 \\ x_{id}^{t+1} = \begin{cases} 1, rand() \leq s(v_{id}^{t+1}) \\ x_{id}, otherwise \end{cases}, v_{id}^{t+1} > 0 \end{cases} \tag{9}$$

In order to improve the algorithm's searching ability, the global searching ability should be enhanced in the first stage of the algorithm and the local searching ability should be enhanced in the later stage. In this thesis, based on NBPSO, an improvement of the position update method based on Sigmoid function is proposed to enhance the global search ability in the first stage. The particle position update formula is adjusted:

$$\begin{cases} x_{in}^{t+1} = \begin{cases} 0, v_{in}^{t+1} \geq 0, gbest_n^t = 0 \\ 1, v_{in}^{t+1} \leq 0, gbest_n^t = 1 \end{cases}, |v_{in}^{t+1}| \geq \theta \\ x_{in}^{t+1} = \begin{cases} 1, rand() \leq s(v_{id}^{t+1}) \\ 0, otherwise \end{cases}, |v_{in}^{t+1}| < \theta \end{cases} \tag{10}$$

In the formula, the θ is a preset threshold value and $\theta \in (0, \&v\_max())$, thus increasing the probability that the particle maintains its original position at a large velocity, which is more favorable to the diversity of the particle swarm.

For each particle, update the velocity and position according to the following method:
if t < *MaxIter

Use of formulas (6), (7), (10)

else

Use of formulas (6), (8), (9)

Step 5: Calculate the adaptation value of each particle and compare it with the particle's current historical optimal solution $pbest_{id}^t$ adaptation value, and if the current solution is better, update the particle's historical optimal solution $pbest_{id}^t = x_i^{t+1}$. The adaptation value function is calculated as:

$$Fit(lnd) = \sum_{i=1}^{m} \sum_{j=1}^{|P|} SIM(d_i(lnd), d_i(j)), j \neq lnd \tag{11}$$

Of thesem The samples are randomly selected. SIM is the cosine similarity, the larger the cosine similarity means the more similar the two texts are, so the larger the value of the fitness value function, the better.

Step 6: Based on the adaptation value calculated in step 5, compare the adaptation value of the optimal individual of the contemporary population with the adaptation value of $gbest_n^t$ the adaptation value of the contemporary population, and if the former is better, then update the global optimal solution $gbest^{t+1}$.

Step 7: Judge the termination condition of the algorithm, if the accuracy required by the algorithm is not reached, or the maximum number of iterations is not reached, then $t = t + 1$, repeat the third, fourth, fifth, sixth and seventh steps, if the termination condition is satisfied then output the optimal solution.

4 Text Clustering and Evaluation of Predictions

4.1 GSBPSO-Based Bisection SKM Clustering Algorithm

After the improved binary particle swarm optimization algorithm obtains the optimal individuals with the advantages of time-consuming break and rapid convergence, the final feature subset T obtained is clustered by the bisection SKM clustering algorithm, which develops a classification criterion to classify the text data on the existing text corpus. The bisection SKM clustering algorithm is much better than K-means, spectral clustering, and SKM algorithm in terms of text clustering accuracy. The following are the steps of the algorithm for text clustering of the final feature subset X using the bisection SKM clustering algorithm:

Input: sample set X, number of clusters k

Output: k clusters

Step 1: Normalize all sample vectors to ensure that the vector modulus is 1

Step 2: Initialize the cluster table to contain a cluster made up of sample points

Step 3: Remove the cluster with the largest SCE value of the objective function from the cluster table

Step 4: Perform a loop for $(t = 1; t < = n; t ++)$ {Cluster the selected clusters using SKM method with k value set to 2}

Step 5: Select the cluster that minimizes the SCE value of the objective function from the experimental results

Step 6: Add the two clusters to the cluster table

Step 7: Determine whether the cluster table contains k clusters, if no, repeat steps 3 to 6, if yes, output the result

Among them, the objective function SCE used in the third to fifth steps is calculated as follows:

$$SCE = \sum_{j=1}^{k} \sum_{x \in C_j} 1 - xc_j^T \tag{12}$$

where C_j denotes the cluster to which the samplex belongs to, in hard clustering, each sample belongs to only one cluster. C_j denotes the cluster center vector of C_j the cluster center vector of the sample, in SKM, the cluster center vector is normalized, i.e., the mode length is 1. Thus xc_j^T denotes the cluster center vector of the samplex and the cluster center vector C_j the cosine similarity of the sample to the cluster center vector, and $(1 - xc_j^T)$ denotes the cosine similarity of x the cosine distance between the sample and the cluster center C_j cosine distance from the cluster center.

4.2   Evaluating the XGBOOST Algorithm for Prediction

As a Boosting algorithm, the XGBOOST algorithm generates a new tree through iteration after iteration, thus strengthening the weak learner into a strong one, which not only solves the problem of low classification performance of the weak learner, but also ensures the high accuracy of the strong learner. The algorithm uses the second-order Taylor expansion of the loss function as an alternative function to solve for its minimization (i.e., the derivative is 0) to determine the optimal cut-off point of the regression tree and the leaf node output values. In addition, XGBoost considers the regularization problem and introduces the number of subtrees and subtree leaf node values, etc. to effectively avoid overfitting. In terms of algorithmic efficiency, XGBoost has a unique approximate regression tree bifurcation point estimation and sub-node parallelization, etc., as well as the second-order convergence property, the modeling efficiency has a substantial improvement.

Using the obtained clustering results, an intelligent clustering scoring model for English text based on XGBOOST can be built, where the text to be scored is passed into the model as an input value, and the input text is evaluated according to the clustering results of the model.

5   Basic Steps of English Text Clustering Based on XGBOOST Algorithm

Step 1: Preprocess the original English text data to obtain a roughly selected feature subset D.

Step 2: Calculate the weights of the text vector formed by D by applying the TF-IDF algorithm

Step 3: Encoding and decoding of text annotations (particles).

Step 4: Feature reselection using improved binary particle swarm optimization algorithm (GSBPSO).

Step 5: Text clustering with binary SKM algorithm on the obtained final feature subset X.

Step 6: According to the above implementation algorithm, and the results of clustering, establish an intelligent clustering scoring model based on XGBOOST English text, pass the text to be scored into the model as an input value, and evaluate the input text according to the clustering results of the model.

Step 7: Further optimize the model and develop the software based on the above implementation algorithm as well as the established model.

6   Simulation Experiments

6.1   Experimental Data

The text dataset used for the experiment is from the English categorized dataset of the official Kaggle website, which contains texts in the categories of business, politics, education, science and sports. Taking this paper as an example, six sets of data are organized from it, containing different numbers of categories and proportions of categories. The specific descriptions of the six sets of data are shown in the following Table 1:

Table 1. Description of the six data sets

	Group 1	Group 2	Group 3	Group 4	Group 5	Group 6
Total number of texts N	200	200	200	200	200	200
Number of categories k	5	7	5	5	5	6
Balance or not	be	clogged	be	be	clogged	be

In order to facilitate the clustering results of the identification of sample categories, the documents will be numbered according to the order of the categories, for example, in the fifth group: number 1–200 for the art category documents and so on, so that the model training data is ready.

6.2 Experimental Flow Simulation

In this section, we use examples to illustrate and validate the textual intelligence assessment model proposed in this paper.

Firstly, the English text dataset is preprocessed, and each English text dataset is processed by NLTK segmentation with simple segmentation and removal of deactivated words to obtain a candidate feature word set. Calculate the document frequency of each candidate feature word in each data set, and eliminate the feature words whose document frequency is lower than 0.03% of the total number of texts in each group and higher than 20% of the total number of texts to obtain the feature subset.

The number of feature words after doing feature coarsening with document frequency is shown in the Table 2 below:

Table 2. Number of feature words after rough selection

	Group 1	Group 2	Group 3	Group 4	Group 5	Group 6
Number of feature words after feature rough selection	2149	2236	2381	2092	2285	2372

Since the rough selection of features removes some of the features that do not contribute much to the clustering, but the text vectors are still highly sparse at this time, this step re-selects the text feature words through the particle swarm optimization algorithm. The particle swarm parameters are set as $\omega = 0.4/0.9$, $c1 = c2 = 2$, $\gamma = 0.9$, $\theta = 5$, size $= 80$. The particle swarm parameters are set as MaxIter Set as 80, the small batch of sample books in the adaptation value calculation m $= 0.1*N$, where N is the total number of texts in the group. BPSO, NBPSO, GSBPSO are used to compare the optimal solution of adaptation value under different particle swarm algorithms, respectively, and this step uses the average of adaptation value of each group of data to compare, that is, using the sum of adaptation value of each group of data divided by the number of texts in the corresponding dataset, and the experimental results are shown in the following Table 3:

Table 3. Performance of different particle swarm algorithms in different datasets

	Group 1	Group 2	Group 3	Group 4	Group 5	Group 6
BPSO	70.089	72.314	70.861	69.206	71.263	70.841
NBPSO	96.632	96.810	97.256	96.382	96.325	96.465
GSBPSO	**97.655**	**97.131**	**97.879**	**97.218**	**97.452**	**97.723**

Table 4. Changes in the number of feature words before and after feature reselection via GSBPSO algorithm

	Group 1	Group 2	Group 3	Group 4	Group 5	Group 6
Number of features after feature roughing	2149	2236	2381	2092	2285	2372
Number of features after feature re-selection	1762	1878	1904	1631	1872	1874
compression ratio	**0.82**	**0.84**	**0.80**	**0.78**	**0.82**	**0.79**

It can be seen that in this experiment, the performance of GSBPSO is better than the performance of the other two particle swarm algorithms, in which the performance of GSBPSO is far superior to the BPSO algorithm before the improvement, and can be optimized on the basis of the original NBPSO algorithm, which proves that the improvement of the particle swarm algorithm in this paper is effective.

The unselected feature terms in the TF-IDF matrix obtained from the Chinese text dataset are deleted by the particle swarm algorithm, and the final text feature matrix is obtained and used for clustering.

The number of text features before and after feature reselection with GSBPSO is shown in the Table 4.

It can be seen that using the particle swarm algorithm proposed in this paper for feature word re-selection can effectively reduce the dimension of the text feature matrix, which in turn improves the efficiency of the subsequent text clustering as well as the accuracy of the prediction of the XGBoost classifier.

Clustering experiments are carried out using the text feature matrix obtained in the previous steps. It is divided into two parts of experiments: comparison of clustering effect between traditional Kmeans algorithm and Skmeans algorithm and comparison between Skmeans algorithm and hybrid clustering algorithm proposed in this paper. The clustering effect is evaluated using three metrics: CH (Calinski-Harabaz) value, Silhouette Coefficient, and SSE (Sum of the Squared Errors) value. The CH metric is used to characterize the degree of closeness of the classes as well as the degree of dispersion among the classes, the Silhouette Coefficient metric is used to characterize the differences between inside and outside of clusters, and the SSE value is used to characterize the differences between inside and outside of clusters through the calculation of the clusters. The CH metric is used to describe the closeness of the classes and the dispersion between classes, the SC metric is used to describe the difference between inside and outside the clusters, and the SSE value describes the clustering error of all samples by calculating the sum of the error squares of the distances between the sample points and the centers of the clusters in which they are located. The following experimental data are obtained from 30 repetitions of independent experiments (Tables 5, 6, 7, 8, 9, 10, 11, 12, 13, 14, 15 and 16).

Table 5. Comparison of the effectiveness of different clustering algorithms using cosine similarity under the first data set

	Skmeans	BSKM	Skmeans++
CH	3.163	2.825	**4.146**
SC	0.03658	0.02493	**0.06241**
SSE	144.271	142.822	**137.816**

Table 6. Comparison of the effectiveness of different clustering algorithms using cosine similarity under the second dataset

	Skmeans	BSKM	Skmeans++
CH	**5.728**	4.614	5.652
SC	**0.04819**	0.03966	0.04571
SSE	364.152	372.478	**363.596**

Table 7. Comparison of the effectiveness of different clustering algorithms using cosine similarity under the third data set

	Skmeans	BSKM	Skmeans++
CH	**9.663**	6.718	8.581
SC	0.03814	0.01941	**0.03962**
SSE	606.182	601.524	**597.122**

From the above table, the improved clustering algorithm proposed in this paper outperforms the other algorithms in SC index and SSE index in the first, third, fourth, and fifth group of experiments, while in the second group of experiments, it slightly underperforms Skmeans algorithm in SC index, and the performance of the clustering algorithm in SC is not much different from that of Skmeans algorithm in the fifth group of experiments. In the fifth group of experiments, the performance of the clustering algorithm proposed in this paper is not much different from Skmeans algorithm in terms of SC, indicating that the clustering algorithm proposed in this paper is more stable and more reliable in the case of uniform text categories, while in each group of experiments,

Table 8. Comparison of the effectiveness of different clustering algorithms using cosine similarity under the fourth data set

	Skmeans	BSKM	Skmeans++
CH	11.228	7.359	**11.523**
SC	0.03715	0.01649	**0.03802**
SSE	759.812	765.717	**757.283**

Table 9. Comparison of the effectiveness of different clustering algorithms using cosine similarity under the fifth data set

	Skmeans	BSKM	Skmeans++
CH	6.197	5.821	**6.469**
SC	0.03692	0.02946	**0.03811**
SSE	729.450	744.899	**725.812**

Table 10. Comparison of the effectiveness of different clustering algorithms using cosine similarity under the sixth data set

	Skmeans	BSKM	Skmeans++
CH	5.797	5.422	**6.148**
SC	0.03728	0.03014	**0.03893**
SSE	730.415	746.129	**724.652**

Table 11. Comparison of clustering effect between Kmeans algorithm and SKM algorithm under the first dataset

	Kmeans (math.)	Skmeans
SC	0.00802	**0.03712**
SSE	159.128	**142.006**

the clustering algorithm using cosine distance as the distance function outperforms the Kmeans algorithm using the Euclidean distance as the distance function, which indicates that cosine distance is more suitable for the comparison of textual data. The cosine distance is more suitable for the comparison of text data.

Table 12. Comparison of clustering effect between Kmeans algorithm and SKM algorithm under the second data set

	Kmeans (math.)	Skmeans
SC	−0.00469	**0.05119**
SSE	425.218	**368.172**

Table 13. Comparison of clustering effect between Kmeans algorithm and SKM algorithm under the third data set

	Kmeans (math.)	Skmeans
SC	−0.00614	**0.03612**
SSE	659.721	**608.928**

Table 14. Comparison of clustering effect of Kmeans algorithm and SKM algorithm under the fourth data set

	Kmeans (math.)	Skmeans
SC	−0.00293	**0.03294**
SSE	839.329	**755.964**

Table 15. Comparison of clustering effect between Kmeans algorithm and SKM algorithm under the fifth data set

	Kmeans (math.)	Skmeans
SC	−0.02015	**0.03521**
SSE	832.746	**730.452**

Table 16. Comparison of clustering effect between Kmeans algorithm and SKM algorithm under the sixth data set

	Kmeans (math.)	Skmeans
SC	−0.02153	**0.04511**
SSE	833.766	**722.745**

The second group, the fifth group for the unbalanced clustering, and the fifth group of text number is larger, the number of categories is more, is the highest difficulty of clustering a group of clustering algorithms proposed in this paper in the performance of these two groups of data does not have a great advantage, and the other algorithms in the performance of the two groups of data compared to the other three groups decreased significantly, indicating that these four algorithms are still not a good solution to the difficult clustering problems, there is still a large room for improvement. However, the clustering algorithm proposed in this paper still has a slight advantage.

The clustering results obtained from the fourth set of data were tested for text category evaluation, where the matrix dimensions within each cluster and the number of different categories contained are shown in the Table 17 below:

Table 17. Dimensions and number of categories for different clusters

Cluster number	Dimension (math.)	Classification
1	(256,2273)	5
2	(170,2273)	3
3	(242, 2273)	3
4	(122,2273)	5
5	(210,2273)	5
6	(224, 2273)	5

The clustering results obtained from the fourth set of data were evaluated for accuracy comparison using XGBoost, BP neural network, Bayesian classification, and K-nearest neighbor algorithm, respectively (Table 18):

Table 18. Comparison of text evaluation accuracy for different classifier algorithms

	Cluster 1	Cluster 2	Cluster 3	Cluster 4	Cluster 5
XGBoost	87.11	**99.41**	**97.93**	**85.25**	**90.48**
BP neural network	67.97	90.0	69.42	61.48	46.19
Bayesian classification	73.44	90.0	93.38	63.93	83.33
K-nearest neighbor	**87.89**	92.94	97.11	79.51	84.28

As can be seen from Table 6.16, except for the first cluster of data, which has a slightly weaker effect than the K nearest neighbor classifier, the rest of the clusters of XGBoost's performance results are significantly better than the BP neural network, Bayesian classifier, and K nearest neighbor algorithm, which indicates that XGBoost is more suitable to be used as a predictor for the text intelligent evaluation model proposed in this paper.

At the end of the model, the particle swarm algorithm described in this paper is used to reselect the features of the roughly selected text data, and each cluster created for the improved clustering method is respectively trained to predict the XGBoost classifier, and after 40 repetitions of the independent experiments, an average prediction correctness rate of about 89.6% is finally obtained, which preliminarily indicates that the approach proposed in this paper has a better effect in the prediction of the class of the unknown text.

7 Conclusions

In this paper, a XGBoost text intelligent evaluation model based on particle swarm optimization algorithm is proposed, and the final experimental results show that:

(1) As if just filtering the feature words with too high or too low DF to roughly select the text, the text feature matrix after roughly selecting is still highly sparse, through the improved particle swarm algorithm proposed in this paper to re-select the text features after roughly selecting, it achieves a high compression ratio, dramatically reduces the sparsity of the text matrix, and the running efficiency as well as the accuracy of the whole model after the simplification is improved to a certain extent.

(2) Based on the traditional Kmeans algorithm, this paper makes certain improvements, changes the distance measure of clustering, and at the same time combines the ideas of bisection clustering algorithm and Kmeans + + clustering algorithm, and proposes an improved hybrid clustering algorithm by changing the process of clustering as well as the selection of the initial value of the center of clustering, and ultimately in the clustering indexes can be seen in the improved clustering algorithm proposed in this paper is more desirable than that of the traditional clustering algorithm. Clustering algorithm, the effect is more ideal.

(3) The final accuracy of the prediction of the category of the unknown text by XGBoost is 89.6%, which can more accurately make a judgment on the category to which the text belongs, and can provide some reference and help to the user.

Computational examples show that the method proposed in this paper is practical and effective. Migration of the model in this paper can be considered in future work for application to fraud detection, intelligent scoring of English essays, and other fields. In addition, the heuristic search algorithm optimizes the feature subset and also brings time overhead, this paper slightly improves the efficiency by introducing the concept of small batch of random samples in the adaptation value, but it still requires a large time overhead. How to further speed up the search process of the optimal feature subset as well as designing more robust clustering algorithms are what needs to be further investigated in this topic.

References

1. Gao, X., Shao, G., Zhang, H., Zhou, Z.: Improved binary sparrow search algorithm for feature selection and text clustering. J. Chongqing Univ. Technol. (Natural Sci.) **37**(8), 166–176 (2023)
2. Dou, X.: Research on Intrusion Detection Method Based on PopulationClustering and Feature Selection. Hebei University (2023)
3. Wang, C., Dong, Y.: Feature selection based on binary grey wolf optimization and text clustering. Comput. Eng. Des. **42**(9), 2526–2535 (2021)

4. Zeng, H.: Research on Text Sentiment Classification Based on Feature Selection and TFIDF. Huazhong University of Science and Technology (2023)
5. Jiang, W.: Research on Imbalanced Classification Algorithm Based on Text Data Augmentation and Feature Selection. University of Science and Technology of China (2023)
6. Tian, X., Xiong, Y.: Text clustering algorithm combined with new feature selection mechanism. Comput. Eng. Des. **42**(3), 734–741 (2021)
7. Abualigah, L., Gandomi, A.H., Elazz, M.A., et al.: Advances in meta-heuristic opfimization algorithms in big data text clustering. Electronics **10**(2), 101–129 (2021)
8. Xu, Z.: Comparative study on the effect of different features on text clustering: take news text as an example. Inf. Stud. Theory Appl. **43**(1), 169–176 (2020)
9. Gao, X., Zhou, Z., Wang, L., Shao, G., Zhang, O.: Feature Selection and Text Clustering Algorithm Based on Binary Mayfly Optimization. J. Jilin Univ. (Sci. Edition) **61**(3), 631–640 (2023)
10. Hao, X., Fang, X., Yang, G.: News text clustering and visualization based on TFIDF+LSA algorithm. Comput. Technol. Dev. **32**(7), 34–38+45 (2022)
11. Ying, F.: Text Clustering Based on Particle Swarm Optimization. South China Agricultural University (2023)
12. Wang, M.: Research on Text Clustering Algorithm Based on Particle Swarm Optimization Algorithm. Guangdong University of Technology (2021)
13. Niu, Y., Wu, B.: Research on text clustering algorithm based on improved particle swarm optimization and K-means. J. Lanzhou Univ. Arts Sci. (Natural Sci.) **33**(4), 44–47 (2019)
14. Yang, J., Zhao, T., Zhao, Y., Gao, M., Chen, H., Yao, W.: Active distribution network optimization method based on improved binary particle swarm optimization. Electr. Automation **44**(3), 48–49+53 (2022)
15. Liu, J.: The Research and Improvement of Particle Swarm Optimization. Harbin Institute of Technology (2006)

Machine Learning-Assisted Optimization of Direction-Finding Antenna Arrays

Qing Zhang[1], Miao Gong[2(✉)], Gouqiong Li[3], Xinyu Ma[2], Yiheng Chen[2], Fei Zhao[3], and Sanyou Zeng[2]

[1] School of Computer Science, Huanggang Normal University, Huanggang 438000, China
[2] Germany School of Mechanical Engineering and Electronic Information, China University of Geosciences, Wuhan 430074, China
miaogong913@qq.com
[3] Science and Technology on Blind Signal Processing Laboratory, Chengdu 610041, China

Abstract. Antenna design is usually a complex optimization problem. It usually includes two stages. The first stage is to formulate it as an optimization problem. And then the second stage is to design an algorithm to solve the optimization problem. This paper aims to design unequally spaced antenna arrays for direction-finding, which requires cooperation among direction-finding users, antenna designers and optimizers. The paper combines the direction finding array with the optimization design and analyzes how to construct a reasonable optimization problem from the practical optimization design problem construction stage which has been neglected. A multi-objective problem is constructed by taking the direction-finding performance as the optimization objective. The mutual coupling effect causes the design of array antennas to require electromagnetic simulation to ensure the reliability of the results, which is a very time-consuming and expensive problem. Gaussian regression model is introduced into the multi-objective optimization algorithm to construct inexpensive surrogate optimization problems that reduce the number of accurate simulation evaluations. The experimental results show that the designed array meets all the requirements with good robustness while the number of simulations required is less than that of previous methods.

Keywords: Direction-finding Antenna Array · Antenna Design · Optimization Problem · Time-consuming Computation

1 Introduction

Array design for antenna systems often requires the evaluation of the radiation characteristics of the array antennas using electromagnetic simulations (EM). The time-consuming nature of these simulations, coupled with the large number of function calls required for optimization algorithms, makes array design an expensive optimization problem [1].

In order to address the challenge of time-consuming evaluation, many machine learning methods (e.g., Gaussian process regression (GPR) [2], support vector machine (SVM) [3]) have been used in the array optimization process for constructing agent

K. Li and Y. Liu (Eds.): ISICA 2023, CCIS 2147, pp. 476–486, 2024.
https://doi.org/10.1007/978-981-97-4396-4_44

optimization function to replace the expensive EM evaluations with computationally inexpensive surrogate problems. These models are trained using a limited number of samples obtained from the electromagnetic simulations, and they are then used to predict the performance of the array for different design configurations.

Designing a non-uniform antenna array for direction finding is the main focus of this paper. Unlike conventional antenna designs, direction-finding array designs have traditionally relied on the expertise of the designer, as the performance of direction-finding arrays requires an analysis of the radiation information obtained from simulation evaluations. This introduces additional challenges in the optimization process. In this paper, the problem of direction-finding array design is approached by formulating it as a multi-objective problem. Two important metrics, the direction-finding error and the correlation peak ratio are selected as the optimization objectives. A Gaussian process model is used to approximate the direction-finding objectives, which saves a significant amount of computational resources by avoiding the need for intermediate calculations.

The rest of this paper is structured as follows: the second section describes the proposed direction-finding metrics. The third part describes the proposed two-stage optimization algorithm in detail. In the fourth part, the experiment and result discussion are carried out. Finally, a conclusion is drawn in the fifth part.

2 Background

2.1 Multi-objective Optimization Problems

Many real-world problems are multi-objective problems (MOPs), consisting of several sub-problems that conflict with each other. A MOP can be denoted as:

$$\min \vec{F}(\vec{x}) = (f_1(\vec{x}), \dots, f_m(\vec{x}))^T$$
$$s.t. \vec{x} \in \Omega \tag{1}$$

where $\vec{x} = (x_1, x_2, \dots, x_n)$ denotes the n-dimensional decision variable and Ω is the variable space. m is the number of objectives, $\vec{F}(\vec{x})$ is the vector of m-dimensional objectives.

2.2 Gaussian Process Model

Gaussian process is a spatial interpolation technique based on stochastic process theory [4]. It can offer information regarding the uncertainty of the predicted objective value.

Consider any x whose single target value $f(x)$ can be expressed as a sample point in $F(x)$, $F(x) \sim N(\mu, \sigma^2)$, μ and σ are two constants independent of x. For any two points x, x' in training data, the correlation $c(x, x')$ between $F(x)$ and $F(x')$, depends only on the correlation between x and x'.

$$c(x, x') = \exp\left[-\sum_{i=1}^{D} \theta_i |x_i - x_i'|^{p_i}\right] \tag{2}$$

where D is the dimension of problems, $\Theta = [\theta_1, \dots, \theta_D]^T$, $\theta_i > 0$ and $1 \le p_i \le 2$. θ_i indicates the importance of x_i on $g(x)$, p_i is related to the smoothness of the prediction function $g(x)$. The known N points $x^1, \ldots, x^N$ and their function values $y^1, \ldots, y^N$ form the training data (X, Y). The likelihood function that $g(x) = y^i$ at $x = x^i (i = 1, 2, \ldots, N)$ is

$$\frac{1}{(2\pi\sigma^2)^{\frac{N}{2}} \sqrt{\det(C)}} \exp\left[-\frac{(y-\mu 1)^T C^{-1}(y-\mu 1)}{2\sigma^2}\right] \tag{3}$$

where C is a $N \times N$ matrix whose (i, j) -element is $c(x^i, x^j)$, $y = (y^1, \ldots, y^N)$ and 1 is a N -dimensional column vector of ones. To maximize the likelihood function (3), the value of μ and σ^2 must be

$$\hat{\mu} = \frac{1^T C^{-1} y}{1^T C^{-1} 1}$$
$$\hat{\sigma}^2 = \frac{(y-1\hat{\mu})^T C^{-1}(y-1\hat{\mu})}{N} \tag{4}$$

Substituting Eq. (4) into Eq. (3) eliminates the unknown parameters μ and σ. Thus the likelihood function Eq. (3) can be translated into a function on θ_i and p_i. The values of θ_i and p_i can be obtained by maximizing Eq. (3), which is multimodal, gradient-based methods could be trapped on its local optima. After deriving these hyperparameters $\hat{\theta}_i$, $\hat{p}_i$, $\hat{\mu}$ and $\hat{\sigma}$, the Gaussian process will be adopted to predict the function value $f(x)$ based on the best linear unbiased prediction function

$$\hat{y}(x) = \hat{\mu} + r^T C^{-1}(y - 1\hat{\mu}) \tag{5}$$

and its mean squared error

$$\hat{s}^2(x) = \hat{\sigma}^2\left[1 - r^T C^{-1} r + \frac{(1-1^T C^{-1} r)^2}{1^T C^{-1} r}\right] \tag{6}$$

where $r = (c(x, x^1), \ldots, c(x, x^N))^T$, $N(\hat{y}(x), \hat{s}^2(x))$ is the corresponding prediction model.

3 Optimization Problem Construction for Direction-Finding Arrays

The direction-finding antenna linear array in this paper is placed horizontally. It finds the direction-of-arrival (DOA) of the source by comparing the phase difference of the received signals of different array elements, while considering only the polar angle θ. Due to the position difference between each element, there is a phase difference in the signal received by each element. Figure 1 shows the Phase difference schematic over the polar angle θ.

When the signal source is located at the polar angle θ, without considering the mutual coupling, the phase difference of the signal received by the two arrays is:

$$\Delta\phi = \frac{2\pi d}{\lambda} sin(\theta) \tag{7}$$

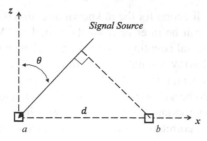

Fig. 1. Phase Difference Schematic

where λ is the wavelength at frequency f, d is distance between the two elements called baseline length. The value of $\Delta\phi$ can be obtained through the phase discriminator, and the measured DOA is

$$\hat{\theta} = argmin\left(\frac{\lambda\Delta\phi}{2\pi d}\right) \tag{8}$$

In the N-dimensional linear antenna array, considering the diversity of received signals, the operating frequency range is divided into K frequency points. M discrete polar angles $\theta_m(1 \le m \le M)$ are evenly selected from the direction finding space of each frequency point. Before starting the direction finding, a data sample set V needs to be constructed to store the corresponding phase difference from the array element at zero generated by each array element in the linear array antenna after receiving the signal from angle θ_m at different frequencies.

$$
\begin{aligned}
v_k(\theta_m) &= [\Delta\phi_{k1}(\theta_m), \dots, \Delta\phi_{kn}(\theta_m), \dots, \Delta\phi_{kN}(\theta_m)]^T, \\
V(\theta_m) &= [v_1(\theta_m), \dots, v_k(\theta_m), \dots, v_K(\theta_m)], \\
V &= [V(\theta_1), \dots, V(\theta_m), \dots, V(\theta_M)].
\end{aligned} \tag{9}
$$

where $\Delta\phi_{kn}(\theta_m)$ denotes the phase difference between the n th element and the element at zero when receiving a signal from the direction θ_m. Therefore, $\Delta\phi_{k1}(\theta_m) = 0$. Gather K sets of phase differences due to the unknown frequency of the signal after receiving the signal to be measured.

$$
\begin{aligned}
x_k(\theta) &= [\Delta\phi_{k1}(\theta), \dots, \Delta\phi_{kn}(\theta), \dots, \Delta\phi_{kN}(\theta)]^T, \\
X(\theta) &= [x_1(\theta), \dots, x_k(\theta), \dots, x_K(\theta)].
\end{aligned} \tag{10}
$$

where $\Delta\phi_{kn}(\theta)$ denotes the phase difference between the n th element and the element at zero when receiving a signal from the unknown direction θ. $X(\theta)$ denotes the set of phase differences at k th frequency.

Calculate the cosine similarity of the data sample with the corresponding phase difference of the signal to be measured and the m th direction Angle of the k th frequency:

$$\rho_k(\theta_m) = \frac{\sum_{n=1}^{N} \cos(\Delta\phi_{kn}(\theta_m) - \Delta\phi_{kn}(\theta))}{N} \tag{11}$$

After $K \times M$ calculations the maximum value ρ_{max} and its direction θ_{max} can be obtained. For θ_{max} the front and rear sampling angles can be quadratic interpolated.

$$\hat{\theta} = \theta_{max} - \frac{(\theta_{max+1} - \theta_{max-1})(\rho_{max+1} - \rho_{max-1})}{4(\rho_{max+1} + \rho_{max-1} - 2\rho_{max})} \tag{12}$$

where $\hat{\theta}$ is the estimated direction for the unknown direction θ signal source.

The estimated DOA can be in error from the actual DOA due to factors such as ambient noise signals, mutual coupling between array elements, and so on. In order to ensure that the designed array antenna can accurately determine the direction of the signal in a noisy environment, technical indicators that can measure the probability of accurate prediction should be added in the design process to ensure that the final design of the array antenna is appropriate [5]. Introduced two direction finding array metrics: direction finding error and correlation peak ratio, to measure direction finding accuracy at different spacings.

A. *Direction-finding Error (DFE):* It refers to the error between the incoming wave orientation and the true orientation of the target measured by the directional measuring equipment, usually using the root mean square value for statistics. The value of directional error is related to the working frequency and the orientation of the incoming signal, so the actual application needs to use different frequencies and different orientations of the measured directional error to indicate the accuracy of the system (shown in (13)), which is actually a technical indicator to measure the credibility of the incoming wave direction. The formula of DFE is

$$f_e = \sqrt{\frac{1}{K \times M} \sum_{k=1}^{K} \sum_{m=1}^{M} \left(\hat{\theta}_{km} - \theta_{km} \right)^2} \tag{13}$$

where θ_{km} is the angle of the m th angle sampling point at the k th discrete frequency point (assuming it is the actual DOA of the signal). And $\hat{\theta}_{km}$ is the estimated direction of the θ_{km}.

B. *Correlation Peak Ratio (CPR):* Interpolation of (11) yields a continuous correlation curve (cosine similarity - polar angle). Combined with (11), the highest peak of the correlation curve corresponds to the estimated direction $\hat{\theta}_{km}$. Define the highest peak of the correlation curve as the first correlation peak and the second highest peak as the second correlation peak. When there are multiple higher peaks in the correlation curve, it is easy to yield phase ambiguity resolution leading to the failure of direction finding. Therefore CPR can quantify the performance of phase ambiguity resolution for linear direction finding array antennas. Under the premise that the estimated DOA does not significantly deviate from the actual DOA, the lower the CPR value, the higher the success rate of direction finding. The calculation formula is as follows:

$$r(\theta) = \frac{\rho_{secmax}(\theta)}{\rho_{max}(\theta)},$$
$$f_r = \max_{\substack{1 \le m \le M \\ 1 \le k \le K}} (r(\theta_{km})) \tag{14}$$

where θ is the direction of the measured signal, $\rho_{max}(\theta)$ is the value of the first correlation peak (maximum value of the correlation curve), and $\rho_{secmax}(\theta)$ is the value of the second correlation peak value. Considering the diversity of signal directions and the fact that the working frequency band can be divided into multiple frequency points, the worst value

r of the second correlation peak in all working airspace and working frequency bands is taken to reflect the overall performance of the direction finding array in the working airspace and working frequency bands.

In order to reduce the DFE, the baseline length d_n in the linear array needs to be extended as much as possible. However, in order to reduce the CPR to improve the probability of successful direction finding, the length of some baselines should be reduced appropriately. The optimization of spacings for linear direction-finding array is constructed as a MOP:

$$
\begin{aligned}
\min \quad & F(x) = (f_e(x), f_r(x)) \\
where \quad & x = (x_1, \ldots, x_i, \ldots, x_D) \\
st \quad & l_i \leq x_i \leq u_i
\end{aligned}
\tag{15}
$$

where x is the baseline length vector between all neighboring array elements, l_i and u_i are the lower bound and upper bound on the nth baseline length, limited by the size of the individual antenna element and the maximum length of the linear array.

4 Machine Learning-Assisted Optimization Algorithm

To solve the constructed MOP (15), it is proposed to decompose the problem using the popular decomposition-based multi-objective evolutionary algorithm (MOEA/D) [6] and introduce a Gaussian process model [7] to construct computationally inexpensive function via modeling the database. The algorithmic framework is shown in Algorithm 1.

In the step of *Initialization*, a total of $11n - 1$ (n is the variable dimension) sample points are obtained through LHS sampling, which is a statistical method for generating a diverse set of samples. These sample points are then evaluated expensively to obtain an initial database. In each iteration, K_E new points are added to the database, and the predictive model is updated accordingly in the next loop. As the number of generations increases, the predictive model becomes progressively more accurate, enabling better predictions of the behavior of the system being modeled.

In the step of *Evolutionary Optimization*, we uesd the MOEA/D algorithm shown in Algorithm 2 and choose the most widely used Tchebycheff aggregation method, whose decomposition mechanism is

$$
\begin{aligned}
\min g^{te}(x|\lambda, z^*) = \max_{1 \leq i \leq m} \left\{ \lambda^i \left| \left(f_i(x) - z_i^* \right) \right| \right\} \\
where z_i^* = min(f_i(x))
\end{aligned}
\tag{16}
$$

where $z^* = \left(z_1^*, \ldots, z_m^* \right)$ is the ideal point in the objective space, λ^i indicates the i th weight vector, $f_i(x)$ is the i th minimized objective function value.

Algorithm 1 Array Optimization with MOEA/D-EGO Algorithm Framework

Input: MOP(15); Computational budget $FEs = 0$; The number of EM simulation evaluations K_E.

Output: Output the Non-dominated solutions in the final database as the best-choice distribution structures.

1: **Initialization:** Use LHS to generate the initial samples for the initial database and evaluate by using EM simulation to obtain expensive function values, get training database;

2: **While** The computational budget FEs is not exhausted **do**

3:    **Models Building:** Use the training database to build a Gaussian Process model for f_e and f_r, respectively.

4:    **Evolutionary Optimization:** Evolve the population by the prediction models with MOEA/D [6].

5:    **Expensive Evaluation:** Select K_E points in the population by a selection scheme to evaluate via expensive EM simulation, $FEs + K_E$.

6:    **Update Database:** Add the solutions to the training database, and update the database.

7: **End while**

Algorithm 2 MOEA/D Framework

Input: A uniform spared of N weight vectors: $\lambda^1, \ldots, \lambda^N$; T: the number of the weight vectors in the neighborhood of each weight vectors.

Output: An external population EP, which is used to store non-domination solutions found during the search.

1: **Initialization:** Generate an initial population $x^1, \ldots, x^N$ randomly or by a problem-specific method and evaluate their objective values $(f_e(x), f_r(x))$; Find the T neighbor vectors for each weight vector; Match the weight vector with N points one by one by a special method; Compute z^*.

2: **For** $i = 1, \ldots, N$ **do**

3:    **Reproduction and repair:** Select two individuals x^j and x^k, then generate a solution y from x^i, x^j and x^k by a DE/rand/1/bin operator [8], and then perform a mutation operator on y.

4:    **Update of z^*:** For each $j = 1, \ldots, m$, if $z_j^* < f_j(y)$, then set $z_j^* = f_j(y)$.

5:    **Update of neighboring solution:** For each individual that matches with the neighbor vector of λ^i, if $g^{te}(y|\lambda, z^*) \leq g^{te}(x|\lambda, z^*)$, then set the point $x = y$ and update the objective values at the same time.

6:    **Update of EP:** Remove from EP all the individuals dominated by $(f_e(x), f_r(x))$; Add to if no individual in EP dominate $(f_e(x), f_r(x))$.

---

5 Experiment and Result Discussion

To validate the application of machine learning-assisted optimization in direction-finding array design, a spacing optimization is performed for a 5-element direction-finding antenna array. The design variables are four-dimensional. The operating frequency is 600 MHz–1200 MHz. Parameters used in this paper are as follows:

1. The variable dimension is 4, the aborting evaluations *FEs* is 200.
2. The number N of subproblems and weight vectors is 300.
3. The operating frequency is 600 MHz–1200 MHz. Sampling at 25 MHz intervals, a total of 25 frequencies are obtained.
4. In order to more closely match the real-world environment, noise (mean is 0, variance is 5°) is added to each incoming angle to simulate the real direction finding. Each objective value is the result of 15 independent repetitions of the calculation, ensuring the robustness of the results.

The Fig. 2 shows the pareto front of the final database. The Table 1 shows the four sets of spacings results randomly selected from the pareto front, and their target values were repeated several times to calculate the display in Figs. 3, 4, 5 and 6. All figures show two curves, the worst values (max) and the mean values (rms) of the objective at the current frequency. All four sets of results have a direction-finding error of less than 0.2° and a correlation peak ratio of less than 0.8, which meets the requirements needed for direction-finding measurement. Also from the fact that the number of times of using expensive simulation in this experiment is less compared to the number of times required by other expensive algorithms, it shows that the method is effective in designing the direction-finding array antenna.

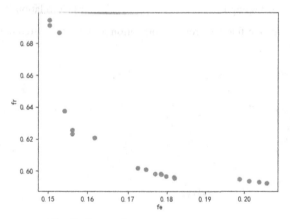

Fig. 2. Pareto front of the final database

Table 1. There are 4 sets of optimal spacings from the pareto front.

	5-element array spacings (m)	f_e	f_r
1	[0.658, 0.391, 0.530, 0.258]	0.1344	0.7754
2	[0.509, 0.846, 0.394, 0.217]	0.1215	0.7555
3	[0.537, 0.375, 0.410, 0.244]	0.1584	0.7923
4	[0.484, 0.920, 0.363, 0.232]	0.1197	0.7444

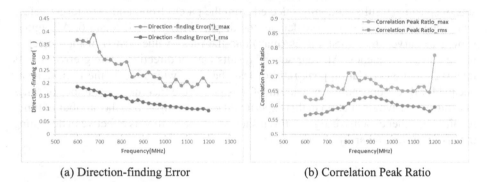

(a) Direction-finding Error　　　　　　(b) Correlation Peak Ratio

Fig. 3. Values of direction-finding error and correlation peak ratio at different frequencies for the first set of arrays

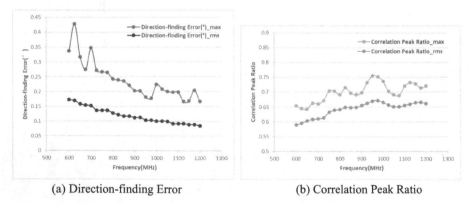

(a) Direction-finding Error　　　　　　(b) Correlation Peak Ratio

Fig. 4. Values of direction-finding error and correlation peak ratio at different frequencies for the second set of arrays

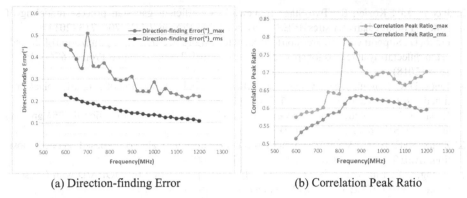

(a) Direction-finding Error   (b) Correlation Peak Ratio

Fig. 5. Values of direction-finding error and correlation peak ratio at different frequencies for the third set of arrays

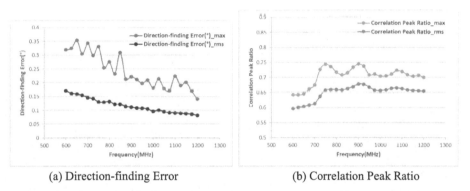

(a) Direction-finding Error   (b) Correlation Peak Ratio

Fig. 6. Values of direction-finding error and correlation peak ratio at different frequencies for the fourth set of arrays

6 Conclusion

In this paper, two technical metrics that can measure the performance of direction-finding arrays are proposed and a multi-objective optimization problem for array spacings is constructed. To solve the complex and expensive antenna linear array design problem, the Gaussian regression model is introduced into the optimization algorithm to complete the construction of the surrogate evaluation model. The experimental results show that the introduction of the predictive model greatly reduces the optimization time with a limited number of expensive EM evaluations, and the designed array still maintains a good direction-finding performance.

References

1. Wu, Q., Cao, Y., Wang, H.M., Hong, W.: Machine-Learning-Assisted optimization and its application to antenna designs: opportunities and challenges. China Commun. **17**, 152–164 (2020)

2. Jacobs, J.P., Koziel, S.: Two-Stage framework for efficient gaussian process modeling of Antenna Input Characteristics. IEEE Trans. Antennas Propag. **62**(2), 706–713 (2014)
3. Prado, D.: Support vector regression to accelerate design and crosspolar optimization of shaped-beam reflectarray antennas for space applications. IEEE Trans. Antennas Propag. **67**(3), 1659–1668 (2018)
4. Jones, D.R., Schonlau, M., Welch, W.J.: Efficient global optimization of expensive black-box functions. J. Global Optim. **13**(4), 455–492 (1998)
5. Kadan, F.E.: Ambiguity and gross error probability calculation for direction finding antenna arrays. In: SIU, pp. 248–251 (2017)
6. Zhang, Q.F., Li, H.: MOEA/D: a multiobjective evolutionary algorithm based on decomposition. IEEE Trans. Evol. Comput. **11**(6), 712–731 (2007)
7. Zhang, Q.F., Liu, W.D.: Expensive multiobjective optimization by MOEA/D with Gaussian process model. IEEE Trans. Evol. Comput. **14**(3), 456–474 (2010)
8. Storn, R., Price, K.: Differential evolution - a simple and efficient heuristic for global optimization over continuous spaces. J. Glob. Optim. **11**(4), 341–359 (1997)

Author Index

C

Cai, Jiahui I-388
Cai, Tie I-388, II-49, II-135
Cai, Xingjuan I-277
Cao, Jiale I-73, II-337
Chai, Lu II-196
Che, Yongdie II-372
Chen, Bing II-196
Chen, Feiyun II-459
Chen, Weicong I-463
Chen, Xuhang I-266
Chen, Yan I-223, II-28
Chen, Yiheng II-476
Chen, Yongxian II-113
Chen, Yu I-137
Chen, Zeming II-135
Chen, Zhiqiang II-103, II-384
Cheng, Hangchi I-369
Cheng, Huabin I-137
Cheng, Shi II-407
Cheng, Ziyu I-409
Chu, Yih Bing I-231
Cui, Dandan I-165
Cui, Xiaojun I-18, II-170, II-187
Cui, Zhihua I-277

D

Damian, Maria Amelia E. II-3
Diao, Zhenya I-152
Dong, Ani I-443
Dong, Qianqian I-178

F

Fang, Wanhan II-196
Feng, Tian II-127

G

Gao, Zihang I-18, II-170, II-187
Gong, Miao II-476
Guan, Jian I-152, I-326
Guan, Jing I-207

Guo, Li I-409
Guo, Xuan I-292

H

He, Dan II-159
He, Fufa I-178
He, Jinfeng II-103
He, Jinrong I-340
He, Kejin II-287
He, Shuizhen I-312
He, Yongqiang I-277
Hu, Kun II-270
Hu, Min I-125
Huang, Peiquan I-3
Huang, Siming I-409
Huang, Weidong I-292, I-302, II-225
Huang, Xing I-302

J

Jalil, Hassan I-49
Ji, Dong I-165
Ji, Qi II-430
Jiang, Chengyu II-214
Jiang, Tian I-231
Jiang, Zifeng II-39
Jin, Xiao II-56
Jing, Furong I-88

K

Kang, Lanlan I-101, I-192
Kangshun, Li I-49, I-369, I-463, II-15
Kong, Yuyan I-3

L

Lai, Luyan II-351, II-372
Lai, Tao II-360
Lai, Yu I-101
Lei, Jiawei II-459
Lei, Yishu I-463
Li, Changrui I-451
Li, Gouqiong II-476

Li, Huade II-337, II-394
Li, Jiahao II-225
Li, Jiawang II-142
Li, Jing I-409
Li, Kangshun I-39, I-266, I-312, II-39,
    II-127, II-360
Li, Mingxing II-430
Li, Shaobo I-420
Li, Wei I-113, I-178
Li, Yuanxiang II-71
Li, Zhaokui I-340
Li, Zhengying I-223
Liang, Bang I-326
Liang, Haiyan I-409
Liang, Zhixun II-71
Liao, Futao I-125
Lin, Chanjuan I-26
Lin, Kexin I-113
Liu, Feng I-409
Liu, Hui I-88
Liu, Yi II-240
Liu, Yue II-142
Liu, Zitu II-142
Lu, Huyuan II-384
Lu, Jintao II-56

M
Ma, Sha II-214, II-287
Ma, Xinyu II-476
Meng, Wen II-278

N
Nie, Huabei I-247, I-443
Niu, Lu I-420
Niu, Yi I-443

O
Ou, Qingrong I-326
Ou, Yangcong I-207
Ou, Yuqi I-113

P
Peng, Hongxing II-28
Peng, Ling I-357

Q
Qian, Haijun I-247
Qiu, Wenbin I-223
Qiu, Zhenzhen I-3

S
Shang, Jingtong II-270
Shao, Peng I-395
Shen, Chao II-430
Shen, Jianqiao I-247, I-443
Shi, Dehao II-28
Shi, Jia I-451
Shi, Yunying I-62
Su, Hongwei II-3
Sun, Jian I-137
Sun, Yuming II-179, II-208, II-262

T
Tan, Xixian II-419
Tang, Bo I-326
Teng, Zi II-71
Tian, Wenjie II-142

W
Wang, Hao II-214, II-287
Wang, Haoliang II-270
Wang, Hua II-179, II-208, II-262, II-297,
    II-319, II-442
Wang, Hui I-125, I-388, II-49, II-135
Wang, Jian'ou I-26
Wang, Jiancong I-312
Wang, Juan I-3
Wang, Junjie II-15
Wang, Lei II-450
Wang, Lili II-270
Wang, Lingwei II-319, II-442
Wang, Ping I-357, II-95, II-326
Wang, Shuai I-125
Wang, Wenjun I-125
Wang, Xianmin I-409
Wang, Xiaofeng I-26
Wang, Yong II-196
Wang, Yuanbing II-253, II-309
Wang, Zhiyong I-62
Wei, Bo II-56
Wei, Yong I-388, II-135
Wei, Yunshan II-113, II-419
Wen, Peihua II-459
Wu, Jian I-137

X
Xiang, Lu I-254
Xiao, Dong I-125
Xiao, Hongyu I-18, II-170, II-187
Xie, Mingchen I-266

Xie, Yutong   I-409
Xiong, Yi   I-429
Xu, Qiner   II-384
Xu, Rui   II-337, II-394
Xu, Yiying   II-240
Xu, Zexin   II-337, II-394

Y
Yan, Yuangao   II-419
Yang, Lei   I-73, II-337, II-394
Yang, Ming   I-207
Yang, Shuxin   II-225
Yang, Zhongxin   I-340
Yao, Jintao   I-3
Yao, Wanyi   II-459
Ye, Chen   I-395
Yi, Yunfei   I-62, II-71
Yu, Chao   II-278
Yu, Fei   I-152, I-326
Yu, Haili   II-240
Yu, Jingkun   II-196
Yuan, Jia   I-254

Z
Zang, Yanhui   I-88
Zeng, Jia   II-459
Zeng, Sanyou   II-476
Zeng, Zhaolian   II-459

Zha, Wentao   II-56
Zhan, Guangsheng   II-407
Zhang, Dongbo   I-254
Zhang, Jiayu   I-39
Zhang, Jibo   I-254
Zhang, Jinbao   II-407
Zhang, Jinen   I-340
Zhang, Jingbo   I-277
Zhang, Jun   II-253, II-309
Zhang, Qing   II-476
Zhang, Shaoping   I-395
Zhang, Shaowei   I-125
Zhang, Xuming   I-357, II-95, II-326
Zhang, Yanjun   I-277
Zhang, Yongcai   I-231, I-443
Zhang, Yuanye   I-73
Zhao, Fei   II-476
Zhao, Jingwen   II-103
Zhao, Nannan   I-18, II-170, II-187
Zhong, Beixin   II-384
Zhong, Yi   I-192
Zhou, Xuesong   I-451
Zhu, Fuyu   II-297
Zhu, Haihua   I-443
Zhu, Tianjin   I-369
Zhu, Wenbin   II-15
Zhu, Zhanyang   I-137
Zong, Xuanyi   II-103

Printed in the United States
by Baker & Taylor Publisher Services